T0181945

Communications in Computer and Information Science 1961

Editorial Board Members

Joaquim Filipe⊙, *Polytechnic Institute of Setúbal, Setúbal, Portugal*
Ashish Ghosh⊙, *Indian Statistical Institute, Kolkata, India*
Raquel Oliveira Prates⊙, *Federal University of Minas Gerais (UFMG), Belo Horizonte, Brazil*
Lizhu Zhou, *Tsinghua University, Beijing, China*

Rationale

The CCIS series is devoted to the publication of proceedings of computer science conferences. Its aim is to efficiently disseminate original research results in informatics in printed and electronic form. While the focus is on publication of peer-reviewed full papers presenting mature work, inclusion of reviewed short papers reporting on work in progress is welcome, too. Besides globally relevant meetings with internationally representative program committees guaranteeing a strict peer-reviewing and paper selection process, conferences run by societies or of high regional or national relevance are also considered for publication.

Topics

The topical scope of CCIS spans the entire spectrum of informatics ranging from foundational topics in the theory of computing to information and communications science and technology and a broad variety of interdisciplinary application fields.

Information for Volume Editors and Authors

Publication in CCIS is free of charge. No royalties are paid, however, we offer registered conference participants temporary free access to the online version of the conference proceedings on SpringerLink (http://link.springer.com) by means of an http referrer from the conference website and/or a number of complimentary printed copies, as specified in the official acceptance email of the event.

CCIS proceedings can be published in time for distribution at conferences or as post-proceedings, and delivered in the form of printed books and/or electronically as USBs and/or e-content licenses for accessing proceedings at SpringerLink. Furthermore, CCIS proceedings are included in the CCIS electronic book series hosted in the SpringerLink digital library at http://link.springer.com/bookseries/7899. Conferences publishing in CCIS are allowed to use Online Conference Service (OCS) for managing the whole proceedings lifecycle (from submission and reviewing to preparing for publication) free of charge.

Publication process

The language of publication is exclusively English. Authors publishing in CCIS have to sign the Springer CCIS copyright transfer form, however, they are free to use their material published in CCIS for substantially changed, more elaborate subsequent publications elsewhere. For the preparation of the camera-ready papers/files, authors have to strictly adhere to the Springer CCIS Authors' Instructions and are strongly encouraged to use the CCIS LaTeX style files or templates.

Abstracting/Indexing

CCIS is abstracted/indexed in DBLP, Google Scholar, EI-Compendex, Mathematical Reviews, SCImago, Scopus. CCIS volumes are also submitted for the inclusion in ISI Proceedings.

How to start

To start the evaluation of your proposal for inclusion in the CCIS series, please send an e-mail to ccis@springer.com.

Biao Luo · Long Cheng · Zheng-Guang Wu ·
Hongyi Li · Chaojie Li

Editors

Neural
Information Processing

30th International Conference, ICONIP 2023
Changsha, China, November 20–23, 2023
Proceedings, Part VII

 Springer

Editors
Biao Luo ⓘ
School of Automation
Central South University
Changsha, China

Zheng-Guang Wu ⓘ
Institute of Cyber-Systems and Control
Zhejiang University
Hangzhou, China

Chaojie Li ⓘ
School of Electrical Engineering
and Telecommunications
UNSW Sydney
Sydney, NSW, Australia

Long Cheng ⓘ
Institute of Automation
Chinese Academy of Sciences
Beijing, China

Hongyi Li ⓘ
School of Automation
Guangdong University of Technology
Guangzhou, China

ISSN 1865-0929   ISSN 1865-0937 (electronic)
Communications in Computer and Information Science
ISBN 978-981-99-8125-0   ISBN 978-981-99-8126-7 (eBook)
https://doi.org/10.1007/978-981-99-8126-7

© The Editor(s) (if applicable) and The Author(s), under exclusive license
to Springer Nature Singapore Pte Ltd. 2024

This work is subject to copyright. All rights are reserved by the Publisher, whether the whole or part of the material is concerned, specifically the rights of translation, reprinting, reuse of illustrations, recitation, broadcasting, reproduction on microfilms or in any other physical way, and transmission or information storage and retrieval, electronic adaptation, computer software, or by similar or dissimilar methodology now known or hereafter developed.
The use of general descriptive names, registered names, trademarks, service marks, etc. in this publication does not imply, even in the absence of a specific statement, that such names are exempt from the relevant protective laws and regulations and therefore free for general use.
The publisher, the authors, and the editors are safe to assume that the advice and information in this book are believed to be true and accurate at the date of publication. Neither the publisher nor the authors or the editors give a warranty, expressed or implied, with respect to the material contained herein or for any errors or omissions that may have been made. The publisher remains neutral with regard to jurisdictional claims in published maps and institutional affiliations.

This Springer imprint is published by the registered company Springer Nature Singapore Pte Ltd.
The registered company address is: 152 Beach Road, #21-01/04 Gateway East, Singapore 189721, Singapore

Paper in this product is recyclable.

Preface

Welcome to the 30th International Conference on Neural Information Processing (ICONIP2023) of the Asia-Pacific Neural Network Society (APNNS), held in Changsha, China, November 20–23, 2023.

The mission of the Asia-Pacific Neural Network Society is to promote active interactions among researchers, scientists, and industry professionals who are working in neural networks and related fields in the Asia-Pacific region. APNNS has Governing Board Members from 13 countries/regions – Australia, China, Hong Kong, India, Japan, Malaysia, New Zealand, Singapore, South Korea, Qatar, Taiwan, Thailand, and Turkey. The society's flagship annual conference is the International Conference of Neural Information Processing (ICONIP). The ICONIP conference aims to provide a leading international forum for researchers, scientists, and industry professionals who are working in neuroscience, neural networks, deep learning, and related fields to share their new ideas, progress, and achievements.

ICONIP2023 received 1274 papers, of which 394 papers were accepted for publication in Communications in Computer and Information Science (CCIS), representing an acceptance rate of 30.93% and reflecting the increasingly high quality of research in neural networks and related areas. The conference focused on four main areas, i.e., "Theory and Algorithms", "Cognitive Neurosciences", "Human-Centered Computing", and "Applications". All the submissions were rigorously reviewed by the conference Program Committee (PC), comprising 258 PC members, and they ensured that every paper had at least two high-quality single-blind reviews. In fact, 5270 reviews were provided by 2145 reviewers. On average, each paper received 4.14 reviews.

We would like to take this opportunity to thank all the authors for submitting their papers to our conference, and our great appreciation goes to the Program Committee members and the reviewers who devoted their time and effort to our rigorous peer-review process; their insightful reviews and timely feedback ensured the high quality of the papers accepted for publication. We hope you enjoyed the research program at the conference.

October 2023

Biao Luo
Long Cheng
Zheng-Guang Wu
Hongyi Li
Chaojie Li

Organization

Honorary Chair

Weihua Gui   Central South University, China

Advisory Chairs

Jonathan Chan	King Mongkut's University of Technology Thonburi, Thailand
Zeng-Guang Hou	Chinese Academy of Sciences, China
Nikola Kasabov	Auckland University of Technology, New Zealand
Derong Liu	Southern University of Science and Technology, China
Seiichi Ozawa	Kobe University, Japan
Kevin Wong	Murdoch University, Australia

General Chairs

Tingwen Huang	Texas A&M University at Qatar, Qatar
Chunhua Yang	Central South University, China

Program Chairs

Biao Luo	Central South University, China
Long Cheng	Chinese Academy of Sciences, China
Zheng-Guang Wu	Zhejiang University, China
Hongyi Li	Guangdong University of Technology, China
Chaojie Li	University of New South Wales, Australia

Technical Chairs

Xing He	Southwest University, China
Keke Huang	Central South University, China
Huaqing Li	Southwest University, China
Qi Zhou	Guangdong University of Technology, China

Local Arrangement Chairs

Wenfeng Hu	Central South University, China
Bei Sun	Central South University, China

Finance Chairs

Fanbiao Li	Central South University, China
Hayaru Shouno	University of Electro-Communications, Japan
Xiaojun Zhou	Central South University, China

Special Session Chairs

Hongjing Liang	University of Electronic Science and Technology, China
Paul S. Pang	Federation University, Australia
Qiankun Song	Chongqing Jiaotong University, China
Lin Xiao	Hunan Normal University, China

Tutorial Chairs

Min Liu	Hunan University, China
M. Tanveer	Indian Institute of Technology Indore, India
Guanghui Wen	Southeast University, China

Publicity Chairs

Sabri Arik	Istanbul University-Cerrahpaşa, Turkey
Sung-Bae Cho	Yonsei University, South Korea
Maryam Doborjeh	Auckland University of Technology, New Zealand
El-Sayed M. El-Alfy	King Fahd University of Petroleum and Minerals, Saudi Arabia
Ashish Ghosh	Indian Statistical Institute, India
Chuandong Li	Southwest University, China
Weng Kin Lai	Tunku Abdul Rahman University of Management & Technology, Malaysia
Chu Kiong Loo	University of Malaya, Malaysia
Qinmin Yang	Zhejiang University, China
Zhigang Zeng	Huazhong University of Science and Technology, China

Publication Chairs

Zhiwen Chen   Central South University, China
Andrew Chi-Sing Leung   City University of Hong Kong, China
Xin Wang   Southwest University, China
Xiaofeng Yuan   Central South University, China

Secretaries

Yun Feng   Hunan University, China
Bingchuan Wang   Central South University, China

Webmasters

Tianmeng Hu   Central South University, China
Xianzhe Liu   Xiangtan University, China

Program Committee

Rohit Agarwal   UiT The Arctic University of Norway, Norway
Hasin Ahmed   Gauhati University, India
Harith Al-Sahaf   Victoria University of Wellington, New Zealand
Brad Alexander   University of Adelaide, Australia
Mashaan Alshammari   Independent Researcher, Saudi Arabia
Sabri Arik   Istanbul University, Turkey
Ravneet Singh Arora   Block Inc., USA
Zeyar Aung   Khalifa University of Science and Technology, UAE
Monowar Bhuyan   Umeå University, Sweden
Jingguo Bi   Beijing University of Posts and Telecommunications, China
Xu Bin   Northwestern Polytechnical University, China
Marcin Blachnik   Silesian University of Technology, Poland
Paul Black   Federation University, Australia
Anoop C. S.   Govt. Engineering College, India
Ning Cai   Beijing University of Posts and Telecommunications, China
Siripinyo Chantamunee   Walailak University, Thailand
Hangjun Che   City University of Hong Kong, China

Wei-Wei Che	Qingdao University, China
Huabin Chen	Nanchang University, China
Jinpeng Chen	Beijing University of Posts & Telecommunications, China
Ke-Jia Chen	Nanjing University of Posts and Telecommunications, China
Lv Chen	Shandong Normal University, China
Qiuyuan Chen	Tencent Technology, China
Wei-Neng Chen	South China University of Technology, China
Yufei Chen	Tongji University, China
Long Cheng	Institute of Automation, China
Yongli Cheng	Fuzhou University, China
Sung-Bae Cho	Yonsei University, South Korea
Ruikai Cui	Australian National University, Australia
Jianhua Dai	Hunan Normal University, China
Tao Dai	Tsinghua University, China
Yuxin Ding	Harbin Institute of Technology, China
Bo Dong	Xi'an Jiaotong University, China
Shanling Dong	Zhejiang University, China
Sidong Feng	Monash University, Australia
Yuming Feng	Chongqing Three Gorges University, China
Yun Feng	Hunan University, China
Junjie Fu	Southeast University, China
Yanggeng Fu	Fuzhou University, China
Ninnart Fuengfusin	Kyushu Institute of Technology, Japan
Thippa Reddy Gadekallu	VIT University, India
Ruobin Gao	Nanyang Technological University, Singapore
Tom Gedeon	Curtin University, Australia
Kam Meng Goh	Tunku Abdul Rahman University of Management and Technology, Malaysia
Zbigniew Gomolka	University of Rzeszow, Poland
Shengrong Gong	Changshu Institute of Technology, China
Xiaodong Gu	Fudan University, China
Zhihao Gu	Shanghai Jiao Tong University, China
Changlu Guo	Budapest University of Technology and Economics, Hungary
Weixin Han	Northwestern Polytechnical University, China
Xing He	Southwest University, China
Akira Hirose	University of Tokyo, Japan
Yin Hongwei	Huzhou Normal University, China
Md Zakir Hossain	Curtin University, Australia
Zengguang Hou	Chinese Academy of Sciences, China

Lu Hu	Jiangsu University, China
Zeke Zexi Hu	University of Sydney, Australia
He Huang	Soochow University, China
Junjian Huang	Chongqing University of Education, China
Kaizhu Huang	Duke Kunshan University, China
David Iclanzan	Sapientia University, Romania
Radu Tudor Ionescu	University of Bucharest, Romania
Asim Iqbal	Cornell University, USA
Syed Islam	Edith Cowan University, Australia
Kazunori Iwata	Hiroshima City University, Japan
Junkai Ji	Shenzhen University, China
Yi Ji	Soochow University, China
Canghong Jin	Zhejiang University, China
Xiaoyang Kang	Fudan University, China
Mutsumi Kimura	Ryukoku University, Japan
Masahiro Kohjima	NTT, Japan
Damian Kordos	Rzeszow University of Technology, Poland
Marek Kraft	Poznań University of Technology, Poland
Lov Kumar	NIT Kurukshetra, India
Weng Kin Lai	Tunku Abdul Rahman University of Management & Technology, Malaysia
Xinyi Le	Shanghai Jiao Tong University, China
Bin Li	University of Science and Technology of China, China
Hongfei Li	Xinjiang University, China
Houcheng Li	Chinese Academy of Sciences, China
Huaqing Li	Southwest University, China
Jianfeng Li	Southwest University, China
Jun Li	Nanjing Normal University, China
Kan Li	Beijing Institute of Technology, China
Peifeng Li	Soochow University, China
Wenye Li	Chinese University of Hong Kong, China
Xiangyu Li	Beijing Jiaotong University, China
Yantao Li	Chongqing University, China
Yaoman Li	Chinese University of Hong Kong, China
Yinlin Li	Chinese Academy of Sciences, China
Yuan Li	Academy of Military Science, China
Yun Li	Nanjing University of Posts and Telecommunications, China
Zhidong Li	University of Technology Sydney, Australia
Zhixin Li	Guangxi Normal University, China
Zhongyi Li	Beihang University, China

Ziqiang Li   University of Tokyo, Japan
Xianghong Lin   Northwest Normal University, China
Yang Lin   University of Sydney, Australia
Huawen Liu   Zhejiang Normal University, China
Jian-Wei Liu   China University of Petroleum, China
Jun Liu   Chengdu University of Information Technology,
                                China
Junxiu Liu   Guangxi Normal University, China
Tommy Liu   Australian National University, Australia
Wen Liu   Chinese University of Hong Kong, China
Yan Liu   Taikang Insurance Group, China
Yang Liu   Guangdong University of Technology, China
Yaozhong Liu   Australian National University, Australia
Yong Liu   Heilongjiang University, China
Yubao Liu   Sun Yat-sen University, China
Yunlong Liu   Xiamen University, China
Zhe Liu   Jiangsu University, China
Zhen Liu   Chinese Academy of Sciences, China
Zhi-Yong Liu   Chinese Academy of Sciences, China
Ma Lizhuang   Shanghai Jiao Tong University, China
Chu-Kiong Loo   University of Malaya, Malaysia
Vasco Lopes   Universidade da Beira Interior, Portugal
Hongtao Lu   Shanghai Jiao Tong University, China
Wenpeng Lu   Qilu University of Technology, China
Biao Luo   Central South University, China
Ye Luo   Tongji University, China
Jiancheng Lv   Sichuan University, China
Yuezu Lv   Beijing Institute of Technology, China
Huifang Ma   Northwest Normal University, China
Jinwen Ma   Peking University, China
Jyoti Maggu   Thapar Institute of Engineering and Technology
                                Patiala, India
Adnan Mahmood   Macquarie University, Australia
Mufti Mahmud   University of Padova, Italy
Krishanu Maity   Indian Institute of Technology Patna, India
Srimanta Mandal   DA-IICT, India
Wang Manning   Fudan University, China
Piotr Milczarski   Lodz University of Technology, Poland
Malek Mouhoub   University of Regina, Canada
Nankun Mu   Chongqing University, China
Wenlong Ni   Jiangxi Normal University, China
Anupiya Nugaliyadde   Murdoch University, Australia

Toshiaki Omori	Kobe University, Japan
Babatunde Onasanya	University of Ibadan, Nigeria
Manisha Padala	Indian Institute of Science, India
Sarbani Palit	Indian Statistical Institute, India
Paul Pang	Federation University, Australia
Rasmita Panigrahi	Giet University, India
Kitsuchart Pasupa	King Mongkut's Institute of Technology Ladkrabang, Thailand
Dipanjyoti Paul	Ohio State University, USA
Hu Peng	Jiujiang University, China
Kebin Peng	University of Texas at San Antonio, USA
Dawid Połap	Silesian University of Technology, Poland
Zhong Qian	Soochow University, China
Sitian Qin	Harbin Institute of Technology at Weihai, China
Toshimichi Saito	Hosei University, Japan
Fumiaki Saitoh	Chiba Institute of Technology, Japan
Naoyuki Sato	Future University Hakodate, Japan
Chandni Saxena	Chinese University of Hong Kong, China
Jiaxing Shang	Chongqing University, China
Lin Shang	Nanjing University, China
Jie Shao	University of Science and Technology of China, China
Yin Sheng	Huazhong University of Science and Technology, China
Liu Sheng-Lan	Dalian University of Technology, China
Hayaru Shouno	University of Electro-Communications, Japan
Gautam Srivastava	Brandon University, Canada
Jianbo Su	Shanghai Jiao Tong University, China
Jianhua Su	Institute of Automation, China
Xiangdong Su	Inner Mongolia University, China
Daiki Suehiro	Kyushu University, Japan
Basem Suleiman	University of New South Wales, Australia
Ning Sun	Shandong Normal University, China
Shiliang Sun	East China Normal University, China
Chunyu Tan	Anhui University, China
Gouhei Tanaka	University of Tokyo, Japan
Maolin Tang	Queensland University of Technology, Australia
Shu Tian	University of Science and Technology Beijing, China
Shikui Tu	Shanghai Jiao Tong University, China
Nancy Victor	Vellore Institute of Technology, India
Petra Vidnerová	Institute of Computer Science, Czech Republic

Shanchuan Wan	University of Tokyo, Japan
Tao Wan	Beihang University, China
Ying Wan	Southeast University, China
Bangjun Wang	Soochow University, China
Hao Wang	Shanghai University, China
Huamin Wang	Southwest University, China
Hui Wang	Nanchang Institute of Technology, China
Huiwei Wang	Southwest University, China
Jianzong Wang	Ping An Technology, China
Lei Wang	National University of Defense Technology, China
Lin Wang	University of Jinan, China
Shi Lin Wang	Shanghai Jiao Tong University, China
Wei Wang	Shenzhen MSU-BIT University, China
Weiqun Wang	Chinese Academy of Sciences, China
Xiaoyu Wang	Tokyo Institute of Technology, Japan
Xin Wang	Southwest University, China
Xin Wang	Southwest University, China
Yan Wang	Chinese Academy of Sciences, China
Yan Wang	Sichuan University, China
Yonghua Wang	Guangdong University of Technology, China
Yongyu Wang	JD Logistics, China
Zhenhua Wang	Northwest A&F University, China
Zi-Peng Wang	Beijing University of Technology, China
Hongxi Wei	Inner Mongolia University, China
Guanghui Wen	Southeast University, China
Guoguang Wen	Beijing Jiaotong University, China
Ka-Chun Wong	City University of Hong Kong, China
Anna Wróblewska	Warsaw University of Technology, Poland
Fengge Wu	Institute of Software, Chinese Academy of Sciences, China
Ji Wu	Tsinghua University, China
Wei Wu	Inner Mongolia University, China
Yue Wu	Shanghai Jiao Tong University, China
Likun Xia	Capital Normal University, China
Lin Xiao	Hunan Normal University, China
Qiang Xiao	Huazhong University of Science and Technology, China
Hao Xiong	Macquarie University, Australia
Dongpo Xu	Northeast Normal University, China
Hua Xu	Tsinghua University, China
Jianhua Xu	Nanjing Normal University, China

Xinyue Xu	Hong Kong University of Science and Technology, China
Yong Xu	Beijing Institute of Technology, China
Ngo Xuan Bach	Posts and Telecommunications Institute of Technology, Vietnam
Hao Xue	University of New South Wales, Australia
Yang Xujun	Chongqing Jiaotong University, China
Haitian Yang	Chinese Academy of Sciences, China
Jie Yang	Shanghai Jiao Tong University, China
Minghao Yang	Chinese Academy of Sciences, China
Peipei Yang	Chinese Academy of Science, China
Zhiyuan Yang	City University of Hong Kong, China
Wangshu Yao	Soochow University, China
Ming Yin	Guangdong University of Technology, China
Qiang Yu	Tianjin University, China
Wenxin Yu	Southwest University of Science and Technology, China
Yun-Hao Yuan	Yangzhou University, China
Xiaodong Yue	Shanghai University, China
Paweł Zawistowski	Warsaw University of Technology, Poland
Hui Zeng	Southwest University of Science and Technology, China
Wang Zengyunwang	Hunan First Normal University, China
Daren Zha	Institute of Information Engineering, China
Zhi-Hui Zhan	South China University of Technology, China
Baojie Zhang	Chongqing Three Gorges University, China
Canlong Zhang	Guangxi Normal University, China
Guixuan Zhang	Chinese Academy of Science, China
Jianming Zhang	Changsha University of Science and Technology, China
Li Zhang	Soochow University, China
Wei Zhang	Southwest University, China
Wenbing Zhang	Yangzhou University, China
Xiang Zhang	National University of Defense Technology, China
Xiaofang Zhang	Soochow University, China
Xiaowang Zhang	Tianjin University, China
Xinglong Zhang	National University of Defense Technology, China
Dongdong Zhao	Wuhan University of Technology, China
Xiang Zhao	National University of Defense Technology, China
Xu Zhao	Shanghai Jiao Tong University, China

Liping Zheng	Hefei University of Technology, China
Yan Zheng	Kyushu University, Japan
Baojiang Zhong	Soochow University, China
Guoqiang Zhong	Ocean University of China, China
Jialing Zhou	Nanjing University of Science and Technology, China
Wenan Zhou	PCN&CAD Center, China
Xiao-Hu Zhou	Institute of Automation, China
Xinyu Zhou	Jiangxi Normal University, China
Quanxin Zhu	Nanjing Normal University, China
Yuanheng Zhu	Chinese Academy of Sciences, China
Xiaotian Zhuang	JD Logistics, China
Dongsheng Zou	Chongqing University, China

Contents – Part VII

Theory and Algorithms

Theory and Algorithms

A 3D UWB Hybrid Localization Method Based on BSR and L-AOA

Bin Shu[1,2](✉), Chuandong Li[1,2](✉), Yawei Shi[1,2], Huiwei Wang[1,2], and Huaqing Li[1,2]

[1] College of Electronic and Information Engineering, Southwest University, Chongqing, China
shubin1234s@163.com, {cdli,tomson,hwwang}@swu.edu.cn
[2] BeiBei, Chongqing 400715, China

Abstract. In this paper, the reliability of the base station(BSR) and the low-cost Angle of arrival(L-AOA) positioning method are proposed to optimize the positioning result of the time difference of arrival(TDOA) positioning method, and then the result is substituted into the Taylor algorithm as the initial value for iterative optimization. The experimental results in non-line-of-sight environment show that the proposed method improves the localization accuracy by about 20% compared with that of TDOA only. In addition, we also apply this algorithm to the mixed algorithms of TDOA and Taylor, under with the same error environment, the positioning accuracy is improved by about 10%.

Keywords: TDOA · Taylor · L-AOA · BSR

1 Introduction

With the rapid development of science and technology, people's demand for indoor wireless technology also presents exponential growth, so the research of this technology has been put on the agenda by more and more scientists and scholars [1]. Traditional positioning technologies include WiFi positioning technology, Bluetooth positioning technology, Radio Frequency Identification(RFID) positioning technology and so on. The main idea of these positioning technologies is to judge the signal reaching strength, and then estimate the location of the positioning point [1]. Such positioning methods are greatly affected by environmental factors, such as temperature, occlusion and so on. The positioning field of Ultra Wide Band(UWB) positioning technology has incomparable advantages over other positioning technologies. The reason is that the signal used by the UWB indoor positioning system has a bandwidth of GHz order, which does not require the carrier wave as the transmission medium for transmission, but transmits data through sending and receiving narrow pulses with nanoseconds or below [2]. Therefore, UWB pulse signal has strong penetration ability, good anti-multipath effect, and can provide more accurate positioning results than the traditional positioning method. At present, the relatively mature

© The Author(s), under exclusive license to Springer Nature Singapore Pte Ltd. 2024
B. Luo et al. (Eds.): ICONIP 2023, CCIS 1961, pp. 3–15, 2024.
https://doi.org/10.1007/978-981-99-8126-7_1

positioning methods of UWB positioning technology include: based on the time difference of signal arrival (TDOA), based on the time of signal arrival (TOA), based on the intensity of signal arrival (RSSI), based on the angle of signal arrival (AOA), Taylor series expansion of the distance function at a given initial point, iterative optimization (Taylor), etc. [2]. Along this line, one of the possible improved direction is the combination of several algorithms, which may combine the advantages of each algorithm and avoiding weaknesses and expect the optimal positioning results [3]. So, in order to continuously improve the positioning accuracy, many scholars have carried out research on hybrid location algorithm. The authors in literature [3] systematically studied the mixed positioning algorithm of Taylor algorithm and chan algorithm. A hybrid positioning algorithm of AOA/TOA was proposed in [9], and In [12], the authors proposed another hybrid location method based on TDOA with single source and multiple base stations. In [13], the inertial measurement unit was applied to the UWB positioning system and a hybrid positioning system with a single base station was proposed. A hybrid positioning algorithm to track and locate indoor moving targets based on kalman filter was stutied both in [14,15].

In order to further improve the positioning accuracy, we propose a new hybrid positioning method in this paper. In this method, BSR is used as the initial weight matrix of the weighted least square method to optimize the initial solution of TDOA, and then L-AOA method is used to optimize the obtained optimization results by using spatial information to obtain an optimal TDOA positioning result, and then the result is substituted into the Taylor algorithm as the initial value for calculation.

More specifically, the main contributions of this paper is summarized as follows: (i) The location of anchor points can be corrected using spatial information at low cost. (ii) The positioning accuracy of chan and taylor hybrid positioning algorithm is further improved. (iii) The concept of base station credibility is proposed and the initial weight matrix of first weighted least squares is optimized.

The arrangement of this paper is as follows: Sect. 2 introduces the principle of basic positioning algorithm in three-dimensional space and the problem to be solved. Section 3 introduces reliability of base station and L-AOA location method in three-dimensional space. In Sect. 4, experimental simulation results are shown and the rationality of the proposed method is proved. Section 5 concludes the conclusion and gives the direction of the follow-up research.

2   Theory of Basic Algorithm

2.1   Basic Chan Algorithm

In the three-dimensional TDOA location system, at least four base stations are required. We only need to measure the arrival time of the signal from the positioning label to each base station, set as t_1, t_2, t_3,......, t_i. The time of signal arrival at each base station is subtracted from the time of arrival at the main base station, and then multiplied by the speed of light c, to obtain some series of distance difference equations, which are solved by the chan algorithm [4]. TDOA positioning principle is shown in Fig. 1.

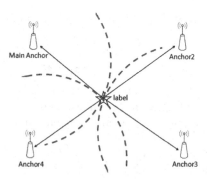

Fig. 1. Theory of TDOA.The four base stations are arranged in a rectangular form (Color figure online)

The schematic diagram of TDOA is shown in Fig. 1. The solution satisfying the three equations at the same time is the registration point to be solved, which is the intersection of the blue dashed lines in the figure. We set the distance between the label and each station as d_i, and it is obtained by the Euclidean distance expression in three dimensional space, like the Eq.(1).

$$d_i = \sqrt{(x - x_i)^2 + (y - y_i)^2 + (z - z_i)^2} \tag{1}$$

(x, y, z) represents unknown tag coordinates,and (x_i, y_i, z_i) represents the coordinates of each anchor where $i = 1, 2, 3, 4$. According to the distance and time differences listed by this theory, the TDOA equations are as follows [4].

$$\begin{cases} d_2 - d_1 = c(t_2 - t_1) \\ d_3 - d_1 = c(t_3 - t_1) \\ \quad \vdots \\ d_i - d_1 = c(t_i - t_1) \end{cases} \tag{2}$$

where t_1, t_2, t_3,... t_i are the arrival time of the signal from the positioning label to each base station, d_1, d_2, d_3,... d_i are the real distance between tag and each anchor calculated by Eq. (1). Then, using the chan algorithm to solve Eq. (2), the coordinates of the unknown label can be obtained.

Chan algorithm can transform the nonlinear hyperbolic equations into linear matrix form, which is easy to solve [4]. Let (x, y, z) be the unknown label position, and (x_i, y_i, z_i) be the known coordinates of the ith base station. Then, the distance between the label and the i-th base station is:

$$\begin{aligned} r_i^2 &= (x - x_i)^2 + (y - y_i)^2 + (z - z_i)^2 \\ &= K_i - 2xx_i - 2yy_i - 2zz_i + x^2 + y^2 + z^2 \end{aligned} \tag{3}$$

where $K_i = x_i^2 + y_i^2 + z_i^2$, let $r_{i,1}$ be the value of the distance from the label to the i-th base station minus the distance from the label to the main base station.

Therefore, it can be expressed as $r_{i,1} = r_i - r_1$. Combining with Eq. (3), we can obtain:

$$r_{i,1}^2 + 2r_{i,1} + r_1^2 = K_i - 2xx_i - 2yy_i - 2zz_i + x^2 + y^2 + z^2 \tag{4}$$

By substituting $i = 1, 2, 3, ..., n$ into Eq. (4) one by one, i equations can be obtained and then subtracting from the equation obtained by $i = 1$ respectively, finally some equations can be obtained and written in matrix form as follows [5].

$$\begin{bmatrix} x_{2,1} & y_{2,1} & z_{2,1} & r_{2,1} \\ x_{3,1} & y_{3,1} & z_{3,1} & r_{3,1} \\ \vdots & \vdots & \vdots & \vdots \\ x_{i,1} & y_{i,1} & z_{i,1} & r_{i,1} \end{bmatrix} \begin{bmatrix} x \\ y \\ z \\ r_1 \end{bmatrix} = \frac{1}{2} \begin{bmatrix} K_2 - K_1 - r_{2,1} \\ K_3 - K_1 - r_{3,1} \\ \vdots \\ K_i - K_1 - r_{i,1} \end{bmatrix} \tag{5}$$

where $x_{i,1} = x_i - x_1$. $r_{i,1}$ is the known value, so only x, y, z, r_1 are unknown. Then the solution of the label coordinate points is the linear equations of Eq. (6). The least squares method is usually used to solve the linear equation in matrix form [5].

$$X = (A^T A)^{-1} A^T B \tag{6}$$

Using the estimated coordinates and additional variables obtained from the first calculation, the second weighted least squares estimation is carried out to obtain the improved estimated coordinates.

$$\Theta_s = \zeta_s Q \zeta_s \tag{7}$$

where $\zeta_s = \begin{bmatrix} D_2 & \cdots & 0 & 0 \\ 0 & D_3 & 0 & 0 \\ \vdots & \vdots & \vdots & \vdots \\ 0 & \cdots & 0 & D_n \end{bmatrix}$, D_n is the distance from the calculated initial coordi-

nates to the base station other than the main base station. Θ_s is the covariance matrix of the second WLS.

$$X' = (A^T \Theta_s^{-1} A)^{-1} A^T \Theta_s^{-1} \zeta_s \tag{8}$$

Perform the third WLS based on the second WLS.

$$Z' = (G^T \Theta_t^{-1} G)^{-1} G^T \Theta_t^{-1} \zeta_t \tag{9}$$

where $\zeta_t = 4\Theta_t \psi \Theta_t$. $\psi = A^T \Theta_s A$. $G = \begin{bmatrix} 1 & 0 & 0 \\ 0 & 1 & 0 \\ 0 & 0 & 1 \\ 1 & 1 & 1 \end{bmatrix}$ and ζ is calculated in a similar

way to the second WLS. The final calculation result of the third WLS is as follows.

$$X_f = -\sqrt{Z'} + X_1 \text{ or } X_f = \sqrt{Z'} + X_1 \tag{10}$$

where X_1 is the coordinates of the main base station.

2.2   Taylor Algorithm

The Taylor series is used to expand the distance function at the given initial point, and the weighted least square method is used to estimate the position [6]. Then the local least square solution of the estimation error is solved, and the position of the label is iteratively updated [7]. Assume that the initial position of the given label is (x_0, y_0, z_0), and the real position of the label is (x, y, z). The result of the distance function expanded at the initial point position and ignore the components above second order using the Taylor series is as follows [7].

$$\varphi_i(x, y, z, x_i, y_i, z_i) = \varphi_i(x_0, y_0, z_0, x_i, y_i, z_i)$$
$$+(\Delta x \frac{\partial}{\partial x} + \Delta y \frac{\partial}{\partial y} + \Delta z \frac{\partial}{\partial z})\varphi_i(x_0, y_0, z_0, x_i, y_i, z_i) \tag{11}$$

Setting a threshold η in advance and the value is 0.01 in this paper. When meet $|\Delta x + \Delta y + \Delta z| < \eta$, stop the iteration [7]. Convert the above formula into matrix form. We can write it as

$$\epsilon = h - G\delta \tag{12}$$

and ϵ is the error vector. In addition:

$$G = \begin{bmatrix} \frac{x_1-x}{r_1} - \frac{x_2-x}{r_2} & \frac{y_1-y}{r_1} - \frac{y_2-y}{r_2} & \frac{z_1-z}{r_1} - \frac{z_2-z}{r_2} \\ \frac{x_1-x}{r_1} - \frac{x_3-x}{r_3} & \frac{y_1-y}{r_1} - \frac{y_3-y}{r_3} & \frac{z_1-z}{r_1} - \frac{z_3-z}{r_3} \\ \vdots & \vdots & \vdots \\ \frac{x_1-x}{r_1} - \frac{x_i-x}{r_i} & \frac{y_1-x}{r_1} - \frac{y_i-y}{r_i} & \frac{z_1-z}{r_1} - \frac{z_i-x}{r_i} \end{bmatrix}$$

$$h = \begin{bmatrix} r_{2,1} - (r_2 - r_1) \\ r_{3,1} - (r_3 - r_1) \\ \vdots \\ r_{i,1} - (r_i - r_1) \end{bmatrix}, \delta = \begin{bmatrix} \Delta x \\ \Delta y \\ \Delta z \end{bmatrix}$$

where i=1, 2, 3, 4. r_i represents the distance between the label and the i-th base station in an iterative calculation and $r_{i,1}$ are given quantities. The weighted least squares solution of the above equation is as:

$$\delta = (G^T Q^{-1} G)^{-1} G^T Q^{-1} h \tag{13}$$

where Q is the measurement covariance matrix. After calculating δ, let: $x = x_0 + \Delta x$, $y = y_0 + \Delta y$, $z = z_0 + \Delta z$. Repeat the above process until the error meets the set threshold value and stop the iterative calculation.

2.3   Problem to Be Solved

Chan algorithm is an effective linearization algorithm for nonlinear hyperbolic equations, but the effect is best when the measurement errors obey the Gaussian distribution. However, in the actual NLOS error environment, the measurement errors are difficult to obey the Gaussian distribution, so the chan algorithm alone will have a large error [8]. Taylor iterative optimization algorithm can be combined to optimize the results calculated by chan algorithm. Since the calculation error of chan algorithm is relatively large in the actual application environment, the initial point provided is not accurate, resulting in the accuracy of Taylor algorithm is not very ideal [8]. The existing solution is to install the antenna array to measure the arrival angle of signal and use the angle information for positioning. The central limit theorem is used to gaussialize the errors and improve the accuracy of chan algorithm. Kalman filter is used to estimate the position of tags in the dynamic case, but these three schemes undoubtedly increase the cost of the system and the cost of computing time [9].

Therefore, in practical applications, how to optimize the calculation results of chan algorithm in NLOS environment at a low cost, so as to provide a more accurate initial point for Taylor algorithm, is a problem that needs to be considered and solved. This paper is to solve this problem.

3   Base Station Reliability and Theory of L-AOA

3.1   Base Station Reliability

In the actual environment, the base station used for communication will become unreliable due to factors such as signal wall reflection, hardware temperature and external weather. Therefore, if every base station is calculated as the same reliability, the calculation results will become unreliable. Therefore, we introduce the concept of base station reliability. It is used to characterize the reliability of the base station involved in positioning.

The calculation process of base station reliability(BSR) is as follows: The labels are placed at three non-collinear points in the experimental environment successively, and 100 groups of distance data from the labels to each base station are collected at each point, that is, the distance value of the labels to each base station is 100 groups of data. Then, the credibility of each base station at each point is calculated by the following formula, so that each base station can obtain three credibility b_1, b_2, b_3. Finally, the average value is the reliability of each base station in the experimental environment.

$$b_j = \frac{1}{\sqrt{\frac{\sum_{i=1}^{N}(d_i - d_r)^2}{N}}}, \varepsilon_k = \frac{b_j}{3}$$

where j = 1, 2, 3. N = 100. k = 1,2,3,...,n,and n is the number of base stations, and d_r is the real distance between ith base station and label.

The weighted least square method adds weight on the basis of the least square method, mainly to take into account that not all the observed values in the system have the same weight, that is, the confidence is not the same. So the question becomes:

$$\|X\|_w^2 = \sum_{i=1}^{L} w_i^2 x_i^2 \tag{14}$$

where w is the weight. We consider using the credibility of the base station to initialize the weight matrix.

$$W = \begin{bmatrix} \varepsilon_1 & \cdots & 0.5 & 0.5 \\ 0.5 & \varepsilon_2 & 0.5 & 0.5 \\ \vdots & \vdots & \vdots & \vdots \\ 0.5 & \cdots & 0.5 & \varepsilon_k \end{bmatrix} \tag{15}$$

Therefore, TDOA can be solved by weighted least squares, and Eq.(7) can be rewritten as:

$$X = (A^T W^+ A)^{-1} A^T W^+ B \tag{16}$$

W^+ is the pseudo-inverse of W.

3.2 Theory of L-AOA

In three-dimensional space, there is spatial information between points, such as angle and distance [10]. Although AOA positioning algorithm considers the use of signal arrival angle information for positioning, due to the need to use antenna array, the cost of the system has been greatly increased, and the positioning accuracy is not high [11]. Therefore, on the basis of not using the antenna array, we consider using the anchor point obtained by the chan algorithm to calculate the elevation angle with the main base station and another base station, and use the obtained degree elevation angle for position solving, which is the core idea of this method [12]. The theory is shown in Fig. 2.

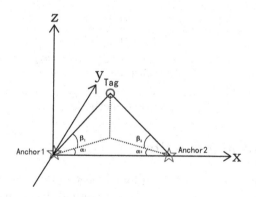

Fig. 2. Schematic diagram of the proposed method

The location of the base station has certain requirements: the label point to be positioned takes the location of the main base station as the reference for position correction, so the main base station should be taken as the origin of coordinates, and another base station should be selected as the angle analysis.

The base station on the same axis as the main base station should be selected as the angle analysis base station. Such selection of the base station can ensure the reliability of spatial information such as angle to a large extent [12]. As shown in Fig. 2, we set the coordinate of the label as (x, y, z), the coordinate of base station 1 as (x_1, y_1, z_1), the distance between the two base stations as d, the distance between the projection point of the label on the xoy plane and base station 1 as L, and the coordinate of base station 2 as (x_2, y_2, z_2). The elevation angle of the label relative to base station 1 is β_1, and the azimuth angle is α_1. The elevation angle of the label relative to base station 2 is β_2 and the azimuth angle is α_2.

First of all, calculate the value of each angle according to the existing coordinates, and the following expressions can be obtained according to the spatial geometric relationship [13].

$$\begin{cases} \alpha_1 = \arctan \dfrac{y - y_1}{x - x_1} \\[2mm] \alpha_2 = \arctan \dfrac{y - y_2}{x - x_2} \\[2mm] \beta_1 = \arctan \dfrac{z - z_1}{\sqrt{(x - x_1)^2 + (y - y_1)^2}} \\[2mm] \beta_2 = \arctan \dfrac{z - z_2}{\sqrt{(x - x_2)^2 + (y - y_2)^2}} \end{cases} \tag{17}$$

Calculate coordinates using angle information. In the triangle composed of dashed lines in the xoy plane, the sine theorem can be used to obtain [13]. The equation is like:

$$L = \frac{d \sin \alpha_2}{\sin (\alpha_1 + \alpha_2)} \tag{18}$$

And by geometric relation, we can get: $\Delta x = L \cos \alpha_1, \Delta y = L \sin \alpha_1, \Delta z = L \tan \beta_1,$. Set the final positioning coordinate as (x_0, y_0, z_0), it can be obtained by the following formula: $x_0 = x + \Delta x$, $y_0 = y + \Delta y$, $z_0 = z + \Delta z$. The above is the calculation process of the proposed method.

4   Experiment and Results

The experiments include the following four aspects: using only Chan algorithm, Chan algorithm combine with the proposed method, Chan algorithm combine with Taylor algorithm, Chan algorithm combine with the proposed method and Taylor algorithm. In the simulation environment, in order to simulate the NLOS environment, we added Gaussian noise with the mean value of zero and the variance of σ^2 to the real distance value [14]. In this experiment, σ^2 values were successively 2, 4, 6, 8. Under each σ^2 value, 100 measurements were made, and then the average error was calculated. In our experiment, four base stations are used for positioning, and the main base station was taken as the origin of coordinates to establish a three-dimensional coordinate system. The coordinates of each base station were set as $(0,0,0)$, $(50,0,0)$, $(50,50,50)$, $(0,50,50)$, and the real coordinates of the labels were $(20,20,20)$. The three-dimensional layout of the simulation environment is shown in the Fig. 3(a).

4.1   Experiments on BSR

First, we tested and proved the feasibility of the BSR. The error coefficients of the error environment of the experiment were set as 2,4,6,8 in turn. For each error environment, the experiment was repeated 100 times at the same point. The experiment was divided into two groups: one group used BSR to initialize the weight matrix, the other did not use BSR to initialize the weight. The solution algorithm used was TDOA-chan algorithm, and the experimental results were as follows. It can be seen from the experimental results that the effect of BSR initializing the weight matrix is very good, and it still has a good effect on reducing the error under the circumstance of relatively large error.

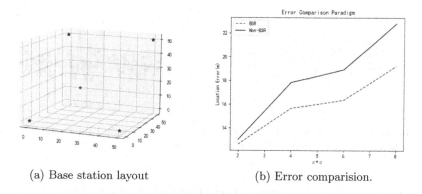

(a) Base station layout   (b) Error comparision.

Fig. 3. (a) is the Base station layout. (b) is the Error comparision on BSR

4.2   Combine Chan with L-AOA and BSR

We applied BSR to chan algorithm, and did comparable experiments. The error coefficients of the error environment of the experiment were set as 2,4,6,8 in turn. For each error environment, the experiment was repeated 100 times at the same point. The solution algorithm used was TDOA-chan algorithm, and the experimental results are shown as Fig. 4.

The red dots in the figure represent the error results of positioning using only the chan method. Compared with the method proposed by the standard, the errors are more divergent and the overall value is larger, which effectively proves the rationality and feasibility of the proposed method.

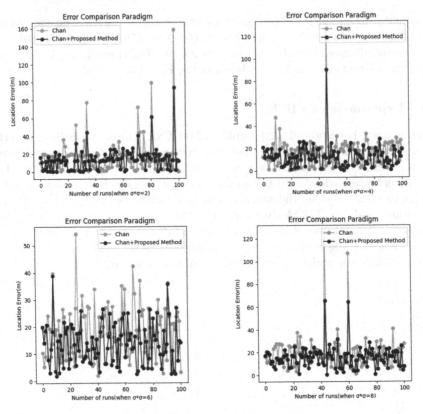

Fig. 4. Error comparision. σ^2 values were successively 2, 4, 6, 8. Under each σ^2 value, 100 measurements were made. (Color figure online)

The results of 100 positioning times show that the combination of the proposed method and the chan algorithm can effectively improve the accuracy. To see the advantage more directly, we calculate the average error. As shown in Fig. 6(a), the proposed method has more obvious convergence, and the overall error has been obviously reduced.

4.3   Combine L-AOA and BSR with Taylor and Chan

The characteristic of Taylor algorithm is that the higher the accuracy of the initial point, the smaller the coordinate error [15]. Therefore, the anchoring points of the chan algorithm optimized by the proposed method were substituted into Taylor as initial values for iteration. In order to see the effect, we did a comparative experiment, that is, we did not use the proposed method to optimize the positioning results of chan algorithm, and directly substituted the results of chan algorithm into the Taylor algorithm. In order to ensure the accuracy of the experiment. The experimental results are as follows (Fig. 5).

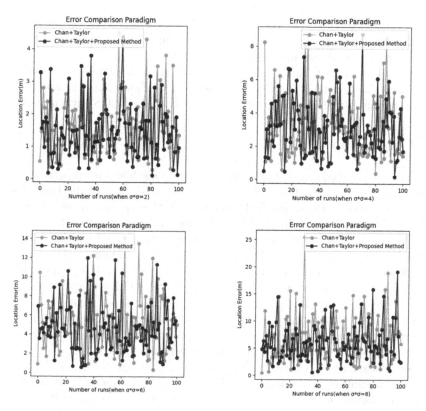

Fig. 5. Error comparision. σ^2 values were successively 2, 4, 6, 8. Under each σ^2 value, 100 measurements were made. (Color figure online)

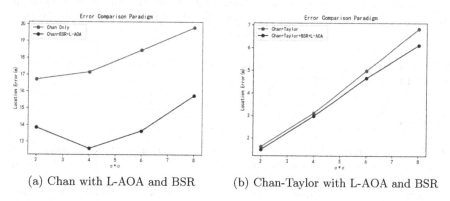

(a) Chan with L-AOA and BSR   (b) Chan-Taylor with L-AOA and BSR

Fig. 6. Error comparision. σ^2 values were successively 2, 4, 6, 8. Under each σ^2 value, 100 measurements were made.

The results of 100 positioning times show that the proposed method combined with chan algorithm provides more accurate initial value for Taylor's algorithm, so the final results are also more accurate. To see the advantage more directly, we calculated the average error. In each error environment, the average error was calculated, as shown in the Fig. 6(b).

As can be seen from the Fig. 6(b), especially when the actual error coefficient is 6 and 8, the advantage is even more obvious. It shows that our method has good accuracy in the environment with large error like NLOS environment.

5   Conclusion

This paper discussed a low-cost AOA positioning method, and proposed to initialize the weight of the first WLS by using the credibility of the base station innovatively, so that the value of TDOA can be further optimized, and the results obtained by Taylor algorithm would be higher, and the experimental proof has also been given. From the overall experimental effect, applying the proposed method in this paper to a scenario with only TDOA positioning algorithm and TDOA and Talyor hybrid positioning algorithms respectively can improve the overall positioning accuracy by about 20% and 10%, which is a great progress in the field of indoor positioning. On the other hand, the experimental results also has shown that the accuracy of this method is improved more obviously when the error environment is large, and it also provides an effective way to reduce the error in NLOS environment. In dynamic personnel positioning system, position estimation using Kalman filter was also carried out on the basis of rough position calculation, so accurate initial value is also very useful in dynamic personnel positioning system. In the future, we will apply the method discussed in this paper to the actual positioning system, and combine Kalman filter, neural network and other technologies for continuous improvement and optimization, which is also our follow-up research plan. In addition, the source of error of the method has not been studied in this paper, which will be the direction of further improvement of the method in the future.

Acknowledgment. This work was supported by the National Natural Science Foundation of China (61873213).

References

1. Mazhar, F., Khan, M.G., Sällberg, B.: Precise indoor positioning using UWB: a review of methods, algorithms and implementations. J. Wirel. Pers. Commun. **97**(3), 4467–4491 (2017)
2. Tian, D., Xiang, Q.: Research on indoor positioning system based on UWB technology. In: 2020 IEEE 5th Information Technology and Mechatronics Engineering Conference (ITOEC), pp. 662–665. IEEE (2020). https://doi.org/10.1109/ITOEC49072.2020.9141707
3. Zang, L., Shen, C., Zhang, K., et al.: Research on hybrid algorithm based on TDOA. In: 2020 IEEE 20th International Conference on Communication Technology (ICCT), pp. 539–542. IEEE (2020). https://doi.org/10.1109/ICCT50939.2020.9295853

4. Díez-González, J., Álvarez, R., Sánchez-González, L., et al.:3D Tdoa problem solution with four receiving nodes. Sensors **19**(13), 2892 (2019)
5. Chan, Y.T., Ho, K.C.: A simple and efficient estimator for hyperbolic location. IEEE Trans. Signal Process. **42**(8), 1905–1915 (1994)
6. Wang, Z., Hu, D., Zhao, Y., et al.: Real-time passive localization of TDOA via neural networks. IEEE Commun. Lett. **25**(10), 3320–3324 (2021)
7. Kocur, D., Vecová, M., Kažimír, P.: Taylor series based localization method of moving persons in 3D space by UWB sensors. In: 2019 IEEE 23rd International Conference on Intelligent Engineering Systems (INES), pp. 000017–000022. IEEE (2019). https://doi.org/10.1109/INES46365.2019.9109468
8. Liao, M., Li, Y., Wang, G., et al.: An indoor positioning algorithm against NLOS error based on UWB. In: 2022 41st Chinese Control Conference (CCC), pp. 5140–5145. IEEE (2022). https://doi.org/10.23919/CCC55666.2022.9902618
9. Yang, B., Li, J., Shao, Z., et al.: Robust UWB indoor localization for NLOS scenes via learning spatial-temporal features. IEEE Sens. J. **22**(8), 7990–8000 (2022)
10. Smaoui, N., Heydariaan, M., Gnawail, O.: Single-antenna AoA estimation with UWB radios. In: 2021 IEEE Wireless Communications and Networking Conference (WCNC), pp. 1–7. IEEE (2021). https://doi.org/10.1109/WCNC49053.2021.9417526
11. Galler, S., Gerok, W., Schroeder, J., et al.: Combined AOA/TOA UWB localization. In: 2007 International Symposium on Communications and Information Technologies, pp. 1049–1053. IEEE (2007). https://doi.org/10.1109/ISCIT.2007.4392171
12. Martaló, M., Ferrari, G., Perri, S., et al.:UWB TDoA-based positioning using a single hotspot with multiple anchors. In: 2019 4th International Conference on Computing, Communications and Security (ICCCS), pp. 1–7. IEEE (2019). https://doi.org/10.1109/CCCS.2019.8888099
13. Liu, Y., Yan, R., Lian, B., et al.: Hybrid IMU/UWB cooperative localization algorithm in single-anchor networks. IEEE Geosci. Remote Sens. Lett. **19**, 1–5 (2022)
14. Pham, N.H.H., Nguyen, M.A., Sun, C.C.: Indoor positioning system using UWB and Kalman filter to increase the accuracy of the localization system. In: 2022 IEEE International Conference on Consumer Electronics-Taiwan, pp. 339–340. IEEE (2022). https://doi.org/10.1109/ICCE-Taiwan55306.2022.9869066
15. Lu, N., Gao, Z.: Ultra wideband indoor positioning method based on Kalman filter and Taylor algorithm. In: 2022 4th International Conference on Intelligent Control, Measurement and Signal Processing (ICMSP), pp. 910–914. IEEE (2022). https://doi.org/10.1109/ICMSP55950.2022.9859227

Unsupervised Feature Selection Using Both Similar and Dissimilar Structures

Tao Xiang[1], Liang Tian[2](✉), Peng Li[1], Jie Liu[1], and Mao Ye[2]

[1] CETC Key Laboratory of Avionics Information System Technology,
Southwest China Institute of Electronic Technology, Chengdu 610036, China
[2] School of Computer Science and Engineering, University of Electronic Science and
Technology of China, Chengdu 611731, China
201911081308@std.uestc.edu.cn

Abstract. Unsupervised feature selection method is widely used to handle the rapid increasing complex and high-dimensional sparse data without labels. Many good methods have been proposed in which the relationships between the similar data points are mainly considered. The graph embedding theory is used which occupies a large proportion. Despite their achievements, the existing methods neglect the information from the most dissimilar data. In this paper, we follow the research line of graph embedding and present a novel method for unsupervised feature selection. Two different viewpoints in the positive and negative are used to keep the data structure after feature selection. Besides a Laplacian matrix by which the most similar data structure is kept, we build an additional Laplacian matrix to keep the least similar data structure. Furthermore, an efficient algorithm is designed by virtue of the existing generalized powered iteration method. Extensive experiments on six benchmark data sets are conducted to verify the state-of-the-art effectiveness and superiority of the proposed method.

Keywords: Unsupervised feature selection · similar matrix · dimensionality reduction

1 Introduction

Along with the development of data analysis, feature selection is still a widely used dimension reduction technique because of its interpretability. There already exist many good feature selection methods. From the viewpoint whether the data set is labeled, there are three kinds of feature selection methods, i.e., supervised feature selection [3,23,26], unsupervised feature selection [5,7], and semi-supervised feature selection [1,4,27]. As it is easy to get a lot of data from the Internet and the work of labeling these data is time consuming and laborious, unsupervised feature selection is required in an increasing number of areas.

Due to the absence of label information, the fundamental issue in unsupervised feature selection is how to model the data structure and produce a faithful

© The Author(s), under exclusive license to Springer Nature Singapore Pte Ltd. 2024
B. Luo et al. (Eds.): ICONIP 2023, CCIS 1961, pp. 16–28, 2024.
https://doi.org/10.1007/978-981-99-8126-7_2

feature subset which preserves the intrinsic structure accurately [17]. Multiple criteria have been proposed to evaluate feature relevance. One common criterion is to choose those features, which could preserve the manifold structure after feature selection. Typically, the structure is characterized in the form of graph. Another frequently used criterion is to seek cluster indicators via clustering algorithms and then leaves a supervised problem [28]. Related to clustering, a work in [5] extend the Random Forests paradigm to unsupervised learning. Without the aid of an affinity graph-based learning method, the method in [8] allows the target matrix to capture latent cluster centers via orthogonal basis clustering and simultaneously select discriminative features guided by these latent cluster centers.

So far, most of methods ignore the information of dissimilarity. Actually, the dissimilar relationship between data pairs can also help keeping the data structure. In this paper, we explore the potential information that dissimilar data contain. It is reasonable to assume that two most dissimilar data point belong to different classes. However, when the graph weights are adaptively reconstructed during the procedure of feature selection, two data points, which far away form each other in original space, may fall into a same class in the selected subspace. To keep the dissimilar property of the data after feature selection, we use an additional Laplacian matrix. During the procedure of feature selection, these data points will approach to different centers by the dissimilar constrains. At last, the data will be more distinguishable in the selected feature space. Based on this motivation, we present an unsupervised feature selection with repulsive effect model(UFSRE). Experimental results confirm the effectiveness of our method.

Our contributions can be summarized as follows. 1) An uncorrelated regression model is presented to perform feature selection and manifold learning simultaneously. In this model, it is the first time that the concept of using the farthest points to assist unsupervised feature selection has been introduced. 2) Two adaptive graph structures based on the nearest and farthest points respectively are incorporated into the feature selection procedure as a graph regularization, which embeds the data structure into manifold learning. 3) Comprehensive experiments on several large scale data sets demonstrate the efficiency and effectiveness of the proposed method.

2   Related Work

In early stage, data reconstruction based methods is mainly used in linear models, such as PCA (Principle Component Analysis) and LDA (Linear Discriminate Analysis). Inspired by the above research route, the following general objective function can be summarized based on many unsupervised feature selection models [2, 7, 8, 10, 15, 28],

$$\min_{\mathbf{W}, \mathbf{F}} \mathcal{F}(\mathbf{W}, \mathbf{F}, \mathbf{X}) + \lambda \mathrm{rank}(\mathbf{W}),$$

where $\mathbf{W} \in \mathbf{R}^{d \times c}$ is a projection matrix, $\mathbf{F}$ contains pseudo labels and $\mathbf{F}^T \mathbf{F} = \mathbf{I}_k$ which ensures the data represented by the selected features could be effectively divided into k parts, and avoids a trivial solution. $\mathcal{F}(\mathbf{W}, \mathbf{F}, \mathbf{X})$ is a data reconstruction function and plays the role of clustering. Data reconstruction based methods have been well studied [8,16,23,24,30]. The second term is a regularization term, which avoids the over-fitting for training data, λ is a parameter and rank($\cdot$) means the rank of a matrix. Usually, the matrix norm is used to limit the rank of matrix. In general, there are l_1-norm [2], $l_{2,p}$-norm [7,15,28], nuclear norm [14] and Frobenius norm [29].

3   Methodology

In this paper, we use boldface uppercase letters represent matrices. Vectors are written as boldface lowercase letters. The i-th row of matrix $\mathbf{M}$ is denoted by $\mathbf{m}^i$, and its j-th column is denoted by $\mathbf{m}_j$. $\mathbf{m}_{ij}$ represents the ij-th element of $\mathbf{M}$. Tr($\mathbf{M}$) denotes the trace of matrix $\mathbf{M}$. The transpose of matrix $\mathbf{M}$ is denoted by $\mathbf{M}^T$. Identity matrix is denoted by $\mathbf{I}$, and $\mathbf{I}_k$ means $k-$dimensional identity matrix. The Frobenius norm of matrix $\mathbf{M}$ is denoted by $\|\mathbf{M}\|_F^2$. The l_2 and $l_{2,1}$-norm of matrix $\mathbf{M}$ is defined as $\|\mathbf{M}\|_2$ and $\|\mathbf{M}\|_{2,1}^2$ respectively. $\mathbf{1}$ is the column vector of all ones. A defined matrix $\mathbf{H} = \mathbf{I} - (1/n)\mathbf{1}\mathbf{1}^T$ and it is referred to as the centering matrix.

Now, we will give a detail introduction of unsupervised feature selection with repulsive effect (UFSRE) algorithm. Given $\mathbf{X} = \{\mathbf{x}_1, \mathbf{x}_2, \cdots, \mathbf{x}_n\} \in \mathbf{R}^{d \times n}$ denote input data set with c classes. We construct two graph laplacian $\mathbf{L}_{\tilde{s}}$ and $\mathbf{L}_{\tilde{t}}$. $\mathbf{L}_{\tilde{s}} = \mathbf{D}_{\tilde{s}} - \tilde{\mathbf{S}}$ is the graph laplacian and $\mathbf{D}_{\tilde{s}}$ is a diagonal matrix whose diagonal entries are defined as $\sum_{j=1}^{n} \tilde{\mathbf{s}}_{ij}$. We define the similarity between $\mathbf{x}_i$ and $\mathbf{x}_j$ as:

$$\tilde{\mathbf{s}}_{ij} = \begin{cases} 1, \forall \mathbf{x}_j \in \langle i_{\tilde{s}} \rangle \\ 0, \text{otherwise} \end{cases} \tag{1}$$

where $\langle i_{\tilde{s}} \rangle$ indicates the k nearest neighbors around $\mathbf{x}_i$. Similarly, we set a dissimilar matrix $\tilde{\mathbf{T}} \in \mathbf{R}^{n \times n}$, and define the similarity between $\mathbf{x}_i$ and $\mathbf{x}_j$ as:

$$\tilde{\mathbf{t}}_{ij} = \begin{cases} 1, \forall \mathbf{x}_j \in \langle i_{\tilde{t}} \rangle, \\ 0, \text{otherwise}. \end{cases} \tag{2}$$

where $\langle i_{\tilde{t}} \rangle$ indicates the l most dissimilar data points around $\mathbf{x}_i$.

Based on the above definitions, the objective function of UFSRE can be written as the following,

$$\min_{\mathbf{W}, \mathbf{F}, \mathbf{E}, \mathbf{S}, \mathbf{T}} \left\| \mathbf{H}(\mathbf{X}^T \mathbf{W} - \mathbf{F}) \right\|_F^2 + \lambda \|\mathbf{W}\|_{2,1}^2 + \alpha_1 (2\text{Tr}(\mathbf{F}^T \mathbf{L}_s \mathbf{F}) + \beta_1 \|\mathbf{S} - \mathbf{A}\|_F^2)$$

$$+ \alpha_2 (-2\text{Tr}(\mathbf{E}^T \mathbf{L}_t \mathbf{E}) + \beta_2 \|\mathbf{T} - \mathbf{B}\|_F^2) + \eta \|\mathbf{F} - \mathbf{E}\|_F^2, \tag{3}$$

$s.t.\ \mathbf{W}^T(\mathbf{Z}_t + \lambda \tilde{\mathbf{D}})\mathbf{W} = \mathbf{I},\ \mathbf{s}_{ij}, \mathbf{t}_{ij} > 0, \mathbf{s}_i^T \mathbf{1} = 1,\ \mathbf{t}_i^T \mathbf{1} = 1,\ \mathbf{F}^T \mathbf{F} = \mathbf{I},\ \mathbf{E}^T \mathbf{E} = \mathbf{I},$

where λ, α_1, α_2, β_1, β_2 and η are nonnegative parameters.

The first two items in Eq. (3) are nonlinear sparsity model which are proposed in [15]. $\mathbf{W} \in \mathbf{R}^{d \times c}$, is a projection function which projects $\mathbf{X}$ from d-dimensional space to c-dimensional space. The orthogonal basis constrain $\mathbf{F}^T\mathbf{F} = \mathbf{I}$ ensure the projected data are discriminable in the new space. $l_{2,1}$-norm of $\mathbf{W}$ with parameter λ ensures the row sparsity of $\mathbf{W}$ for feature selection [4]. A generalized uncorrelated constraint $\mathbf{W}^T(\mathbf{Z}_t + \lambda\widetilde{\mathbf{D}})\mathbf{W} = \mathbf{I}$ is utilized to avoid trivial solution [15] where $\mathbf{Z}_t = \mathbf{X}\mathbf{H}\mathbf{X}^T$ is a total scatter matrix [19]. Because $\mathbf{Z}_t$ may encounter the singularity problem when the number of samples is less than the features size, a $d \times d$ diagonal matrix $\widetilde{\mathbf{D}}$ is used in which $\widetilde{\mathbf{d}}_{ii}$ is defined as follows:

$$\widetilde{\mathbf{d}}_{ii} = \frac{1}{2\sqrt{\|\mathbf{w}^i\|_2^2 + \varepsilon}}(\varepsilon \to 0, i = 1, 2, 3, \cdots, d) \tag{4}$$

where ε is added for preventing zero-denominator.

The third item is to exploit the nearest relationship structure. The initial affinity matrix $\mathbf{A}$ is calculated by Eq. (1). As indicated in [19], an ideal similarity structure for clustering should have the property that the number of connected components is equal to the number of clusters. So we use a matrix $\mathbf{S}$ to approximate the initial affinity matrix $\mathbf{A}$ while keeping this similarity structure. For satisfying the symmetric requirement, we define the Laplacian matrix $\mathbf{L}_s = \mathbf{D}_s - (\mathbf{S}^T + \mathbf{S})/2$. A theorem in [17] shows that the ideal similarity structure could be achieved by making the matrix $\mathbf{L}_s$ has k zero eigenvalues. Hence, we compute $\sum_{i=1}^k \delta_i(\mathbf{L}_s) = \min_{\mathbf{F} \in R^{k\times n}, \mathbf{F}^T\mathbf{F}=\mathbf{I}_k} \text{Tr}(\mathbf{F}^T\mathbf{L}_s\mathbf{F})$, where $\delta_i(\mathbf{L}_s)$ is the ith smallest eigenvalue of $\mathbf{L}_s$. The rank constraint $rank(\mathbf{L}_s) = n - k$ can be satisfied when $\sum_{i=1}^k \delta_i(\mathbf{L}_s) = 0$.

The forth item is to exploit the farthest relationship structure. Similarly, let $\mathbf{B}$ be the initial dissimilar matrix, and $\mathbf{L}_t = \mathbf{D}_t - (\mathbf{T}^T + \mathbf{T})/2$. The matrix $\mathbf{T}$ is also used to approximate $\mathbf{B}$, such that the dissimilar structure information can be considered. Since we want the k-largest eigenvalues of $\mathbf{L}_t$, the negative sign is used in the first item. For keeping the nearest and farthest information, the cluster indicator $\mathbf{F}$ is required to be the same as $\mathbf{E}$ which are guaranteed in the last item.

In UFSRE, when the parameters α_2, β_2 and η are set to zeros, our model is to

$$\min_{\mathbf{W},\mathbf{F},\mathbf{S}} \left\|\mathbf{H}(\mathbf{X}^T\mathbf{W} - \mathbf{F})\right\|_F^2 + \lambda\|\mathbf{W}\|_{2,1}^2 + \beta_1\|\mathbf{S} - \mathbf{A}\|_F^2 + \alpha_1\text{Tr}(\mathbf{F}^T\mathbf{L}_s\mathbf{F}), \tag{5}$$

$$s.t.\ \mathbf{W}^T(\mathbf{Z}_t + \lambda\widetilde{\mathbf{D}})\mathbf{W} = \mathbf{I},\ s_{ij} > 0, s_i^T\mathbf{1} = 1,\ \mathbf{F}^T\mathbf{F} = \mathbf{I},$$

We compare the experimental results of the two models later. The process of determination of super parameters could be laborious. Here we use the model (5) to fix three parameters, λ, α_1 and β_1 at first. These parameter values make the model (5) get the best value will be a good start for the model (3). If α_2, β_2 and η get close to zeros, the result of model (5) approaches to the result of model (3). Therefor we set the parameters λ, α_1 and β_1 to constants when we do optimization of the model (3), and the good results are confirmed in the experiments.

4 Optimization

The optimization of the model (3) involves five variables: $\mathbf{W}$, $\mathbf{F}$, $\mathbf{E}$, $\mathbf{S}$ and $\mathbf{T}$. So we adopt an iterative optimization method via fixing other variables and choose only one to optimize alternately, i.e., the coordinate blocking method.

4.1 Updating W

The optimization of $\mathbf{W}$ while other variables are fixed is equal to solving the following problem:

$$\min_{\mathbf{W}^T(\mathbf{Z}_t+\lambda\widetilde{\mathbf{D}})\mathbf{W}=\mathbf{I}} \left\|\mathbf{H}(\mathbf{X}^T\mathbf{W}-\mathbf{F})\right\|_F^2 + \lambda\left\|\mathbf{W}\right\|_{2,1}^2. \tag{6}$$

Lemma 1. *As $\mathbf{W}$ approaches the optimal value $\mathbf{W}^*$, and $\varepsilon \to 0$, the value of $\frac{\partial\|\mathbf{W}\|_{2,1}^2}{\partial\mathbf{w}^i}$ approaches the value of $\frac{\partial\mathrm{Tr}(\mathbf{W}^T\widetilde{\mathbf{D}}\mathbf{W})}{\partial\mathbf{w}^i}$.*

The proof of Lemma (1) can get from [15]. According to Lemma 1, Eq. (6) is equivalent to the following,

$$\min_{\mathbf{W}^T(\mathbf{Z}_t+\lambda\widetilde{\mathbf{D}})\mathbf{W}=\mathbf{I}} \left\|\mathbf{H}(\mathbf{X}^T\mathbf{W}-\mathbf{F})\right\|_F^2 + \lambda\mathrm{Tr}(\mathbf{W}^T\widetilde{\mathbf{D}}\mathbf{W}) \tag{7}$$

$$\Leftrightarrow \min_{\mathbf{W}^T(\mathbf{Z}_t+\lambda\widetilde{\mathbf{D}})\mathbf{W}=\mathbf{I}} \mathrm{Tr}(\mathbf{W}^T(\mathbf{Z}_t+\lambda\widetilde{\mathbf{D}})\mathbf{W} - 2\mathbf{W}^T\mathbf{X}\mathbf{H}\mathbf{F} + \mathbf{F}^T\mathbf{H}\mathbf{F})$$

$$\Leftrightarrow \max_{\mathbf{W}} \mathrm{Tr}(\mathbf{W}^T\mathbf{X}\mathbf{H}\mathbf{F}) \quad s.t. \quad \mathbf{W}^T(\mathbf{Z}_t+\lambda\widetilde{\mathbf{D}})\mathbf{W} = \mathbf{I}$$

$$\Leftrightarrow \max \mathrm{Tr}(\mathbf{Q}^T\mathbf{G}) \tag{8}$$

where

$$\mathbf{Q} = (\mathbf{Z}_t+\lambda\widetilde{\mathbf{D}})^{\frac{1}{2}}\mathbf{W}, \qquad \mathbf{G} = (\mathbf{Z}_t+\lambda\widetilde{\mathbf{D}})^{-\frac{1}{2}}\mathbf{X}\mathbf{H}\mathbf{F}. \tag{9}$$

By the virtue of the definition above, the optimal solution to model (7) can be obtained as:

$$\mathbf{W} = (\mathbf{Z}_t+\lambda\widetilde{\mathbf{D}})^{-\frac{1}{2}}\mathbf{Q} \tag{10}$$

where $\mathbf{Q}$ can effectively achieved in a closed form [20].

4.2 Updating F

The optimization of $\mathbf{F}$ while other variables are fixed is equal to solving the following equation,

$$\min_{\mathbf{F}^T\mathbf{F}=\mathbf{I}} \left\|\mathbf{H}(\mathbf{X}^T\mathbf{W}-\mathbf{F})\right\|_F^2 + 2\alpha_1\mathrm{Tr}(\mathbf{F}^T\mathbf{L}_s\mathbf{F}) + \eta\left\|\mathbf{F}-\mathbf{E}\right\|_F^2 \tag{11}$$

$$\Leftrightarrow \min_{\mathbf{F}^T\mathbf{F}=\mathbf{I}} \mathrm{Tr}(\mathbf{F}^T\mathbf{P}\mathbf{F} - 2\mathbf{F}^T\mathbf{C}) \tag{12}$$

where

$$\mathbf{P} = \mathbf{H} + 2\alpha_1 \mathbf{L}_s, \quad \mathbf{C} = \mathbf{H}\mathbf{X}^T\mathbf{W} + \mathbf{E}. \tag{13}$$

Based on Lemma 2, [20] has proved problem (12) could be closed by problem (14).

$$\max_{\mathbf{F}^T\mathbf{F}=\mathbf{I}} \mathrm{Tr}(\mathbf{F}^T\mathbf{M}) \tag{14}$$

where $\mathbf{M} = 2\widetilde{\mathbf{P}}\mathbf{F} + 2\mathbf{C}$.

Lemma 2. *The optimal solution $\mathbf{F}$ to problem (11) is defined as $\mathbf{F} = \mathbf{U}\mathbf{V}^T$ where $\mathbf{U}$ and $\mathbf{V}$ are the left and right singular matrices of compact SVD decomposition of $\mathbf{M}$ defined above, respectively [12].*

4.3 Updating S

The optimization of $\mathbf{S}$ while other variables are fixed is equal to solving the following problem,

$$\min_{\mathbf{S}} 2\mathrm{Tr}(\mathbf{F}^T\mathbf{L}_s\mathbf{F}) + \beta_1 \|\mathbf{S} - \mathbf{A}\|_F^2, \, s.t. \, \mathbf{s}_{ij} > 0, \, \mathbf{s}_i^T\mathbf{1} = 1. \tag{15}$$

Eq. (15) could be rewritten as the following

$$\min_{\sum_j \mathbf{s}_{ij}=1, \mathbf{s}_{ij}\geq 0} \sum_j \|\mathbf{f}^i - \mathbf{f}^j\|_2^2 \mathbf{s}_{ij} + \beta_1 \sum_j (\mathbf{s}_{ij} - \mathbf{a}_{ij})^2, \tag{16}$$

Denoting $\mathbf{v}_{ij} = \|\mathbf{f}^i - \mathbf{f}^j\|_2^2$, and $\mathbf{v}_i$ as a vector with the j-th element equal to $\mathbf{v}_{ij}$ (and similarly for $\mathbf{s}_i$ and $\mathbf{a}_i$), we write the problem (16) in vector form as follows,

$$\min_{\mathbf{s}_i^T\mathbf{1}=1, \mathbf{s}_i\geq 0} \left\|\mathbf{s}_i - \left(\mathbf{a}_i - \frac{1}{2\beta_1}\mathbf{v}_i\right)\right\|_2^2. \tag{17}$$

The method in [11] has given an efficient iterative algorithm to solve Eq. (17). If $\mathbf{y}$ represents $\mathbf{a}_i - \frac{1}{2\beta_1}\mathbf{v}_i$, $\hat{\mathbf{y}} = \mathbf{y} - \frac{\mathbf{1}^T}{n}\mathbf{y} + \frac{1}{n}\mathbf{1}$, and $\mathbf{s}^*$ represents the optimal solution to the proximal problem (15), $\mathbf{s}_j^*$ can be solved as follows,

$$\mathbf{s}_j^* = (\hat{\mathbf{y}}_j - \overline{\lambda}^*)_+,$$

where $x_+ = \max(x, 0)$, $\overline{\lambda}^*$ is optimal Lagrangian coefficients. For details, please refer to [11].

4.4 Updating E

The optimization of $\mathbf{E}$ while other variables are fixed is equal to solving the following problem,

$$\max_{\mathbf{E}^T\mathbf{E}=\mathbf{I}} 2\alpha_2\mathrm{Tr}(\mathbf{E}^T\mathbf{L}_t\mathbf{E}) - \eta\|\mathbf{F} - \mathbf{E}\|_F^2 \Leftrightarrow \max_{\mathbf{E}^T\mathbf{E}=\mathbf{I}} \mathrm{Tr}(\alpha_2\mathbf{E}^T\mathbf{L}_t\mathbf{E} + \eta\mathbf{E}^T\mathbf{F}). \tag{18}$$

The Lagrangian function for the problem (18) can be written as

$$\mathbf{L}_2(\mathbf{E}, \Lambda_1) = \text{Tr}(\mathbf{E}^T \mathbf{L}_t \mathbf{E}) + 2\text{Tr}(\frac{\eta}{2\alpha_2} \mathbf{E}^T \mathbf{F}) - \text{Tr}(\Lambda_2(\mathbf{E}^T \mathbf{E} - \mathbf{I})),$$

where Λ_2 is a lagrange multiplier. Similar as the method of updating $\mathbf{F}$, [20] has proposed an efficient algorithm for Eq. (18) as follows,

1) update $\mathbf{R} \leftarrow 2\mathbf{L}_t\mathbf{E} + \frac{\eta}{\alpha_2}\mathbf{F}$;
2) update $\mathbf{E}$ by solving $\max_{\mathbf{E}^T\mathbf{E}=\mathbf{I}} \text{Tr}(\mathbf{E}^T\mathbf{R})$ according to Lemma 2.
until convergence.

4.5   Updating T

The optimization of $\mathbf{T}$ while other variables are fixed is equal to solving the following problem,

$$\min_{\mathbf{T}} -2\text{Tr}(\mathbf{E}^T\mathbf{L}_t\mathbf{E}) + \beta_2 \|\mathbf{T} - \mathbf{B}\|_F^2 , s.t. \ t_{ij} > 0, \ \mathbf{t}_i^T\mathbf{1} = 1, \tag{19}$$

The first part of Eq. (19) could be rewritten as follows,

$$- 2\text{Tr}(\mathbf{E}^T\mathbf{L}_t\mathbf{E}) = -\sum_{i=1}^{n}\sum_{j=1}^{n} \left\|\mathbf{e}^i - \mathbf{e}^j\right\|_2^2 \mathbf{t}_{ij}. \tag{20}$$

Noted that the problem Eq. (20) is independent to the index i, so we can solve the following problem separately for each i,

$$\min_{\sum_j t_{ij}=1, t_{ij}\geq 0} -\sum_{j} \left\|\mathbf{e}^i - \mathbf{e}^j\right\|_2^2 \mathbf{t}_{ij} + \beta_2 \sum_{j}(\mathbf{t}_{ij} - \mathbf{b}_{ij})^2. \tag{21}$$

Denoting $\mathbf{u}_{ij} = \left\|\mathbf{e}^i - \mathbf{e}^j\right\|_2^2$, and $\mathbf{u}_i$ as a vector with the j-th element equal to $\mathbf{u}_{ij}$ (and similarly for $\mathbf{t}_i$ and $\mathbf{b}_i$), we write the problem (21) as the following,

$$\min_{\mathbf{t}_i^T\mathbf{1}=1, t_i\geq 0} \left\|\mathbf{t}_i - (\mathbf{b}_i + \frac{1}{2\beta_2}\mathbf{u}_i)\right\|_2^2. \tag{22}$$

Similar to Eq. (17), Eq. (22) could be solved as follow:

$$\mathbf{t}_j^* = (\hat{\mathbf{u}}_j - \overline{\lambda}_2^*)_+,$$

where $\mathbf{t}_i^*$ to represent the optimal solution to the proximal problem (22), $\overline{\lambda}_2^*$ is the optimal Lagrangian coefficients. For details, please refer to [11].

5   Experiments

5.1   Experimental Setup

Data Sets. The proposed UFERE is evaluated on six real-world data sets, including one handwritten digit data set, i.e., USPS [13], three object image data set, i.e., FEI [25], warpPIE10P [22] and COIL20 [18], a spoken letter recognition, i.e., Isolet[1], and one biological data set, i.e., Lung[2].

Table 1. Accuracy and NMI of various unsupervised feature selection methods on several data sets.

	Datasets	UFSRE	Model (5)	URAFS	USCFS	JELSR	MCFS	LS
ACC (%)	COIL20	**67.18**	65.47	66.54	66.07	66.94	64.45	62.17
	Isolet	**70.44**	69.29	65.73	68.97	68.13	60.60	57.56
	USPS	**76.97**	76.13	74.59	74.42	75.56	73.09	72.15
	lung	**91.18**	90.64	90.25	90.59	90.89	91.13	86.26
	FEI	**51.74**	51.41	43.23	47.63	49.97	37.61	17.12
	warpPIE	**47.81**	46.47	44.38	43.10	43.01	37.81	30.95
NMI (%)	COIL20	**77.17**	76.01	76.51	76.54	77.03	73.52	72.63
	Isolet	**80.34**	79.79	77.09	80.01	78.77	74.08	71.18
	USPS	**65.09**	64.46	62.71	62.92	64.26	61.66	61.60
	lung	65.33	**65.83**	64.75	65.58	65.56	65.28	53.52
	FEI	78.50	**78.87**	73.96	77.05	77.96	70.34	52.32
	warpPIE	**46.92**	45.39	45.68	44.58	46.56	37.44	25.53

Implementation Details. In terms of parameter settings, for the sake of simplicity, we fix the number of neighboring parameters to 15 in affinity graphs which are constructed on the basis on similarity matrixes S and T. We planed to tune the six parameters in Eq. (3) by a grid search strategy from $\{10^{-3}, 10^{-2}, 10^{-1}, 1, 10, 10^2, 10^3\}$. However, it's turned out quite time-consuming. Therefore, We first train the model (5), and obtain three corresponding parameters: λ, α_1 and β_1. Then we fix them, and obtain another three parameters. We use k-means clustering method to verify the results. To eliminate the occasionality in k-means, k-means clustering with random starting points is performed 10 times, and the mean of ACC and NMI are reported. The number of selected features is commonly set to $\{20, 40, 80, 120, 160, 200\}$ for all the data sets. We follow the initializations of the variables of each unsupervised feature selection methods recommended in the corresponding papers, and give the best average results of 10 times experiments.

---

[1] http://archive.ics.uci.edu/dataset/54/isolet.
[2] https://jundongl.github.io/scikit-feature/datasets.html.

In accordance with the experimental setups of previous methods, we assumed that the number of classes of each data set is known. Two metrics are employed in the experiments. One is the clustering accuracy (ACC) [21], another is normalized mutual information (NMI) [6].

On each dataset, the proposed method (UFSRE) and the model (5) mentioned above are compared with the state-of-the-art unsupervised feature selection approaches which have open source codes: URAFS [15], USCFS [8], JELSR [10], MCFS [2], and LS [9].

5.2   Experimental Results and Analysis

We put the ACC results of our experiments in first part of Table 1 along with the results of other state-of-the-art models, and the NMI results in the second par of Table 1. As it can be seen that UFSRE selects the most discriminative features under different multilabel conditions. On all datasets, our method is superior to other methods in accuracy. And except for two datasets, our method is superior to other methods in NMI. Regardless of the results of model (5), our method have more than 3% advantage on warpPIE10P and more than 1% advantage on Isolet and FEI in term of ACC. Our method keep the edge on five datasets in term of NMI, and is only slightly lower than USCFS and JELSR on one datasets (lung).

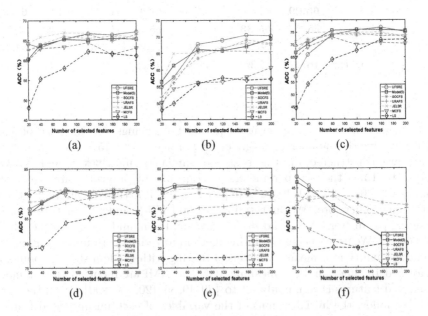

Fig. 1. ACC mean (%) of each selected feature set from various unsupervised feature selection methods. (a) COIL20. (b) Isolet. (c) USPS. (d) lung. (e) FEI. (f) warpPIE10P. (Color figure online)

Figures (1–2) show the best results versus the number of selected features. The number of selected features ranges from 20 to 200 following the experimental setup shown earlier. We use solid line represent the results of two methods proposed in this paper. The read line represents UFSRE and blue one represents model (5). Other methods showed by dash lines.

It shows that the clustering results of UFSRE coincide the trend of the clustering results of model (5). This is because that we fix three parameters by model (5) when we optimize UFSRE. The clustering results of UFSRE has outperformed the results of model (5) in most cases, and is only slightly lower on two datasets (lung and FEI) in term of NMI. The results of these experiments prove the least similarity data also contain some unexplored information.

5.3 Parameter Sensitivity

We investigate the impacts of parameters α_1, α_2, β_1, β_2, λ and η in Eq. (3) on the performance of UFSRE. Specifically, we vary one parameter by fixing the others. ACC is employed to evaluate the performance of clustering with each parameter alternatively searched in the grid of $\{10^{-3}, 10^{-2}, 10^{-1}, 1, 10, 10^2, 10^3\}$. The number of selected features varies in $\{20, 40, 80, 120, 160, 200\}$. From the results in Fig. 3, we can conclude that the performance of UFSRE is insensitive to α_1, α_2, β_1 and β_2, while it is sensitive to λ especially in the condition that λ is small. As the coefficient of $\mathbf{W}$, λ impacts on the sparsity of $\mathbf{W}$ actually.

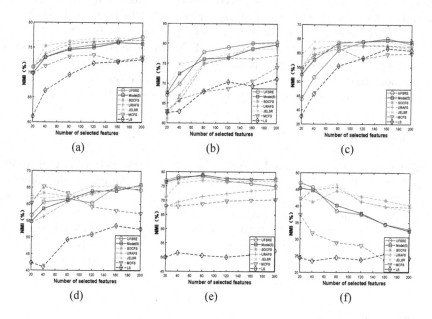

Fig. 2. NMI mean (%) of each selected feature set from various unsupervised feature selection methods. (a) COIL20. (b) Isolet. (c) USPS. (d) lung. (e) FEI. (f) warpPIE10P. (Color figure online)

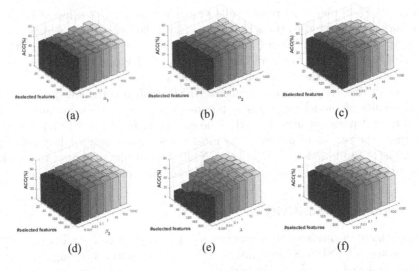

Fig. 3. Parameter sensitivity demonstration on COIL20. The x-axis represents selected parameter while y-axis represents the number of selected features, and z-axis represents the clustering accuracy. Except the selected parameter, other parameters are assigned as follows: $\alpha_1 = 0.001$, $\alpha_2 = 10$, $\beta_1 = 1000$, $\beta_2 = 0.1$, $\lambda = 10$, $\eta = 0.1$

When λ is larger, a sparser $\mathbf{W}$ is acquired, and thus a more exact feature ranking is acquired to some extent. Figure 3 also shows little fluctuations in (f) which indicates that there exist certain sensitives to η. As the coefficient of $\mathbf{F} - \mathbf{E}$, η reflects the impaction of dissimilar matrix $\mathbf{T}$ on the indicator matrix $\mathbf{F}$. When η is smaller, $\mathbf{T}$ has lesser work on $\mathbf{F}$. However, if η is too large, $\mathbf{F}$ affected by $\mathbf{E}$ even more than affected by $\mathbf{L}_s$ in the process of iteration. This could lead to damage to the structure of similar matrix $\mathbf{S}$ when we update $\mathbf{S}$ with $\mathbf{F}$.

6   Conclusion

In this paper, we have presented a method named Feature Selection with Repulsive Effect (FSRE), which constructs the structure of data with the information of the nearest and farthest points. The selected features are obtained by the analysis of projection matrix $\mathbf{W}$. An efficient optimization algorithm is proposed to solve this problem. Comprehensive experiments on six benchmark datasets demonstrate the effectiveness of our approach. In the future work, we will improve this method to design a more effective model combining the most similarity and the least similarity matrices.

References

1. Bellal, F., Elghazel, H., Aussem, A.: A semi-supervised feature ranking method with ensemble learning. Pattern Recogn. Lett. **33**(10), 1426–1433 (2012)

2. Cai, D., Zhang, C., He, X.: Unsupervised feature selection for multi-cluster data. In: Proceedings of the 16th ACM SIGKDD International Conference on Knowledge Discovery and Data Mining, pp. 333–342 (2010)
3. Chen, R., Sun, N., Chen, X., Yang, M., Wu, Q.: Supervised feature selection with a stratified feature weighting method. IEEE Access **6**, 15087–15098 (2018)
4. Chen, X., Yuan, G., Nie, F., Huang, J.Z.: Semi-supervised feature selection via rescaled linear regression. In: IJCAI, vol. 2017, pp. 1525–1531 (2017)
5. Elghazel, H., Aussem, A.: Unsupervised feature selection with ensemble learning. Mach. Learn. **98**(1–2), 157–180 (2015)
6. Fan, K.: On a theorem of Weyl concerning eigenvalues of linear transformations I. Proc. Natl. Acad. Sci. **35**(11), 652 (1949)
7. Han, D., Kim, J.: Unsupervised simultaneous orthogonal basis clustering feature selection. In: Proceedings of the IEEE Conference on Computer Vision and Pattern Recognition, pp. 5016–5023 (2015)
8. Han, D., Kim, J.: Unified simultaneous clustering and feature selection for unlabeled and labeled data. IEEE Trans. Neural Netw. Learn. Syst. **29**(12), 6083–6098 (2018)
9. He, X., Niyogi, P.: Locality preserving projections. In: Advances in Neural Information Processing Systems, pp. 153–160 (2004)
10. Hou, C., Nie, F., Li, X., Yi, D., Wu, Y.: Joint embedding learning and sparse regression: a framework for unsupervised feature selection. IEEE Trans. Cybern. **44**(6), 793–804 (2013)
11. Huang, J., Nie, F., Huang, H.: A new simplex sparse learning model to measure data similarity for clustering. In: Twenty-Fourth International Joint Conference on Artificial Intelligence (2015)
12. Huang, J., Nie, F., Huang, H., Ding, C.: Robust manifold nonnegative matrix factorization. ACM Trans. Knowl. Disc. Data (TKDD) **8**(3), 1–21 (2014)
13. Hull, J.J.: A database for handwritten text recognition research. IEEE Trans. Pattern Anal. Mach. Intell. **16**(5), 550–554 (1994)
14. Li, S., Fu, Y.: Robust subspace discovery through supervised low-rank constraints. In: Proceedings of the 2014 SIAM International Conference on Data Mining, pp. 163–171. SIAM (2014)
15. Li, X., Zhang, H., Zhang, R., Liu, Y., Nie, F.: Generalized uncorrelated regression with adaptive graph for unsupervised feature selection. IEEE Trans. Neural Netw. Learn. Syst. **30**(5), 1587–1595 (2018)
16. Li, Z., Yang, Y., Liu, J., Zhou, X., Lu, H.: Unsupervised feature selection using nonnegative spectral analysis. In: Twenty-Sixth AAAI Conference on Artificial Intelligence (2012)
17. Luo, M., Nie, F., Chang, X., Yang, Y., Hauptmann, A.G., Zheng, Q.: Adaptive unsupervised feature selection with structure regularization. IEEE Trans. Neural Netw. Learn. Syst. **29**(4), 944–956 (2017)
18. Nene, S.A., Nayar, S.K., Murase, H., et al.: Columbia object image library (coil-20) (1996)
19. Nie, F., Wang, X., Huang, H.: Clustering and projected clustering with adaptive neighbors. In: Proceedings of the 20th ACM SIGKDD International Conference on Knowledge Discovery and Data Mining, pp. 977–986 (2014)
20. Nie, F., Zhang, R., Li, X.: A generalized power iteration method for solving quadratic problem on the Stiefel manifold. Sci. China Inf. Sci. **60**(11), 112101 (2017)
21. Papadimitriou, C.H., Steiglitz, K.: Combinatorial optimization: algorithms and complexity. Courier Corporation (1998)

22. Strehl, A., Ghosh, J.: Cluster ensembles–a knowledge reuse framework for combining multiple partitions. J. Mach. Learn. Res. **3**(Dec), 583–617 (2002)
23. Tang, J., Alelyani, S., Liu, H.: Feature selection for classification: a review. Data Class.: Algorithms Appl. **37** (2014)
24. Tang, J., Hu, X., Gao, H., Liu, H.: Unsupervised feature selection for multi-view data in social media. In: Proceedings of the 2013 SIAM International Conference on Data Mining, pp. 270–278. SIAM (2013)
25. Thomaz, C.E., Giraldi, G.A.: A new ranking method for principal components analysis and its application to face image analysis. Image Vis. Comput. **28**(6), 902–913 (2010)
26. Xue, Y., Xue, B., Zhang, M.: Self-adaptive particle swarm optimization for large-scale feature selection in classification. ACM Trans. Knowl. Disc. Data (TKDD) **13**(5), 1–27 (2019)
27. Yu, Z., et al.: Incremental semi-supervised clustering ensemble for high dimensional data clustering. IEEE Trans. Knowl. Data Eng. **28**(3), 701–714 (2015)
28. Zhang, R., Nie, F., Wang, Y., Li, X.: Unsupervised feature selection via adaptive multimeasure fusion. IEEE Trans. Neural Netw. Learn. Syst. **30**(9), 2886–2892 (2019)
29. Zhang, Z., Lai, Z., Xu, Y., Shao, L., Wu, J., Xie, G.S.: Discriminative elastic-net regularized linear regression. IEEE Trans. Image Process. **26**(3), 1466–1481 (2017)
30. Zhao, Z., He, X., Cai, D., Zhang, L., Ng, W., Zhuang, Y.: Graph regularized feature selection with data reconstruction. IEEE Trans. Knowl. Data Eng. **28**(3), 689–700 (2015)

STA-Net: Reconstruct Missing Temperature Data of Meteorological Stations Using a Spatiotemporal Attention Neural Network

Tianrui Hou[1] , Li Wu[1(✉)] , Xinzhong Zhang[2], Xiaoying Wang[1], and Jianqiang Huang[1]

[1] Department of Computer Technology and Applications, Qinghai University, No. 251, Ningda Road, Xining 810016, Qinghai, China
{wuli777,wxy_cta,hjq}@qhu.edu.cn
[2] Research and Development Department, Beijing PRESKY Co., Ltd., Beijing, China

Abstract. Reconstructing the missing meteorological site temperature data is of great significance for analyzing climate change and predicting related natural disasters, but is a trickily and urgently solved problem. In the past, various interpolation methods were used to solve this problem, but these methods basically ignored the temporal correlation of the site itself. Recently, the methods based on machine learning have been widely studied to solve this problem. However, these methods tend to handle the missing value situation of single site, neglecting spatial correlation between sites. Hence, we put forward a new spatiotemporal attention neural network (STA-Net) for reconstructing missing data in multiple meteorological sites. The STA-Net utilizes the currently state-of-the-art encoder-decoder deep learning architecture and is composed of two sub-networks which include local spatial attention mechanism (LSAM) and multidimensional temporal self-attention mechanism (MTSAM), respectively. Moreover, a multiple-meteorological-site data processing method is developed to generate matrix datasets containing spatiotemporal information so the STA-Net can be trained and tested. To evaluate the STA-Net, a large number of experiments on real Tibet and Qamdo datasets with the missing rates of 25%, 50% and 75%, respectively, are conducted, meanwhile compared with U-Net, PConvU-Net and BiLSTM. Experimental results have showed that our data processing method is effective and meantime and our STA-Net achieves greater reconstruction effect. In the case with the missing rate of 25% on Tibet test datasets and compared to the other three methods, the MAE declines by 60.21%, 36.42% and 12.70%; the RMSE declines by 56.28%, 32.03% and 14.17%; the R2 increases by 0.75%, 0.20% and 0.07%.

Keywords: Attention mechanism · Deep learning · Neural networks · Missing data imputation · Meteorological station data · Time series

Supplementary Information The online version contains supplementary material available at https://doi.org/10.1007/978-981-99-8126-7_3.

© The Author(s), under exclusive license to Springer Nature Singapore Pte Ltd. 2024
B. Luo et al. (Eds.): ICONIP 2023, CCIS 1961, pp. 29–52, 2024.
https://doi.org/10.1007/978-981-99-8126-7_3

1   Introduction

In meteorological field, temperature data are mostly obtained from corresponding meteorological stations. Complete temperature data are an important basis for meteorologists to conduct the relevant weather forecast and climate analysis; meanwhile, it also plays a critical role in agriculture and ecological research, and is also an important data source for relevant scientific research [1–4]. But the data obtained by meteorological sites are not always complete due to various issues such as electromagnetic interference, equipment failure, harsh environmental conditions or manual operation errors, etc. [2,5,6]. Thus, reconstructing the missing temperature data is an essential previous work when ones conduct relevant scientific studies, and is also a trickily and urgently solved problem.

In order to impute missing meteorological site temperature data, related researchers have undertaken a lot of work, and proposed various kinds of methods. Generally, these methods can be divided into two categories: traditional space-based interpolation method and data-driven based machine learning method. The methods based on traditional interpolation often use the mathematical or physical properties in space to restore missing data [7]. For example, researchers have already reconstructed the missing meteorological station temperature data using interpolation based methods such as inverse-distance weighting (IDW) [8], kriging [9,10], thin-plate splines [11] and multiple regressions [12]. However, these methods are inevitably affected by the spatial distance and the number of meteorological stations [2,5]. Besides, they seldom consider the specific geographical status of station position, being prone to causing that the theoretical models used by such methods may not match the complexity of the actual data. Such limitations would restrict the performance of the methods based on interpolation.

On the other hand, the data-driven methods represented by machine learning can dig out the complex relationship or potential distribution that the data given exist and thus reconstruct the missing meteorological station data. These methods can also be divided into two categories: traditional machine learning and the most currently popular deep learning. Examples based on traditional machine learning to reconstruct various missing meteorological station data have k-nearest neighbor [13], logistic regression [14] and autoregressive integrated moving average [6,15]. Although these methods have a relatively stronger imputation ability in datasets with a low missing rate, they may show a poor performance on datasets with a relatively high missing rate [16].

Meantime, due to meteorological station data are typically time-series data, in recent years, people have used deep learning models based sequence-to-sequence to impute various missing meteorological station data. For instance, Cao et al. [17] applied bidirectional recurrent neural network (BiRNN) to reconstruct multiple types of time series data, such as air quality data collected by stations, etc. Zhang et al. [18,19] and Dagtekin et al. [20] used long short-term memory (LSTM) [21] or gated recurrent units (GRUs) to structure network architectures to recover missing water quality data collected by stations. Xie et al. [22] applied LSTM based model to impute missing temperature station data. These models flexibly designed the corresponding network architecture based on the data used and achieved better reconstruction effects. However, they only considered the problem that the missing value existed in long-term time series data collected by single station, neglecting spatial correlation between multiple local sits.

To address against the faced problem, in this paper, we firstly propose a weather station point data processing method. The datasets produced by this method is a format of the numerical matrix, which can be input to the convolutional neural network (CNN). That is, we solve a tricky problem that previously people often use traditional interpolation method but not CNN-based network to impute missing data in space. Secondly, we design a completely new spatiotemporal attention neural network (STA-Net) which can effectively handle spatiotemporal information through two attention mechanism: local spatial attention mechanism (LSAM) and multidimensional temporal self-attention mechanism (MTSAM). In addition, this paper also discusses reconstruction effects of the proposed model on Tibet and Qamdo datasets with different missing rate 25%, 50%, and 75%, which provides a new technical means for other similar missing data imputation work.

The main contributions of this work include: (1) We design a simple multiple-meteorological-station data processing method. The method can generate matrix datasets containing spatiotemporal information. These generated datasets can be input into the CNN. Hence, we now can dig out spatial correlation between multiple local stations by taking advantage of CNN's characteristics; (2) Propose a novel reconstruction neural network (STA-Net) with a new spatiotemporal attention based on encoder-decoder network architecture. Our model is able to recover missing data based on the information from both the spatial and temporal relation; (3) Design two new attention mechanism: local spatial attention mechanism (LSAM) and multidimensional temporal self-attention mechanism (MTSAM), respectively. The LSAM mainly handles spatial correlation between multiple local meteorological site data. The MTSAM mainly solves the time sequence relationship of single meteorological site data itself.

2   Related Works

2.1   Missing Data Reconstruction

In many research areas, missing data reconstruction is an important and necessarily previous beforehand work when conducting relevant scientific research. Especially in recent years, with the continuous significant progress in deep learning, people are more and more inclined to use deep learning knowledge to solve this problem. For example, in the field of image repair, people have applied based on CNN network structure for image repair [23–26]. Similarly, in geography and meteorology, related researchers have begun to use deep learning networks to repair a variety of missing data. For example, relevant researchers used CNN based network model to reconstruct the global average temperature [27], recover the land water storage data [28], repair the remote sensing sea surface temperature image [29], and reconstruct the missing seismic data [7,30]. In addition, for the missing weather station data, in the past, people had used the interpolation method for recovery. However, at present, based on meteorological station data is a typical timing data, meteorologists often use the LSTM based network model to repair the missing various meteorological data [17–20, 22]. However, so far, no one has used CNN based networks to reconstruct missing meteorological station data.

2.2   Attention

At present, much research has been done on attention. Here, we have selected several representative models for a brief description. On spatial attention, Yu et al. [31] first proposed contextual attention (CA) model which is used in image repair tasks. However, the model is faced with a regular missing region data reconstruction problem and is therefore not suitable for irregular ones. Although Shin et al. [25] and Peng et al. [26] made some improvements to CA, having not addressed the problem of irregular missing region data repair. On temporal attention, Bahdanau et al. [32] first proposed the concept of attention, having been applied to related temporal tasks and natural language translation tasks [33]. Meanwhile, this attention explicitly models global dependencies, which has been successfully applied in multiple models [34]. These tasks motivate us to try using attention to improve network performance.

3   Methods and Network Architecture

3.1   A Data Processing Method of Meteorological Stations

The original used meteorological datasets in this paper are provided by China Meteorological Administration (http://data.cma.cn/) which is publicly accessible. They are Tibet regional meteorological datasets and Qamdo regional meteorological datasets, respectively. Both two datasets hourly store the observed meteorological data in text format.

According to the actual situation, any meteorological station data have its own idiosyncratic time sequence. Hence, we can apply LSTM-based neural network to impute missing data. However, these network models hardly deal with the spatial connection of multiple meteorological stations. Although the interpolation methods can reconstruct missing values between multiple meteorological stations in space, these methods are usually driven by a mathematical or physical model [7], which is prone to cause incompatible between data and model. In addition, these methods ignore the temporal relationship of the weather station itself.

To overcome the above plights, we design a simple data processing method. It generate a numerical matrix data set including spatiotemporal relationship. The proposed method first selects s meteorological stations as the research object, and then integrates the meteorological data observed within 24 h in chronological order (from 0:00 to 23:00). Finally, we will get a matrix data set with dimensions of $(24, s)$, where s is determined by the specific data set used. In addition, the ranking of s meteorological sites is based on the order in which they appear in the original data set.

Obviously, the new matrix data set can be used as the input of CNN, which effectively alleviates the deficiencies of the interpolation methods, better coping with spatial correlation between multiple local sites. At the same time, since each column of data is still a time series data, we can still use the LSTM-based network models. More importantly, this data set allows us to design a new network architecture to reconstruct the missing value from the space and time dimensions.

3.2 Attention Mechanism and Network Architecture

Local Spatial Attention Mechanism. According to the characteristics of CNN, convolution operation only simply aggregates local information [30], neglecting the spatial correlation or similarity between adjacent local areas. Hence, when we apply the neural network model only containing CNN to reconstruct the missing data, a large missing rate will bring a great impact on the reconstruction effect. In addition, our input data is two-dimensional numerical matrix mathematically and basically can be equivalent to single channel image data. Meanwhile, previous studies showed that the nearby pixels between adjacent local areas in images could keep consecutive and consistent in most cases [35–38]. Particularly, Yu et al. [31] proposed contextual attention (CA) to repair the images including missing pixels by utilizing the principle. Therefore, based on the foregoing discussed theories and methods, we design a local spatial attention mechanism, drawing and exploiting the correlation in neighboring local areas to resolve the problem encountered at the beginning of this paragraph.

In the original paper, the CA model only processed the situation that missing pixels or part was regular. And it did not utilize all attention scores, either [26]. In contrast, the proposed local spatial attention model makes use of all the corresponding masks to judge whether the filled missing values are valid.

Figure 1 shows how the local spatial attention mechanism works. The input feature map which concurrently can be regarded the as foreground feature map and the background feature map. The foreground feature map matches with the patches $b_{x',y'}$ extracted from the background feature map by the way of sliding window $b_{x',y'}$ with the step acquiescently set to 1, which is similar to the convolution operation. That is, the patches $b_{x',y'}$ can be treated as convolutional kernel whose size is set to $(3, 3)$ in this paper, where $\left(x', y'\right)$ denotes the central coordinate of the patches in the background feature map. Let $f_{x,y}$ denotes the sub feature map corresponding to the current sliding (convolution) window. Then, the attention score can be computed as:

$$s_{(x,y),(x',y')} = \langle \frac{f_{x,y}}{\|f_{x,y}\|}, \frac{b_{x',y'}}{\left\|b_{x',y'}\right\|} \rangle \tag{1}$$

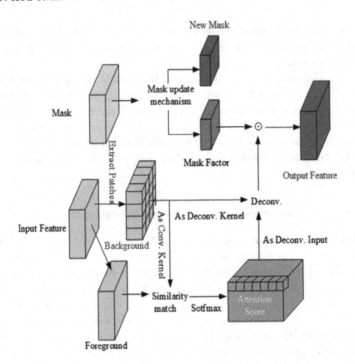

Fig. 1. Illustration of the local spatial attention mechanism. Firstly after the patches extracted from background feature and foreground feature go through the similarity matching calculation and *softmax*, we will get the attention scores. Meanwhile we attain mask factor by utilizing mask update mechanism with corresponding mask. Next, the result that deconvoluting attention score outputs and the mask factor execute hadamard product (element-wise multiplication). Finally, we get the output feature map.

$$a_{(x,y),(x',y')} = softmax(\alpha s_{(x,y),(x',y')}) \tag{2}$$

wherein $s_{(x,y),(x',y')}$ denotes cosine similarity between $f_{x,y}$ and $b_{x',y'}$, $a_{(x,y),(x',y')}$ denotes the attention scores for $f_{x,y}$ corresponding to $b_{x',y'}$, and α is a hyperparameter that avoids smaller values in performing *softmax* operations and is set to 10.

By processing mask M corresponding to input feature map using the mask update mechanism, we get the new mask and mask factor f_m whose element values can only be 1 or 0 which indicates whether the reconstructed missing values are valid. The f_m is computed as:

$$f_m = G(F(m, w^1)) \tag{3}$$

wherein w^1 is a matrix with elements of all 1 whose size is same as $b_{x',y'}$, and F and G denote convolutional function and cropping function that limits the value to $[0, 1]$, respectively.

Eventually, through the result output by the deconvolution operation $Deconv$ [39,40] whose input is $a_{(x,y),(x',y')}$ and $b_{x',y'}$, and combines with f_m, we can get final outcome o, which is computed as:

$$o = f_m \odot Deconv(a_{(x,y),(x',y')}, b_{x',y'}) \tag{4}$$

wherein $\odot$ denotes element-wise multiplication.

Multidimensional Temporal Self-attention Mechanism. According to the data processing method proposed, per column in input data has its own temporal relationship which can be acquired by making use of LSTM based network models. In this paper, We used bidirectional LSTM (BiLSTM), an extended version of BiRNN [17], to cope with the temporal relationship.

However, existing BiLSTM simply concatenates forward hidden $\overrightarrow{y_i}$ and backward hidden state $\overleftarrow{y_i}$, not taking into account that the elements corresponding to sub dimensions between $\overrightarrow{y_i}$ and $\overleftarrow{y_i}$ have different importance to the s_i, where $\overrightarrow{y_i}$ and $\overleftarrow{y_i}$ denote high-dimensional vectors, as shown in Fig. 2. In order to weigh the significance of different sub elements values, we propose a multidimensional temporal self-attention mechanism (MTSAM).

The MTSAM proposed works as Fig. 3 shows. For the input time series data $X = (x_1, x_2, \cdots, x_i, \cdots, x_{n-1}, x_n)$, when they are input into the BiLSTM, we will obtain vector sequences of two states: forward hidden states

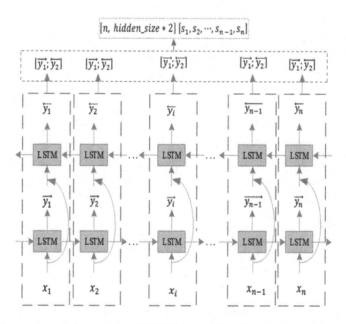

Fig. 2. Illustration of the existing BiLSTM model. The after input sequence data $X = (x_1, x_2, \cdots, x_i, \cdots, x_{n-1}, x_n)$ go through BiLSTM, we get a new sequence $S = (s_1, s_2, \cdots, s_i, \cdots, s_{n-1}, s_n)$. For each s_i, it is attained by concatenating forward hidden states $\overrightarrow{y_i}$ and backward hidden states $\overleftarrow{y_i}$, namely which can be expressed as $[\overrightarrow{y_i}; \overleftarrow{y_i}]$.

$\vec{y} = (\vec{y_1}, \vec{y_2}, \cdots, \vec{y_i}, \cdots, \overrightarrow{y_n - 1}, \vec{y_n})$ and backward hidden states $\overleftarrow{y} = (\overleftarrow{y_1}, \overleftarrow{y_2}, \cdots, \overleftarrow{y_i}, \cdots, \overleftarrow{y_n - 1}, \overleftarrow{y_n})$. Here, we stack the states corresponding subscript between $\vec{y}$ and $\overleftarrow{y}$, obtaining the output $S = (s_1, s_2, \cdots, s_i, \cdots, s_{n-1}, s_n)$ whose size is $[n, 2, hidden\_size]$ contrary to the output of existing BiLSTM.

To figure out the influence degree between the element values in each sub dimensions of the status s_i in S on itself, we first transpose S and get the result $S^T = (s_1^T, s_2^T, \cdots, s_i^T, \cdots, s_{n-1}^T, s_n^T)$ whose size is $[n, hidden\_size, 2]$, computed as follows:

$$S^T = T(S) \tag{5}$$

wherein T denotes transpose function. Next S^T enters a single layer perceptron [41] and *softmax* operation layer, in turn. After this step, getting attention scores s_a, which is computed as follows:

$$s_w = L(S^T; \theta) \tag{6}$$

$$s_a = softmax(s_w) \tag{7}$$

wherein L is a linear neural network, θ denotes learnable parameters and s_w shows importance of the element values which are in each sub dimensions of each sub status s_i in S.

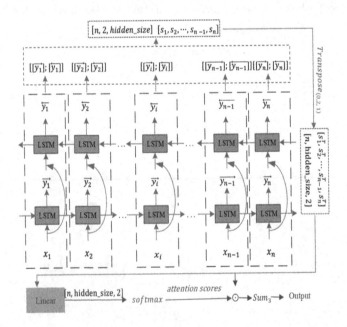

Fig. 3. Illustration of the multidimensional temporal self-attention mechanism. After input sequence data $X = (x_1, x_2, \cdots, x_i, \cdots, x_{n-1}, x_n)$ go through BiLSTM, the output sequence data $S = (s_1, s_2, \cdots, s_i, \cdots, s_{n-1}, s_n)$ is obtained by stacking forward hidden states $\vec{y_i}$ and backward hidden states $\overleftarrow{y_i}$ not concatenating them. Then the output S^T after transposing S executes the linear and *softmax* operations to get self-attention scores s^*. Finally after summing the hadamard product of S^T and s^*, getting ultimate result.

Finally, we calculate the weighted sum S_w^T of the attention scores s_a and S^T. Now each dimension of S_w^T fuses the forward and backward information, which makes the amount of messages implied in S_w^T unchanged, but reduces data size by half compared with the previous BiLSTM model output. Hence, the cost of subsequent operations can be reduced. The S_w^T is computed as:

$$S_w^T = Sum(s_a \odot S^T) \tag{8}$$

wherein $\odot$ denotes element-wise multiplication and Sum represents summarizing function.

Variable Factor Analysis. Intuitively, the temperature of a specific geographical environment is affected by a variety of factors. These factors usually include solar altitude (SA), latitude (Lat), longitude (Lon), cloud thickness (CT), cloud range (CR), wind speed (WS), and other related factors. Apparently, the influence focus of these factors on temperature varies in different regions and in different time periods. Here we express this relation by applying a function, which is defined as:

$$Tem = F_c(SA, Lat, Lon, CT, CR, WS, \cdots) \tag{9}$$

wherein Tem is multivariate function value, F_c denotes a multivariate function and SA et al. is variable.

When performing temporal regularity analysis of a certain meteorological site, the temperature data need to be decomposed and had better to be expressed by many factors, which is defined as:

$$F_m = NN_d(Tem; \theta) \tag{10}$$

wherein NN_d denotes a neural network layer with learnable parameter θ, and F_m is a multivariable representation symbol.

Specifically, by studying changes in the number of variables impacted on Tem, finding that the size of hyperparameter *hidden_size* has an extremely similar meaning as F_m, and NN_d can be understood as the function consisting of the BiLSTM with learnable parameters θ. We will further elaborate this in detail in ablation experiments to clarify their relationship.

Network Architecture. In order to extract the spatiotemporal information of used data as much as possible to make the filled missing data more accurate, we carefully design a new network architecture containing spatiotemporal attention mechanism. It is named as spatiotemporal attention network (STA-Net), as shown in Fig. 4. STA-Net is made up of two encoder-decoder based sub network frameworks: the PConvU-Net [24] based on local spatial attention model (LSAPConvU-Net) and the PConvU-Net based on multidimensional temporal self-attention model (MTSAPConvU-Net). And both two sub-networks are based on encoder-decoder architecture [23].

The encoder layers and decoder layers of both LSAPConvU-Net and MTSAPConvU-Net adopt partial convolution (PConv) [24] block as shown in

Fig. 7. By using the PConv in encoder layers, more abstract and higher representation or feature map [23] could be obtained. In decoder layers, utilize such features and the output of encoder layers to reconstruct the missing information [24]. Meanwhile, in both LSAPConvU-Net and MTSAPConvU-Net, we use the Hardswish activation function [42] but not the ReLU activation function [43] in encoder layers and the LeakyReLu activation function [44,45] in decoder layers to accelerate network training. Following the activation function, the batch normalization [46] is also applied to accelerate training and improve the network performance. Besides, the architectures based on PConvU-Net apply the skip connections to the two corresponding feature maps in both encoder and decoder layers, which not only promotes the network performance and the training speed, but also alleviates the phenomenon of gradient disappearance or explosion, which is similar as the residual net [47]. In skip connections, the upsampling keeps the size of the two corresponding feature maps same. In the whole network architecture, the convolutional kernel size in both encoder and decoder layers is set to. The stride is set to 2 in encoder layers and set to 1 in decoder layers. In the last layer of network, kernel size, out-channels and stride are set to, 1 and 1, separately. The detailed parameters could be found in Table 1.

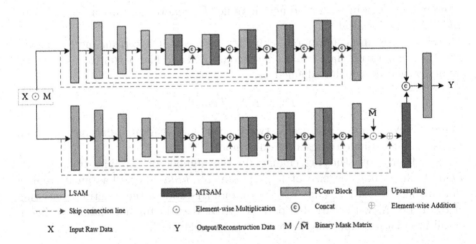

Fig. 4. The architecture of STA-Net which is consisting of two sub-networks: the PConvU-Net based on local spatial attention model on top (LSAPConvU-Net) and the PConvU-Net based on multidimensional temporal self-attention model beneath (MTSAPConvU-Net). First, we input the mask data ($X \odot M$) into both sub-networks, in turns. In MTSAPConvU-Net, the result obtained through PConvU-Net will be first multiplied by $\widetilde{M}$ (which equals to $1 - M$), secondly added by the mask data and input into multidimensional temporal self-attention. Next we concatenate the output of the two sub-networks. Finally, after the concatenated result goes through the last layer, we can get the final reconstruction data Y. The local spatial attention model (LSAM), multidimensional temporal self-attention model (MTSAM) and PConv Block are shown as Fig. 5, Fig. 6 and Fig. 7, respectively.

3.3   Loss Function

When training a neural network, we need a suitable loss function to guide the network to convergence [30]. In meteorological field, root mean squared error (RMSE) is mostly used as the evaluation index for the reconstruction algorithm of missing data. Concurrently, due to the data input to proposed network is a numerical matrix which can be regarded as image data, we can also use the loss function in the field of image repair. According to the above reason, here we chose the L_2 function with a mask [23] as our loss function. For each ground truth input data X with corresponding mask M, our reconstruction network R can output corresponding imputation result $R(X \odot M)$. M is a binary mask with the same size as X, where 0 corresponds to the data is discarded while 1 means the opposite. Then the loss function is expressed as:

$$Loss_{mse} = ||(X - R(X \odot M)) \odot (1 - M)||_2^2 \qquad (11)$$

wherein $\odot$ is the element-wise product operation. In our experiments, mimicking the preprocessing method of the image data in the computer vision (CV) field

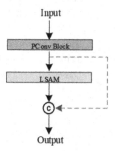

Fig. 5. The local spatial attention model. Taking the output from the upper layer of neural network as input, then passing through PConv Block and the local spatial attention mechanism, concatenating the output of two networks by channels to obtain result with local spatial information.

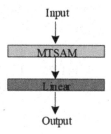

Fig. 6. The multidimensional temporal self-attention model. After the output of the previous layer goes through multidimensional temporal self-attention mechanism, expression containing timing information could be obtained. Since multidimensional temporal self-attention will change the size of input data, we apply the linear network to deal with this issue and get the output with temporal correlation.

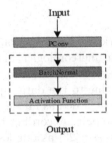

Fig. 7. The partial convolution block.

Table 1. The details of STA-Net

subnetwork_name	net_name	kernel_size	in_channels	out_channels	stride	activation function
LSAPConvU-Net	Net1	3	1	32	2	Hardswish
	Net2	3	64	128	2	Hardswish
	Net3	3	256	512	2	Hardswish
	Net4	3	1024	1024	2	Hardswish
	Net5	3	2048	2048	2	Hardswish
	Net6	3	6144	1536	1	LeakyReLU(0.3)
	Net7	3	4096	1024	1	LeakyReLU(0.3)
	Net8	3	2304	576	1	LeakyReLU(0.3)
	Net9	3	1216	304	1	LeakyReLU(0.3)
	Net10	3	609	1	1	LeakyReLU(1.)
MTSAPConvU-Net	Net1	3	1	32	2	Hardswish
	Net2	3	32	64	2	Hardswish
	Net3	3	64	128	2	Hardswish
	Net4	3	128	256	2	Hardswish
	Net5	3	256	512	2	Hardswish
	Net6	3	768	384	1	LeakyReLU(0.3)
	Net7	3	512	256	1	LeakyReLU(0.3)
	Net8	3	320	160	1	LeakyReLU(0.3)
	Net9	3	192	96	1	LeakyReLU(0.3)
	Net10	3	97	1	1	LeakyReLU(1.)
last layer	Net	3	2	1	1	LeakyReLU(1.)

before being input to the network [23–26], the data used was normalized to minimum-maximum normalization between $[-1, 1]$.

4   Experiments

To evaluate the imputation effectiveness of the spatiotemporal attention network (STA-Net) in the meteorological missing data reconstruction task, in three different missing rates: 25%, 50% and 75%, we performed a large number of experiments on two meteorological station datasets: Tibet and Qamdo. Meanwhile,

STA-net was compared with other three classical models: U-Net [48], BiLSTM [17] and PConv-Net [24]. Besides, to examine the effectiveness of each part and design attention, we also performed ablation experiments on the Tibet dataset with three missing rates: 25%, 50% and 75%.

4.1 Experimental Details and Evaluation Indicators

In training networks, the Adam algorithm [49] was used as the optimizer with an initial learning rate $1e^{-3}$. When the test rmse remained unchanged for 10 consecutive times, learning rate was reduced to one-tenth of the last time until $1e^{-8}$. The stochastic gradient descent (SGD) [50] was used to train our models where the batch size was set to 32, and training epoch was set to 300, 400 or 600 as required. Our all models have been trained and tested upon a server configured with Nvidia Tesla T4 GPUs, intel(R) Xeon(R) Gold 6242 CPU @ 2.80 GHz and 386G memory. Meantime, in order to measure the reconstruction performance of four models, we selected three different indicators, including mean absolute error (MAE) and root mean squared error (RMSE) [2,5,17,18], coefficient of determination (R^2) [51], computed as follows:

$$MAE = \frac{1}{n} \sum_{i=1}^{n} |x_i - y_i| \tag{12}$$

$$RMSE = \sqrt{\frac{1}{n} \sum_{i=1}^{n} (x_i - y_i)^2} \tag{13}$$

$$R^2 = 1 - \frac{\sum_{i=1}^{n} (x_i - y_i)^2}{\sum_{i=1}^{n} (x_i - \overline{x})^2} \tag{14}$$

wherein x_i and y_i are true and imputation values, respectively. $\overline{x}$ is mean value.

4.2 Datasets and Mask

TibetDataset. The original Tibet meteorological data set recorded the meteorological data observed hourly from 2009 to 2019. Used this data set, s was set to 61. After it went through the proposed data processing method, missing data was eliminated to avoid the impact of missing values contained in the data on the experimental accuracy. The non-missing data used in the experiment was obtained, named as TND. The TND contains 2918 numeric matrices and is divided into two parts: training set (80% of the total) and test set (20% of the total). The size of each numeric matrix is $(24, 61)$.

QamdoDataset. The original Qamdo meteorological data set records the meteorological data observed hourly from 1978 to 2020. We chose 37 meteorological observation stations in Qamdo area to use in the following experiment, that is, s is set to 37. The relevant data processing process is same as TND. Finally, we acquired the non-missing data used in the experiment, named as QND, in which 80% was used for training and 20% for testing and validation. It contained 4980 numeric matrices whose size was $(24, 37)$.

Mask. To simulate the actual situation, we also analyzed the change rule of missing values of meteorological stations and found that the missing values were random in time. Based on the findings, by randomly masking at time, we made binary mask matrix with different missing ratios whose dimension was the same as used data.

4.3   Experimental Results

Analysis of TND Experimental Results. To compare the performance of four models in meteorological missing data reconstruction, we conducted experiments on the TND dataset when missing rates were 25%, 50% and 75%, respectively. We use three evaluation metrics for comparison: MAE, RMSE, and R^2. Table 2 lists the performance comparison of STA-Net with other methods on the TND dataset, and the best results are shown in bold. Moreover, by filling the missing data of a single site within different missing rates, we try to show the performance difference between models from the side. Here, we select Lhasa weather station as a representative example, the experimental results are shown in Fig. 8, Fig. 9 and Fig. 10.

Table 2. Performance comparison results of different models with 25%, 50% and 75% missing rate on TND dataset.

Missing rate	Method	MAE($\downarrow$)	RMSE($\downarrow$)	R^2($\uparrow$)
25%	U-Net [48]	0.3594	0.9362	0.9980
	PConv-Net [24]	0.2249	0.6022	0.9962
	BiLSTM [17]	0.1638	0.4769	0.9975
	STA-Net(ours)	**0.1430**	**0.4093**	**0.9982**
50%	U-Net [48]	0.8113	1.4799	0.9770
	PConv-Net [24]	0.5648	1.0530	0.9883
	BiLSTM [17]	0.4248	0.8599	0.9920
	STA-Net(ours)	**0.3874**	**0.7536**	**0.9939**
75%	U-Net [48]	1.2252	1.8140	0.9654
	PConv-Net [24]	1.1098	1.6656	0.9707
	BiLSTM [17]	0.8915	1.4296	0.9778
	STA-Net(ours)	**0.8819**	**1.3496**	**0.9794**

As it can be observed from the results in Table 2 and figures, the performance of STA-Net is the best compared to other methods. For example, in the case with

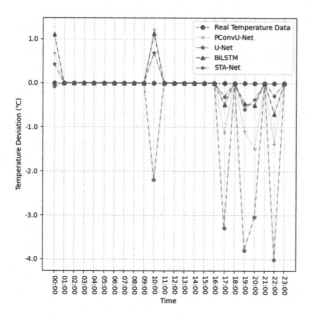

Fig. 8. The differences between reconstruction values Y and real values X and when the missing rate is 25% on TND test datasets, which are obtained by different models in Lhasa meteorological station.

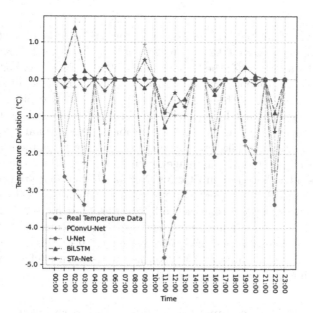

Fig. 9. The differences between reconstruction values Y and real values X and when the missing rate is 50% on TND test datasets, which are obtained by different models in Lhasa meteorological station.

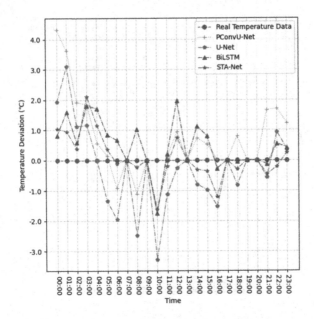

Fig. 10. The differences between reconstruction values Y and real values X and when the missing rate is 75% on TND test datasets, which are obtained by different models in Lhasa meteorological station.

the missing rate of 25% on test datasets and compared to the other three methods, the MAE declines by 60.21%, 36.42% and 12.70%; the RMSE declines by 56.28%, 32.03% and 14.17%; the R2 increases by 0.75%, 0.20% and 0.07%. At the same time, the experimental results also show that the multiple-meteorological-station data processing method proposed is effective and workable. Furthermore, the results also reflect that PConvU-Net performs better than U-Net, since partial convolution is proposed to solve the problem that ordinary convolutional neural network does not take into account the effect of the missing part on the extracted spatial features [24] in image repair field. Meanwhile, the performance of BiLSTM is much better than PConvU-Net and U-Net, since temporal dependence of site data self is stronger than spatial correlation reflected by matrix data generated by our data processing method. However, these models do not perform as well as STA-Net, since it makes full use of space-time correlation through the two attention mechanisms designed. The experimental results demonstrate that our constructed spatiotemporal attention neural network is effective in meteorological missing data reconstruction tasks.

Analysis of QND Experimental Results. To further evaluate the performance of our method in meteorological missing data reconstruction, we conducted experiments with 25%, 50% and 75% missing rate on QND dataset. Table 3 lists the performance comparison of STA-Net with other methods on the QND dataset, and the best results are shown in bold. It can be again observed that the designed data preprocessing method is effective and feasible, and generated datasets not only utilizes CNN-based networks to reconstruct missing data in space but also makes use of LSTM-based models to fill missing data in time sequence.

Table 3. Performance comparison results of different models with 25%, 50% and 75% missing rate on QND dataset

Missing rate	Method	MAE($\downarrow$)	RMSE($\downarrow$)	R^2($\uparrow$)
25%	U-Net [48]	0.2252	0.6088	0.9954
	PConv-Net [24]	0.1670	0.4642	0.9973
	BiLSTM [17]	0.1509	0.4400	0.9976
	STA-Net(ours)	**0.1427**	**0.4054**	**0.9980**
50%	U-Net [48]	0.5420	1.0274	0.9869
	PConv-Net [24]	0.4232	0.8135	0.9918
	BiLSTM [17]	0.3937	0.7945	0.9923
	STA-Net(ours)	**0.3303**	**0.6574**	**0.9946**
75%	U-Net [48]	0.9463	1.4563	0.9738
	PConv-Net [24]	0.9508	1.4454	0.9743
	BiLSTM [17]	0.8543	1.3749	0.9769
	STA-Net(ours)	**0.6480**	**1.0326**	**0.9869**

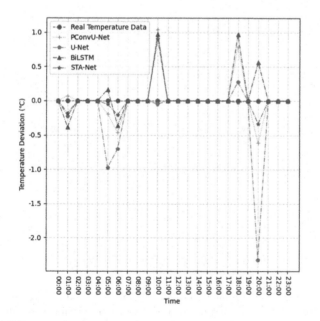

Fig. 11. The differences between reconstruction values Y and real values X and when the missing rate is 25% on QND test datasets, which are obtained by different models in Qamdo meteorological station.

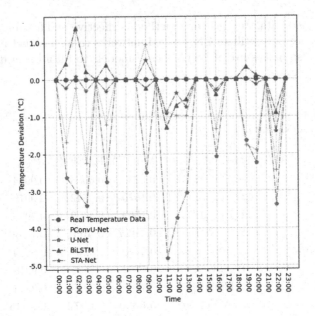

Fig. 12. The differences between reconstruction values Y and real values X and when the missing rate is 50% on QND test datasets, which are obtained by different models in Qamdo meteorological station.

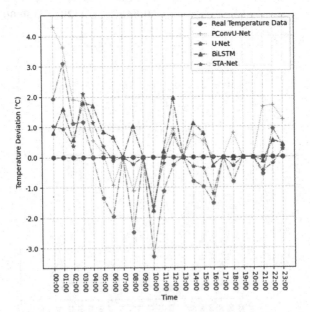

Fig. 13. The differences between reconstruction values Y and real values X and when the missing rate is 75% on QND test datasets, which are obtained by different models in Qamdo meteorological station.

Besides, due to the meteorological stations in Qamdo area are relatively closer geographically compared with TND, the QND has stronger spatial correlation. Hence the performance gaps between PConvU-Net and BiLSTM are smaller. However, since these two models only consider space or time correlation, their reconstruction effects are not as good as STA-Net. As the same as previous experiments, here, Qamdo meteorological station is selected to side-on show analogous conclusions, the corresponding experimental results are shown in Fig. 11, Fig. 12 and Fig. 13. To sum up, the experimental results again show that our constructed spatiotemporal attention neural network is effective in meteorological missing data reconstruction tasks.

4.4   Ablation Experiments

We conduct the ablation experiments on the TND dataset with the 25%, 50% and 75% missing rate to evaluate that two different factors have an impact on network performance.

Effectiveness of the Spatiotemporal Attention Model. To validate the effectiveness and superiority of the local spatial attention model (LSAM) and multidimensional temporal self-attention model (MTSAM), we take the PConvU-Net as baseline network architecture. Then conduct the corresponding ablation experiments with the 25%, 50% and 75% missing rate, respectively. The results of ablation experiment are shown Table 4.

Table 4. The results of ablation experiments on STA-Net

Missing rate	Method	MAE($\downarrow$)	RMSE($\downarrow$)	R^2($\uparrow$)
25%	Baseline	0.2249	0.6022	0.9962
	Baseline+LSAM	0.1897	0.5117	0.9972
	Baseline+MTSAM	0.1487	0.4216	0.9981
	STA-Net(ours)	**0.1430**	**0.4093**	**0.9982**
50%	Baseline	0.5648	1.0530	0.9883
	Baseline+LSAM	0.5201	0.9703	0.9899
	Baseline+MTSAM	0.4154	0.7989	0.9931
	STA-Net(ours)	**0.3874**	**0.7536**	**0.9939**
75%	Baseline	1.1098	1.6656	0.9707
	Baseline+LSAM	1.0011	1.5165	0.9753
	Baseline+MTSAM	0.9363	1.4252	0.9781
	STA-Net(ours)	**0.8819**	**1.3496**	**0.9794**

Obviously, since PConvU-Net does not consider local spatial and temporal relations, its performance is worst. When LSAM or MTSAM is combined, due to the using attention mechanisms, spatial or temporal information is obtained so that the effect of reconstruction can get improved. When both LSAM and MTSAM are integrated, the performance will be the best because of obtaining spatiotemporal information.

Effectiveness of the Hyperparameter Hidden_Size. The hyperparameter hidden_size in LSTM is closely related to the types of factors affecting the temperature of meteorological stations. Typically, the performance of LSTM based models is influenced by it. Here, for simplicity, we conduct the ablation experiments with the 25%, 50% and 75% missing rate on MTSAPConvU-Net. The hidden size is set to 1, 32, 64, 128, and 256, respectively, to observe the impact on MTSAPConvU-Net. The results of ablation experiment are shown in Table 5, which show the impact from hidden size into MTSAPConvU-Net.

Table 5. The results of ablation experiments on MTSAPConvU-Net

Missing rate	Hidden_size	MAE(↓)	RMSE(↓)	R^2(↑)
25%	1	0.1991	0.5523	0.9968
	32	0.1774	0.4883	0.9975
	64	0.1773	0.4886	0.9975
	128	**0.1487**	**0.4216**	**0.9981**
	256	0.1731	0.4836	0.9975
50%	1	0.5208	0.9826	0.9898
	32	0.4549	0.8655	0.9921
	64	0.4750	0.8990	0.9913
	128	**0.4154**	**0.7989**	**0.9931**
	256	0.4186	0.8024	0.9929
75%	1	1.0223	1.5736	0.9744
	32	**0.0.9363**	**1.4252**	**0.9781**
	64	0.9928	1.4497	0.9762
	128	0.9535	1.4491	0.9774
	256	0.9438	1.4340	0.9779

It is observed from the experimental results in Table 5 that the performance comparison in different settings of the hidden size. Roughly, as the hidden size becomes larger, the performance of the model increases first and then drop. We infer that when the hidden size is small, multiple factors decomposed cannot fully express the data, which suppresses the performance of the model. Especially when it is set to 1 as an extreme setting, only one constraint has an impact on the data and the model performance theoretically is the worst in this case. However, if the size is too big, a huge number of factors may lead to over expression of the data, which will also degrade the performance of the model due to resource waste and reduce the time efficiency. Hence, the hidden size should be cautiously set in a reasonable range to achieve the best performance.

5   Conclusion and Future Work

In this research, we mainly designed a now network architecture STA-Net to fill missing data in multiple meteorological station data. We first proposed a simply novel a multiple-meteorological-station data processing method, making generated data able to utilize currently the most popular CNN to fill the missing values. At the same time, in order to exploit the spatio-temporal correlation in data as much as possible, we designed two attention mechanisms: local spatial attention mechanism and multidimensional temporal self-attention mechanism. Through a large number of experiments, it has been proved that the proposed data processing method is feasible. By conducting the ablation study, we further confirmed that the two attention models could greatly improve the effect of missing value reconstruction. In this way, our work can provide a meaningful reference for the repair of missing values of multiple meteorological station data, even other kinds of data. In the future, we will further extend our current framework to be applicable to support other types of meteorological data, such as wind speed, relative humidity and so on. In addition, better network models could be further exploited to improve the accuracy and time.

Acknowledgment. This work is funded by the National Natural Science Foundation of China (No. 42265010, No. 62162053, No. 62062059, No. 62166032), Natural Science Foundation of Qinghai Province (No. 2023-ZJ-906M), Youth Scientific Research Foundation of Qinghai University (No. 2022-QGY-6) and the Open Project of State Key Laboratory of Plateau Ecology and Agriculture, Qinghai University (No. 2020-ZZ-03).

References

1. Peterson, T.C., Vose, R.S.: An overview of the global historical climatology network temperature database. Bull. Am. Meteor. Soc. **78**(12), 2837–2850 (1997)
2. Lompar, M., Lalić, B., Dekić, L., Petrić, M.: Filling gaps in hourly air temperature data using debiased ERA5 data. Atmosphere **10**(1), 13 (2019). Publisher: MDPI
3. Lara-Estrada, L., Rasche, L., Sucar, L.E., Schneider, U.A.: Inferring missing climate data for agricultural planning using Bayesian networks. Land **7**(1), 4 (2018). Publisher: MDPI
4. Huang, M., et al.: Air temperature optima of vegetation productivity across global biomes. Nat. Ecol. Evol. **3**(5), 772–779 (2019). Publisher: Nature Publishing Group UK London
5. Henn, B., Raleigh, M.S., Fisher, A., Lundquist, J.D.: A comparison of methods for filling gaps in hourly near-surface air temperature data. J. Hydrometeorol. **14**(3), 929–945 (2013). Publisher: American Meteorological Society

6. Afrifa-Yamoah, E., Mueller, U.A., Taylor, S., Fisher, A.: Missing data imputation of high-resolution temporal climate time series data. Meteorol. Appl. **27**(1), 1873 (2020). Publisher: Wiley Online Library
7. Park, J., Yoon, D., Seol, S.J., Byun, J.: Reconstruction of seismic field data with convolutional U-Net considering the optimal training input data. In: SEG International Exposition and Annual Meeting. OnePetro (2019)
8. Daly, S., Davis, R., Ochs, E., Pangburn, T.: An approach to spatially distributed snow modelling of the Sacramento and San Joaquin Basins, California. Hydrol. Processes **14**(18), 3257–3271 (2000). Publisher: Wiley Online Library
9. Tobin, C., Nicotina, L., Parlange, M.B., Berne, A., Rinaldo, A.: Improved interpolation of meteorological forcings for hydrologic applications in a swiss alpine region. J. Hydrol. **401**(1–2), 77–89 (2011). Publisher: Elsevier
10. Garen, D.C., Johnson, G.L., Hanson, C.L.: Mean areal precipitation for daily hydrologic modeling in mountainous regions 1. JAWRA J. Am. Water Resourc. Assoc. **30**(3), 481–491 (1994). Publisher: Wiley Online Library
11. Pape, R., Wundram, D., Löffler, J.: Modelling near-surface temperature conditions in high mountain environments: an appraisal. Climate Res. **39**(2), 99–109 (2009)
12. Stahl, K., Moore, R., Floyer, J., Asplin, M., McKendry, I.: Comparison of approaches for spatial interpolation of daily air temperature in a large region with complex topography and highly variable station density. Agric. Forest Meteorol. **139**(3–4), 224–236 (2006). Publisher: Elsevier
13. Belachsen, I., Broday, D.M.: Imputation of missing pm2. 5 observations in a network of air quality monitoring stations by a new knn method. Atmosphere **13**(11), 1934 (2022). Publisher: MDPI
14. Chen, M., Zhu, H., Chen, Y., Wang, Y.: A novel missing data imputation approach for time series air quality data based on logistic regression. Atmosphere **13**(7), 1044 (2022). Publisher: MDPI
15. Kihoro, J., Athiany, K., et al.: Imputation of incomplete nonstationary seasonal time series data. Math. Theory Model. **3**(12), 142–154 (2013)
16. Wang, H., Yuan, Z., Chen, Y., Shen, B., Wu, A.: An industrial missing values processing method based on generating model. Comput. Netw. **158**, 61–68 (2019). Publisher: Elsevier
17. Cao, W., Wang, D., Li, J., Zhou, H., Li, L., Li, Y.: Brits: bidirectional recurrent imputation for time series. In: Advances in Neural Information Processing Systems, vol. 31 (2018)
18. Zhang, Y.F., Thorburn, P.J., Xiang, W., Fitch, P.: SSIM - A deep learning approach for recovering missing time series sensor data. IEEE Internet Things J. **6**(4), 6618–6628 (2019). Publisher: IEEE
19. Zhang, Y., Thorburn, P.J.: A dual-head attention model for time series data imputation. Comput. Electron. Agric. **189**, 106377 (2021). Publisher: Elsevier
20. Dagtekin, O., Dethlefs, N.: Imputation of partially observed water quality data using self-attention LSTM. In: 2022 International Joint Conference on Neural Networks (IJCNN), pp. 1–8 (2022). Publisher: IEEE
21. Hochreiter, S., Schmidhuber, J.: Long short-term memory. Neural Comput. **9**(8), 1735–1780 (1997). Publisher: MIT Press
22. Xie, C., Huang, C., Zhang, D., He, W.: BiLSTM-I: a deep learning-based long interval gap-filling method for meteorological observation data. Int. J. Environ. Res. Publ. Health **18**(19), 10321 (2021). Publisher: MDPI
23. Pathak, D., Krahenbuhl, P., Donahue, J., Darrell, T., Efros, A.A.: Context encoders: feature learning by inpainting. In: Proceedings of the IEEE Conference on Computer Vision and Pattern Recognition, pp. 2536–2544 (2016)

24. Liu, G., Reda, F.A., Shih, K.J., Wang, T.-C., Tao, A., Catanzaro, B.: Image inpainting for irregular holes using partial convolutions. In: Ferrari, V., Hebert, M., Sminchisescu, C., Weiss, Y. (eds.) ECCV 2018. LNCS, vol. 11215, pp. 89–105. Springer, Cham (2018). https://doi.org/10.1007/978-3-030-01252-6_6

25. Shin, Y.G., Sagong, M.C., Yeo, Y.J., Kim, S.W., Ko, S.J.: Pepsi++: fast and lightweight network for image inpainting. IEEE Trans. Neural Netw. Learn. Syst. **32**(1), 252–265 (2020). Publisher: IEEE

26. Peng, J., Liu, D., Xu, S., Li, H.: Generating diverse structure for image inpainting with hierarchical VQ-VAE. In: Proceedings of the IEEE/CVF Conference on Computer Vision and Pattern Recognition, pp. 10775–10784 (2021)

27. Kadow, C., Hall, D.M., Ulbrich, U.: Artificial intelligence reconstructs missing climate information. Nat. Geosci. **13**(6), 408–413 (2020). Publisher: Nature Publishing Group UK London

28. Irrgang, C., Saynisch-Wagner, J., Dill, R., Boergens, E., Thomas, M.: Self-validating deep learning for recovering terrestrial water storage from gravity and altimetry measurements. Geophys. Res. Lette. **47**(17), 2020–089258 (2020). Publisher: Wiley Online Library

29. Dong, J., Yin, R., Sun, X., Li, Q., Yang, Y., Qin, X.: Inpainting of remote sensing SST images with deep convolutional generative adversarial network. IEEE Geosci. Remote Sens. Lett. **16**(2), 173–177 (2018). Publisher: IEEE

30. Yu, J., Wu, B.: Attention and hybrid loss guided deep learning for consecutively missing seismic data reconstruction. IEEE Trans. Geosci. Remote Sens. **60**, 1–8 (2021). Publisher: IEEE

31. Yu, J., Lin, Z., Yang, J., Shen, X., Lu, X., Huang, T.S.: Generative image inpainting with contextual attention. In: Proceedings of the IEEE Conference on Computer Vision and Pattern Recognition, pp. 5505–5514 (2018)

32. Bahdanau, D., Cho, K., Bengio, Y.: Neural machine translation by jointly learning to align and translate. arXiv preprint arXiv:1409.0473 (2014)

33. Vaswani, A., et al.: Attention is all you need. In: Advances in Neural Information Processing Systems, vol. 30, pp. 5998–6008 (2017)

34. Chaudhari, S., Mithal, V., Polatkan, G., Ramanath, R.: An attentive survey of attention models. ACM Trans. Intell. Syst. Technol. (TIST) **12**(5), 1–32 (2021). Publisher: ACM, New York, NY, USA

35. Chen, Y., Liu, S., Wang, X.: Learning continuous image representation with local implicit image function. In: Proceedings of the IEEE/CVF Conference on Computer Vision and Pattern Recognition, pp. 8628–8638 (2021)

36. Karras, T., et al.: Alias-free generative adversarial networks. In: Advances in Neural Information Processing Systems, vol. 34, pp. 852–863 (2021)

37. Liu, H., et al.: Video super-resolution based on deep learning: a comprehensive survey. Artif. Intell. Rev. **55**(8), 5981–6035 (2022). Publisher: Springer

38. Hays, J., Efros, A.A.: Scene completion using millions of photographs. ACM Tran. Graph. (ToG) **26**(3), 4 (2007). Publisher: ACM, New York, NY, USA

39. Zeiler, M.D., Taylor, G.W., Fergus, R.: Adaptive deconvolutional networks for mid and high level feature learning. In: 2011 International Conference on Computer Vision, pp. 2018–2025 (2011). Publisher: IEEE

40. Shi, W., et al.: Is the deconvolution layer the same as a convolutional layer? arXiv preprint arXiv:1609.07009 (2016)

41. Rosenbaltt, F.: The perceptron - a perceiving and recognizing automation. Cornell Aeronautical Laboratory (1957)

42. Howard, A., et al.: Searching for mobilenetv3. In: Proceedings of the IEEE/CVF International Conference on Computer Vision, pp. 1314–1324 (2019)

43. Krizhevsky, A., Sutskever, I., Hinton, G.E.: Imagenet classification with deep convolutional neural networks. Commun. ACM **60**(6), 84–90 (2017). Publisher: ACM, New York, NY, USA
44. Maas, A.L., Hannun, A.Y., Ng, A.Y., et al.: Rectifier nonlinearities improve neural network acoustic models. In: Proceedings of ICML, Atlanta, Georgia, USA, vol. 30, p. 3 (2013)
45. Xu, B., Wang, N., Chen, T., Li, M.: Empirical evaluation of rectified activations in convolutional network. arXiv preprint arXiv:1505.00853 (2015)
46. Ioffe, S., Szegedy, C.: Batch normalization: accelerating deep network training by reducing internal covariate shift. In: International Conference on Machine Learning, pp. 448–456 (2015). Publisher: PMLR
47. He, K., Zhang, X., Ren, S., Sun, J.: Deep residual learning for image recognition. In: Proceedings of the IEEE Conference on Computer Vision and Pattern Recognition, pp. 770–778 (2016). Publisher: IEEE
48. Ronneberger, O., Fischer, P., Brox, T.: U-net: convolutional networks for biomedical image segmentation. In: Navab, N., Hornegger, J., Wells, W.M., Frangi, A.F. (eds.) MICCAI 2015. LNCS, vol. 9351, pp. 234–241. Springer, Cham (2015). https://doi.org/10.1007/978-3-319-24574-4_28
49. Kingma, D.P., Ba, J.: Adam: a method for stochastic optimization. arXiv preprint arXiv:1412.6980 (2014)
50. Bottou, L.: Stochastic gradient descent tricks. In: Montavon, G., Orr, G.B., Müller, K.-R. (eds.) Neural Networks: Tricks of the Trade. LNCS, vol. 7700, pp. 421–436. Springer, Heidelberg (2012). https://doi.org/10.1007/978-3-642-35289-8_25
51. Junninen, H., Niska, H., Tuppurainen, K., Ruuskanen, J., Kolehmainen, M.: Methods for imputation of missing values in air quality data sets. Atmos. Environ. **38**(18), 2895–2907 (2004). Publisher: Elsevier

Embedding Entity and Relation for Knowledge Graph by Probability Directed Graph

Chunming Yang[1,2]($\boxtimes$), Xinghao Song[3], Yue Luo[1], and Litao Zhang[1]

[1] School of Computer Science and Technology,
Southwest University of Science and Technology, Mianyang 621010, Sichuan, China
[2] Sichuan Big Data and Intelligent Systems Engineering Technology Research Center,
Mianyang 621010, Sichuan, China
yangchunming@swust.edu.cn
[3] Sichuan Branch of China Telecom Co., Ltd., Chengdu 610015, Sichuan, China

Abstract. Knowledge graph embedding (KGE) represents entities and relationship as low dimensional dense vectors in knowledge graphs (KGs), and to improve the computational efficiency of downstream tasks. This paper regards the semantic relations between the head and tail entities in the KGs as semantically probability transfers, and proposes a method based on the structured probability model for KGE. This method considers the relations as nodes and uses probability-directed graphs to model the KGs. The scoring function is defined as a probability distribution that represents the directed transitivity between the entities and relations. Finally, the function is used to infer the probability that the triples are true. Experimental results on several standard datasets show that this method achieves better results in complex relations and uneven distribution of triples.

Keywords: Knowledge graph embedding · Directed Relational Transitivity · Probability Graph Model · Scoring Function

1 Introduction

Knowledge graphs(KGs) are collections of triples (head entity, relation, and tail entity) which represent the real-world facts. However, there are a large number of missing relations in KGs. It makes the data sparse and increases the computational difficulty of downstream tasks, such as recommendation systems and automatic question answering [1,2]. Knowledge Graph Embedding(KGE) maps entities and relations to a low-dimensional continuous vector space by representation learning. It can quickly predicts the semantic relations between entities.

Most of the existing KGE methods use the scoring function $f(h, r, t)$ to describe the possibility of a candidate triple (h, r, t), where h, r and t denote the head entity, relation and tail entity in the triples respectively. The goal of optimization is to score the true triple $f(h, r, t)$ higher than false triple $f(h', r, t)$

© The Author(s), under exclusive license to Springer Nature Singapore Pte Ltd. 2024
B. Luo et al. (Eds.): ICONIP 2023, CCIS 1961, pp. 53–65, 2024.
https://doi.org/10.1007/978-981-99-8126-7_4

or $f(h, r, t')$. The existing scoring function describe the connection between the head entities h, the tail entities t and relations r in the triples. Such as, TransE [3] series regard the relation r as the "translation distance" from h to t, RESCAl [4] uses a bi-linear relation matrix to represents the semantic connection between entities and relations, and deep learning method replace the bi-linear relation matrix with neural network. In addition to the distance, matrix and neural network, the semantic relationship between the entities of the triplet can also be regarded as the probability transfer between the head and tail entities. That is to say, h, r and t are regarded as the three random variables in KG, h may affect the r, and r also affects the t. For example, in the triple composed of singer "Taylor Swift" as the h, the r will most likely be "singing", "arranging", etc., rather than "programming" or "therapy", When the r is "arrangement", the t will most likely be the "song".

Therefore, this paper regards the semantic relationship between entities in KGs as the probability transfer of the probability-directed graph, and regards the relationship as a node in it. And it proposes a KGE method based on the structured probability model – Probability Graph Model Embedding(PGME). This method defines the scoring function $f(h, r, t)$ as the probability $p(h, r, t)$ of the triple is true, and expresses the directed transitivity between entities and relations through the decomposition of the probability distribution, so as to calculate the probability that the whole triples are true. Experimental results on the benchmark datasets show that PGME has higher prediction accuracy.

The rest of the paper is organized as follows: Section "Related Work" introduces the basic concepts and embedding approach of KG. Section "Embedding Entity and Relation for Knowledge Graph Completion by Probability Directed Graph Model" describes the proposed PGME in detail. Section "Experiments and analysis of results" compares PGME with several embedding models followed by a conclusion in Section "Conclusion".

2   Related Work

KGE maps the entities and relations in KG into the low-dimensional and dense vectors while keeping the semantic relations unchanged, so as to improve the computational efficiency of downstream tasks. The mapping methods mainly include translation models, bi-linear models and deep learning models.

Translation models map the relations and entities in KG to the same vector space, and regard the relations as translation operations from head to tail entities. TransE [3] is the most representative that shows good results in the simple relation completion of KG. However, TransE does not perform well on 1-N, N-1 and N-N relations. TransH [5] solves the problem about of the complex relations by modeling the relations as a hyperplane. TransR [6] supposes that the semantics and attributes of the same entities should be different for different relations, so the entities and relations are mapped into different vector spaces to express different semantics. TransD [7] constructs a dynamic mapping matrix for each entity-relation pair by considering the diversity of entities and relations at the same time, so as to have fewer parameters and no matrix vector multiplication. In addition, there are TransF [8], ManifoldE [9], TransM [10], TranSpare [11] and other translation-based models, and these improved models have all achieved good performance. However, as the translation model is projected by different matrices of the head and tail entities, it is unable to accurately reflect the semantic relations between entities, so the coordination between entities is poor, and it is still difficult to deal with the KG of complex relations.

Bi-linear models hold that the connection between entities and relations in KG can be represented by the relation matrix. RESCAL [4] uses vectors to capture the underlying semantics of entities and each relation is represented as a matrix, which models heap interactions between potential factors. Since each relation is a matrix, the computational complexity of this model is very high. DistMult [11] limits the relation matrix to a diagonal matrix, which is equivalent to expressing the relation as a vector, reducing the complexity of the model. ComplEx [12] represents entities and relations as complex vectors to capture antisymmetric relations. Analogy [13] expresses the relation as a normal matrix B_r to capture more relations. SimplE [14] learns two independent embedding vectors for each relation, one is for the normal relation and the other is for the inverse relation. TuckER [15] treats KG as a third-order binary tensor, each element corresponds to a fact triple, and decomposes the tensor into a core tensor multiplied by the product of three factor matrices in three modes, which is fully expressed. However, the core tensor is not constrained in TuckER, and the correlation between entities and relations is not better handled. So it is prone to the risk of over-fitting. In this regard, CP decomposition can be used to solve this problem [16].

Deep learning models use neural networks instead of bi-linear transformations to correlate entities and relations in different dimensions. NTN [17] uses tensors to express complex semantic relations between entities and relations, it requires a large number of triples to be fully studied because of the high complexity of tension calculation. Some models like R-GCN [18], ConvE [19] and ConvKB [20] use convolution to learn the vectored representation of entities and relations in KG. R-GCN uses relational graph convolutions networks to model relational paths in KG. ConvE converts a one-dimensional entity and relation vector into a two-dimensional matrix for convolution. ConvKB only uses one-dimensional convolution. KBGAN [21] uses generative adversarial networks to enhance existing KGE models.

In KGs, the relation reflects the semantic transmission path between entities, with interactions between the head entities, relations, and the tail entities in triples. In addition to using distances, matrices, and neural networks to reflect this interaction, it can also be characterized as a probabilistic transfer of head-tail entities. That is, KG can be modeled using the probability graph, and the head entity, tail entity and relation are regarded as three random variables. The relation are nodes in the probability graph, with directed edges indicating causal relationships between the random variables. Therefore, the scoring function $f(h, r, t)$ is defined as the probability $p(h, r, t)$ that the triple is true. And the probability is calculated through the probability-distribution. The relationship in KGs is regarded as the node in the probability directed graph, and the probability is used to represent the semantic transfer between the head and tail entities, which can well express the complex relationship(1-N, N-N, etc.) [22], and can learn more accurate embedded representation.

3   Embedding Entity and Relation for Knowledge Graph by Probability Directed Graph

3.1   Problem Definition

Knowledge Graph Embedding(KGE): For a given KG $G = (\mathcal{E}, \mathcal{R})$, let $\mathcal{E}$ denote the set of all entities, $\mathcal{R}$ the set of all relations, and the triple (h, r, t) denotes a fact in KG. The KGE learns vectors $\boldsymbol{h}, \boldsymbol{t} \in \mathbb{R}^{d_e}$ for $h, t \in \mathcal{E}$, and $\boldsymbol{r} \in \mathbb{R}^{d_r}$ for $r \in \mathcal{R}$, others learn projection matrices $M_r \in \mathbb{R}^{d_r \times d_e}$ for relations, and true triples are evaluated by specific scoring functions. The embedding of entities and relations is optimized and learned through the score function $f(h, r, t)$ $\mathcal{E} \times \mathcal{R} \times \mathcal{E} \to \mathbb{R}$ and cost function $\mathcal{L}(\Delta, \Delta', \Theta)$, defined over a set of positive triples Δ, set of negative triples Δ', and the parameters Θ is optimized.

Probabilistic Directed Graph: Set $G = (V, E)$ be a directed graph, $V = \{v_1, v_2, \ldots, v_n\}$ is the set of nodes, E the set of edges, and G is Acyclic Graph. Each node v_i in G represents a random variable X_i in the probability digraph. Let π_i denote the set of parent nodes of node v_i, then X_{π_i} denotes the parent of random variable X_i. The arrows connecting the two nodes indicate that two random variables have a causal relationship, or unconditionally independent. The probability directed graph model defines the joint probability $P(X_1, X_2, \ldots, X_n)$ as:

$$P(X_1, X_2, \ldots, X_n) = \prod_{i=1}^{n} P(X_i | X_{\pi_i}) \tag{1}$$

3.2   PGME Model

The structured information of triples in KG reflects semantic connections between entities pass on the relation path. In order to represent this transfer relation, we regard the entities and relations of triples as three random variables, and model them with a probability graph, and propose a KGE method based on structured probability model–PGME. PGME defines the scoring function $f(h, r, t)$ as the probability $p(h, r, t)$ of the triple (h, r, t) is true, that is, $f(h, r, t) = p(h, r, t)$. In order to model KG more accurately with probabilistic directed graph models, we denote the directed relation in KG as nodes in the graph, as shown in Fig. 1 (a), and the probability distribution $p(h, r, t)$ is modeled by the constructed probability model. Considering that the three random variables h, r and t have order-dependent relations, and the probability distribution of the triples h, r, t are also dependent on h, r and t respectively, as shown in Fig. 1 (b). Therefore, we use a probabilistic directed graph to model the probability of true triples and use this probability to represent the scoring function of triples: $f(h, r, t) = p(h, r, t) = p(h) p(r|h) p(t|h, r)$.

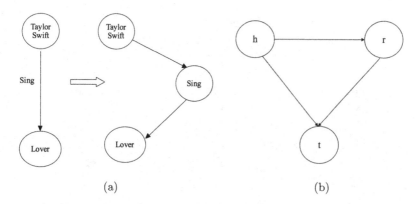

Fig. 1. Embedding entity and relation by probability directed graph model

In the specific model, we use the single-layer fully connected neural network to represent the probability distribution of these three factors, as follows:

$$p(h) = \sigma\left(\boldsymbol{W}_1\boldsymbol{h}^T + b_1\right) \tag{2}$$

$$p(r|h) = \sigma[\boldsymbol{W}_2\,[\boldsymbol{h},\boldsymbol{r}]^T + b_2] \tag{3}$$

$$p(t|h,r) = \sigma[\boldsymbol{W}_3\,[\boldsymbol{h},\boldsymbol{r},\boldsymbol{t}]^T + b_3] \tag{4}$$

KGE maps entities and relations to a continuous and low-dimensional vector space, with the learning vectors $\boldsymbol{h}, \boldsymbol{t} \in \mathbb{R}^{d_e}$ denote entities, and the learning vector $\boldsymbol{r} \in \mathbb{R}^{d_r}$ denotes relations.

$$\sigma(x) = \frac{1}{1 + e^{-x}} \tag{5}$$

$[\boldsymbol{x}, \boldsymbol{y}]$ denotes the "join" calculations between vector $\boldsymbol{x}$ and vector $\boldsymbol{y}$, where the "join" calculation can take many forms. For example, the product: $[\boldsymbol{x}, \boldsymbol{y}] = \boldsymbol{x} \odot \boldsymbol{y}$, where $\odot$ means the multiplication of elements at the corresponding positions of two vectors, adding: $[\boldsymbol{x}, \boldsymbol{y}] = \boldsymbol{x} + \boldsymbol{y}$, and splicing: $[\boldsymbol{x}, \boldsymbol{y}] = (\boldsymbol{x}, \boldsymbol{y})$, $(\boldsymbol{x}, \boldsymbol{y})$ denotes a 2d vector, and we tested these three "join" calculation in the experiment.

The overall structure of the PGME is shown in Fig. 2, which does not add additional parameters in addition to the three fully connected networks compared to other knowledge representation learning models. It is related to the scoring function $f_{dist}(h, r, t) = \boldsymbol{h}^T\boldsymbol{M}_r\boldsymbol{t}$ of DistMult [17]. DistMult uses a diagonal matrix $\boldsymbol{M}_r$ to denote each relation or a vector $\boldsymbol{r}$. So the scoring function is represented as $f_{dist}(h, r, t) = \boldsymbol{1}\,(\boldsymbol{h} \odot \boldsymbol{r} \odot \boldsymbol{t})^T$, where $\boldsymbol{1}$ means a matrix which $1 \times d$ that values are all 1. It can be seen that when there is no sigmoid activation function when $\boldsymbol{W}_3 = \boldsymbol{1}$, $b_3 = 0$, $f_{dist}(h, r, t)$ is a special case of $p(t|h, r)$ in Eq. 4. And the scoring function of DistMult is a symmetric function that $f_{dist}(h, r, t) = f_{dist}(t, r, h)$, the quadrant transformations that DisMult performs

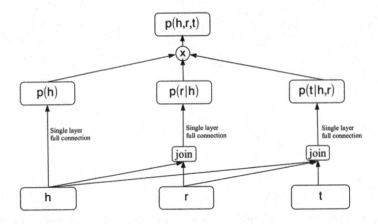

Fig. 2. Structure illustration of PGME

on entities and relations embedding vectors are stretched only and cannot model asymmetric relations. However, most of relations in KG are complex (1-N, N-1, multi-level relations, etc.), data sparseness, uneven distribution of triples, etc. Which can easily affect the learning of the model and reduce the accuracy of link prediction, so it is difficult to capture the structural information in KGs. PGME denotes entities and relations as nodes in the directed probability graph model, not only reflects the directed transmission relations in the triples, but also can use the probability directed graph to deal with the complex relations well when there are more and more 1-N, N-1 relations in KG, reflects fully the structured information of KG, and makes it more expressive.

3.3 Model Learning

To learn PGME, we define the optimized objective function as:

$$\mathcal{L} = -\sum_{(h,r,t)\in\mathcal{B}(h,r,t)} l\,(h,r,t)\,logf\,(h,r,t) + (1 - l\,(h,r,t))\log\,(1 - f\,(h,r,t))$$

(6)

where $\mathcal{B}(h,r,t) = \Delta\bigcup\Delta'$, Δ denotes the set of fact triples, Δ' gets the set of negatively sample triples from positive triples, i.e., $\mathcal{B}(h,r,t)$ is the set composed of positive triples and negative sampled triples.

$$l\,(h,r,t) = \begin{cases} 0, & if\ (h,r,t)\in\Delta' \\ 1, & if\ (h,r,t)\in\Delta \end{cases}$$

(7)

where $f\,(h,r,t) = p\,(h,r,t) = p\,(h)\,p\,(r|h)\,p(t|h,r)$, the negative sampling triples set Δ' is generated by the true triples randomly replacing the head or tail entities with the same number of samples as the true triples.

Table 1. Details and partitioning of each data set.

Dataset	Rel	Ent	Train	Test	Valid
FB15K237	237	14,541	272,115	20,466	17,535
WN18RR	11	40,934	86,835	3,134	3,034

4 Experiments and Analysis of Results

4.1 Datasets

In order to verify the effect of the model, two standard data sets, FB15K237 [23] and WN18RR [24], were used to evaluate PGME. The two data sets were obtained by deleting one of the reversible relation pairs on the basis of FB15K and WN18. There was no reversible relation pair and no test leakage [25] would occur due to the redundant reverse relation. The data set is shown in Table 1.

4.2 Evaluation Metrics

We evaluate the accuracy of the model using link prediction, as measured by mean reciprocal rank (MRR) and Hits@N. For a triples (h, r, t) in the test set, the scores of all candidate triples (h', r, t) and (h, r, t') are calculated by scoring function $f(h, r, t)$ in the experiment, and the scores were sorted in ascending order. The ranking of the correct answer triplet was denoted as R_h, and the ranking of (h, r, t) in the candidate set (h, r, t') is denoted as R_t. $h', t' \in \mathcal{E}$ means all entities, when $t' \neq t, h' \neq h$, the candidate triples (h, r, t') and (h', r, t) may contain other true triples, these true triples do not participate in the calculation and only the unreal triples and the correct answers participate in the calculation.

MRR is to take the reciprocal of the ranking to make the result fall between [0,1]. The closer the result is to 1, the better the model effect. The calculation method is as follows:

$$MRR = \frac{1}{2n_{test}} \sum_{(h,r,t) \in test} \frac{1}{R_h} + \frac{1}{R_t} \tag{8}$$

where n_{test} means the number of test triples. Hit@N represents the proportion of the top N correct answers in all candidate sets, and its calculation method can be expressed as:

$$Hit@N = \frac{count\{R_h \leq N\} + count R_t \leq N}{2n_{test}} \tag{9}$$

4.3 Experiment Setting

We implemented PGME on OpenKE [26], train the model using Adam and set the batch size $b = n_{train}/100$, where n_{train} represents the number of samples the

Table 2. Experimental parameters setting.

	Train Times	Learning rate	Number of batches	dimensions
FB15K237	1,500	0.0005	100	150
WN18RR	1,500	0.01	100	100

base train epochs $e = 1000$. During the training, evaluate the model with validation set every 100 epochs and save the best model for testing, if the best result is in the last 100 epochs, add 500 epochs on the base epochs to retrain the model. We search and select the learning rate $\gamma \in \{0.0005, 0.001, 0.003, 0.005, 0.01\}$, vector dimension d from $\{100, 150, 200, 250, 300\}$, the product "join" operation is used between vectors in the experiment. Selecting the best results for each dataset by tuning parameters is shown in Table 2.

4.4 Experimental Results and Analysis

In order to verify the performance of the algorithm, the proposed model is compared with the classical models based on transfer idea, tensor decomposition, deep learning and graph neural networks, as well as the recent models. The experimental results of the comparison model are from the original literature, and the missing results of some indicators are replaced by '-'. The experimental comparison results are shown in Table 3.

As can be seen from Table 3, PGME has only 0.002 difference with QuatE in Hit@3 of FB15K237 data set, which is superior to the optimal value in the comparison model in other indicators. In the MRR of the WN8RR dataset,

Table 3. Link prediction results on FB15K237 and WN18RR

Method	FB15K237				WN18RR			
	MRR	Hit@10	Hit@3	Hit@1	MRR	Hit@10	Hit@3	Hit@1
TransE	0.249	0.465	–	–	0.226	0.501	–	–
PairRE [27]	0.351	0.544	0.387	0.256	0.452	0.546	0.467	0.41
RotatE [28]	0.338	0.533	0.375	0.241	0.476	0.571	0.492	0.428
QuatE	0.366	0.556	**0.401**	0.271	0.488	0.582	0.508	0.438
DistMult	0.241	0.419	0.263	0.155	0.43	0.49	0.44	0.39
ComplEx	0.247	0.428	0.275	0.158	0.44	0.51	0.46	0.41
TuckER	0.358	0.544	0.394	0.266	0.47	0.526	0.482	0.443
CP [16]	0.371	0.552	0.399	0.272	0.482	0.547	0.484	0.455
ConvE	0.316	0.491	0.35	0.239	0.46	0.48	0.43	0.39
Rotat-GCN	0.356	0.555	0.388	0.252	0.485	0.578	0.51	0.438
InteractE [29]	0.354	0.535	–	0.263	0.463	0.528	–	0.43
PGME	**0.383**	**0.583**	0.399	**0.294**	**0.532**	**0.601**	**0.511**	**0.498**

compared with the optimal value in the comparison model, the value increased by 4.4%. The experimental results show that using the probability graph model to represent the semantic relations between entities in KG can better deal with the complex relations, data sparse and other problems.

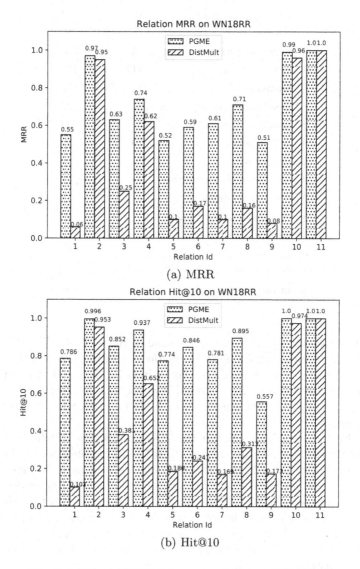

(a) MRR

(b) Hit@10

Fig. 3. MRR and Hit@10 for each relation on WN18RR

In order to further illustrate PGME model can better express the complex relations and deal with the data sparseness problem, on the WN18RR dataset for each relation compared with DistMult, as shown in Fig 3.

Table 4. Number of relationships between WN18 and WN18RR

ID	relation	WN18RR	WN18
1	hypernym	1251	1251
2	derivationally_related_form	1074	1074
3	instance_hypernym	122	122
4	also_see	56	56
5	member_meronym	253	253
6	synset_domain_topic_of	114	114
7	has_part	172	172
8	member_of_domain_usage	24	24
9	member_of_domain_region	8	26
10	verb_group	39	39
11	similar_to	3	3
12	member_of_domain_topic	–	111
13	member_holonym	–	278
14	synset_domain_usage_of	-	14
15	part_of	–	165
16	hyponym	–	1153
17	synset_domain_region_of	–	37
18	instance_hyponym	–	108

As can be seen from the figure, among the 11 relations of WN18RR, PGME is better to DistMult, and among the 7 relations (relation 1, 3, 5, 6, 7, 8, 9), the model in this paper is much higher to DistMult on MMR and Hit@10. WN18RR is obtained by deleting one of the reversible relation pairs on the basis of WN18. The relations in the datasets are shown in Table 4, among which relations 12 to 18 are reversible.

It can be seen from Table 4 that in the WN18, the reversible relationship pairs of "has_part" and "par_of" account for 76.56% of the whole data set. Combined with the conclusion of Fig. 3, the seven relations with the most obvious enhancement effect of the model in this paper are exactly the seven deleted reversible relations, which account for 37.72%. This indicates that when the relations in KG is deleted, its structure changes and the data becomes sparse. The model in this paper can still well express the semantic relations among entities in the graph, and better solve the problems such as complex relations, sparseness data and uneven distribution of triples.

5   Conclusion

In this work, the KGs is regarded as a kind of directed structured data, and the head, tail entity and relation are respectively regarded as three random variables. The probabilistic directed graph model is constructed to express the structured information of KGs, to capture the semantic transfer between entities. It can be deal with the complex relations of the KG, sparseness data, uneven distribution of triples and other problems. Then, we verify the validity of the proposed model on two publicly available data set. Due to the large differences in KGs in different fields, sparseness data and complex relations are still the difficulties. How to better learn long-tail entities and integrate additional text, pictures and other information in KGs into the model are the key issues to be considered in the later research.

Acknowledgment. This study is supported by the Sichuan Science and Technology Program (NO.2021YFG0031, 22YSZH0021) and Advanced Jet Propulsion Creativity Center (Projects HKCX2022-01-022).

References

1. Wang, Q., Mao, Z., Wang, B., Guo, L.: Knowledge graph embedding: a survey of approaches and applications. IEEE Trans. Knowl. Data Eng. **29**(12), 2724–2743 (2017)
2. Zhou, Z., et al.: Knowledge-based recommendation with hierarchical collaborative embedding. In: Phung, D., Tseng, V.S., Webb, G.I., Ho, B., Ganji, M., Rashidi, L. (eds.) PAKDD 2018. LNCS (LNAI), vol. 10938, pp. 222–234. Springer, Cham (2018). https://doi.org/10.1007/978-3-319-93037-4_18
3. Bordes, A., Usunier, N., Garcia-Duran, A., Weston, J., Yakhnenko, O.: Translating embeddings for modeling multi-relational data. In: Advances in Neural Information Processing Systems, vol. 26 (2013)
4. Nickel, M., Tresp, V., Kriegel, H.P.: A three-way model for collective learning on multi-relational data. In: ICML (2011)
5. Wang, Z., Zhang, J., Feng, J., Chen, Z.: Knowledge graph embedding by translating on hyperplanes. In: Proceedings of the AAAI Conference on Artificial Intelligence, vol. 28 (2014)
6. Fan, M., Zhou, Q., Chang, E., Zheng, F.: Transition-based knowledge graph embedding with relational mapping properties. In: Proceedings of the 28th Pacific Asia Conference on Language, Information and Computing, pp. 328–337 (2014)
7. Xiong, S., Huang, W., Duan, P.: Knowledge graph embedding via relation paths and dynamic mapping matrix. In: Woo, C., Lu, J., Li, Z., Ling, T.W., Li, G., Lee, M.L. (eds.) ER 2018. LNCS, vol. 11158, pp. 106–118. Springer, Cham (2018). https://doi.org/10.1007/978-3-030-01391-2_18
8. Feng, J., Huang, M., Wang, M., Zhou, M., Hao, Y., Zhu, X.: Knowledge graph embedding by flexible translation. In: Fifteenth International Conference on the Principles of Knowledge Representation and Reasoning (2016)
9. Xiao, H., Huang, M., Zhu, X.: From one point to a manifold: knowledge graph embedding for precise link prediction. arXiv preprint arXiv:1512.04792 (2015)

10. Ji, G., Liu, K., He, S., Zhao, J.: Knowledge graph completion with adaptive sparse transfer matrix. In: Thirtieth AAAI Conference on Artificial Intelligence (2016)
11. Yang, B., Yih, W.T., He, X., Gao, J., Deng, L.: Embedding entities and relations for learning and inference in knowledge bases. arXiv preprint arXiv:1412.6575 (2014)
12. Trouillon, T., Welbl, J., Riedel, S., Gaussier, É., Bouchard, G.: Complex embeddings for simple link prediction. In: International Conference on Machine Learning, pp. 2071–2080. PMLR (2016)
13. Liu, H., Wu, Y., Yang, Y.: Analogical inference for multi-relational embeddings. In: International Conference on Machine Learning, pp. 2168–2178. PMLR (2017)
14. Kazemi, S.M., Poole, D.: Simple embedding for link prediction in knowledge graphs. In: Advances in Neural Information Processing Systems, vol. 31 (2018)
15. Balažević, I., Allen, C., Hospedales, T.M.: Tucker: tensor factorization for knowledge graph completion. arXiv preprint arXiv:1901.09590 (2019)
16. Luo, Y., Yang, C., Li, B., Zhao, X., Zhang, H.: CP tensor factorization for knowledge graph completion. In: Memmi, G., Yang, B., Kong, L., Zhang, T., Qiu, M. (eds.) KSEM 2022. LNCS, vol. 13368, pp. 240–254. Springer, Cham (2022). https://doi.org/10.1007/978-3-031-10983-6_19
17. Socher, R., Chen, D., Manning, C.D., Ng, A.: Reasoning with neural tensor networks for knowledge base completion. In: Advances in Neural Information Processing Systems, vol. 26 (2013)
18. Schlichtkrull, M., Kipf, T.N., Bloem, P., van den Berg, R., Titov, I., Welling, M.: Modeling relational data with graph convolutional networks. In: Gangemi, A., et al. (eds.) ESWC 2018. LNCS, vol. 10843, pp. 593–607. Springer, Cham (2018). https://doi.org/10.1007/978-3-319-93417-4_38
19. Dettmers, T., Minervini, P., Stenetorp, P., Riedel, S.: Convolutional 2D knowledge graph embeddings. In: Proceedings of the AAAI Conference on Artificial Intelligence, vol. 32 (2018)
20. Nguyen, D.Q., Nguyen, T.D., Nguyen, D.Q., Phung, D.: A novel embedding model for knowledge base completion based on convolutional neural network. arXiv preprint arXiv:1712.02121 (2017)
21. Cai, L., Wang, W.Y.: KbGAN: adversarial learning for knowledge graph embeddings. arXiv preprint arXiv:1711.04071 (2017)
22. Wang, Z., Lai, K.P., Li, P., Bing, L., Lam, W.: Tackling long-tailed relations and uncommon entities in knowledge graph completion. arXiv preprint arXiv:1909.11359 (2019)
23. Bollacker, K., Evans, C., Paritosh, P., Sturge, T., Taylor, J.: Freebase: a collaboratively created graph database for structuring human knowledge. In: Proceedings of the 2008 ACM SIGMOD International Conference on Management of Data, pp. 1247–1250 (2008)
24. Miller, G.A.: Wordnet: a lexical database for English. Commun. ACM **38**(11), 39–41 (1995)
25. Akrami, F., Saeef, M.S., Zhang, Q., Hu, W., Li, C.: Realistic re-evaluation of knowledge graph completion methods: an experimental study. In: Proceedings of the 2020 ACM SIGMOD International Conference on Management of Data, pp. 1995–2010 (2020)
26. Han, X., et al.: OpenKE: an open toolkit for knowledge embedding. In: Proceedings of the 2018 Conference on Empirical Methods in Natural Language Processing: System Demonstrations, pp. 139–144 (2018)
27. Chao, L., He, J., Wang, T., Chu, W.: PairRE: knowledge graph embeddings via paired relation vectors. arXiv preprint arXiv:2011.03798 (2020)

28. Sun, Z., Deng, Z.H., Nie, J.Y., Tang, J.: Rotate: knowledge graph embedding by relational rotation in complex space. arXiv preprint arXiv:1902.10197 (2019)
29. Vashishth, S., Sanyal, S., Nitin, V., Agrawal, N., Talukdar, P.: InteractE: improving convolution-based knowledge graph embeddings by increasing feature interactions. In: Proceedings of the AAAI Conference on Artificial Intelligence, vol. 34, pp. 3009–3016 (2020)

Solving the Inverse Problem of Laser with Complex-Valued Field by Physics-Informed Neural Networks

Naiwen Chang[1,2] , Ying Huai[1(✉)] , and Hui Li[3]

[1] Key Laboratory of Chemical Lasers, Dalian Institute of Chemical Physics, Chinese Academy of Sciences, Zhongshan Road 457, Dalian 116023, China
huaiying@dicp.ac.cn
[2] University of Chinese Academy of Sciences, Beijing 100049, China
[3] Zhengzhou Tobacco Research Institute of CNTC, Zhengzhou 450001, China

Abstract. In the resonator of an actual laser oscillator, the complex-valued laser field is extracted from the gain. The inverse problem of the laser is to construct the gain utilizing the given complex-valued laser field, which is essential for the design purpose. However, it is a challenge for conventional numerical methods because the governing equations cannot be solved inversely. In this paper, a deep learning method based on physics-informed neural networks is introduced to solve the inverse laser problem. The complex-valued laser field and partial differential equation are divided into real and imaginary parts because the optimizer of neural networks cannot deal with the derivation of complex values. A given paraxial wave equation is used as an example to validate the performance of the method. The comparison between the predictions of PINNs and fast Fourier transform numerical solutions shows the average relative error of gain is 6.78%. This method can be generalized to laser design and optimal problems.

Keywords: Inverse problem · Physics-informed neural networks · Complex-valued laser

1 Introduction

Laser is generated through stimulated emission by gain media within the laser cavity. The distribution of solid-state or gaseous gain media typically affects the laser output. Therefore, determining gain in the laser cavity is critical in achieving the desired output laser. Researchers usually rely on empirical assumptions for gain distribution and subsequently calculate the intensity and phase of the output laser. This series of calculations is commonly involved in conventional numerical methods but is time-consuming and computationally expensive. The inverse laser problem involves constructing the gain from a given complex-valued laser field. The governing equations cannot be solved in reverse, making it difficult for conventional numerical methods to solve the inverse problem. Fortunately, recent advancements in AI for science provide a solution to inverse laser problems. By AI-based approaches, researchers can simplify the design process and achieve accurate results.

© The Author(s), under exclusive license to Springer Nature Singapore Pte Ltd. 2024
B. Luo et al. (Eds.): ICONIP 2023, CCIS 1961, pp. 66–75, 2024.
https://doi.org/10.1007/978-981-99-8126-7_5

In recent years, neural network development has seen significant progress. Physics-informed Neural Networks (PINNs) have emerged as a promising approach in the field of NNs [1]. PINNs incorporate the governing equation into the loss function to enable the neural network to predict values that meet the constraints of the governing equations.

Raissi Maziar presented the first paper using neural networks to solve differential equations in 2017 [2]. In his study, a computation framework for learning generic linear equations was developed. They infer the solution of the linear differential equation from the data with noises. After that, Raissi et al. developed "Hidden physics models" in 2018, in which neural networks were effectively used to solve nonlinear partial differential equations [3]. Data with noises and boundary conditions are used to obtain solutions to equations including Burger equation, Kuramoto-Sivashinsky equation, Navier-Stocks equations. Both the forward problem and the inverse problem are proved equivalent to PINNs. Many studies have also proved the effectiveness of the PINNs method in solving the inverse problem [4–6]. Then Raissi et al. presented the "Physics-informed Neural Networks" (PINNs) in 2019 and explored the forward and inverse issues in the Schrodinger equation, Allen-Cahn equation, Navier-Stokes equation, and Korteweg–de Vries equation [7].

Since then, many studies have been conducted to broaden PINNs theoretical under-pinnings and practical applications. Guofei Pang et al. proposed fPINNs and successfully applied PINNs to solve fractional partial differential equations [8]. But the convergence cannot be guaranteed, and a large number of optimization iterations are required. Ameya D. Jagtap et al. conducted a comprehensive study on different activation functions in PINNs and introduced an adaptive activation function [9]. By introducing a hyperparameter in the activation function and dynamically changing the loss function, the best performance of PINNs is achieved. Qizhi He et al. applied PINNs to groundwater problems and studied the effectiveness of MPINNs in the coupling system of the static flow field and electric field [10]. Raissi et al. used PINNs to address the problem of 2D and 3D flow through a cylinder in the year 2020 [11]. And they used it to undertake three-dimensional numerical research on patient-specific cerebral aneurysms.

These advancements highlight the potential of PINNs in addressing complex challenges across diverse fields. However, incorporating PINNs into laser research has encountered several challenges. Propagating laser light requires computing light amplitudes by considering their phase, which necessitates complex-valued lasers. The optimizer of neural networks cannot compute complex number gradients, making integrating them into artificial neural networks unfeasible. To overcome this challenge, we propose a solution by splitting the complex number into real and imaginary parts and training them separately within the neural network. We extend this approach to address the complex-valued partial differential equation within the control equation of the loss function. Nonetheless, the application of this approach in the laser has some limitations that are discussed in Sect. 3.

The paper is organized as follows. Section 2 outlines the methodology, including governing equations and data preparation. The predictions of PINNs are compared to the numerical results in Sect. 3. The accuracy and limitations of the method are discussed. The research conclusion is given in Sect. 4.

2 Methodology

In this paper, we discuss the problem in the context of chemical lasers. Let us consider the propagating complex-valued field $U(x, y)$ along axis z in the laser cavity with gain distribution $g(x, y)$ which is uniform along axis z. The field $U(x, y)$ can be expanded as $U = U_{real} + i \cdot U_{imag}$. This light propagation is described by the paraxial wave equation [12].

$$\frac{i}{2k}\left(\frac{\partial^2}{\partial x^2} + \frac{\partial^2}{\partial y^2}\right)U + \frac{g}{2} \cdot U = \frac{\partial U}{\partial z}, \tag{1}$$

where k is the wave number.

In our notation, a symbol with a tilde is a non-dimensional quantity, the symbol with a star is the actual value. In our case, we scale the complex-valued field by $U_0 = 10^3 \cdot m$, $g_0 = 1/m$, $l_0 = 1 \cdot m$. And the dimensionless variables $\tilde{U} = \frac{U}{U_0}$, $\tilde{g} = \frac{g}{g_0}$, $\tilde{x} = \frac{x}{l_0}$, $\tilde{y} = \frac{y}{l_0}$ are obtained. The dimensionless paraxial wave equation is,

$$\frac{i \cdot U_0}{2kl_0^2}\left(\frac{\partial^2}{\partial \tilde{x}^2} + \frac{\partial^2}{\partial \tilde{y}^2}\right)\tilde{U} + \frac{U_0 g_0}{2}\tilde{g} \cdot \tilde{U} = \frac{U_0}{l_0}\frac{\partial \tilde{U}}{\partial \tilde{z}}. \tag{2}$$

Since numerous studies have addressed PINNs in detail [13], the mathematical principles of PINNs are not presented. In our research, the complex-valued field data points are used as labels to construct gain distribution through PINNs.

The total loss function consists of two components. The loss of data is expressed as,

$$Loss_{data} = \frac{1}{N}\left[\sum_N \left(U_{real}^* - \tilde{U}_{real}\right)^2 + \sum_N \left(U_{imag}^* - \tilde{U}_{imag}\right)^2\right] \tag{3}$$

where N is the size of a minibatch in each training iteration. U_{real} and U_{imag} are the real and imaginary parts of the field. The asterisk mark indicates that the quantity is the predicted value of the neural network. The loss of governing equations is expressed as,

$$Loss_{eqns} = \frac{1}{N}\sum_N\left(10^3 \cdot \left(\frac{\partial \tilde{U}_{real}}{\partial \tilde{z}} - \frac{1}{2k}\left(\frac{\partial^2}{\partial \tilde{x}^2} + \frac{\partial^2}{\partial \tilde{y}^2}\right)\tilde{U}_{imag} - \frac{\tilde{g}}{2} \cdot \tilde{U}_{real}\right)\right)^2$$

$$+ \frac{1}{N}\sum_N\left(10^3 \cdot \left(\frac{\partial \tilde{U}_{imag}}{\partial \tilde{z}} + \frac{1}{2k}\left(\frac{\partial^2}{\partial \tilde{x}^2} + \frac{\partial^2}{\partial \tilde{y}^2}\right)\tilde{U}_{real} - \frac{\tilde{g}}{2} \cdot \tilde{U}_{imag}\right)\right)^2 \tag{4}$$

The total loss function of PINNs is,

$$Loss = Loss_{data} + Loss_{eqns} \tag{5}$$

Figure 1 shows the schematic diagram of the PINNs in our laser model. The input of PINNs consists of spatial coordinates x, y, and z. And the output consists of U_{real}, U_{imag} and g. We employ a 12-layer neural network. The neural network comprises 1 input layer, 10 hidden layers, and 1 output layer. Each of the 10 hidden layers has 100 neurons. The activation function for the hidden layer is set as Swish, while no activation function is added for the output layer. The Adam optimizer is utilized for achieving the convergence of PINNs in this study. As for the training procedure, we make the learning rate change with the total loss. When loss is greater than 10^{-3}, the learning rate is set to 10^{-3}. When loss is less than 10^{-3}, the learning rate is set to 10^{-4}. The number of data points in our dataset is around 400 thousand. The mini-batch size is 10000. When the maximum number of iterations is reached, the gain distribution can be obtained by feeding the spatiotemporal coordinates to the neural network. The training runs on a NVIDIA Tesla V100 GPU card.

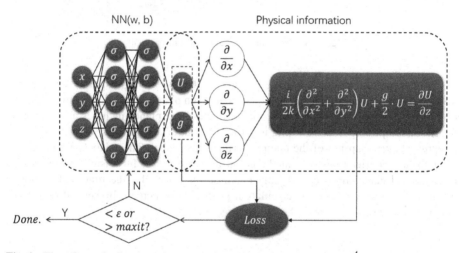

Fig. 1. The schematic diagram of PINNs. ε is a minute quantity (e.g. 10^{-4}). Upon reduction of the loss function to a value lower than ε, one may regard the neural network to have converged. Moreover, *maxit* is defined as a ceiling for the number of iterations performed by the neural network to avoid interminable training.

3   Results and Discussions

Figure 2 shows a schematic diagram of a laser cavity. The laser is generated by stimulated emission within the gain medium, oscillates continuously within the cavity, and is emitted from the side of a semi-transparent mirror. In this study, the cavity mirror is circular, resulting in a circular beam profile. The figure also shows the real and imaginary parts of the complex-valued laser field at the interface indicated by the dashed line.

The laser cavity

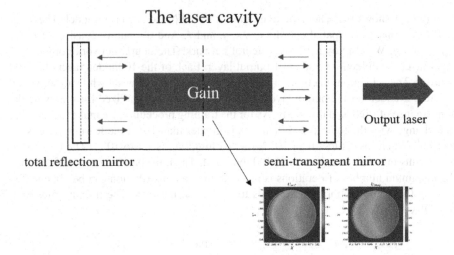

Fig. 2. The schematic diagram of a laser cavity.

We selected the inscribed rectangular area of the circular light spot as the training domain for our analysis. The dataset consists of seven planes perpendicular to the direction of light propagation (along the z-axis). Each of the planes includes a 241×241 matrix of sampling points, resulting in a total of 406,567 data points. Figure 3 depicts a representative snapshot of the complex-valued field label. These labels are obtained through fast Fourier transform methods but are with noise. Increasing the precision of numerical calculations can slightly decrease the error. But the time and computing resources consumed in the calculation process will increase by several orders of magnitude. Figure 4 illustrates the distribution of noise.

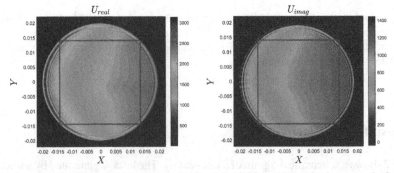

Fig. 3. A schematic diagram of the complex-valued field. The real part is shown on the left and the imaginary part is shown on the right. A data collection area enclosed within an inscribed rectangular (regions marked in red) is set to prevent the adverse impact of zero values outside the circular spot on the neural network training.

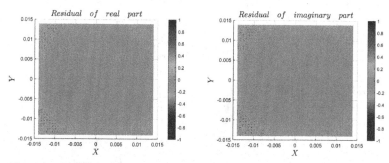

Fig. 4. The noise distributions of the real and imaginary parts of numerical results. The average residual of real part and imaginary part are 13.71% and 14.23%.

The gain distribution within the rectangular area is constructed by the joint constraints associated with the label and the corresponding physics constraints. Figure 5 shows the comparison between the exact values and PINNs' predictions of complex-valued field and gain distribution. It can be seen that our algorithm's predictions roughly align with the precise data within the training domain. The utilization of complex-valued data as a label ensures accurate PINNs predictions for the real and imaginary amplitude components. However, there is a divergence between the predicted and actual values of gain distribution in sections where noise distribution is relatively high.

This current level of accuracy is not satisfactory, and further improvements are needed. Notably, we have observed that the distribution of gain reaches a constant value at the extreme boundary of "x" (regions marked in red in Fig. 6). Therefore, let us modify the loss function by including the mean squared error between the predicted and true values of gain at the boundary. The $Loss_{bc}$ can be written as,

$$Loss_{bc} = \frac{1}{N}\left[\sum\nolimits_{N}\left(g^{*}_{x=x_{max}} - \tilde{g}_{x=x_{max}}\right)^{2}\right], \tag{6}$$

and the total loss function transform to,

$$Loss = Loss_{data} + Loss_{eqns} + Loss_{bc}. \tag{7}$$

Figure 6 shows the comparison of gain distribution between the exact values and PINNs' predictions after adding labels at the extreme end of x. Including these supplementary labels has shown remarkable improvements in accuracy. The average relative error of gain is reduced to 6.78%, which is a notable improvement.

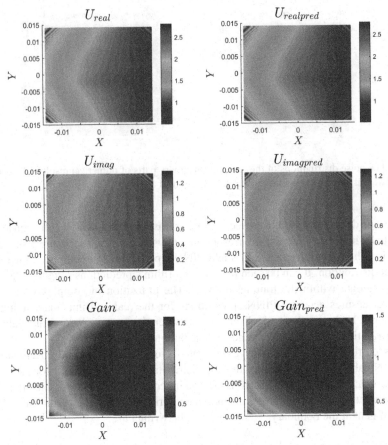

Fig. 5. Comparisons between the exact values and PINNs' predictions of complex-valued field and gain distribution. The utilization of complex-valued data as a label ensures accurate PINNs predictions for the real and imaginary amplitude components. However, it should be noted that the average relative error between the predictions generated by the PINNs and the exact gain is 17.98%.

It should be emphasized that there is a limitation in the approach. The distance between the seven sample surfaces is required to be a multiple of the wavelength. It is necessary to fulfill this requirement to guarantee that the phase differences between waves on these surfaces are whole number multiples of 2π. Failure to meet this condition could result in substantial residual errors between the laser field and the gain distributions.

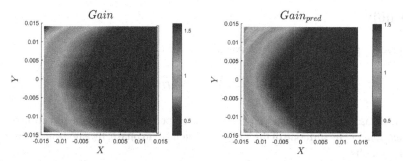

Fig. 6. The comparison of gain distribution between the exact values and PINNs' predictions after adding labels at the extreme end of x (regions marked in red). The average relative error is reduced to 6.78%.

According to our analysis, the limitation imposed by $\frac{g}{2} \cdot U$ in Eq. 1 is the explanation for this phenomenon. This term describes the same impact from the gain on the real and imaginary components of U, which means that it acts only on the amplitude but is limited in its ability to affect the phase. Let us consider the effect of shifting one of the 7 planes, which are in phase, by a non-integer number of wavelengths. Note that the $\frac{i}{2k}\left(\frac{\partial^2}{\partial x^2} + \frac{\partial^2}{\partial y^2}\right)U$ and $\frac{\partial U}{\partial z}$ items provide insight into the amplitude changes resulting from the phase shift, while the $\frac{g}{2} \cdot U$ item does not provide relevant information. The specific mathematical derivation is detailed in the appendix. Therefore, it is crucial to adhere to this requirement to obtain accurate results.

4 Conclusion

In this study, we develop a method to solve the inverse laser problem using the Physics-informed neural networks (PINNs) approach. The motivation of this method is to construct gain from the complex-valued laser field. However, the general neural network optimizers cannot compute complex number derivation. To deal with this challenge, the complex-valued laser field and complex partial differential equation in the loss function are split into real and imaginary components within the neural network. The comparison between the predictions of the PINNs and the fast Fourier transformation method shows a good agreement. To improve the accuracy of the results, we included the boundary data as the supplementary label. The relative error of gain reduces from 17.98% to 6.78% by comparing PINNs' results and numerical solutions. This numerical method can preliminarily solve the inverse laser problem.

Acknowledgements. This work was financially supported by the National Natural Science Foundation of China (Grant No. 22073095), Science Planning Fund of Dalian (2021JJ11CG006), Research foundation of Yulin Laboratory for Clean Energy (DNL-YLA202208).

Appendix

The governing equation of laser propagation in the cavity is as follows,

$$\frac{i}{2k}\left(\frac{\partial^2}{\partial x^2} + \frac{\partial^2}{\partial y^2}\right)U + \frac{g}{2} \cdot U - \frac{\partial U}{\partial z} = 0$$

The dataset comprises of seven planes, placed perpendicular to the direction of light propagation. Let us shift one of the 7 planes, which are in phase, by a non-integer number of wavelengths.

The complex-valued field can be written as $U = A \cdot e^{i\phi}$. The complex-valued field on the shifted plane is $U' = (A + \Delta A) \cdot e^{i(\phi + \Delta\phi)}$. Substituting it into the governing equation, the expansion of each term is as follows.

The expansion of the first term:

$$\frac{i}{2k}\left(\frac{\partial^2\left((A + \Delta A) \cdot e^{i(\phi + \Delta\phi)}\right)}{\partial x^2} + \frac{\partial^2\left((A + \Delta A) \cdot e^{i(\phi + \Delta\phi)}\right)}{\partial y^2}\right)$$

$$= \frac{i}{2k}\left[\begin{array}{l} e^{i\Delta\phi} \cdot \left(\frac{\partial^2 U}{\partial x^2} + \frac{\partial^2 U}{\partial y^2}\right) + U \cdot \left(\frac{\partial^2 e^{i\Delta\phi}}{\partial x^2} + \frac{\partial^2 e^{i\Delta\phi}}{\partial y^2}\right) + 2 \cdot \left(\frac{\partial U}{\partial x} \cdot \frac{\partial e^{i\Delta\phi}}{\partial x} + \frac{\partial U}{\partial y} \cdot \frac{\partial e^{i\Delta\phi}}{\partial y}\right) \\ + e^{i\Delta\phi} \cdot \left(\frac{\partial^2 \Delta A e^{i\phi}}{\partial x^2} + \frac{\partial^2 \Delta A e^{i\phi}}{\partial y^2}\right) + \Delta A e^{i\phi} \cdot \left(\frac{\partial^2 e^{i\Delta\phi}}{\partial x^2} + \frac{\partial^2 e^{i\Delta\phi}}{\partial y^2}\right) + 2 \cdot \left(\frac{\partial \Delta A e^{i\phi}}{\partial x} \cdot \frac{\partial e^{i\Delta\phi}}{\partial x} + \frac{\partial \Delta A e^{i\phi}}{\partial y} \cdot \frac{\partial e^{i\Delta\phi}}{\partial y}\right) \end{array}\right]$$

The expansion of the second term:

$$\frac{g}{2} \cdot (A + \Delta A) \cdot e^{i(\phi + \Delta\phi)} = \frac{g}{2} \cdot U \cdot e^{i\Delta\phi} + \frac{g}{2} \cdot \Delta A e^{i\phi} \cdot e^{i\Delta\phi}$$

The expansion of the third term:

$$-\frac{\partial (A + \Delta A) \cdot e^{i(\phi + \Delta\phi)}}{\partial z} = -\frac{\partial U}{\partial z} \cdot e^{i\Delta\phi} - \frac{\partial \Delta A e^{i\phi}}{\partial z} \cdot e^{i\Delta\phi} - U \cdot \frac{\partial e^{i\Delta\phi}}{\partial z} - \Delta A e^{i\phi} \cdot \frac{\partial e^{i\Delta\phi}}{\partial z}$$

The terms denoted by underscores and double underscores are consistently equal to zero during summation. The remaining terms cannot be eliminated through any means. Based on our knowledge, these residual terms embody the amplitude changes resulting from the phase shift.

References

1. Karniadakis, G.E., Kevrekidis, I.G., Lu, L., et al.: Physics-informed machine learning. Nat. Rev. Phys. 3(6), 422–440 (2021)
2. Raissi, M., Perdikaris, P., Karniadakis, G.E.: Machine learning of linear differential equations using Gaussian processes. J. Comput. Phys. 348, 683–693 (2017)
3. Raissi, M., Karniadakis, G.E.: Hidden physics models: machine learning of nonlinear partial differential equations. J. Comput. Phys. 357, 125–141 (2018)
4. Cheng, C., Zhang, G.T.: Deep learning method based on physics informed neural network with resnet block for solving fluid flow problems. Water 13(4) (2021)
5. Dwivedi, V., Parashar, N., Srinivasan, B.: Distributed learning machines for solving forward and inverse problems in partial differential equations. Neurocomputing 420, 299–316 (2021)

6. Mishra, S., Molinaro, R.: Estimates on the generalization error of physics informed neural networks (PINNs) for approximating PDEs II: a class of inverse problems arXiv. arXiv (USA), 35 p. (2020)

7. Raissi, M., Perdikaris, P., Karniadakis, G.E.: Physics-informed neural networks: a deep learning framework for solving forward and inverse problems involving nonlinear partial differential equations. J. Comput. Phys. **378**, 686–707 (2019)

8. Pang, G., Lu, L., Karniadakis, G.E.: FPINNs: fractional physics-informed neural networks. SIAM J. Sci. Comput. **41**(4), A2603–A2626 (2019)

9. Jagtap, A.D., Kawaguchi, K., Karniadakis, G.E.: Adaptive activation functions accelerate convergence in deep and physics-informed neural networks. J. Comput. Phys., **404** (2020)

10. He, Q., Barajas-Solano, D., Tartakovsky, G., et al.: Physics-informed neural networks for multiphysics data assimilation with application to subsurface transport. Adv. Water Resour. **141** (2020)

11. Raissi, M., Yazdani, A., Karniadakis, G.E.: Hidden fluid mechanics: learning velocity and pressure fields from flow visualizations. Science 367(6481), 1026–1030 (2020)

12. Wu, K., Huai, Y., Jia, S., et al.: Coupled simulation of chemical lasers based on intracavity partially coherent light model and 3D CFD model. Opt. Express **19**(27), 26295–26307 (2011)

13. Cai, S.Z., Mao, Z.P., Wang, Z.C., Yin, M.L., Karniadakis, G.E.: Physics-informed neural networks (PINNs) for fluid mechanics: a review. Acta Mech. Sin. **37**(12), 1727–1738 (2021). https://doi.org/10.1007/s10409-021-01148-1

Efficient Hierarchical Reinforcement Learning via Mutual Information Constrained Subgoal Discovery

Kaishen Wang[1,2], Jingqing Ruan[2], Qingyang Zhang[2], and Dengpeng Xing[1,2(✉)]

[1] University of Chinese Academy of Sciences, Beijing 100049, China
[2] Institute of Automation, Chinese Academy of Sciences, Beijing 100190, China
{wangkaishen2021,ruanjingqing2019,zhangqingyang2019,
dengpeng.xing}@ia.ac.cn

Abstract. Goal-conditioned hierarchical reinforcement learning has demonstrated impressive capabilities in addressing complex and long-horizon tasks. However, the extensive subgoal space often results in low sample efficiency and challenging exploration. To address this issue, we extract informative subgoals by constraining their generation range in mutual information distance space. Specifically, we impose two constraints on the high-level policy during off-policy training: the generated subgoals should be reached with less effort by the low-level policy, and the realization of these subgoals can facilitate achieving the desired goals. These two constraints enable subgoals to act as critical links between the current states and the desired goals, providing more effective guidance to the low-level policy. The empirical results on continuous control tasks demonstrate that our proposed method significantly enhances the training efficiency, regardless of the dimensions of the state and action spaces, while ensuring comparable performance to state-of-the-art methods.

Keywords: Hierarchical reinforcement learning · Subgoal discovery · Mutual information

1 Introduction

Goal-conditioned hierarchical reinforcement learning (HRL) [9,11,12,19] has received much attention due to its significant performance in solving complex and long-term tasks. Among HRL frameworks, goal-conditioned HRL typically consists of a high-level policy and a low-level policy. The high-level policy decomposes the desired goal into simpler subgoals, allowing the low-level policy to learn and explore more effectively. However, identifying informative and reachable subgoals in the extensive subgoal space to enhance sample and training efficiency still remains a major challenge.

This work is supported by the Program for National Nature Science Foundation of China (62073324).

© The Author(s), under exclusive license to Springer Nature Singapore Pte Ltd. 2024
B. Luo et al. (Eds.): ICONIP 2023, CCIS 1961, pp. 76–87, 2024.
https://doi.org/10.1007/978-981-99-8126-7_6

Over the past few years, several works [8,9,12,20] have been proposed to improve sample efficiency in HRL. Kulkarni et al. [9] predefine a set of key states as the subgoal space, which is efficient but requires task-relevant knowledge. Nachum et al. [12] propose the off-policy correction method that enables the high-level policy to be trained in an off-policy manner. However, its training cost increases considerably when dealing with larger state and action spaces. Zhang et al. [20] leverage the concept of adjacency distance to confine subgoals within a reachable range of k steps, which reduces the subgoal space and enhances training efficiency. Nonetheless, maintaining an adjacency matrix to calculate the distances between subgoals can incur additional training expense and storage requirement. Kim et al. [8] construct a landmark graph and select subgoals by planning on the graph based on prior work [20]. Although these approaches reduce the subgoal space, they also entail an increase in training time.

In this paper, we propose a novel method called **M**utual **I**nformation-based **S**ubgoal **D**iscovery (MISD) to improve training efficiency while maintaining sample efficiency. Concretely, by maximizing the mutual information between the subgoal and the actually achieved goal, this subgoal can be reached easily by the low-level policy, as illustrated in Fig. 1. Analogously, by maximizing the mutual information between the subgoal and the desired goal, the probability of achieving the desired goal will increase after accomplishing this subgoal. Our main contributions are outlined below: 1) We introduce mutual information as a metric of distance between subgoals and identify the most informative subgoals for the agent to explore in HRL. 2) Our method can take advantage of the correlation between subgoals to avoid costly goal-relabeling and reduce the need for extensive exploration, resulting in improved sample efficiency. 3) The experimental results demonstrate that our method is dimension-agnostic with respect to the state and action spaces and significantly improves training efficiency in diverse continuous control tasks.

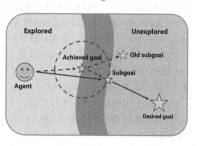

Fig. 1. The dashed blue line denotes the historical transition, the solid blue line denotes the new transition, and the dashed orange line denotes the subgoal constraint range. The old subgoal may be unachievable in reality due to being too far away. (Color figure online)

The remaining of the paper is structured as follows: Sect. 2 discusses related works in HRL. Section 3 briefly introduces goal-conditioned HRL and mutual information estimator. Section 4 presents the framework of our method. Section 5 includes experimental settings, empirical results, ablation studies and visualizations. Last, Sect. 6 concludes this paper with a summary.

2   Related Works

Subgoal Discovery. Goal-conditioned HRL[9,13], which incorporates high-level and low-level policies, has demonstrated immense potential in solving diverse complex tasks. By combining the hindsight technique [1], Levy et al. [11] can train multiple levels of policies concurrently. However, the vast subgoal space limits the identification of effective subgoals. In order to tackle this issue, some methods [5,13] utilize online planning to select feasible subgoals. Hafner et al. [5] apply a world model to generate an imagined trajectory and select subgoals. However, this approach needs additional training of the world model for planning purpose. Some graph-based methods [7,19] have also been proposed to address this challenge. Nevertheless, these approaches require creating a graph and performing online planning over it, which can be computationally expensive and time-consuming.

Mutual Information. In recent years, mutual information is applied in skill-based HRL [3,18] to generate diverse skills, increase the exploration ability of the agent, and enhance the transferability of the skills. Eysenbach et al. [3] can train various skills using mutual information without relying on environmental rewards, but these skills need to be pre-trained before they can be transferred to downstream tasks. In practice, due to unknown data distributions, calculating the mutual information accurately is often difficult [15]. Oord et al. [14] propose the InfoNCE loss, a method of contrastive learning [4,10], to estimate the lower bound on mutual information. In our paper, we utilize mutual information as a distance metric [6,16] between subgoals by learning a representation function which maps these subgoals to the mutual information distance space.

3   Preliminaries

Goal-Conditioned HRL can be expressed as a finite horizon Markov Decision Process (MDP) with tuple $(\mathcal{S}, \mathcal{A}, \mathcal{G}, P, R, \gamma)$, where $\mathcal{S}$ is the state space, $\mathcal{A}$ is the action space, $\mathcal{G}$ is the subgoal space which is mapped from $\mathcal{S}$ by the function $\varphi : \mathcal{S} \rightarrow \mathcal{G}$, $P : \mathcal{S} \times \mathcal{A} \times \mathcal{S} \rightarrow [0,1]$ is the transition function, $R : \mathcal{S} \times \mathcal{A} \rightarrow \mathbb{R}$ is the reward function, and $\gamma \in [0,1)$ is the discount factor. Following prior works [12], we formulate the framework composed of two hierarchies: a high-level policy $\pi_{\theta_h}^h$ and a low-level policy $\pi_{\theta_l}^l$ parameterized by θ_h and θ_l, respectively. The high-level policy generates subgoal $g_t \sim \pi_{\theta_h}^h(s_t, g_d)$ every k steps until the episode terminates at step T, where $g_d \in \mathcal{G}$ is the desired goal that the agent needs to achieve. When $t \equiv 0$ (mode k), the low-level policy receives subgoal $g_t \in \mathcal{G}$ from the high-level policy, otherwise, it resorts to using a fixed subgoal transition function:

$$g_{t+1} = h(s_t, g_t, s_{t+1}) = \varphi(s_t) + g_t - \varphi(s_{t+1}). \tag{1}$$

Then, the low-level policy performs a primitive action $a_t \sim \pi_{\theta_l}^l(s_t, g_t)$, which results in the environment transferring to the next state according to the transition function $s_{t+1} \sim P(s_{t+1}|s_t, a_t)$ and giving reward $r_t \sim R(s_t, a_t)$. Without involving environmental rewards, the low-level policy is motivated by the high-level policy with intrinsic reward $r_t^l = -\|\varphi(s_t) + g_t - \varphi(s_{t+1})\|_2$.

Based on the above setups, the transitions of high-level and low-level policies can be denoted as $(s_t, g_d, g_t, r_t^h, s_{t+k})$ and $(s_t, g_t, a_t, r_t^l, s_{t+1})$, respectively, where

$r_t^h = \sum_{t:t+k-1} r_t$, and s_{t+k} is the achieved state by the low-level policy. The objective of the high-level policy is to maximize the expected cumulative reward provided by the environment:

$$\mathcal{L}_{rew}(\theta_h) = -\mathbb{E}_{\pi_{\theta_h}^h}\left[\sum_{t=0}^{T-1} \gamma^t r_t^h\right]. \tag{2}$$

Mutual Information Estimator [14,17] is a technique used to estimate the mutual information between two random variables. In our method, we employ the InfoNCE loss [14] to estimate the lower bound on mutual information, which learns representations by maximizing the similarity between positive samples and minimizing the similarity between negative samples:

$$\mathcal{L}_{InfoNCE} = -\mathbb{E}_{s \in \mathcal{S}}\left[\log \frac{\exp(\psi_\phi(s_i)^T \cdot \psi_\phi(s_j)/\tau)}{\sum_{n=0}^{N} \exp(\psi_\phi(s_i)^T \cdot \psi_\phi(s_n)/\tau)}\right] \tag{3}$$
$$\geq \log(N) - I(s_i; s_j),$$

where ψ_ϕ is a encoder network parameterized by ϕ, s_i and s_j are different states, τ is a temperature scale factor, s_n is the state sample, N is the number of samples, and I is the mutual information function.

4   Methodology

In this section, we present MISD: Mutual Infomation-based Subgoal Discovery, a simple and effective method for training the high-level policy with mutual information distance constraints, as shown in Fig. 2.

4.1   Mutual Information Distance Space

Previous works [8,20] have extensively studied the measurement of the distance between different subgoals, utilizing the shortest transition steps. However, these approaches overlook the correlation between subgoals which can be used to gauge the difficulty of achieving them. In contrast, we propose the concept of the mutual information distance space to estimate the distance between subgoals without considering the transition steps. In the mutual information distance space, a smaller distance corresponds to a higher mutual information, indicating a stronger correlation and a higher likelihood of achieving the subgoals jointly, even with a random policy. Therefore, we define the distance between the subgoals g_i and g_j as follows:

$$d_{st}(g_i, g_j) := -\mathbb{E}_{\pi \in \prod}[I(g_i; g_j | \pi)], \tag{4}$$

where $\prod$ is the set of policy π used by the agent, $g_i = \varphi(s_i)$, and $g_j = \varphi(s_j)$.

Minimizing the distance between subgoals can facilitate their successful realization with less effort. However, accurately and directly calculating mutual

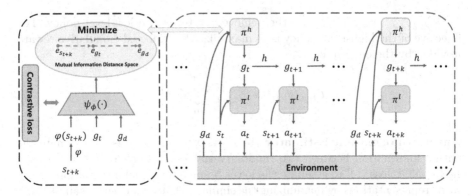

Fig. 2. The MISD framework with mutual information distance constraints implemented by the subgoal encoder ψ_ϕ (dashed red box), where $e_{s_{t+k}} = \psi_\phi(\varphi(s_{t+k})), e_{g_t} = \psi_\phi(g_t)$, and $e_{g_d} = \psi_\phi(g_d)$. The encoder, trained using a contrastive loss, maps the subgoals to the mutual information distance space which is then utilized to train the high-level policy to constrain subgoals generation. (Color figure online)

information is often impractical due to intractable data distributions [15]. Instead, we estimate a lower bound on mutual information by using the InfoNCE [14] along with a limited set of policies that the agent has employed in recent C episodes. Based on the fact that the adjacent states have relatively higher mutual information in MDP, we select the next achieved goal reached by the low-level policy as the positive sample, and randomly sampled achieved goals from the current episode as the negative samples. Consequently, we derive the following optimization objective function:

$$\mathcal{L}_{dis}(\phi) = -\mathbb{E}_{s_i \in \mathcal{S}}\left[\log\frac{\exp(\psi_\phi(\varphi(s_i))^T \cdot \psi_\phi(\varphi(s_{i+1}))/\tau)}{\sum_{n=0}^{N}\exp(\psi_\phi(\varphi(s_i))^T \cdot \psi_\phi(\varphi(s_n))/\tau)}\right], \quad (5)$$

where ψ_ϕ is the subgoal encoder parameterized by ϕ, s_{i+1} is the next state following state s_i, τ is a temperature scale, s_n is the state sample, and N is the number of samples.

By optimizing the objective function, we can obtain a subgoal encoder $\psi_\phi(\cdot)$, which allows for the mapping of subgoals to the mutual information distance space and simplifies the calculation of distances between them. For subgoals g_i and g_j, the minimization of distance is equivalent to the maximization of the numerator of Eq. 5, as they demonstrate a negative correlation:

$$\begin{aligned} I(g_i; g_j) &\propto \frac{\exp(sim(\psi_\phi(g_i), \psi_\phi(g_j)))}{\sum_{n=0}^{N}\exp(sim(\psi_\phi(g_i), \psi_\phi(g_n)))} \\ &\propto sim(\psi_\phi(g_i), \psi_\phi(g_j)) \\ &\propto -d_{st}(g_i, g_j), \end{aligned} \quad (6)$$

where g_i and g_j are different subgoals, g_n is the subgoal sample, N is the number of samples, and sim denotes the similarity scoring function. Inspired by [14], we choose the cosine similarity function defined as follows:

$$f_{cs}(\psi_\phi(g_i), \psi_\phi(g_j)) = \frac{\psi_\phi(g_i)^T \cdot \psi_\phi(g_j)}{||\psi_\phi(g_i)||_2 \cdot ||\psi_\phi(g_j)||_2}. \tag{7}$$

Alternatively, the minimization of mutual information distance is equivalent to the maximization of Eq. 7, which we then employ to enforce the aforementioned distance constraints.

4.2   Efficient Subgoal Discovery with Distance Constraint

To identify informative subgoals that efficiently guide the low-level policy to accomplish the desired goal, we introduce two mutual information distance constraints on the high-level policy.

Constrain with the Achieved Goal. In HRL, goal-relabeling [1,12] is an effective technique to help the high-level policy more quickly learn to choose achievable subgoals. However, this approach has a limitation in that the high-level policy may only be aware of specific subgoals that the low-level policy can achieve while remaining unaware of other subgoals that may be more relevant. We address this limitation by constraining the distance between subgoals and achieved goals to make these subgoals representation more informative and relevant, allowing for more effective guidance of the low-level policy. Specifically, the subgoals proposed by the high-level policy are highly correlated with the achieved goals that have already been reached by the low-level policy, making them easier to achieve. Moreover, the ability of the low-level policy can be fed back to the high-level policy to generate more effective subgoals, maintaining the consistency between hierarchical policies and enhancing the stability of the learning process. Therefore, transition samples can be used directly for training to improve data efficiency without goal-relabeling. In summary, the objective function can be written as follows:

$$\mathcal{L}_{ag}(\theta_h) = -\mathbb{E}_{\pi_{\theta_h}^h} [f_{cs}(\psi_\phi(g_t), \psi_\phi(\varphi(s_{t+k})))], \tag{8}$$

where $g_t \sim \pi_{\theta_h}^h(s_t, g_d)$, and $\varphi(s_{t+k})$ is the achieved goal reached by the low-level policy after k steps.

Constrain with the Desired Goal. After imposing the distance constraint mentioned above, the subgoals generated by the high-level policy may be too simple for the low-level policy, resulting in limited exploration capability. To address this issue, we introduce another constraint with the desired goal, aiming to ensure that the desired goal can be easily achieved once the subgoal is accomplished by the low-level policy. For this purpose, we minimize the distance between the subgoal and the desired goal, thereby expanding the exploration area around the achieved goal by the low-level policy, formulated as follows:

$$\mathcal{L}_{dg}(\theta_h) = -\mathbb{E}_{\pi_{\theta_h}^h} [f_{cs}(\psi_\phi(g_t), \psi_\phi(g_d))], \tag{9}$$

where $g_t \sim \pi_{\theta_h}^h(s_t, g_d)$, and g_d is the desired goal.

Algorithm 1. MISD algorithm

Initialize: the trajectory buffer $\mathcal{B} \leftarrow \emptyset$;
Initialize: θ_h, θ_l, and ϕ for $\pi_{\theta_h}^h$, $\pi_{\theta_l}^l$, and ψ_ϕ;
1: **for** $n = 1$ to $num\_episodes$ **do**
2:     $t = 0$;
3:     Reset the environment, sample the initial state s_0 and the desired goal g_d;
4:     **repeat**
5:         **if** $t \equiv 0 \pmod{k}$ **then**
6:             Generate subgoal $g_t \sim \pi_{\theta_h}^h(s_t, g_d)$;
7:         **else**
8:             Perform subgoal transition $g_t = h(s_{t-1}, g_{t-1}, s_t)$;
9:         **end if**
10:         Execute low-level action $a_t \sim \pi_{\theta_l}^l(s_t, g_t)$;
11:         Sample next state $s_{t+1} \sim \mathcal{P}(s_{t+1}|s_t, a_t)$ and reward $r_t \sim \mathcal{R}(s_t, a_t)$;
12:         $t = t + 1$;
13:     **until** episode terminates;
14:     Store the trajectory in buffer $\mathcal{B}$;
15:     Train high-level policy $\pi_{\theta_h}^h$ according to Equation 10;
16:     Train low-level policy $\pi_{\theta_l}^l$;
17:     **if** $n \equiv 0 \pmod{C}$ **then**
18:         Train the subgoal encoder ψ_ϕ with Equation 5 using buffer $\mathcal{B}$;
19:         Clear $\mathcal{B}$;
20:     **end if**
21: **end for**

4.3   The Policy Optimization

By enforcing mutual information distance constraints on both parts as previously stated, we can obtain the final objective function of the high-level policy:

$$\mathcal{L}_{high}(\theta_h) = \mathcal{L}_{rew}(\theta_h) + \alpha \cdot [(1 - \beta) \cdot \mathcal{L}_{ag}(\theta_h) + \beta \cdot \mathcal{L}_{dg}(\theta_h)], \tag{10}$$

where $\alpha \in [0, +\infty)$ is a scale factor, and $\beta \in [0, 1]$ is the distance coefficient used to determine the exploration range. In practice, we incorporate $\mathcal{L}_{ag}$ and $\mathcal{L}_{dg}$ as additional terms into the original loss function $\mathcal{L}_{rew}$ of the high-level policy. For the low-level policy, we utilize a common reinforcement learning algorithm, e.g., temporal-difference learning methods, to train it as usual without modification. The main process of our method is presented in Algorithm 1.

5   Experiments

We design experiments to answer the following questions: 1) How does MISD perform compared to state-of-the-art methods on various continuous control tasks? 2) Can MISD enhance sample and training efficiency in HRL? 3) Does the dimensionality of the state and action spaces affect the training efficiency of MISD? 4) What is the impact of hyperparameters on the performance of MISD?

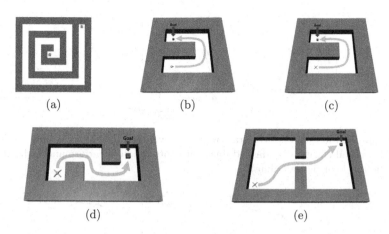

(a)   (b)   (c)

(d)   (e)

Fig. 3. Environment descriptions. (a) Spiral: navigate from the edge (denoted as 'S') to the center (denoted as 'G') with sparse rewards. (b) Point Maze (U-shape) and (c) Ant Maze (U-shape): navigate from a fixed position to a fixed target location with dense rewards. (d) Ant Maze (S-shape): navigate from a random position to a fixed target location with sparse rewards. (e) Ant TwoRooms: navigate from a fixed position in one room to a fixed target position in another room with dense rewards. (Color figure online)

5.1 Environment Setup

We evaluate our method on diverse control tasks with continuous state and action spaces based on the Mujoco simulator [2], as illustrated in Fig. 3. In all tasks, the agent needs to achieve goals with either sparse or dense rewards, where the subgoal space consists of two dimensions that correspond to the position (x, y) of the agent. The code of our method is available at https://github.com/RandyButters/MISD.

5.2 Comparative Experiments

We compare MISD with the following baselines: 1) HIRO [12]: a baseline that proposes the off-policy correction method to improve data efficiency. 2) HRAC [20]: a baseline that employs k-step reachability to constrain the range of subgoal and reduce the subgoal space. 3) HIGL [8]: a baseline that utilizes a graph of landmarks for online planning and to guide the generation of subgoals.

We present the learning curves of success rate plotted against both training time and steps, as shown in Figs. 4 and 5, respectively. The results indicate that our proposed MISD algorithm demonstrates a significant improvement in training efficiency by reducing the overall training time while achieving a comparable performance and sample efficiency to other baselines.

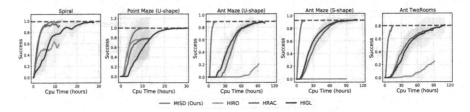

Fig. 4. The learning curves over training time, averaged over 5 trials and smoothed equally for visual clarity. The dashed line represents the best performance achieved by our method. (Color figure online)

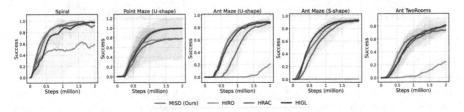

Fig. 5. The learning curves over training steps, averaged over 5 trials and smoothed equally for visual clarity. (Color figure online)

Additionally, we generate a bar chart of the training time across the Mujoco environments utilized in our experiments, as illustrated in Fig. 6. The results demonstrate that the training efficiency of MISD remains stable, while other baselines experience a significant increase in training time as the state and action spaces become larger. For example, the Ant environment has larger state and action spaces compared to the simpler Point environment, additional training time is required due to its increased complexity, resulting in lower training efficiency for the baseline methods.

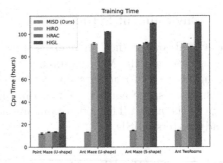

Fig. 6. Training time on different environments within the same benchmark, averaged over 5 trials. (Color figure online)

5.3 Visualizations

We visualize both achieved goals and subgoals at different training stages in the Ant Maze (U-shape) environment in Fig. 7. These subgoals generated by the high-level policy are consistently located near the achieved goals by the low-level policy, effectively guiding the low-level policy to reach the desired goal and improving training efficiency. Figure 8 displays the features of subgoals extracted by the subgoal encoder ψ_ϕ after dimension reduction by t-SNE. Initially, the

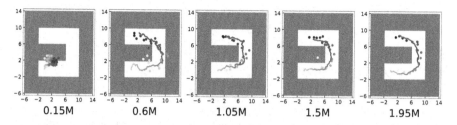

Fig. 7. Visualization of subgoals generated by the high-level policy and achieved goals accomplished by the low-level policy within one episode at a series of training steps in the Ant Maze (U-shape) environment (red for the subgoals, blue for the achieved goals, light for the start, dark for the end). (Color figure online)

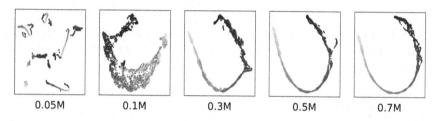

Fig. 8. Subgoal representation learning process in the Ant Maze (U-shape) environment at different training steps. Each subfigure contains 5 trajectories with the start positions highlighted in blue and the end positions in yellow. The visualization only shows the learning phase and excludes the stable phase that occurs afterward. (Color figure online)

feature map is disordered at 0.05M steps but becomes gradually more ordered over time, reaching stability at 0.3M steps and remaining stable thereafter. This indicates that the features can be learned quickly and become stable as the environment is explored extensively, facilitating the generation of reliable and consistent subgoals by the high-level policy.

5.4   Ablation Studies

We perform ablation studies on the Ant Maze (U-shape) environment to analyze the impact of parameter α, parameter β, and component loss in Fig. 9.

Parameters α and β: The result of Fig. 9a indicates that $\alpha = 20$ is better for promotion in the learning process. Our algorithm MISD is robust against β due to the comparable performance with various values, as shown in Fig. 9b. Notably, we set $\alpha = 20$ and $\beta = 0.1$ as default values based on our ablation study results.

Component loss: The high-level policy employs three loss functions: $\mathcal{L}_{rew}$, $\mathcal{L}_{ag}$ and $\mathcal{L}_{dg}$, represented as RL, AGL, and DGL in the Fig. 9c, respectively. Compared to the algorithm with only loss $\mathcal{L}_{rew}$, the addition of loss $\mathcal{L}_{ag}$ can significantly improve performance by generating efficient samples that can be directly

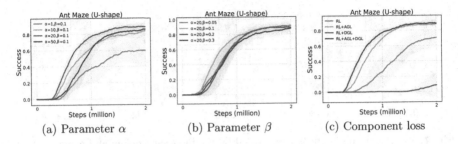

(a) Parameter α (b) Parameter β (c) Component loss

Fig. 9. Ablation studies on different parameters and components: (a) parameter α, (b) parameter β, and (c) component loss. All curves are averaged over 5 trials and smoothed equally for visual clarity. (Color figure online)

used for training without goal-relabeling. By incorporating loss $\mathcal{L}_{dg}$ under the constraint of loss $\mathcal{L}_{ag}$, the algorithm can enhance exploration capability and speed up learning process.

6   Conclusion

We propose a novel method MISD, Mutual Information-based Subgoal Discovery, which utilizes mutual information constraint to identify informative subgoals in hierarchical reinforcement learning. By minimizing the mutual information distance between subgoals and both the achieved and desired goals, MISD effectively enhances sample and training efficiency. The experimental results demonstrate that our method achieves comparable performance to state-of-the-art methods while significantly reducing the training time in various continuous control tasks. Moreover, MISD is dimension-agnostic with respect to the state and action spaces, highlighting its potential applicability in real-world scenarios.

References

1. Andrychowicz, M., et al.: Hindsight experience replay. In: Advances in Neural Information Processing Systems, vol. 30, pp. 5048–5058. Curran Associates, Inc. (2017)
2. Duan, Y., Chen, X., Houthooft, R., Schulman, J., Abbeel, P.: Benchmarking deep reinforcement learning for continuous control. In: Proceedings of The 33rd International Conference on Machine Learning, vol. 48, pp. 1329–1338. PMLR (2016)
3. Eysenbach, B., Gupta, A., Ibarz, J., Levine, S.: Diversity is all you need: learning skills without a reward function. In: International Conference on Learning Representations (2018)
4. Eysenbach, B., Zhang, T., Levine, S., Salakhutdinov, R.R.: Contrastive learning as goal-conditioned reinforcement learning. In: Advances in Neural Information Processing Systems, vol. 35, pp. 35603–35620. Curran Associates, Inc. (2022)
5. Hafner, D., Lee, K.H., Fischer, I., Abbeel, P.: Deep hierarchical planning from pixels. In: Advances in Neural Information Processing Systems. vol. 35, pp. 26091–26104. Curran Associates, Inc. (2022)

6. Hartikainen, K., Geng, X., Haarnoja, T., Levine, S.: Dynamical distance learning for semi-supervised and unsupervised skill discovery. In: International Conference on Learning Representations (2020)

7. Huang, Z., Liu, F., Su, H.: Mapping state space using landmarks for universal goal reaching. In: Advances in Neural Information Processing Systems, vol. 32, p. 1942–1952. Curran Associates, Inc. (2019)

8. Kim, J., Seo, Y., Shin, J.: Landmark-guided subgoal generation in hierarchical reinforcement learning. In: Advances in Neural Information Processing Systems, vol. 34, pp. 28336–28349. Curran Associates, Inc. (2021)

9. Kulkarni, T.D., Narasimhan, K., Saeedi, A., Tenenbaum, J.: Hierarchical deep reinforcement learning: integrating temporal abstraction and intrinsic motivation. In: Advances in Neural Information Processing Systems, vol. 29, pp. 3675–3683. Curran Associates, Inc. (2016)

10. Laskin, M., Srinivas, A., Abbeel, P.: CURL: Contrastive unsupervised representations for reinforcement learning. In: Proceedings of the 37th International Conference on Machine Learning, vol. 119, pp. 5639–5650. PMLR (2020)

11. Levy, A., Konidaris, G., Platt, R., Saenko, K.: Learning multi-level hierarchies with hindsight. In: International Conference on Learning Representations (2019)

12. Nachum, O., Gu, S.S., Lee, H., Levine, S.: Data-efficient hierarchical reinforcement learning. In: Advances in Neural Information Processing Systems, vol. 31, p. 3303–3313. Curran Associates, Inc. (2018)

13. Nasiriany, S., Pong, V., Lin, S., Levine, S.: Planning with goal-conditioned policies. In: Advances in Neural Information Processing Systems, vol. 32, p. 14843–14854. Curran Associates, Inc. (2019)

14. Van Den Oord, A., Li, Y., Vinyals, O.: Representation learning with contrastive predictive coding. arXiv preprint arXiv:1807.03748 (2018)

15. Paninski, L.: Estimation of entropy and mutual information. Neural Comput. **15**(6), 1191–1253 (2003)

16. Pong, V., Gu, S., Dalal, M., Levine, S.: Temporal difference models: model-free deep RL for model-based control. In: International Conference on Learning Representations (2018)

17. Poole, B., Ozair, S., Van Den Oord, A., Alemi, A., Tucker, G.: On variational bounds of mutual information. In: Proceedings of the 36th International Conference on Machine Learning, vol. 97, pp. 5171–5180. PMLR (2019)

18. Sharma, A., Gu, S., Levine, S., Kumar, V., Hausman, K.: Dynamics-aware unsupervised discovery of skills. In: International Conference on Learning Representations (2020)

19. Zhang, Q., Yang, Y., Ruan, J., Xiong, X., Xing, D., Xu, B.: Balancing exploration and exploitation in hierarchical reinforcement learning via latent landmark graphs. In: 2023 International Joint Conference on Neural Networks (IJCNN), pp. 1–8. IEEE (2023)

20. Zhang, T., Guo, S., Tan, T., Hu, X., Chen, F.: Generating adjacency-constrained subgoals in hierarchical reinforcement learning. In: Advances in Neural Information Processing Systems, vol. 33, pp. 21579–21590. Curran Associates, Inc. (2020)

Accelerate Support Vector Clustering via Spectral Data Compression

Yongyu Wang$^{(\boxtimes)}$ ⓘ and Yuxuan Song ⓘ

JD Logistics, Beijing 101111, China
wangyongyu1@jd.com

Abstract. This paper proposes a novel framework for accelerating support vector clustering. The proposed method first computes much smaller compressed data sets while preserving the key cluster properties of the original data sets based on a novel spectral data compression approach. Then, the resultant spectrally-compressed data sets are leveraged for the development of fast and high quality algorithm for support vector clustering. We conducted extensive experiments using real-world data sets and obtained very promising results. The proposed method allows us to achieve 100X and 115X speedups over the state of the art SVC method on the Pendigits and USPS data sets, respectively, while achieving even better clustering quality. To the best of our knowledge, this represents the first practical method for high-quality and fast SVC on large-scale real-world data sets.

Keywords: Support Vector Clustering · Spectral Theory · Data Compression

1 Introduction

Support vector-based algorithms are among the most widely used machine learning methods. To solve the supervised classification problem, support vector machine (SVM) constructs a hyperplane in a high-dimensional space via support vectors. Extended from SVM, [1] proposed support vector clustering (SVC) method, employing support vectors to discover the boundaries of clusters. Compared to traditional clustering methods such as k-means, SVC can well-detect linearly inseparable clusters by mapping the samples into a high-dimensional feature space. Compared to advanced clustering methods such as spectral clustering [2], it does not require people to specify the number of clusters in advance, which is highly desired for unseen data sets. When comparing with the DBSCAN clustering method [3], SVC can well-detect clusters from any density scenario.

Due to the rigorous theoretical foundations and many desired characteristics, SVC has drawn great attention from both the research and industrial communities. However, the cluster labeling step of the original SVC has a $O(N^2)$ time

---

Y. Wang and Y. Song—These authors contributed equally and are co-first authors.

ⓒ The Author(s), under exclusive license to Springer Nature Singapore Pte Ltd. 2024
B. Luo et al. (Eds.): ICONIP 2023, CCIS 1961, pp. 88–97, 2024.
https://doi.org/10.1007/978-981-99-8126-7_7

complexity which can immediately hinder its applications in emerging big data analytics tasks. Another main drawback of the original SVC is that it cannot handle boundary support vectors (BSVs). In large-scale real-world data sets where a lot of BSVs exist, only a small portion of points can be clustered by the original SVC method and a large amount of BSVs are mis-recognized as outliers.

To solve the above problems, substantial effort has been devoted and many methods have been developed. Among them, two representative works are most important: [4] attempted to accelerate SVC through proximity graphs. However, it cannot solve the BSV problem. [5] proposed a stable equilibrium point (SEP)-based method that can solve the BSV problem very well. However, the SEP calculation process requires to perform gradient descent on each data point which is an extremely time-consuming process for large-scale data set. So as acknowledge in [5], it is impractical to do it on thousands of data points. [6] proposed a rough-fuzzy SVC based on fuzzy clustering, however, it cannot handle multiple clusters. [7] proposed a convex-decomposed-based cluster labeling method, but the BSV problem still exists. Generally speaking, there was almost no meaningful progress on SVC since the SEP-based SVC method was proposed.

However, in spectral clustering field, some important progresses have been made. [9] proposed a general framework for accelerating spectral clustering with representative points and [10] further improved it by using spectrum-preserving node aggregation. Inspired by [9,10], in this paper, we propose to fundamentally address the scalability and BSV issues of SVC simultaneously via spectral data compression. Our method first computes a much smaller compressed data set while preserving the key cluster properties of the original data set, by exploiting a nearly-linear time spectral data compression approach. The compressed data set enables to dramatically reduce the size of search space for SEP searching process. Then, the SEPs found from the compressed data set can be directly applied to the SEP-based SVC algorithm. Due to the guaranteed preservation of the original spectrum for clustering purpose, the proposed method can dramatically improve the efficiency of the SEP-based SVC without loss of solution quality. To the best of our knowledge, this represents the first practical method for high-quality and fast SVC on large-scale real-world data sets. Another distinguished characteristic of the proposed method is that it allows users to set the compression ratio based on their needs. Experimental results show that with 2X, 5X and 10X reductions, the proposed method consistently produces high-quality clustering results on real-world large-scale data sets.

2   Preliminary

2.1   Support Vector Clustering

Given a data set D with N samples $\mathbf{x}_1, ..., \mathbf{x}_N$, the original SVC algorithm find clusters with two main steps, namely SVM training step and cluster labeling step. The SVM training step constructs a trained kernel radius function by first mapping the samples to high dimensional feature space with a nonlinear transformation Φ, then finding the minimal hypersphere of radius R with the following constraint:

$$\|\Phi(\mathbf{x}_j) - \mathbf{a}\|^2 \le R^2 + \xi_j, \tag{1}$$

where $\mathbf{a}$ denotes the center of the hypersphere and $\xi_j \ge 0$ denotes the slack variable.

To form the dual problem for solving the problem (1), the following Lagrangian is introduced:

$$L = R^2 - \sum_j \left(R^2 + \xi_j - \|\Phi(\mathbf{x}_j) - \mathbf{a}\|^2\right) \beta_j$$
$$- \sum \xi_j \mu_j + C \sum \xi_j, \tag{2}$$

where β_j and μ_j are Lagrangian multipliers, C is a manually set constant with the constraint: $0 \le \beta_j \le C$. Then, the solution of (1) can be found by solving the following optimization problem:

$$\text{Maximize: } W = \sum_j \Phi(\mathbf{x}_j)^2 \beta_j - \sum_{i,j} \beta_i \beta_j \Phi(\mathbf{x}_i)\Phi(\mathbf{x}_j)$$
$$\text{Subject to: } 0 \le \beta_j \le C, \sum_j \beta_j = 1, j = 1, ..., N. \tag{3}$$

The points that satisfy $0 < \beta_j < C$ are called support vectors (SVs) and the boundaries of clusters are formed by the SVs. The points that satisfy $\beta_j = 0$ lie inside clusters. The points satisfy $\beta_j = C$ are called bounded support vectors (BSVs) which cannot be clustered by the original SVC method.

By using the Mercer kernel ($K(\mathbf{x}_i, \mathbf{x}_j) = e^{-q\|\mathbf{x}_i - \mathbf{x}_j\|^2}$) to replace the inner product of $\Phi(\mathbf{x}_i) \cdot \Phi(\mathbf{x}_j)$, the following trained kernel radius function can be obtained:

$$f(\mathbf{x}) = R^2(\mathbf{x}) = \|\Phi(\mathbf{x}) - \mathbf{a}\|^2$$
$$= K(\mathbf{x}, \mathbf{x}) - 2\sum_j K(\mathbf{x}_j, \mathbf{x}_j)\beta_j + \sum_{i,j} \beta_i \beta_j K(\mathbf{x}_i, \mathbf{x}_j) \tag{4}$$

Then, in the cluster labeling step, for each pair of points, SVC needs to check whether any segment point $\mathbf{y}$ between them satisfy $R(\mathbf{y}) < R$, which is a very time consuming step.

2.2   Graph Laplacian Matrices

Consider a graph $G = (V, E, w)$, where V is the vertex set of the graph, E is the edge set of the graph, and w is a weight function that assign positive weights to all edges. The Laplacian matrix of graph G is a symmetric diagonally dominant (SDD) matrix defined as follows:

$$L_G(p,q) = \begin{cases} -w(p,q) & \text{if } (p,q) \in E \\ \displaystyle\sum_{(p,t)\in E} w(p,t) & \text{if } (p=q) \\ 0 & \text{if } otherwise. \end{cases} \tag{5}$$

According to spectral graph theory and recent progresses in the field [8], the cluster properties are embedded in the spectrum of the graph Laplacian. For example, for the well-known artificially generated two-moons and two-circles data sets, clustering results of k-means on the original feature spaces and the spectrally-embedded feature spaces are shown in Fig. 1. It can be seen that in the original feature spaces, k-means clustering algorithm fails to discover the correct clusters. In contrast, by integrating spectral information into the feature space, correct clustering results can be generated. Recent research progresses shows that using well-designed spectral algorithm can significantly improve clustering performance [10].

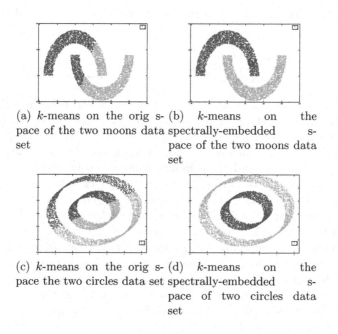

(a) k-means on the orig s-pace of the two moons data set

(b) k-means on the spectrally-embedded s-pace of the two moons data set

(c) k-means on the orig s-pace the two circles data set

(d) k-means on the spectrally-embedded s-pace of two circles data set

Fig. 1. Clustering results of k-means on the original and spectrally-embedded feature spaces.

3   Methods

3.1   Algorithmic Framework

Given a set of data samples, we first construct a k-NN graph G to capture the basic manifold of the data set. Then we can obtain its corresponding graph Laplacian matrix L_G accordingly. The spectral similarity between two points can be measured with the following steps:

- Calculate the bottom K nontrivial graph Laplacian eigenvectors;

- Construct a matrix U with the calculated K eigenvectors stored as column vectors;

- Use each row of U as the embedded feature vector of a point in low-dimensional space;

- The spectral similarity between two points u and v can be calculated as:

$$s_{uv} := \frac{|(\mathbf{X}_u, \mathbf{X}_v)|^2}{(\mathbf{X}_u, \mathbf{X}_u)(\mathbf{X}_v, \mathbf{X}_v)}, \quad (\mathbf{X}_u, \mathbf{X}_v) := \sum_{k=1}^{K} \left(\mathbf{x}_u^{(k)} \cdot \mathbf{x}_v^{(k)} \right). \tag{6}$$

where $\mathbf{x}_u$ and $\mathbf{x}_v$ are the embedded feature vectors of point u and point v, respectively. The statistical interpretation of s_{uv} is that it is the fraction of explained variance of linearly regressing x_u on x_v from spectral perspective. A higher value of s_{uv} indicates stronger spectral correlation between the two points. In practice, to avoid the computational expensive eigen-decomposition procedure, Gauss-Seidel relaxation can be applied to generated smoothed vectors for spectral embedding [11], which can be implemented in linear time.

Based on the spectral similarity, samples can be divided into different subsets. Samples in the same subset are highly correlated. We calculate the average of the feature vectors in each subset as its spectrally-representative pseudo-sample and put the representative pseudo-samples together to form the compressed data set. If the desired compression ratio is not reached, the representative pseudo-samples can be further compressed. Figure 2 and Fig. 3 show the manifolds of the original USPS data set and its compressed data set in 2D space, respectively, by visualizing their corresponding k-NN graphs. It can be seen that the compressed data set leads to a good approximation of the manifold of the original data set.

Then, the SEP searching process can be performed on the compressed data set to efficiently find the SEPs and the downstream cluster labeling step can be applied on SEPs to assign them into different clusters.

[5] demonstrated that each SEP represent a set of original point that belong to the same connected component, so the cluster-membership of each SEP can be assigned to all of its associated original data points. The complete algorithm flow is shown in Algorithm 1.

Fig. 2. The manifold of the original data set (USPS).

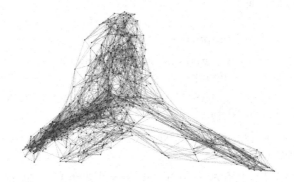

Fig. 3. The manifold of the compressed data set (USPS).

Algorithm 1. Spectral data compression-based fast SVC

Input: A data set D with N samples $x_1, ..., x_N \in R^d$, number of clusters k. **Output:** Clusters $C_1,...,C_k$.

    Construct a k-nearest neighbor (k-NN) graph G from the input data ;

    Compute the Laplacian matrix corresponding to graph G;

    Perform spectral data compression on the original data set ;

    Search SEPs from the compressed data set;

    Perform cluster-labeling on the SEPs;

    Map the cluster-memberships of SEPs back to obtain clustering result of the original data set.

3.2 Algorithm Complexity

The spectral data compression takes $O(|E_G|log(|V|))$ time, where $|E_G|$ is the number of edges in the original graph and $|V|$ is the number of data points. The SEP-based SVC on the compressed data set takes $O(P^2)$ time, where P is the number of compressed data points. The complexity of cluster membership retrieving for the original data points is $O(N)$, where N is the number of samples in the original data set.

4   Experiment

In this section, extensive experiments have been conducted to evaluate the performance of the proposed SVC method. Algorithms are performed using MATLAB running on Laptop.

4.1   Data Sets

We use two large-scale real-world data sets to evaluate the methods: **PenDigits**: A data set consists of 7,494 images of handwritten digits from 44 writers, using the sampled coordination information. Each image is represented by 16 features; **USPS**: A data set with 9,298 scanned images from the envelops of U.S. Postal Service. All the images are deslanted and size normalized, and each of them is a 16-by-16 grayscale image.

4.2   Algorithms for Comparison

We compare the proposed method against both the baseline and the state-of-the-art SVC methods, including:

- Original SVC method (Orig SVC) [1]: It Performs clustering labeling on the complete graph of the data set.

- Proximity method [4]: It uses proximity graph to reduce the computational cost of the original SVC.

- SEP-based SVC [5]: It first searches SEPs and then perform complete-graph based cluster labeling on the SEPs.

4.3   Evaluation Metric

The clustering quality is measured by comparing the label generated by algorithm with the ground-truth label provided by the data set. The normalized mutual information (NMI) is the most widely used metric. The NMI metric can be calculated as follows [12]:

$$NMI = \frac{\sum_{i=1}^{k} \sum_{j=1}^{k} n_{i,j} \log(\frac{n \cdot n_{i,j}}{n_i \cdot n_j})}{\sqrt{(\sum_{i=1}^{k} n_i \log \frac{n_i}{n})(\sum_{j=1}^{k} n_j \log \frac{n_j}{n})}}, \tag{7}$$

where n is the number of data points in the data set, k is the number of clusters, n_i is the number of data points in cluster C_i according to the clustering result generated by algorithm, n_j is the number of data points in class C_j according to the ground truth labels provided by the data set, and $n_{i,j}$ is the number of data points in cluster C_i according to the clustering result as well as in class C_j according to the ground truth labels. The NMI value is in the range of [0, 1], while a higher NMI value indicates a better matching between the algorithm generated result and ground truth result.

4.4   Experimental Results

Clustering quality results of the proposed method and the compared methods are provided in Table 1. The runtime results are reported in Table 2.

Table 1. Clustering Quality Results (NMI)

Data Set	Orig SVC	Proximity	SEP-based SVC	Ours
Pedigits	0.56	0.59	0.79	**0.82**
USPS	0.30	0.31	0.67	**0.68**

Table 2. Runtime Results (s)

Data Set	Orig SVC	Proximity	SEP-based SVC	Ours
Pedigits	4308.5	357.8	1044.6	**10.0**
USPS	245.4	47.4	4665.3	**40.4**

It can be seen from Table 2 that for the original SEP-based SVC method, the SEP searching step and the downstream cluster labeling step take more than 1000 s and more than 4000 s for the Pedigits and USPS data sets, respectively. While in our method, these two steps take only 10.0 s and 40.4 s for the two data sets, respectively. Meanwhile, the proposed method achieved a even better clustering quality. Such a high performance is mainly due to the guaranteed preservation of key spectrum of the original data set. The clustering quality improvement is potentially because our spectral data compression method allows to keep the most significant relations among data points, while avoiding noisy and misleading relationships among data points for clustering tasks. This phenomenon is also observed from the results of the Proximity method. The NMI results of the Proximity method on the two data sets are better than the results of the original SVC, mainly due to its denoising effect.

As shown in Fig. 4, with increasing compression ratio, the proposed method consistently produces high-quality clustering results, demonstrating its robustness. As shown in Fig. 5, our method leads to dramatic speedups for SVC. Moreover, this progress can also lead to dramatically improved memory/storage efficiency for SVC tasks. It is expected that the proposed method will be a key enabler for storing and processing much bigger data sets on more energy-efficient computing platforms, such as FPGAs or even hand-held devices.

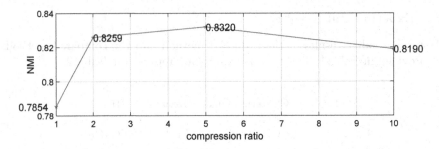

Fig. 4. Clustering quality VS compression ratio for the Pendigits data set.

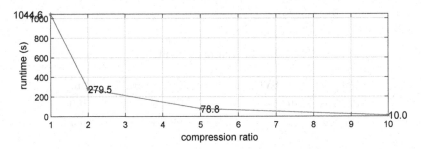

Fig. 5. Runtime VS compression ratio for the Pendigits data set.

5   Conclusion

To fundamentally address the computational challenge in SVC, this work applies a novel spectral data compression method to SEP-based SVC that enables to construct a very small compressed data set with guaranteed preservation of the original spectrum for clustering purpose. Experimental results on large scale real-world data sets show dramatically improved clustering performance when compared with state-of-the-art SVC methods.

6   Contributions

Yongyu Wang and Yuxuan Song conceived the idea. Yongyu Wang, as the corresponding author, supervised, led and guided the entire project. Yongyu Wang and Yuxuan Song designed and conducted the experiments and wrote the paper. All authors discussed the results and implications and commented on the manuscript at all stages.

References

1. Ben-Hur, A., Horn, D., Siegelmann, H.T., Vapnik, V.: Support vector clustering. J. Mach. Learn. Res. **2**(Dec), 125–137 (2001)

2. Von Luxburg, U.: A tutorial on spectral clustering. Stat. Comput. **17**, 395–416 (2007)
3. Schubert, E., Sander, J., Ester, M., Kriegel, H.P., Xu, X.: DBscan revisited, revisited: why and how you should (still) use DBscan. ACM Trans. Database Syst. (TODS) **42**(3), 1–21 (2017)
4. Yang, J., Estivill-Castro, V., Chalup, S.K.: Support vector clustering through proximity graph modelling. In: Proceedings of the 9th International Conference on Neural Information Processing. ICONIP'02, vol. 2, pp. 898–903. IEEE (2002)
5. Lee, J., Lee, D.: An improved cluster labeling method for support vector clustering. IEEE Trans. Pattern Anal. Mach. Intell. **27**(3), 461–464 (2005)
6. Saltos, R., Weber, R.: A rough-fuzzy approach for support vector clustering. Inf. Sci. **339**, 353–368 (2016)
7. Ping, Y., Chang, Y.F., Zhou, Y., Tian, Y.J., Yang, Y.X., Zhang, Z.: Fast and scalable support vector clustering for large-scale data analysis. Knowl. Inf. Syst. **43**, 281–310 (2015)
8. Chung, F.R.: Spectral Graph Theory, vol. 92. American Mathematical Soc. (1997)
9. Yan, D., Huang, L., Jordan, M.I.: Fast approximate spectral clustering. In: Proceedings of the 15th ACM SIGKDD International Conference on Knowledge Discovery and Data Mining, pp. 907–916 (2009)
10. Wang, Y.: Improving spectral clustering using spectrum-preserving node aggregation. In: 2022 26th International Conference on Pattern Recognition (ICPR), pp. 3063–3068. IEEE (2022)
11. Livne, O.E., Brandt, A.: Lean algebraic multigrid (LAMG): fast graph Laplacian linear solver. SIAM J. Sci. Comput. **34**(4), B499–B522 (2012)
12. Strehl, A., Ghosh, J.: Cluster ensembles–a knowledge reuse framework for combining multiple partitions. J. Mach. Learn. Res. **3**(Dec), 583–617 (2002)

A Novel Iterative Fusion Multi-task Learning Framework for Solving Dense Prediction

Jiaqi Wang [ID] and Jianping Luo[(✉)] [ID]

Guangdong Key Laboratory of Intelligent Information Processing,
Shenzhen Key Laboratory of Media Security and Guangdong Laboratory of Artificial
Intelligence and Digital Economy (SZ), Shenzhen University, Shenzhen, China
wangjiaqi2021@email.szu.edu.cn, ljp@szu.edu.cn

Abstract. Dense prediction tasks are hot topics in computer vision that aim to predict each input image pixel, such as Semantic Segmentation, Monocular Depth Estimation, Edge Estimation, etc. With advanced deep learning, many dense prediction tasks have been greatly improved. Multi-task learning is one of the top research lines to boost task performance further. Properly designed multi-task model architectures have better performance and minor memory usage than single-task models. This paper proposes a novel Multi-task Learning (MTL) framework with a Task Pair Interaction Module (TPIM) to tackle several dense prediction tasks. Different from most widely used MTL structures which share features on some specific layer and branch to task-specific layer, the output task-specific features are remixed via a TPIM to get more shared features in this paper. Due to joint learning, tasks are mutually supervised and provide rich shared information to each other for improving final results. The TPIM includes a novel Cross-task Interaction Block (CIB) which comprises two attention mechanisms, self-attention and pixel-wise global attention. In contrast with the commonly used global attention mechanism, an Iterative Fusion Block (IFB) is introduced to effectively fuse affinity information between task pairs. Extensive experiments on two benchmark datasets (NYUD-v2 and PASCAL) demonstrate that our proposal is effective in comparison to existing methods.

Keywords: dense prediction · multi-task learning · cross-task interaction · iterative fusion

1 Introduction

Real-world is inundated with many complex problems that must tackle multiple tasks simultaneously. For example, if an automated vehicle wants to drive safely on the road, one must strictly detect and locate all the objects around and accurately understand traffic signs, lane lines, etc., in the scene [12]. Similarly, in human face detection [4], not only should the landmarks on the human face

© The Author(s), under exclusive license to Springer Nature Singapore Pte Ltd. 2024
B. Luo et al. (Eds.): ICONIP 2023, CCIS 1961, pp. 98–112, 2024.
https://doi.org/10.1007/978-981-99-8126-7_8

be located, but expression recognition, such as smiling and crying, is needed to detect human faces accurately. There are countless instances. Multi-task learning (MTL) [3] is proposed for computationally efficiently solving multiple related tasks.

Traditional MTL methods fall into two categories [18,22]. One is that most parameters are shared among all the tasks with a tiny part of the private parameters, called the hard parameter sharing method. Shared information is propagated via the shared parameters, and task-specific outputs are obtained by independent private parameters [1,13]. The other is that each task has a complete and separate network, and features are fused by specific means such as weighted sum across tasks, named soft parameter sharing [9,16,19]. Recently, a variant of hard parameter sharing MTL has been proposed, i.e., to append a multimodal distillation module at the end of typical hard parameter sharing MTL to improve the information exchange across tasks [24]. It is well-known that multimodal data improve the performance of deep predictions [24]. For example, a Convolutional Neural Network (CNN) trained with RGB-D data perform better than trained with RGB data. However, obtaining depth data requires additional cost. An economical approach is to use a CNN to predict the depth maps and use them as input. Besides the depth maps, we can also use CNN to predict more related information. Inspired by this, previous works [2,14,21,24–27] propose to use a CNN-based MTL to obtain several related information, including but not limited to depth, semantic information, etc., and then use them as multi-modal input which are fed into the following CNN named multi-modal distillation module to fuse features from different tasks to have better MTL performance. It gains improvement with few parameters. However, we found that the previously proposed feature fusion methods are relatively simple and there is much room for improvement. In this paper, we propose a new type of attention-driven multimodal distillation scheme for better cross-task information fusion.

PAP-Net [26] models relationships between pairs of pixels and uses obtained affinity maps to perform interaction across tasks. PAD-Net [24] and MTI-Net [21] prove the effectiveness of self-attention on MTL, i.e., one task can further mine meaningful representations by applying self-attention to the task itself to help the other tasks. DenseMTL [14] introduces an attention module based on PAD-Net, called correlation-guided attention, which calculates the correlation between features from two tasks to guide the construction of exchanged messages. ATRC [2] explores four attention-based contexts dependent on tasks' relations and use Neural Architecture Search (NAS) to find optimal context type for each sourcetarget task pair. We partially follow the same direction as in works mentioned above, using self-attention and global attention in our model. In contrast to them, we propose an iterative mechanism called Iterative Fusion Block (IFB) that further fuses the pixel-wise affinity maps with the original features.

In summary, our contributions are threefold: (i) We propose a novel multimodal distillation design, named Task Pair Interaction Module (TPIM) (Sec. 3.2) for MTL comprising several Cross-task Interaction Blocks (CIB). (ii) We introduce a new mechanism to fuse further the pixel-wise affinity maps between task-pair (Sect. 3.4), which we call Iterative Fusion Block (IFB). IFB adaptively integrates shared and task-specific features and retains the original features to the greatest extent. (iii) Extensive experiments on the challenging NYUD-v2 [20] and PASCAL [8] datasets validate the effectiveness of the proposed method (Sect. 4.2). Our method achieves state-of-the-art results on both NYUD-v2 and PASCAL datasets. More importantly, the proposed method remarkably outperforms state-of-the-art works that optimize different tasks jointly.

2   Related Works

2.1   Multi-task Learning (MTL)

To learn common representations, MTL methods are classified into two paradigms, hard parameter sharing MTL and soft parameter sharing MTL [18,22]. The former typically comprises two stages. Architectures share the intermediate representations among the tasks at the first stage, usually a shared feature extractor, and branch to the independent task-specific representations layer in the second stage. Tasks used in soft parameter sharing have their network; cross-task interaction is conducted by bridging these networks. For example, [16] proposed to use a "cross-stitch" unit to combine features from different independent networks to adaptively learn a proper combination of shared and task-specific representations. Though all the previous works show great multi-task learning potential, they uncover a few challenges. Most notable is the negative transfer phenomenon [11], where learning some less related tasks jointly leads to degrading task performance. Some works [6,10] attribute negative transfer to not balancing the losses among independent tasks and introduce mechanisms to weigh the loss terms carefully. Kendall et al. [10] proposed to use each task's homoscedastic uncertainty to balance the losses. Chen et al. [6] proposed an algorithm named GradNorm to tune the magnitude of each task's gradients dynamically. Liu et al. [13] proposed Dynamic Weight Averaging (DWA) to weigh the tasks based on the task-specific losses dynamically.

2.2   Cross-Task Interaction Mechanisms

Close to our work are methods that distill shared features from task-specific features. Inspired by the acknowledgment that multi-modal data improves the performance of dense predictions, PAD-Net [24] introduced a multi-modal distillation module to refine information across multiple tasks. Vandenhende et al. [21] extended PAD-Net [24] to multi-scale level to better utilize multi-scale cross-task interaction. Zhang et al. [26] proposed to obtain pixel-wise affinity maps of all tasks, which are then diffused to other tasks to perform cross-task interaction. Similarly, Zhou et al. [27] further proposed Pattern Structure Diffusion (PSD) to mine and propagate patch-wise affinities via graphlets. Lopes et al. [14] introduced a cross-task attention mechanism comprising correlation-guided attention and self-attention to carry out multi-modal distillation. ATRC [2] explores four

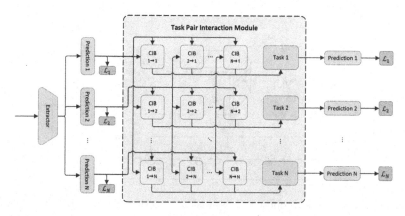

Fig. 1. Overview of our MTL framework with the proposed Task Pair Interaction Module (TPIM). Input features are first through a shared Extractor and several prediction modules. Consequently, the features from prediction modules are fed into the TPIM and fusion in pairs. Within TPIM, each task is routed as a target task to N Cross-task Interaction Blocks (CIB) (n-th row of CIBs) and as a source task to N CIBs (n-th column). The outputs of CIBs are summed together for each task independently (denoted as 'Task n') and fed through the consequent prediction module ('Prediction n'). **Legend**: Green blocks denote modules with learned weights, and orange blocks denote loss functions. Best viewed in color.

attention-based contexts dependent on tasks' relations and use Neural Architecture Search (NAS) to find optimal context type for each source-target task pair. In this paper, we propose a novel cross-task attention mechanism to refine task-specific features. We use a self-attention to explore related features from the source task and a pixel-wise global attention to construct affinity information between task pairs. Besides, an iterative mechanism named Iterative Fusion Block (IFB) is introduced to deeply fuse cross-task affinity information with original task-specific information and combine the cross-task affinities with self-attention by addition operation to learn complementary representations for the target task.

3   Methods

This section will describe the proposed framework used to simultaneously figure out related dense prediction tasks. We first present an overview of the proposed framework and then introduce our framework's details.

3.1   Overall Structure

Figure 1 shows our overall MTL structure for dense prediction tasks. The proposed MTL model consists of four main modules. The first is a shared feature extraction module that extracts shared information among tasks. The second is

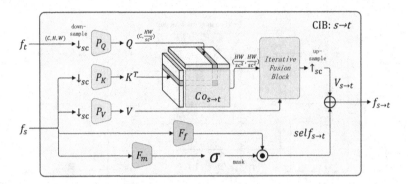

Fig. 2. Overview of our proposed Cross-task Interaction Block (CIB). CIB enables cross-task interaction between task pairs (s, t). It relies on two attention mechanisms. First, pixel-wise global attention (yellow, upper part) to discover $V_{s \to t}$, the features from task s contributing to task t. Second, a self-attention (green, lower part) to discover complementary features $self_{s \to t}$ from s. $\downarrow_{sc}$ is a sc-times downscale for memory efficiency and reversely $\uparrow_{sc}$ is a sc-times upscale operation. Best viewed in color. (Color figure online)

an intermediate prediction module, which takes the shared features extracted by the previous module as input and outputs the prediction of the corresponding task. The third is a Task Pair Interaction Module (TPIM) which uses the predictions from the intermediate prediction module to carry out pair-wise feature interactions. The fourth is a final prediction module, consisting of N task-specific heads to decode the distilled information, to obtain task-specific predictions, where N is the number of tasks. The input of the proposed MTL structure is RGB images during training and testing, and the final output is N maps with the exact resolution as the input RGB images.

3.2   Task Pair Interaction Module (TPIM)

Building on recent works [2,14,21,24–27] illustrating the effectiveness of cross-task interaction, we propose a module that can capture spatial correlation of features at the pixel level while maintaining dedicated task-specific information. Figure 1 depicts our TPIM, which helps to blend task-specific features to enhance cross-task information communication. TPIM comprises N^2 Cross-task Interaction Blocks (CIB) and N feature aggregation blocks. Considering i as the primary task, we perform an element-wise addition on the refined features $f_{j \to i | j \in N}$, contributing towards task i, in the feature aggregation block marked 'Task i'.

3.3   Cross-Task Interaction Block (CIB)

This block seeks to capture the shared pair-wise task knowledge while preserving their non-shared knowledge. We are committed to exploiting features from pair of tasks denoted (s, t). Figure 2 illustrates our components, which help knowledge distillation between tasks. CIB takes as input the task features (f_s, f_t) and returning the corresponding complementary features $f_{s \to t}$.

Note that, for task pair (i, i), CIB is an identity transformation, i.e. $f_{i \to i} = f_i$.

Considering here t as the target task, and s as the source task, we aim to capture related features from task s to improve the performance of task t. For achieving this, two attentions are employed: (i) pixel-wise global attention, which is used to obtain the spatial correlation between two tasks, and (ii) self-attention on source task s to self-discover supplementary features for target task t. We fuse two attentions via an equally element-wise addition operation. Visualizations of the two attentions are colored in yellow and green in Fig. 2, respectively.

Note that each attention contributes differently to the target task t. The pixel-wise global attention relies on identifying shared s and t knowledge, while the other relies on exclusive s knowledge.

Self-attention. We employ a spatial self-attention, see green blocks in Fig. 2, which aims to self-discover important information of source task s that may be helpful to solve relative task t. We formulate self-attention as follows:

$$self_{s \to t} = F_s(f_s) \odot \sigma(F_m(f_s)) \tag{1}$$

where $F_*(\cdot)$ is a 3×3 convolution supervised by the target task t to learn to extract related information from features f_s, $\sigma(\cdot)$ the sigmoid function to normalize the attention map to 0–1, and $\odot$ denotes element-wise product. The self-attention features $self_{s \to t}$ is produced by calculating multiplication between the features coming from F_s and the normalized attention mask from F_m.

Pixel-Wise Global Attention. We rely on the spatial correlation between tasks for pixel-wise global attention; see yellow blocks in Fig. 2. We follow Non-local Block [23] to obtain affinity maps between task s and t. We perform downscale and dimension reduction before calculating for smaller memory footprints and faster inference speed. The correlation maps between task t and task s can be formulated as:

$$Q = P_Q(\downarrow_{sc} f_t), K = P_K(\downarrow_{sc} f_s), V = P_V(\downarrow_{sc} f_s)$$
$$Co_{s \to t} = softmax(\frac{K^T Q}{\sqrt{d}}) \tag{2}$$

where $P_*(\cdot)$ here denotes a 1×1 convolution, following a BatchNorm and ReLu function. $\downarrow_{sc}$ the downscale operator with sc the scale factor. The spatial-correlation matrix $Co_{s \to t}$ is then obtained by applying a softmax on the matric multiplication normalized by $\sqrt{d}$ where d is the length of vector K. The softmax function is used to generate probabilities. Intuitively, $Co_{s \to t}$ has high values where features from s and t are highly correlated and low values otherwise.

Subsequently, Iterative Fusion Block (IFB) takes as input the obtained matrix $Co_{s \to t}$ and the vector V, and we obtain our pixel-wise global attention features by upsampling the output of IFB:

$$V_{s \to t} = \uparrow_{sc} IFB(V, Co_{s \to t}) \tag{3}$$

where $\uparrow_{sc}$ the upscale operator with sc the scale factor. Details of IFB will be discussed in Sec. 3.4 later.

Feature Aggregations. The final features $f_{s \to t}$ are built by combining the features from two attention blocks as:

$$f_{s \to t} = self_s + V_{s \to t} \tag{4}$$

Finally, the corresponding features of task t from multiple source tasks are aggregated as:

$$f_t^o = \sum_{i=0}^{N} f_{i \to t} \tag{5}$$

where $f_{t \to t} = f_t^i$. Output feature f_t^o is consequently fed into the prediction module.

3.4   Iterative Fusion Block (IFB)

Usually, after obtaining the correlation matrix $Co_{s \to t}$, it is multiplied with the matrix V to get global attention. Considering that a single matrix multiplication may not be able to fully integrate the information in the affinity map with the original features V, we introduce an iterative mechanism to fuse the original feature information with details in affinity maps profoundly and effectively. In this paper, we develop and investigate three different iteration block designs, as shown in Fig. 3. The IFB A represents a naive iteration of matrix multiplication of affinity map $Co_{s \to t}$ and features V. The IFB B introduces a residual term to avoid gradient vanishing and a learnable parameter α to adaptively adjust the attention term's weights. The IFB C, based on the IFB B, adds a weight $1 - \alpha$ on the residual term to adaptively balance the weights of the attention term and residual term and keep the magnitude of the input and output consistent.

Iterative Fusion Block A. A common way to iterate is to repeat the operation several times. We also consider this simple scheme as our basic iteration block, which can be formulated as follows:

$$V^{t+1} = CoV^t, t \geq 0 \tag{6}$$

where t is the number of iterations.

However, the values in the affinity map are all probability values, which means the values are between 0 and 1. Multiplication with the affinity map,

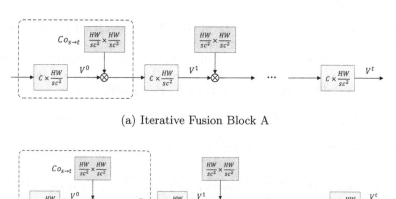

(a) Iterative Fusion Block A

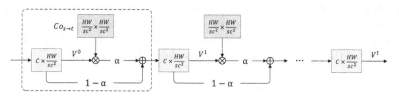

(b) Iterative Fusion Block B

(c) Iterative Fusion Block C

Fig. 3. Illustration of the designed different Iterative Fusion Blocks. The green blocks denote affinity map $Co_{s \to t}$ of task pairs (s, t), and yellow blocks V indicate features from source task s. The IFB A is a naive iteration of matrix multiplication of affinity map $Co_{s \to t}$ and features V; the IFB B introduces a residual term and a learned weight α to adjust the weight of the fused feature adaptively; the IFB C adds a weight $1 - \alpha$ to the residual term to maintain the magnitude of the feature. (Color figure online)

again and again, will cause the values in the feature map to gradually decrease. As the number of iterations increases, the values in the feature map approach 0, resulting in performance degradation.

Iterative Fusion Block B. To overcome the abovementioned problem, we introduce a residual term to maintain the original feature information and stop the values in the feature map from reaching 0.

Here comes another question: How to combine residual term and fused term? Simply adding them together? As the number of iterations continues to increase, simply adding the two terms will lead to an increasing proportion of the original information, losing the meaning of fusion. To tackle this, a learnable parameter α is introduced to adaptively balance the weight of residual and fused terms, and details can be seen in Fig. 3(b).

The learnable parameter α is one of the CNN model parameters, which are dynamically updated with training and supervised by the corresponding ground truth labels. There's no need to set the value of α manually. Besides, proper initialization of α may bring about the improvement; we discuss it in Sect. 4.3.

With a learnable parameter α, IFB B can be formulated as:

$$V^{t+1} = \alpha CoV^t + V^t, t \geq 0 \qquad (7)$$

where t is the number of iterations.

Iterative Fusion Block C. However, IFB B still has problems. With the iteration increases, the output features V^t are significantly larger than the input feature V^0, causing some distortion. To solve this, we add another weight $1 - \alpha$ to the residual term to keep the magnitude of the input and output consistent; see Fig. 3(c).

As mentioned above, we proposed an Iterative Fusion Block to fuse the information from affinity map $Co_{s \to t}$ and features V. Facing the problem of feature vanishing, we append a residual term. To tackle the imbalance of residual and fused terms, we introduce a learnable parameter α and $1 - \alpha$ to adaptively adjust the weights of the two terms. IFB C can be formulated as follows:

$$V^{t+1} = \alpha CoV^t + (1 - \alpha) V^t, t \geq 0 \qquad (8)$$

where t is the number of iterations.

4   Experiments

To demonstrate the effectiveness of the proposed method for simultaneously solving dense prediction tasks, we conduct experiments on two publicly available benchmark datasets, NYUD-v2 and PASCAL. Sect. 4.1 describes our experimental setups, including datasets, task setups, baseline models, etc. Then we show the performance of our method on two benchmark datasets compared to state-of-the-art approaches in Sect. 4.2. Finally, we present three ablation studies in Sect. 4.3, including (i) three designs of IFB, (ii) the number of iterations, and (iii) different initial values for α on the NYUD-v2 dataset.

Table 1. Training details of our experiments.

Dataset	Model	Epoch	LR	Optimizer	Scheduler
NYUD-v2	STL	100	10e-4	SGD	Poly
	Others	100	10e-4	Adam	Poly
PASCAL	STL	60	10e-2	SGD	Poly
	Others	100	10e-4	Adam	Poly

4.1   Experimental Setups

Datasets. We conduct experiments on NYUD-v2 and PASCAL datasets, which are widely used in dense predictions. The NYUD-v2 dataset contains 1449 indoor RGB depth images captured from the Microsoft Kinect and is split into a training subset and a testing subset. The former contains 795 image pairs of images and annotated images for semantic segmentation and monocular depth estimation task; the latter contains 654 pairs. The RGB and annotated images are randomly flipped horizontally and scaled for data augmentation. The PASCAL we used is a split of PASCAL-Context with dense annotations for semantic segmentation, human parts segmentation, and edge detection. It contains 4998 pairs of images, including RGB images and the corresponding ground truth labels for training and 5105 for testing. For data augmentation, the RGB and annotated images are randomly flipped horizontally, rotated, and scaled on the training set.

Task Setup. Since the two datasets provide ground truth labels of different tasks, we use two sets of tasks. The first one is a two-task setup: $T = \{$Semantic Segmentation (**SemSeg**), Monocular Depth Estimation (**Depth**)$\}$ on NYUD-v2 dataset. The second one is a three-task set-up: $T = \{$Semantic Segmentation (**SemSeg**), Human Parts Segmentation (**PartSeg**), Saliency Estimation (**Sal**)$\}$ on PASCAL. Note that, PASCAL does not provide annotations for saliency estimation task, we use the labels from [15], that distilled them from pre-trained state-of-the-art model [5]. As far as we know, there is no necessary relationship between the number of tasks and the performance of the multi-task model. Therefore, different task setups may have different results [21].

Evaluation Metrics. For evaluating the performance of the semantic segmentation, human parts segmentation, and saliency estimation, pixel-level mean Intersection over Union ($mIoU$) is used. The root mean square error in meters ($RMSE$) is used for monocular depth estimation. We formulate *multi-task performance* of model m [15]: $\Delta_m = \frac{1}{N} \sum_{i=1}^{N} (-1)^{\gamma_i} (M_{m,i} - M_{b,i})/M_{b,i}$ as the average of gained performance w.r.t single-task baseline b, where $\gamma_i = 1$ if a lower value means better performance for metric M_i of task i, and 0 otherwise. In our cases, $\gamma_i = 1$ only when the evaluation metric is $RMSE$.

Baseline. We compare the proposed framework against a single-task learning baseline (STL), which is predicted separately by several independent networks without any cross-task interaction and a typical multi-task learning baseline (MTL) consisting of a shared encoder and several task-specific decoders. Moreover, the proposed model is compared against state-of-the-art PAD-Net [24], MTI-Net [21], DenseMTL [14], and ATRC [2]. We replace our TPIM with the distillation modules proposed by the above-mentioned works before the final prediction module. Neural Architecture Search (NAS) is needed to search optimal architecture for ATRC; we use their published search results for simplicity.

Table 2. Comparison with the state-of-the-arts on two validation sets.

(a) Comparison with the state-of-thearts on NYUD-v2 validation set

Model	SemSeg ↑	Depth ↓	$\Delta_m(\%)$ ↑
STL	35.0091	0.6610	0.00
MTL	35.0475	0.6679	-0.59
PAD-Net[24]	35.8012	0.6571	1.30
MTI-Net[21]	37.6181	0.6066	7.71
DenseMTL[14]	37.986	0.6027	8.53
ATRC[2]	38.4576	0.6098	8.66
Ours	39.1596	0.6011	10.32

(b) Comparison with the state-of-the-arts on PASCAL validation set.

Model	SemSeg ↑	PartSeg ↑	Sal ↑	$\Delta_m(\%)$ ↑
STL	59.3427	60.3365	66.892	0.00
MTL	56.2943	60.1417	65.682	-2.42
PAD-Net[24]	51.9943	60.5255	65.894	-4.52
MTI-Net[21]	63.3808	62.1468	67.413	3.53
DenseMTL[14]	63.8954	65.0074	67.519	3.79
ATRC[2]	64.9036	62.0583	66.986	4.12
Ours	65.5761	63.1563	66.944	5.08

Loss Scheme. All the loss schemes are reused from [21]. Specifically, We use the L1 loss for depth estimation and the cross-entropy loss for semantic segmentation on NYUD-v2. On PASCAL, we use the balanced cross-entropy loss for saliency estimation and the cross-entropy loss for others. We do not adopt a particular loss-weighing strategy but sum the losses together with solid weights as in [22], i.e., $\mathcal{L} = \sum_{i=0}^{N} w_i \mathcal{L}_i$.

Training Details. The proposed network structure is implemented base on Pytorch library [17] and on Nvidia GeForce RTX 3090. The backbone model HRNet18 is pre-trained with ImageNet [7]. The training configuration of all models is shown in Table 1 following [22]. A poly learning rate scheduler: $lr = lr \times (1 - \frac{epoch}{TotalEpoch})^{0.9}$ is used to adjust the learning rate.

4.2   Comparison with State-of-the-Arts

Table 2(a) reports our experimental results compared to baseline models on NYUD-v2, while Table 2(b) reports our experimental results on PASCAL. On indoor densely labeled NYUD-v2, our model remarkably outperforms all baselines. Our model on the PASCAL dataset achieves the best results except for the Saliency Estimation task. A likely explanation of low performance is that the ground truth labels for saliency are distilled from pre-trained state-of-the-art model [5] as in [15]. The annotations we used are biased from the ground truth. The improvement of PAD-Net over the MTL baseline confirms the necessity to remix the task-specific features and the efficiency of the self-attention mechanism. MTI-Net retains the multi-modal distillation module of PAD-Net and adds the FPM module to fuse features at multiple scales. The remarkable improvement of MTI-Net over PAD-Net suggests that task interaction varies at different scales and emphasizes the effectiveness of multi-scale cross-task interaction. DenseMTL proposed a correlation-guided attention module, adaptively combined with a self-attention module. The improvement of DenseMTL over PAD-Net validates the importance of cross-task correlation attention. Unlike the aforementioned methods of constructing task correlation on feature space, ATRC explored task relationships on both feature and label space. The high performance of ATRC uncovers the potential of exploring task relationships on label space.

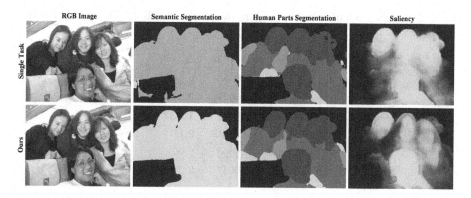

Fig. 4. Qualitative results on PASCAL dataset. We compare the predictions made by a set of single-task models against the predictions made by our model. Our model produces more precise results and smoother boundaries. And the attention distribution is more concentrated, which can be seen from Saliency.

Note that our model outperforms all baseline models on both multi-task performance and single-task performance on the NYUD-v2 dataset and obtains the best on three out of four metrics on the PASCAL dataset, which not only demonstrates the effectiveness of our proposed method but indicates the benefits of jointly solving multiple related tasks and shows the excellent potential of multi-task learning.

Figure 4 shows qualitative results on the PASCAL validation dataset. We can see the advantage of our multitask learning approach over vanilla single-task learning, where we separate objects better especially in the details.

4.3   Ablation Studies

In Table 3, we visualize the results of our ablation studies on NYUD-v2 and PASCAL to verify how IFB contributes to the multi-task improvements.

Table 3. Influence of with and without IFB on the validation set.

Ablation on NYUD-v2 validation set. $n_{iter} = 18$ for all three IFB.

Method	SemSeg ↑	Depth ↓	$\Delta_m(\%)$ ↑
Ours (w/o IFB)	38.4499	0.6016	9.27
Ours (w/ IFB A)	38.1003	0.6125	7.95
Ours (w/ IFB B)	37.9603	**0.5996**	8.73
Ours (w/ IFB C)	**39.1596**	0.6011	**10.32**

Ablation on PASCAL validation set. $n_{iter} = 4$.

Method	SemSeg ↑	PartSeg ↑	Sal ↑	$\Delta_m(\%)$ ↑
Ours (w/o IFB)	64.0299	62.4933	**67.194**	3.97
Ours (w/ IFB A)	64.9804	62.5252	66.590	4.23
Ours (w/ IFB B)	65.4426	62.8790	66.791	4.78
Ours (w/ IFB C)	**65.5761**	**63.1563**	66.944	**5.08**

We focus on the smaller NYUD-v2 dataset first. We alter our method using three IFBs and without IFB. As seen in Table 3(a), IFB A and IFB B lead to decreased performance, -1.32% and -0.54%, respectively. With the iteration increases, models with IFB A may degenerate into MTI-Net, where $V_{s \to t} = 0$

Table 4. Ablating the initialization of α in the proposed IFB on NYUD-v2 validation set. $n_{iter} = 18$.

α	SemSeg ↑	Depth ↓	$\Delta_m(\%)$ ↑
0	<u>38.8510</u>	0.6030	<u>9.74</u>
0.1	38.5180	0.6021	9.33
0.3	**39.1596**	0.6011	**10.32**
0.5	38.2666	**0.6002**	9.11
0.7	38.1899	<u>0.6003</u>	9.00
0.9	38.5608	0.6047	9.20

and CIB contains only self-attention. When $n_{iter} = 18$, The model using IFB A performs only +0.24% better than MTI-Net, proving that pixel-wise global attention plays little role in the whole model. Using IFB B tackles the problem but has a trial that compared with the input features, the output feature is prominent, which may be several times larger than the input features. It gets more prominent with the iteration increases. Instead, improvements (+1.05%) gained by IFB C confirm the rationality of our design. IFB C tackles the above problems and fully uses the cross-task information in the affinity map. A similar trend still appears in the PASCAL, but because of the smaller n_{iter}, the gap is not as obvious as in the NYUD-v2, and we will not conduct specific analysis due to space limitations.

Besides, we vary the number of iterations for better performance. 20 iterations are tested on both NYUD-v2 and PASCAL dataset, resulting in $n_{iter} = 18$ on NYUD-v2 and $n_{iter} = 4$ on PASCAL. Due to limited space, we only show the results of the ablation experiments. The experimental results show the effectiveness of further aggregate affinity information.

We believe that good initialization of α brings better performance. Therefore we perform ablation experiments on the initial value of α. We range it from 0.1 to 0.9 with $n_{iter} = 18$ on the NYUD-v2 dataset. Table 4 shows the performance with different initializations of α. Through experiments, we find that if the initial value of α is set to 0.3, the performance of the overall model is optimal, and we apply $\alpha = 0.3$ to all the above-mentioned experiments unless expressly stated. The results show that with proper initialization, the performance of our framework can be further improved.

5   Conclusions

In this paper, we proposed a novel cross-task interaction block for multi-task learning, which employs two types of attention mechanisms to build cross-task interactions to refine and distill task-specific features. One is the commonly used self-attention, and the other is pixel-wise global attention with an iterative fusion block. Three different designs of IFB are developed to enhance cross-task interaction more effectively. Extensive experiments on two benchmark datasets demonstrate the effectiveness of our proposed framework over state-of-the-art MTL baselines.

Acknowledgments. This work was supported by the National Natural Science Foundation of China under Grant 62176161, and the Scientific Research and Development Foundations of Shenzhen under Grant JCYJ20220818100005011

References

1. Bilen, H., Vedaldi, A.: Integrated perception with recurrent multi-task neural networks. Advances in neural information processing systems 29 (2016)
2. Brüggemann, D., Kanakis, M., Obukhov, A., Georgoulis, S., Van Gool, L.: Exploring relational context for multi-task dense prediction. In: Proceedings of the IEEE/CVF International Conference on Computer Vision, pp. 15869–15878 (2021)
3. Caruana, R.: Multitask learning. Mach. Learn. **28**(1), 41–75 (1997)
4. Chen, B., Guan, W., Li, P., Ikeda, N., Hirasawa, K., Lu, H.: Residual multi-task learning for facial landmark localization and expression recognition. Pattern Recogn. **115**, 107893 (2021)
5. Chen, L.-C., Zhu, Y., Papandreou, G., Schroff, F., Adam, H.: Encoder-decoder with atrous separable convolution for semantic image segmentation. In: Ferrari, V., Hebert, M., Sminchisescu, C., Weiss, Y. (eds.) ECCV 2018. LNCS, vol. 11211, pp. 833–851. Springer, Cham (2018). https://doi.org/10.1007/978-3-030-01234-2_49
6. Chen, Z., Badrinarayanan, V., Lee, C.Y., Rabinovich, A.: Gradnorm: Gradient normalization for adaptive loss balancing in deep multitask networks. In: International Conference on Machine Learning, pp. 794–803. PMLR (2018)
7. Deng, J., Dong, W., Socher, R., Li, L.J., Li, K., Fei-Fei, L.: Imagenet: a large-scale hierarchical image database. In: 2009 IEEE Conference on Computer Vision and Pattern Recognition, pp. 248–255. IEEE (2009)
8. Everingham, M., Van Gool, L., Williams, C.K., Winn, J., Zisserman, A.: The pascal visual object classes (voc) challenge. Int. J. Comput. Vision **88**(2), 303–338 (2010)
9. Gao, Y., Ma, J., Zhao, M., Liu, W., Yuille, A.L.: Nddr-cnn: Layerwise feature fusing in multi-task cnns by neural discriminative dimensionality reduction. In: Proceedings of the IEEE/CVF conference on computer vision and pattern recognition. pp. 3205–3214 (2019)
10. Kendall, A., Gal, Y., Cipolla, R.: Multi-task learning using uncertainty to weigh losses for scene geometry and semantics. In: Proceedings of the IEEE Conference on Computer Vision and Pattern Recognition, pp. 7482–7491 (2018)
11. Kokkinos, I.: Ubernet: Training a universal convolutional neural network for low-, mid-, and high-level vision using diverse datasets and limited memory. In: Proceedings of the IEEE Conference on Computer Vision and Pattern Recognition, pp. 6129–6138 (2017)
12. Lee, S., et al.: Vpgnet: vanishing point guided network for lane and road marking detection and recognition. In: Proceedings of the IEEE International Conference on Computer Vision, pp. 1947–1955 (2017)
13. Liu, S., Johns, E., Davison, A.J.: End-to-end multi-task learning with attention. In: Proceedings of the IEEE/CVF Conference on Computer Vision and Pattern Recognition, pp. 1871–1880 (2019)

14. Lopes, I., Vu, T.H., de Charette, R.: Cross-task attention mechanism for dense multi-task learning. arXiv preprint arXiv:2206.08927 (2022)
15. Maninis, K.K., Radosavovic, I., Kokkinos, I.: Attentive single-tasking of multiple tasks. In: Proceedings of the IEEE/CVF Conference on Computer Vision and Pattern Recognition, pp. 1851–1860 (2019)
16. Misra, I., Shrivastava, A., Gupta, A., Hebert, M.: Cross-stitch networks for multi-task learning. In: Proceedings of the IEEE Conference on Computer Vision and Pattern Recognition, pp. 3994–4003 (2016)
17. Paszke, A., et al.: Pytorch: an imperative style, high-performance deep learning library. Advances in neural information processing systems 32 (2019)
18. Ruder, S.: An overview of multi-task learning in deep neural networks. arXiv preprint arXiv:1706.05098 (2017)
19. Ruder, S., Bingel, J., Augenstein, I., Søgaard, A.: Latent multi-task architecture learning. In: Proceedings of the AAAI Conference on Artificial Intelligence, vol. 33, pp. 4822–4829 (2019)
20. Silberman, N., Hoiem, D., Kohli, P., Fergus, R.: Indoor segmentation and support inference from RGBD images. In: Fitzgibbon, A., Lazebnik, S., Perona, P., Sato, Y., Schmid, C. (eds.) ECCV 2012. LNCS, vol. 7576, pp. 746–760. Springer, Heidelberg (2012). https://doi.org/10.1007/978-3-642-33715-4_54
21. Vandenhende, S., Georgoulis, S., Van Gool, L.: MTI-Net: multi-scale task interaction networks for multi-task learning. In: Vedaldi, A., Bischof, H., Brox, T., Frahm, J.-M. (eds.) ECCV 2020. LNCS, vol. 12349, pp. 527–543. Springer, Cham (2020). https://doi.org/10.1007/978-3-030-58548-8_31
22. Vandenhende, S., Georgoulis, S., Van Gansbeke, W., Proesmans, M., Dai, D., Van Gool, L.: Multi-task learning for dense prediction tasks: a survey. IEEE Trans. Pattern Anal. Mach. Intell. (2021)
23. Wang, X., Girshick, R., Gupta, A., He, K.: Non-local neural networks. In: Proceedings of the IEEE Conference on Computer Vision and Pattern Recognition, pp. 7794–7803 (2018)
24. Xu, D., Ouyang, W., Wang, X., Sebe, N.: Pad-net: multi-tasks guided prediction-and-distillation network for simultaneous depth estimation and scene parsing. In: Proceedings of the IEEE Conference on Computer Vision and Pattern Recognition, pp. 675–684 (2018)
25. Zhang, Z., Cui, Z., Xu, C., Jie, Z., Li, X., Yang, J.: Joint task-recursive learning for semantic segmentation and depth estimation. In: Ferrari, V., Hebert, M., Sminchisescu, C., Weiss, Y. (eds.) ECCV 2018. LNCS, vol. 11214, pp. 238–255. Springer, Cham (2018). https://doi.org/10.1007/978-3-030-01249-6_15
26. Zhang, Z., Cui, Z., Xu, C., Yan, Y., Sebe, N., Yang, J.: Pattern-affinitive propagation across depth, surface normal and semantic segmentation. In: Proceedings of the IEEE/CVF Conference on Computer Vision and Pattern Recognition, pp. 4106–4115 (2019)
27. Zhou, L., et al.: Pattern-structure diffusion for multi-task learning. In: Proceedings of the IEEE/CVF Conference on Computer Vision and Pattern Recognition, pp. 4514–4523 (2020)

Anti-interference Zeroing Neural Network Model for Time-Varying Tensor Square Root Finding

Jiajie Luo[1,2], Lin Xiao[2], Ping Tan[2], Jiguang Li[3], Wei Yao[3],
and Jichun Li[1(✉)]

[1] ICOS Research Group, Newcastle University, Newcastle upon Tyne NE4 5TG, UK
jichun.li@ncl.ac.uk
[2] Hunan Provincial Key Laboratory of Intelligent Computing and Language
Information Processing, Hunan Normal University, Changsha 410081, China
[3] North of England Robotics Innovation Centre,
University of Salford, Manchester M5 4WT, UK

Abstract. Square root finding plays an important role in many scientific and engineering fields, such as optimization, signal processing and state estimation, but existing research mainly focuses on solving the time-invariant matrix square root problem. So far, few researchers have studied the time-varying tensor square root (TVTSR) problem. In this study, a novel anti-interference zeroing neural network (AIZNN) model is proposed to solve TVTSR problem online. With the activation of the advanced power activation function (APAF), the AIZNN model is robust in solving the TVTSR problem in the presence of the vanishing and non-vanishing disturbances. We present detailed theoretical analysis to show that, with the AIZNN model, the trajectory of error will converge to zero within a fixed time, and we also calculate the upper bound of the convergence time. Numerical experiments are presented to further verify the robustness of the proposed AIZNN model. Both the theoretical analysis and numerical experiments show that, the proposed AIZNN model provides a novel and noise-tolerant way to solve the TVTSR problem online.

Keywords: Zeroing neural network · Tensor · Square root finding · Time varying

1 Introduction

The matrix square root (MSR) problem plays an important role in industry. For example, Geoff Pleiss et al. proposed a Gaussian processes and a Bayesian optimization algorithm based on fast MSRs [17]. In [9], an adaptive square root

Supported by Newcastle University seed funding "AI for Synthetic Biology and Brain Health research".

© The Author(s), under exclusive license to Springer Nature Singapore Pte Ltd. 2024
B. Luo et al. (Eds.): ICONIP 2023, CCIS 1961, pp. 113–124, 2024.
https://doi.org/10.1007/978-981-99-8126-7_9

extended Kalman filter is used to estimate the state-of-charge. Generally, traditional algorithms, such as Newton's method [2], can only solve the time-invariant matrix square root (TIMSR) problem. To solve the time-varying matrix square root (TVMSR) problem, a novel zeroing neural network model is required [8]. It has been proved to be an effective method to solve the TVMSR problem [22].

Tensor theory is a branch of applied mathematics, which plays an important role in fields such as artificial intelligence [18], engineering [14], and physics [19]. The term tensor originated in mechanics [15], and was originally used to represent the stress state in an elastic medium [1]. Later, tensor theory developed into a powerful mathematical tool in both mechanics [4] and physics [20]. In recent decades, with the development of artificial intelligence, tensor theory has become a research hotspot again [16]. Tensor is a generalization of scalar, vector and matrix. The vector can be deemed as a first-order tensor, while the matrix can be deemed as a second-order tensor [7].

As a type of recursive neural network, zeroing neural network is an effective tool in solving time-varying problems [24]. It has been intensively studied in both scientific research and engineering fields, including robotics [12], optimization [6], and intelligent computing [10]. According to different activation functions, convergence parameter and design formula, zeroing neural network can be divided into many sub-types, e.g. fixed-time zeroing neural network [21], robust zeroing neural network [11], and fuzzy zeroing neural network [3].

Various zeroing neural network models have been proposed to solve the TVMSR problem. For example, Xiao et al. proposed a fixed-time zeroing neural network model to solve the TVMSR problem in [23]. A novel zeroing neural network is presented in [8] to solve the discrete TVMSR problem. Although the TVMSR problem has been extensively studied, there is very little research about the time-varying tensor square root (TVTSR) problem.

In this study, we aim to develop a fast and robust algorithm based on ZNN, and then to use it to solve the TVTSR problem online. The rest of the paper is organized as follows. Firstly, basic concepts of tensor and the TVTSR problem are presented in Sect. 2. Then we illustrate the novel anti-interference zeroing neural network (AIZNN) model in Sect. 3. Next, a detailed theoretical analysis is also presented to prove the robustness of the proposed AIZNN model. Finally, we show numerical experiments to further verify the robustness of the proposed AIZNN model.

2   Preliminaries

2.1   Tensor Theory

For the convenience of subsequent discussion, we need to understand some basic concepts of tensor product. There are many methods to obtain the product of two tensors, e.g., t-product, face-wise product and Einstein product. In this study, we mainly use the Einstein product, which is defined as follows.

Definition 1. ([5]) *Let* $\mathbb{R}$ *represents the set of real numbers, if* $\mathcal{X} = (x_{p_1 \cdots p_n k_1 \cdots k_n})$ $\in \mathbb{R}^{P_1 \times \cdots \times P_n \times K_1 \times \cdots \times K_n}$, $\mathcal{Y} = (y_{k_1 \cdots k_n q_1 \cdots q_m}) \in \mathbb{R}^{K_1 \times \cdots \times K_n \times Q_1 \times \cdots \times Q_m}$ *are two tensors, then we can define their Einstein product* $\mathcal{X} *_N \mathcal{Y}$ *as follows:*

$$(\mathcal{X} *_N \mathcal{Y})_{p_1 \cdots p_n q_1 \cdots q_m} = \sum_{k_1 \cdots k_n} x_{p_1 \cdots p_n k_1 \cdots k_n} y_{k_1 \cdots k_n q_1 \cdots q_m}. \tag{1}$$

It is obvious that $\mathcal{X} *_N \mathcal{Y} \in \mathbb{R}^{P_1 \times \cdots \times P_n \times Q_1 \times \cdots \times Q_m}$.

1-order and 2-order tensors, i.e. vectors and matrixes, have been extensively studied by mathematicians. For the convenience of subsequent analysis, we can transform tensors into 2-order matrices by defining a suitable mapping.

Definition 2. ([25]) *Considering the mapping* $\Lambda(t)$ *transforming tensors into matrixes:*

$$\mathcal{Z}_{p_1 \cdots p_n q_1 \cdots q_m} \xrightarrow{\Lambda(\cdot)} Z_{ij},$$
$$\mathbb{R}^{P_1 \times \cdots \times P_n \times Q_1 \times \cdots \times Q_m} \xrightarrow{\Lambda(\cdot)} \mathbb{R}^{P' \times Q'}, \tag{2}$$

where i *and* j *are positive integers,* $P' = P_1 \times \cdots \times P_n$, $Q' = Q_1 \times \cdots \times Q_m$, *we have:*

$$i = p_n + \sum_{\gamma=1}^{n-1} \left((p_\gamma - 1) \prod_{o=\gamma+1}^{n} P_o \right),$$
$$j = q_m + \sum_{\gamma=1}^{m-1} \left((q_\gamma - 1) \prod_{o=\gamma+1}^{m} Q_o \right). \tag{3}$$

For the convenience of subsequent discussion, we call the mapping $\Lambda(t)$ *the tensor transformation mapping (TTM).*

TTM provides a powerful tool to study tensor equations. With TTM, we can transform tensor equations into matrix ones. The following lemma illustrates the relationship between matrix multiplication and Einstein product.

Lemma 1. ([13]) *Considering two tensors* $\mathcal{X} = (x_{p_1 \cdots p_n k_1 \cdots k_n}) \in$ $\mathbb{R}^{P_1 \times \cdots \times P_n \times K_1 \times \cdots \times K_n}$, $\mathcal{Y} = (y_{k_1 \cdots k_n q_1 \cdots q_m}) \in \mathbb{R}^{K_1 \times \cdots \times K_n \times Q_1 \times \cdots \times Q_m}$ *and their Einstein product* $\mathcal{Z} = (z_{p_1 \cdots p_n q_1 \cdots q_m}) \in \mathbb{R}^{P_1 \times \cdots \times P_n \times Q_1 \times \cdots \times Q_m}$, *we have:*

$$\mathcal{X} *_N \mathcal{Y} = \mathcal{Z},$$
$$\Leftrightarrow \Lambda(\mathcal{X})\Lambda(\mathcal{Y}) = \Lambda(\mathcal{Z}), \tag{4}$$
$$\Leftrightarrow XY = Z,$$

where $X = \Lambda(\mathcal{X})$, $Y = \Lambda(\mathcal{Y})$ *and* $Z = \Lambda(\mathcal{Z})$.

2.2   Time-Varying Tensor Square Root

In this study, we consider the following time-varying tensor square root (TVTSR) problem:

$$\mathcal{X}(t) *_N \mathcal{X}(t) = \mathcal{A}(t), \tag{5}$$

where $\mathcal{A}(t) \in \mathbb{R}^{K_1 \times \cdots \times K_n \times K_1 \times \cdots \times K_n}$ is a given tensor flow, $\mathcal{X}(t) \in \mathbb{R}^{K_1 \times \cdots \times K_n \times K_1 \times \cdots \times K_n}$ is the square root of $\mathcal{A}(t)$, and $*_N$ denotes the Einstein product.

Based on Definition 2 and Lemma 1, we can transform the TVTSR problem (5) into the following time-varying matrix square root (TVMSR) problem:

$$X^2(t) = A(t), \tag{6}$$

where $X(t) = \Lambda(\mathcal{X}(t))$ and $A(t) = \Lambda(\mathcal{A}(t))$.

3   Anti-interference Zeroing Neural Network

Considering the TVTSR problem (5), we realize it can be reduced to the TVMSR problem by Definition 2 and Lemma 1, after which then we can use the zeroing neural network to solve the TVTSR problem indirectly.

Generally, three steps are required to obtain the dynamics equation of the zeroing neural network for a certain problem. If we want to solve the TVTSR problem (5), we can transform it into the TVMSR problem (6) first, and then perform the following steps:

(1) We need to determine the error function of the problem, and the problem is solved when the error function converges to zero. Naturally, we can define the following error function:

$$\mathfrak{E}(t) = X^2(t) - A(t), \tag{7}$$

where $X(t) = \Lambda(\mathcal{X}(t))$ and $A(t) = \Lambda(\mathcal{A}(t))$.

(2) There are many types of design formula. In this study, we will choose the traditional zeroing neural network design formula [11]:

$$\dot{\mathfrak{E}}(t) = -\Phi(\mathfrak{E}(t)), \tag{8}$$

where $\Phi(\cdot)$ is the activation function, and $\mathfrak{E}(t)$ is the error function defined in the previous step.

(3) Combining the error function (7) and the traditional zeroing neural network design formula (8), the traditional ZNN (TZNN) model for TVTSR problem can be obtained as follows:

$$X(t)\dot{X}(t) + \dot{X}(t)X(t) = -\Phi(X^2(t) - A(t)) + \dot{A}(t) \tag{9}$$

where $X(t) = \Lambda(\mathcal{X}(t))$, $A(t) = \Lambda(\mathcal{A}(t))$ and $\Phi(\cdot)$ is the activation function.

We call (9) the traditional zeroing neural network (TZNN) model. To enhance the performance of the TZNN model, we will use the advanced power activation function (APAF) to activate it. The APAF can be expressed as:

$$\Xi(e) = \left(|e|^{\frac{p}{q}} + |e|^{\frac{q}{p}}\right) \text{sign}(e) + k_1 e + k_2 \text{sign}(e), \tag{10}$$

where $p, q, k_1, k_2 > 0$ are parameters satisfying $p \neq q$, and $\text{sign}(\cdot)$ denotes the sign function:

$$\text{sign}(e) = \begin{cases} 1 & \text{if } e > 0 \\ 0 & \text{if } e = 0. \\ -1 & \text{if } e < 0 \end{cases} \tag{11}$$

In this study, we call the TZNN model activated by the APAF the anti-interference zeroing neural network (AIZNN). For the convenience of subsequent discussion, we can express the AIZNN with noise as:

$$\dot{\mathfrak{E}}(t) = -\Xi(\mathfrak{E}(t)) + D(t). \tag{12}$$

Specifically, the disturbances $D(t)$ in this study satisfy the following conditions:

(1) For vanishing disturbances (VD): $\exists \gamma \leq k_1, \forall i, j \in 1, 2, \cdots, n$, we have:

$$|d_{ij}(t)| \leq \gamma |e_{ij}(t)|. \tag{13}$$

Here, $d_{ij}(t)$, $e_{ij}(t)$ denotes the ijth element of $D(t)$ and $\mathfrak{E}(t)$ respectively, n denotes the number of rows in matrix $D(t)$, and k_1 is the parameter defined in (10).

(2) For non-vanishing disturbances (ND): $\exists \gamma \leq k_2, \forall i, j \in 1, 2, \cdots, n$, we have:

$$|d_{ij}(t)| \leq \gamma. \tag{14}$$

Here, $d_{ij}(t)$ denotes the ijth element of $D(t)$, n denotes the number of rows in matrix $D(t)$, and k_2 is the parameter defined in (10).

4 Theoretical Analysis

In this section, we will prove that the proposed AIZNN is both fixed-time stable and robust in the presence of both vanishing disturbances and non-vanishing disturbances. To begin with, we need to understand some useful definitions and lemmas.

Definition 3. *Consider the following nonlinear system:*

$$\dot{x}(t) = f(x, t), \quad x(0) = x_0, \tag{15}$$

where $x(t) \in \mathbb{R}^n$, $f : U \times \mathbb{R} \to \mathbb{R}^n$ is a continuous function, U_0 is an open neighborhood of the origin x_0 satisfying $f(0, t) = 0$. We say nonlinear system

(15) *is fixed-time stable in the local equilibrium point $x = 0$ if and only if, for any t_0 and initial state $x_0 = x(t_0) \in U_0 \subset U$, there exists an bounded settling-time function $T = T(x_0)$ satisfying:*

$$\begin{cases} \lim_{t \to T(x_0)} \phi(t, t_0, x_0) = 0 \\ \phi(t, t_0, x_0) = 0, \ t > T(x_0) \end{cases}, \tag{16}$$

where $\phi(t, t_0, x_0)$ is the solution $x(t)$ of (15) with initial state x_0.

Lemma 2. *([12]) Suppose $\mathfrak{E}(t)$ satisfies the following inequality:*

$$\dot{\mathfrak{E}}(t) \leq - \left(\mathfrak{E}^{\frac{p}{q}}(t) + \mathfrak{E}^{\frac{q}{p}}(t) \right), \tag{17}$$

then the zeroing neural network model is fixed-time stable for any given initial state, and $\mathfrak{E}(t)$ converge to zero within $T_{\max}$:

$$\begin{aligned} T_{\max} &\leq \left| \frac{1}{1 - \frac{p}{q}} + \frac{1}{1 - \frac{q}{p}} \right| \\ &= \left| \frac{p + q}{p - q} \right| \end{aligned} \tag{18}$$

where p and q are parameters defined in (17).

4.1   No Disturbances

Theorem 1. *Supposing the AIZNN model (12) is used to solve the TVTSR (5) problem, and there are no other disturbances, then the AIZNN model (12) is fixed-time stable, and the error function $\mathfrak{E}(t)$ converges to zero within $T_{\max}$:*

$$T_{\max} \leq \left| \frac{p + q}{p - q} \right|, \tag{19}$$

where $p > 0$ and $q > 0$ are parameters defined in (10).

Proof. The AIZNN model without disturbances can be transformed into the following n^2 subsystems:

$$\dot{\mathbf{e}}_{ij}(t) = -\xi(\mathbf{e}_{ij}(t)), \tag{20}$$

where $i, j \in 1, 2, \cdots, n$ are positive integers, $\xi(\cdot)$ is the element-wise form of the APAF (10) $\Xi(\cdot)$, and $\mathbf{e}_{ij}(t)$ is the ijth element of $\mathfrak{E}(t)$.

Next, we can construct the Lyapunov function candidate $\mathbf{v}_{ij}(t) = |\mathbf{e}_{ij}(t)|$. Obviously, we have:

$$\begin{aligned} \dot{\mathbf{v}}_{ij}(t) &= \dot{\mathbf{e}}_{ij}(t)\text{sign}(\mathbf{e}_{ij}(t)) \\ &= -\xi(\mathbf{e}_{ij}(t))\text{sign}(\mathbf{e}_{ij}(t)) \\ &= - \left(|\mathbf{e}_{ij}|^{\frac{p}{q}} + |\mathbf{e}_{ij}|^{\frac{q}{p}} + k_1 \mathbf{e}_{ij} + k_2 \right) \\ &\leq - \left(|\mathbf{e}_{ij}|^{\frac{p}{q}} + |\mathbf{e}_{ij}|^{\frac{q}{p}} \right). \end{aligned} \tag{21}$$

According to Lemma 2, we know that the ijth element of $\mathfrak{E}(t)$ converges to zero within T_{ij}:

$$T_{ij} \leq \left| \frac{p+q}{p-q} \right|, \tag{22}$$

where $p > 0$ and $q > 0$ are parameters defined in (10).

From previous analysis, we know that the error function $\mathfrak{E}(t)$ converges to zero within $T_{\max}$:

$$T_{\max} = \max(T_{ij}) \leq \left| \frac{p+q}{p-q} \right|, \tag{23}$$

where $p > 0$ and $q > 0$ are parameters defined in (10).

4.2 Vanishing Disturbances

Theorem 2. *Suppose the AIZNN model (12) is at present used to solve the TVTSR (5) problem, and there exist the vanishing disturbances:*

$$|d_{ij}(t)| \leq \gamma |\mathbf{e}_{ij}(t)|, \tag{24}$$

where $\gamma \leq k_1$, $\forall i, j \in 1, 2, \cdots, n$, and k_1 is the parameter defined in (10), then the AIZNN model (12) is fixed-time stable, and the error function $\mathfrak{E}(t)$ converge to zero within $T_{\max}$:

$$T_{\max} \leq \left| \frac{p+q}{p-q} \right|, \tag{25}$$

where $p > 0$ and $q > 0$ are parameters defined in (10).

Proof. The AIZNN model with vanishing disturbances can be transformed into the following n^2 subsystems:

$$\dot{\mathbf{e}}_{ij}(t) = -\xi(\mathbf{e}_{ij}(t)) + d_{ij}(t), \tag{26}$$

where $i, j \in 1, 2, \cdots, n$ are positive integers, $\xi(\cdot)$ is the element-wise form of the APAF (10) $\Xi(\cdot)$, $\mathbf{e}_{ij}(t)$ is the ijth element of $\mathfrak{E}(t)$, and $d_{ij}(t)$ is the ijth element of the disturbances term $D(t)$.

Next, we can construct the Lyapunov function candidate $\mathbf{v}_{ij}(t) = |\mathbf{e}_{ij}(t)|$. Obviously, we have:

$$\begin{aligned}
\dot{\mathbf{v}}_{ij}(t) &= \dot{\mathbf{e}}_{ij}(t)\mathrm{sign}(\mathbf{e}_{ij}(t)) + d_{ij}(t) \\
&= -\xi(\mathbf{e}_{ij}(t))\mathrm{sign}(\mathbf{e}_{ij}(t)) + d_{ij}(t) \\
&= -\left(|\mathbf{e}_{ij}|^{\frac{p}{q}} + |\mathbf{e}_{ij}|^{\frac{q}{p}} + k_1 \mathbf{e}_{ij} + k_2 \right) + d_{ij}(t) \\
&\leq -\left(|\mathbf{e}_{ij}|^{\frac{p}{q}} + |\mathbf{e}_{ij}|^{\frac{q}{p}} \right) - (k_1 - \gamma)\mathbf{e}_{ij} - k_2 \\
&\leq -\left(|\mathbf{e}_{ij}|^{\frac{p}{q}} + |\mathbf{e}_{ij}|^{\frac{q}{p}} \right).
\end{aligned} \tag{27}$$

According to Lemma 2, we know that the ijth element of $\mathfrak{E}(t)$ converges to zero within T_{ij}:

$$T_{ij} \leq \left|\frac{p+q}{p-q}\right|, \tag{28}$$

where $p > 0$ and $q > 0$ are parameters defined in (10).

From previous analysis, we know that the error function $\mathfrak{E}(t)$ converge to zero within T_{max}:

$$T_{max} = \max(T_{ij}) \leq \left|\frac{p+q}{p-q}\right|, \tag{29}$$

where $p > 0$ and $q > 0$ are parameters defined in (10).

4.3   Non-vanishing Disturbances

Theorem 3. *Suppose the AIZNN model* (12) *is at present used to solve the TVTSR* (5) *problem, and there exist the non-vanishing disturbances:*

$$|d_{ij}(t)| \leq \gamma, \tag{30}$$

where $\gamma \leq k_2$, $\forall i,j \in 1,2,\cdots,n$, k_2 is the parameter defined in (10), *then the AIZNN model* (12) *is fixed-time stable, and the error function $\mathfrak{E}(t)$ converge to zero within T_{max}:*

$$T_{max} \leq \left|\frac{p+q}{p-q}\right|, \tag{31}$$

where $p > 0$ and $q > 0$ are parameters defined in (10).

Proof. The AIZNN model with non-vanishing disturbances can be transformed into the following n^2 subsystems:

$$\dot{\mathbf{e}}_{ij}(t) = -\xi(\mathbf{e}_{ij}(t)) + d_{ij}(t), \tag{32}$$

where $i,j \in 1,2,\cdots,n$ are positive integers, $\xi(\cdot)$ is the element-wise form of the APAF (10) $\Xi(\cdot)$, $\mathbf{e}_{ij}(t)$ is the ijth element of $\mathfrak{E}(t)$, and $d_{ij}(t)$ is the ijth element of the disturbances term $D(t)$.

Next, we can construct the Lyapunov function candidate $\mathbf{v}_{ij}(t) = |\mathbf{e}_{ij}(t)|$. Obviously, we have:

$$
\begin{aligned}
\dot{\mathbf{v}}_{ij}(t) &= \dot{\mathbf{e}}_{ij}(t)\mathrm{sign}(\mathbf{e}_{ij}(t)) + d_{ij}(t) \\
&= -\xi(\mathbf{e}_{ij}(t))\mathrm{sign}(\mathbf{e}_{ij}(t)) + d_{ij}(t) \\
&= -\left(|\mathbf{e}_{ij}|^{\frac{p}{q}} + |\mathbf{e}_{ij}|^{\frac{q}{p}} + k_1\mathbf{e}_{ij} + k_2\right) + d_{ij}(t) \\
&\leq -\left(|\mathbf{e}_{ij}|^{\frac{p}{q}} + |\mathbf{e}_{ij}|^{\frac{q}{p}}\right) - k_1\mathbf{e}_{ij} - (k_2 - \gamma) \\
&\leq -\left(|\mathbf{e}_{ij}|^{\frac{p}{q}} + |\mathbf{e}_{ij}|^{\frac{q}{p}}\right).
\end{aligned} \tag{33}
$$

According to Lemma 2, we know that the ijth element of $\mathfrak{E}(t)$ converges to zero within T_{ij}:

$$T_{ij} \leq \left| \frac{p+q}{p-q} \right|, \tag{34}$$

where $p > 0$ and $q > 0$ are parameters defined in (10).

From previous analysis, we know that the error function $\mathfrak{E}(t)$ converge to zero within $T_{\max}$:

$$T_{\max} = \max(T_{ij}) \leq \left| \frac{p+q}{p-q} \right|, \tag{35}$$

where $p > 0$ and $q > 0$ are parameters defined in (10).

5 Numerical Experiments

To further verify the proposed AIZNN model (12), we consider the following TVTSR problem:

$$\mathcal{X}(t) *_N \mathcal{X}(t) = \mathcal{A}(t), \tag{36}$$

where $\mathcal{A}(t)$ is a tensor flow that can be represented as follows:

$$\mathcal{A}(:,:,1,1) = \begin{bmatrix} \sin t \cos t + 1 & 2\sin t \\ 0 & 0 \end{bmatrix} \mathcal{A}(:,:,1,2) = \begin{bmatrix} 2\cos t & \sin t \cos t + 1 \\ 0 & 0 \end{bmatrix}$$

$$\mathcal{A}(:,:,2,1) = \begin{bmatrix} 0 & 0 \\ e^{\sin t}(\sin 2t + 1) + 9 & 7\sin 2t + 7 \end{bmatrix} \tag{37}$$

$$\mathcal{A}(:,:,2,2) = \begin{bmatrix} 0 & 0 \\ 7e^{\sin t} & e^{\sin t}(\sin 2t + 1) + 16 \end{bmatrix}$$

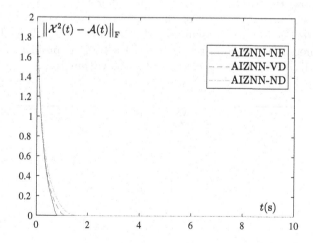

Fig. 1. The trajectory of error

In this study, we will verify the robustness of the proposed AIZNN model (12) in the presence of vanishing and non-vanishing disturbances. By adding different disturbances, the AIZNN models (12) can be divided into the following models:

(1) The anti-interference zeroing neural network-noise free (AIZNN-NF) model:

$$\mathfrak{E}(t) = -\varXi(\mathfrak{E}(t)).\tag{38}$$

(2) The anti-interference zeroing neural network-vanishing disturbances (AIZNN-VD) model:

$$\mathfrak{E}(t) = -\varXi(\mathfrak{E}(t)) + \sin t.\tag{39}$$

(3) The anti-interference zeroing neural network-non-vanishing disturbances (AIZNN-ND) model:

$$\mathfrak{E}(t) = -\varXi(\mathfrak{E}(t)) + 1.\tag{40}$$

Here, we set the parameters of the APAF as $p = 1$, $q = 2$, and the initial state $\mathcal{X}(0)$ is:

$$\mathcal{X}(:,:,1,1) = \begin{bmatrix} 1.5688 & 0.4694 \\ 0.0119 & 0.3371 \end{bmatrix} \mathcal{X}(:,:,1,2) = \begin{bmatrix} 1.1622 & 1.7943 \\ 0.3112 & 0.5285 \end{bmatrix}$$
$$\mathcal{X}(:,:,2,1) = \begin{bmatrix} 0.1656 & 0.6020 \\ 3.2630 & 1.6541 \end{bmatrix} \mathcal{X}(:,:,2,1) = \begin{bmatrix} 0.6892 & 0.7482 \\ 1.4505 & 4.0838 \end{bmatrix}.\tag{41}$$

According to Theorems 1, 2 and 3, the error of the AIZNN model (12) should converge to zero within $T_{\max}$:

$$T_{\max} \leq \left| \frac{p+q}{p-q} \right| = 3.\tag{42}$$

Figure 1 shows the trajectory of error under different disturbances, where $\|\cdot\|_F$ denotes the Frobenius norm. We also the illustrate the trajectory of $\mathcal{X}(1,2,1,1)$ in Fig. 2, where T-solution represents the theoretical solution. It is obvious that the AIZNN model (12) is robust in solving the TVTSR problem (5), and the error of the AIZNN models (12) converge to zero within $T_{\max}$.

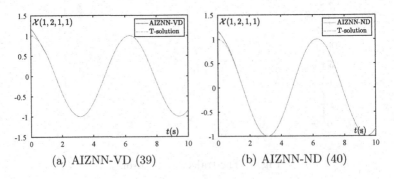

(a) AIZNN-VD (39)   (b) AIZNN-ND (40)

Fig. 2. The trajectory of $\mathcal{X}(1,2,1,1)$

6 Conclusion

In this study, a novel anti-interference zeroing neural network model is proposed to solve the time-varying tensor square root problem online. Both theoretical analysis and numerical experiments illustrate that the anti-interference zeroing neural network model activated by the novel advanced power activation function is robust in the presence of both vanishing and non-vanishing disturbances. This study provides a new perspective to explore the square root problem, and future research may mainly focus on the application of this algorithm.

References

1. Araújo, J.A.F.: A micromechanical analysis of strain concentration tensor for elastoplastic medium containing aligned and misaligned pores. Mech. Res. Commun. **125**, 103989 (2022)
2. Berahas, A.S., Bollapragada, R., Nocedal, J.: An investigation of Newton-sketch and subsampled Newton methods. Optim. Methods Softw. **35**(4), 661–680 (2020)
3. Dai, J., Luo, L., Xiao, L., Jia, L., Li, X.: An intelligent fuzzy robustness ZNN model with fixed-time convergence for time-variant Stein matrix equation. Int. J. Intell. Syst. **37**(12), 11670–11691 (2022)
4. Eggersmann, R., Stainier, L., Ortiz, M., Reese, S.: Model-free data-driven computational mechanics enhanced by tensor voting. Comput. Methods Appl. Mech. Eng. **373**, 113499 (2021)
5. El Guide, M., El Ichi, A., Jbilou, K., Beik, F.: Tensor Krylov subspace methods via the Einstein product with applications to image and video processing. Appl. Numer. Math. **181**, 347–363 (2022)
6. Fu, Z., Zhang, Y., Tan, N.: Gradient-feedback ZNN for unconstrained time-variant convex optimization and robot manipulator application. IEEE Trans. Ind. Inform. (2023)
7. Haynes, M.S., Fenni, I.: T-matrix backprojection imaging for scalar and vector electromagnetic waves. IEEE Trans. Antennas Propagation (2023)
8. Jiang, C., Zhang, Y., Mou, C., Li, B., Sun, X., Shi, Y.: A new ZNN model for finding discrete time-variant matrix square root: From model design to parameter analysis. Journal of Computational and Applied Mathematics, p. 115260 (2023)
9. Jiang, C., Wang, S., Wu, B., Fernandez, C., Xiong, X., Coffie-Ken, J.: A state-of-charge estimation method of the power lithium-ion battery in complex conditions based on adaptive square root extended Kalman filter. Energy **219**, 119603 (2021)
10. Jin, J., Zhu, J., Gong, J., Chen, W.: Novel activation functions-based ZNN models for fixed-time solving dynamic Sylvester equation. Neural Comput. Appl. **34**(17), 14297–14315 (2022)
11. Jin, J., Zhu, J., Zhao, L., Chen, L.: A fixed-time convergent and noise-tolerant zeroing neural network for online solution of time-varying matrix inversion. Appl. Soft Comput. **130**, 109691 (2022)
12. Kong, Y., Hu, T., Lei, J., Han, R.: A finite-time convergent neural network for solving time-varying linear equations with inequality constraints applied to redundant manipulator. Neural Process. Lett. **54**(1), 125–144 (2022)
13. Liu, Y., Liu, J., Long, Z., Zhu, C.: Tensor computation for data analysis. Springer (2022)

14. Metzler, J., Coley, C.: Evaluation of knowledge-guided tensor decomposition in engineering applications. In: AIAA SCITECH 2023 Forum, p. 1433 (2023)

15. Okunishi, K., Nishino, T., Ueda, H.: Developments in the tensor network-from statistical mechanics to quantum entanglement. J. Phys. Soc. Jpn. **91**(6), 062001 (2022)

16. Pang, B., Nijkamp, E., Wu, Y.N.: Deep learning with tensorflow: a review. J. Educ. Behav. Stat. **45**(2), 227–248 (2020)

17. Pleiss, G., Jankowiak, M., Eriksson, D., Damle, A., Gardner, J.: Fast matrix square roots with applications to Gaussian processes and Bayesian optimization. Adv. Neural. Inf. Process. Syst. **33**, 22268–22281 (2020)

18. Rahardja, U., Aini, Q., Manongga, D., Sembiring, I., Girinzio, I.D.: Implementation of tensor flow in air quality monitoring based on artificial intelligence. Int. J. Artif. Intel. Res. **6**(1) (2023)

19. Sheppard, J.Z.: Psychoeducation tensor as a model of cognitive physics. ScienceOpen Posters (2023)

20. Speranza, E., Weickgenannt, N.: Spin tensor and pseudo-gauges: from nuclear collisions to gravitational physics. Europ. Phys. J. A **57**(5), 155 (2021)

21. Tan, Z.: Fixed-time convergent gradient neural network for solving online Sylvester equation. Mathematics **10**(17), 3090 (2022)

22. Xiao, L.: A finite-time convergent Zhang neural network and its application to real-time matrix square root finding. Neural Comput. Appl. **31**, 793–800 (2019)

23. Xiao, L., Jia, L.: FTZNN for time-varying matrix square root (2023)

24. Xiao, L., Li, L., Tao, J., Li, W.: A predefined-time and anti-noise varying-parameter ZNN model for solving time-varying complex Stein equations. Neurocomputing (2023)

25. Xiao, L., Li, X., Jia, L., Liu, S.: Improved finite-time solutions to time-varying Sylvester tensor equation via zeroing neural networks. Appl. Math. Comput. **416**, 126760 (2022)

CLF-AIAD: A Contrastive Learning Framework for Acoustic Industrial Anomaly Detection

Zhaoyi Liu[1(✉)], Yuanbo Hou[2], Haoyu Tang[3], Álvaro López-Chilet[4],
Sam Michiels[1(✉)], Dick Botteldooren[2], Jon Ander Gómez[4],
and Danny Hughes[1(✉)]

[1] imec-Distrinet, Computer Science, KU Leuven, Leuven, Belgium
{zhaoyi.liu,sam.michiels,danny.hughes}@kuleuven.be
[2] WAVES Research Group, Ghent University, Ghent, Belgium
[3] Technical Platform Department, United-Imaging Microelectronics, Shanghai, China
[4] PRHLT Research Center, Universitat Politècnica de València, Valencia, Spain

Abstract. Acoustic Industrial Anomaly Detection (AIAD) has received a great deal of attention as a technique to discover faults or malicious activity, allowing for preventive measures to be more effectively targeted. The essence of AIAD is to learn the compact distribution of normal acoustic data and detect outliers as anomalies during testing. However, recent AIAD work does not capture the dependencies and dynamics of Acoustic Industrial Data (AID). To address this issue, we propose a novel Contrastive Learning Framework (CLF) for AIAD, known as CLF-AIAD. Our method introduces a multi-grained contrastive learning-based framework to extract robust normal AID representations. Specifically, we first employ a projection layer and a novel context-based contrast method to learn robust temporal vectors. Building upon this, we then introduce a sample-wise contrasting-based module to capture local invariant characteristics, improving the discriminative capabilities of the model. Finally, a transformation classifier is introduced to bolster the performance of the primary task under a self-supervised learning framework. Extensive experiments on two typical industrial datasets, MIMII and ToyADMOS, demonstrate that our proposed CLF-AIAD effectively detects various real-world defects and improves upon the state-of-the-art in unsupervised industrial anomaly detection.

Keywords: Acoustic industrial anomaly detection · Contrastive learning · Anomaly detection · Unsupervised learning

1 Introduction

Faults or failures in industrial machines significantly degrade industrial production efficiency and quality. The Industrial Internet of Things (IIoT) tackles this problem by enabling the detection of anomalies in near real time using wireless

© The Author(s), under exclusive license to Springer Nature Singapore Pte Ltd. 2024
B. Luo et al. (Eds.): ICONIP 2023, CCIS 1961, pp. 125–137, 2024.
https://doi.org/10.1007/978-981-99-8126-7_10

sensors and actuators. The exponential growth of the IIoT has led to the generation of vast volumes of Acoustic Industrial Data (AID) from industrial machines that frequently handle critical operations [19,34]. Consequently, Acoustic Industrial Anomaly Detection (AIAD) has attracted substantial attention and holds significant importance as it may serve as an indicator of faults, failures or malicious activities, thereby enabling targeted intervention to avert potential issues [6,15,17].

AIAD tasks can be categorized into supervised and unsupervised approaches. On the one hand, supervised AIAD emphasizes detecting known abnormal noises [29], which represent rare acoustic events. On the other hand, unsupervised AIAD detects unseen anomalous noises that have not been previously observed [3,23]. Due to the nature of anomalous behaviours in industrial equipment, collecting an extensive set of anomalous sounds is impossible. This naturally leads us to focus on unsupervised AIAD.

In unsupervised AIAD, models are trained solely on normal AID, and during inference, anomalies are identified as those significantly deviating from the normal AID. For this purpose, the acoustic feature acquisition and latent representation of normal acoustic data are essential to distinguish the normal and unknown sounds. A range of autoencoder-based (AE-based) models have been proposed for learning better acoustic representations [6,12,25,27]. Typically, the AE-based methods perform a frame-to-frame reconstruction hypothesizing that the error will be higher on abnormal data, allowing to use it as an anomaly detector [9]. However, temporal dependencies in AID are complex and cannot comprehensively characterize the temporal pattern of AID. In addition, the normal AID may change dynamically over time. Thus, existing approaches need to be improved to construct a precise profile for normal AID and hence acquire a robust representation, that delivers good performance in real-world industrial applications.

In this paper, we propose a novel Contrastive Learning Framework for AIAD, namely CLF-AIAD, which aims to capture the dependencies and dynamic features of AIAD. CLF-AIAD presents a multi-grained contrastive learning-based framework, combining contextual-wise and sample-wise contrasting information to construct a comprehensive normal profile in AID. More precisely, we first employ a projection layer combined with a novel acoustic contextual contrasting method. This allows us to effectively learn the temporal feature vectors that capture essential position information within the AID data. Furthermore, we investigate Data Augmentation (DA) in the time and frequency domain. After that, we introduce a sample-level contrasting mechanism to effectively capture the local invariant characteristics of latent representations. Finally, a transformation classifier is introduced as an auxiliary task to improve the performance of the primary task under an unsupervised learning framework. Extensive experiments are conducted to validate the effectiveness of the proposed method. Moreover, our approach achieves state-of-the-art anomaly detection performance in MIMII [24] and ToyAdmos [18] real-record AIAD datasets.

Contributions: The key scientific contributions of this paper are

1. We propose a novel Contrastive Learning Framework for AIAD, called CLF-AIAD. Contrastive learning methods are employed to improve the construction of a robust profile for normal AID, thereby enhancing the detectability of abnormal AID.
2. Multi-grained contrasting approaches are proposed to discover robust AID representations, encompassing both the novel contextual and contrasting acoustic modules.
3. We contribute in-depth comparative experiments to evaluate the performance of CLT-AIAD on two real-world datasets. Our experimental results demonstrate that CLT-AIAD outperforms state-of-the-art baselines.

The remainder of this paper is structured as follows: Sect. 2 provides an overview of relevant work. Section 3 describes the proposed CLF-AIAD technique and key implementation details. Experimental setup, results, and analysis appear in Sect. 4. Finally, Sect. 5 draws conclusions.

2   Related Work

Existing unsupervised Anomaly Detection (AD) methods can be grouped into two categories.

Reconstruction-based methods assume that a generalization gap with anomalies exists in the reconstruction model trained on normal acoustic data, preventing it from reconstructing anomalies. Autoencoders (AE) [15,33], Variational Autoencoders (VAE) [16], and Generative Adversarial Networks (GAN) [14] are typical models belonging to this category. [9] designed a sample-level anomaly score using an AE, calculated by aggregating pixel-wise errors. [35] proposed a masked multi-scale reconstruction method, which enhances the model's ability to infer causality among patches in normal samples and improves performance on different domain shifts.

Representation-based methods involve transforming input data into a feature domain to obtain representative feature embeddings; then, classification or clustering tasks are performed based on these learned representations. Since the embeddings include representative input data characteristics, it is much easier to determine anomalies using normal data. Gaussian Mixture Models(GMMs) based methods [20], k-means [28], and One-Class Support Vector Machines [1] are the common approaches within this category. In [21] combined two classification models as feature extractors and local representations to detect anomalies. PaDiM [4] learns the parameters of multivariate Gaussian distribution from different CNN layers.

Recently, self-supervised learning techniques that do not require handcrafted labels have been applied to detect anomalous data [8,13]. **Contrastive learning-based methods** mainly aim to learn transformation-invariant data representations in a self-supervised setting. These methods seek to minimize the distance between similar samples in the feature space while maximizing the distance between different samples within the same space. SimCLR [2] emphasized the importance of data augmentation in contrastive learning and proposed a learnable nonlinear transformation module to enhance the quality of contrastive representation vectors. He et al. [10] proposed the MoCo to address the challenge of obtaining negative samples by constructing a dynamic dictionary to facilitate the retrieval of sample pairs for contrastive learning. For time series data, Eldele et al. [7] proposed a self-supervised framework with multiple contrasting methods to learn informative representations of time series. Yue et al. [32] introduced a hierarchical contrasting method to capture multi-grained contextual information in time series.

In this paper, we extend these ideas to develop a contrastive learning framework for representation learning and anomaly detection in AID.

3   Proposed Method

In this section, we first present the standard problem description of unsupervised AIAD. Then, we give an overview of the proposed CLF-AIAD and explain its main nodules. Finally, we describe in detail the proposed model and AIAD.

3.1   Problem Description

Unsupervised AIAD is an identification problem of determining whether the acoustic signals emitted from a target are normal or anomalous.

Let us define $\mathbf{x}_\tau \in \mathbb{C}^T$ is the input spectrogram calculated from the observed acoustic signal. Here $\tau = \{1, 2, ..., T\}$ denote the time frames. To detect anomalies, a set of low-dimensional representations is extracted as:

$$\mathbf{z} = \mathcal{F}_{dr}(\mathbf{x}_\tau) \tag{1}$$

where $\mathcal{F}_{dr}$ is a dimensional reduction model usually utilizing a Deep Neural Network (DNN). To determine the spectro-temporal characteristics of observed acoustic feature, the input vector $\mathbf{x}_\tau$ is obtained by concatenating several frames of observation while accounting for previous and future frames, as $\mathbf{x}_\tau = (X_{\omega,\tau-p}, X_{\omega,\tau-p+1}, ..., X_{\omega,\tau+f})^\mathsf{T}$, where T denotes transposition, and p and f are the context window size of previous and future indexes, respectively. Next, an anomaly score $\mathcal{A}_\theta(\mathbf{z})$ is calculated using a normal model as:

$$\mathcal{A}_\theta(\mathbf{z}) = -\ln p(\mathbf{z}|\theta, s = 0) \tag{2}$$

where s indicates the state of observed data, $s = 0$ is normal, and $s \neq 0$ is abnormal. $p(\mathbf{z}|\theta, s = 0)$ is a normal model generally constructed with a Probability Density Function (PDF) of normal acoustic data.

Finally, $\mathbf{x}_\tau$ can be identified as an abnormal acoustic signal when a score of $\mathcal{A}_\theta(\mathbf{z})$ surpasses the pre-defined threshold value ϕ:

$$\mathcal{H}(\mathbf{z}, \phi) = \begin{cases} 0(Normal) & \text{if } \mathcal{A}_\theta(\mathbf{z}) < \phi, \\ 1(Anomaly) & \text{if } \mathcal{A}_\theta(\mathbf{z}) \geq \phi. \end{cases} \tag{3}$$

3.2   Proposed Contrastive Learning Framework

Following the procedure of unsupervised AIAD described in Sect. 3.1, we note that the hidden representation **z** of acoustic data is critical to obtaining good performance to determine anomaly detection. Therefore, we propose a Contrastive Learning Framework for AIAD (CLF-AIAD), which utilizes multi-grained contrasting methods to capture AID's temporal dependency and dynamic variabilities. An overview of our anomaly detection framework is depicted in Fig. 1.

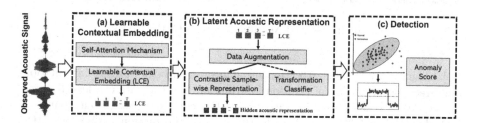

Fig. 1. Overview of proposed CLF-AIAD. (a) Learnable Contextual Embedding (LCE): A novel LCE is applied to learn the robust temporal dependency of AID. (b) Latent acoustic representation: ① LCE undergoes transformations into multiple perspectives through the utilization of both time-domain and frequency-domain time-series-specific augmentations; ② An acoustic sample contrasting method is proposed to learn invariant local hidden acoustic representation; ③ A transformation classifier is used as an auxiliary task to classify the applied augmentations in step ①. (c) Detection: An anomaly score is calculated based on the Mahalanobis distance between the hidden acoustic representations of the input sample and the normal training acoustic instances.

Learnable Contextual Embedding. While simple transformations are commonly employed in contrastive learning to generate diverse views of acoustic data, direct transformation of the original AID may not always be appropriate, especially when certain patterns or trends exist, such as upward trends in machine sounds. To address this issue, in the CLF-AIAD, we first employ a self-attention mechanism [31] as the projection layer to capture the temporal relationships and dependencies of the AID. A mask is applied to assign zero attention scores to future frames to prevent future data information leakage. For an acoustic input $\mathbf{x} = [x_1, x_2, ..., x_T]$, linear mappings are applied to obtain the query matrix $\mathbf{Q}$, key matrix $\mathbf{K}$, and value matrix $\mathbf{V}$. The output LCE $(\boldsymbol{f}_{1:T})$ obtained after applying the self-attention mechanism can be expressed as below.

$$\boldsymbol{f} = Softmax((\mathbf{QK}^{\mathsf{T}}/\sqrt{d_k}) \odot \boldsymbol{Mask})\mathbf{V} \tag{4}$$

where symbols $\odot$ and T denote element-wise and the transpose operations of the matrix, respectively. The $\boldsymbol{f} = [f_1, f_2, ..., f_T], f \in \mathbb{R}^T$ denotes the sequence of the LCE. The $\frac{1}{\sqrt{d_k}}$ is the sacle factor. The matrix $\boldsymbol{Mask}$ is a lower triangular matrix, indicating that the data at frame τ only focuses on current and historical information. That is, the embedding f_τ only attends to the information of $\mathbf{x}_{1:\tau}$.

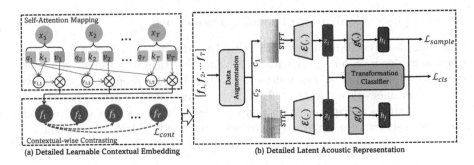

Fig. 2. Detailed Neural Network of CLF-AIAD.

Utilizing only the self-attention mechanism overlooks position information, and direct incorporation of position information into the data can lead to interference [30]. To address this, we propose the contextual-wise contrasting for AID, depicted in (a) of Fig. 2, to learn position information within the AID's embedding space and capture its temporal relationship.

The time frame τ in LCE shows higher similarity to neighbouring frames $[\tau - 1, \tau + 1]$ and lower similarity to distant frames. Contextual-wise contrasting minimizes the distance between neighbouring frames while maximizing the distance between different frames in the embedding space. This is expressed as below:

$$\ell_{cont} = -log \frac{exp(sim(\boldsymbol{f}_\tau, \boldsymbol{f}_{\tau-1}/t)) + exp(sim(\boldsymbol{f}_\tau, \boldsymbol{f}_{\tau+1}/t))}{\sum_{k=1}^{T} \mathbb{1}_{[k \neq \tau]} exp(sim(\boldsymbol{f}_\tau, \boldsymbol{f}_k)/t)}, \tag{5}$$

$$\mathcal{L}_{cont} = \frac{1}{N} \sum_{i=1}^{N} \sum_{\tau=1}^{T} \ell_{cont}. \tag{6}$$

where ℓ_{cont} indicates the loss function at frame τ. The function $\mathbb{1}_{[k \neq \tau]} \in \{0, 1\}$ is an indicator function that equals 1 if $k \neq \tau$, and 0 otherwise. t represents the temperature parameter. The $\mathcal{L}_{cont}$ shows the total contextual-wise contrastive loss of each batch. Besides, $sim(.)$ denotes cosine similarity among embeddings and is equal to $sim(x, y) = \frac{x \cdot y}{\|x\| \|y\|}$.

Latent Acoustic Representation. To leverage the characteristics of the AID fully, we investigate data augmentation transformations for the learnable time-series LCE in both the time and frequency domains. We explore the following transformations to create augmented versions of a given sample:

(a) **Time Shifting:** The time-series embeddings are shifted forwards or backwards, with the degree of shift randomly chosen from the range of zero to half of the embeddings' length.

(b) **Time-domain Gaussian Noise Injection:** In the time domain, we introduce Gaussian noise $N(0, \Sigma_\tau)$ to the embeddings to generate similar samples, where Σ_τ is a diagonal matrix with elements $\sigma_\tau = 0.5$ as the deviation.

(c) **Frequency-domain Gaussian Noise Injection:** In the frequency domain, we perform a short-time Fourier transform (STFT) on the embeddings and add the Gaussian noise $N(0, \Sigma_f)$ for the STFT. The $\Sigma_f = diag(\sigma_f)$ and $\sigma_f = 0.5$ control the deviation of the noise.

(d) **Time Masking:** This transformation randomly selects a segment of the LCE time axis and sets it equal to zero or the mean of the whole embedding. The size of the masked portion of the embedding is randomly chosen to be less than 8% of the LCE's length.

(e) **Frequency Masking:** It applies random masking to the STFT spectrum of the LCE, randomly setting to zeros a segment of frequencies from the embedding. The length of the masked segment is randomly chosen to be less than 5% of the number of frequency bands in the LCE.

In these data augmentations, we define $\mathcal{C}$ as the family of possible augmentation operators. We stochastically apply two operators, $c_1 \sim \mathcal{C}$ and $c_2 \sim \mathcal{C}$, to generate two correlated views from each sample in the batch: $\tilde{f}_1^{(m)} = c_1(f^{(m)})$ and $\tilde{f}_2^{(m)} = c_2(f^{(m)})$.

After applying the transformations, we extract the two-dimensional STFT spectrogram of the embeddings. As shown in Fig. 2 (b), sample-wise contrasting is used to learn the latent variables' local invariant information. The augmented data is then fed into the siamese encoders $\mathcal{E}(.)$, which are DNNs that map the input data to lower-dimensional latent acoustic representations $\mathbf{z}^{(m)} = \mathcal{E}(\tilde{f}^{(m)})$. The Siamese encoders consist of two neural networks with shared weights.

Following the SimCLR paper [2], we introduce a projection head $g(\cdot)$ to map the latent representation $\mathbf{z}$ to a subspace for calculating the sample-wise contrastive loss. This projection head is implemented as a one-hidden-layer Multi-Layer Perceptron (MLP) with ReLU activation. Note that the projection head is not used during inference.

The contrastive loss is computed on the output of the projection head, denoted as $\mathbf{h}^{(m)} = g(\mathbf{z}^{(m)})$. The loss function promotes the proximity of latent representations for positive pairs $(\mathbf{h}_i^{(m)}, \mathbf{h}_i^{(m)+})$ while encouraging separation from negative pairs. We treat all other samples in the batch as negative pairs. In particular, we used Normalized Temperature-scaled Cross-Entropy loss (NT-Xent) as our contrastive loss function [26]. The contrastive loss for the positive pair instance is calculated as follows:

$$\ell_{(i,i+)} = -log \frac{exp(sim(\mathbf{h}_i, \mathbf{h}_i^+/t))}{\sum_{j=1}^{2N} \mathbb{1}_{[j \neq i]} exp(sim(\mathbf{h}_i, \mathbf{h}_j)/t)}, \tag{7}$$

$$\mathcal{L}_{sample} = \frac{1}{2N} \sum_{i=1}^{N} \ell_{(i,i+)} + \ell_{(i+,i)} \tag{8}$$

In self-supervised learning, incorporating an auxiliary task can significantly enhance the performance of the model on the main task [5]. One commonly used auxiliary task involves training a simple classifier on the latent representation to predict the type of transformation applied to the data. In our CLF-AIAD, we utilize a linear classifier, denoted as $p_{cls}(\mathcal{C} = c_i | \tilde{f})$, which operates on the latent representation $\mathbf{z}$ to predict the applied transformation. The loss function employed for this classifier is the Multi-class Cross-Entropy Loss ($\mathcal{L}_{cls}$).

The final loss of the proposed CLF-AIAD is a weighted sum of the contextual-wise and sample-wise contrastive loss and the loss of the transformation classifier:

$$\mathcal{L} = \lambda_1 \mathcal{L}_{cont} + \mathcal{L}_{sample} + \lambda_2 \mathcal{L}_{cls}, \tag{9}$$

where λ_1 and λ_2 are balancing hyperparameters.

Detection. Finally, we require an anomaly scoring function that maps the latent embedding to a scalar to quantify the degree of the AID's abnormality. In this study, we utilize the Mahalanobis distance [22], which is a metric for measuring the distance between a point and a distribution. During the test stage, we compute the distance between the latent embedding of the query sample and the representations of normal training instances. The anomaly score $S_\mathbf{x}$ is calculated as follows for a given input $\mathbf{x}$:

$$S_\mathbf{x} = (\mathbf{z}_\mathbf{x} - \mu)^T \Sigma^{-1} (\mathbf{z}_\mathbf{x} - \mu). \tag{10}$$

where Σ means their covariance matrix, μ denotes the mean vector of normal training AID, and $\mathbf{z}_\mathbf{x}$ denotes the latent representation of the acoustic clip x.

4   Experiments

4.1   Datasets

To evaluate the proposed method, we conducted experiments using two public datasets of machine-operating acoustic data, MIMII [24] and ToyADMOS [18].

MIMII is designed to support the automation of facility maintenance tasks for the machine learning and signal processing communities. It consists of normal and anomalous acoustic data recordings from the four most common industrial machines. The anomalous acoustic data simulate real-life scenarios such as contamination, leakage, rotating unbalance, and rail damage. The acoustic data includes background factory noise to provide a realistic environment. Each recording in the corpus lasts 10 s with a sampling rate of 16 kHz.

ToyADMOS is a dataset created for anomaly detection in machine operating sounds (ADMOS). It includes recordings of anomalous operating sounds from miniature machines (toys) deliberately damaged for this purpose. The dataset comprises three sub-datasets: machine-condition inspection, fault diagnosis of machines with fixed tasks, and fault diagnosis of machines with moving tasks. The recordings were made using four microphones at a sampling rate of 48 kHz.

In our study, we focus on the valves, pumps, fans, and sliders sub-datasets from MIMII as well as toy car and toy conveyor sub-datasets from ToyADMOS.

4.2   Implementation

The STFT spectrogram is obtained using a Blackman window with a length of 1024 and 50% overlap, resulting in 513 frequency bands for the STFT. The self-attention mechanism from [31] is adopted as the contextual projection layer. The

ResNet-18 architecture from [11] is used as the encoder $\mathcal{E}(\cdot)$, with the 256 units in the linear output layer. The projection head consists of an MLP with a hidden layer of 256 units, and an output layer of 128 units for the representation vector. The Adam optimizer is employed with an initial learning rate of 0.01. A batch size of 128 is used, and the network is trained for 600 epochs. The temperature parameter t is empirically set to 0.08, and throughout all our experiments, we set $\lambda_1 = 0.2$ and $\lambda_2 = 0.1$ in Eq.(9).

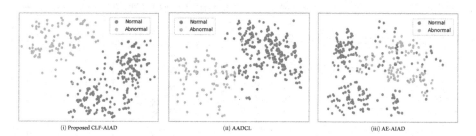

Fig. 3. The latent representations t-SNE visualization of proposed CLF-AIAD, AADCL and AE-AIAD for the machine type Fan.

4.3   Results and Analysis

We evaluate our proposed model using the area under the receiver operating characteristic (ROC) curve (AUC) and the partial-AUC (pAUC) scores [6,15] for each machine type.

Comparison Results. We compared our CLF-AIAD with recent unsupervised AIAD baseline and state-of-the-art (SOTA) methods on DCASE2021/2022 Task2, including AE-AIAD [15], MobileNetV2 [15], IDNN [27], FREAK [21], AADCL [13]. Here, AE-AIAD acts as a baseline, IDNN represents the recent SOTA in reconstruction-based methods, MobileNetV2 and FREAK lead in representation-based classification methods, and AADCL stands as the SOTA self-supervised method utilizing contrastive learning.

As shown in Table 1, our proposed method achieves the best performance in terms of average AUC and average pAUC across six machine types. The proposed CLF-AIAD provides improvements of 1.7% and 3.6% in terms of average AUC and average pAUC, respectively, compared to the current SOTA contrastive learning-based method AADCL.

In order to provide a visual depiction of the efficacy of the learned audio feature representations in anomaly detection, we present in Fig. 3 the t-distributed Stochastic Neighbor Embedding (t-SNE) cluster visualization of the latent acoustic representations obtained from the proposed CLF-AIAD, AADCL, and baseline AE-AIAD methods. Through this visualization, we can discern that our method exhibits better discriminatory capabilities, effectively distinguishing anomalies in the audio data compared to the other approaches.

Table 1. Performance comparison in terms of AUC (%) and pAUC (%) on the test data of MIMII and ToyADMOS datasets.

Methods	Fan		Pump		Slider		Valve		ToyCar		ToyConveyor		Average	
	AUC	pAUC	AUC	pAUC	AUC	pAUC	AUC	pAUC	AUC	pAUC	AUC	pAUC	AUC	pAUC
AE-AIAD [15]	63.24	50.37	61.92	56.74	66.74	50.83	53.41	51.52	63.3	55.21	64.16	56.07	62.29	53.46
IDNN [27]	67.71	52.9	73.71	61.05	86.41	67.85	81.99	64.97	78.33	69.42	71.03	59.7	75.27	62.65
MobileNetV2 [15]	80.19	74.4	82.53	76.5	95.27	85.22	88.65	87.98	87.66	85.92	69.71	56.43	83.46	77.74
FREAK [21]	62.2	53.4	62.4	54.7	66.4	55.1	56.5	50.3	66	54	65.4	52.6	63.01	53.35
AADCL [13]	85.27	79.93	87.75	70.85	95.74	89.62	89.62	87.03	88.79	85.95	71.26	57.14	86.41	78.42
Proposed CLF-AIAD	**87.11**	**81.27**	**90.79**	**83.29**	**96.57**	**90.23**	**90.89**	**89.54**	**92.2**	**87.16**	**72.08**	**60.6**	**88.11**	**82.02**

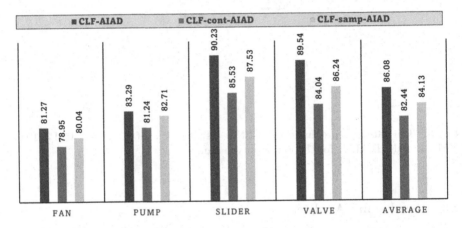

Fig. 4. The pAUC-best (%) results of proposed CLF-AIAD and its variants of CLF-cont-AIAD and CLF-samp-AIAD on four different machine types' sub-datasets

Ablation Study. We evaluated the performance of CLF-AIAD along with two variants: CLF-cont-AIAD and CLF-samp-AIAD. Specifically, CLF-cont-AIAD represents the variant that retains the contextual-wise contrastive loss while eliminating the sample-wise contrastive loss. However, CLF-samp-AIAD retains the sample-wise contrastive loss while discarding the contextual contrastive loss.

As demonstrated in Fig. 4, CLF-AIAD performs best on the four machine types' sub-datasets. CLF-samp-AIAD outperforms CLF-cont-AIAD due to the inherent periodic nature of machine operation, which leads to similarities between distant time frames. The contextual contrastive loss in CLF-cont-AIAD focuses on fine-grained contextual information by maximizing the similarity between neighboring frames and minimizing the similarity between distant frames. However, the generation of periodic acoustic data by machines can lead to similarities between not only neighboring but also more distant frames. And the instance contrastive loss in CLF-samp-AIAD optimizes similarity among the same samples and minimizes similarity between different samples. This allows CLF-samp-AIAD to capture coarse-grained contextual information and local invariance within AID, effectively incorporating the periodic information present in AID.

In summary, using the multi-grained contrasting method in the CLF-AIAD facilitates the acquisition of multi-granularity temporal-dependent and local invariant representations, allowing the model to generate a normal profile accurately and improving the discriminative ability of aberrant AID.

5   Conclusion

In this paper, we propose a novel multi-grained Contrastive Learning-based Framework for AIAD (CLF-AIAD), which can capture dependencies and dynamics in normal AID. CLF-AIAD first incorporates a self-attention layer and a context-based contrast method to learn robust temporal embeddings. Additionally, a sample-wise contrasting method is employed to capture local invariant characteristics, enhancing the discriminative capabilities of the model. We also introduce a transformation classifier within a self-supervised learning framework to improve performance on the primary task. Extensive experiments on the MIMII and ToyADMOS datasets establish the proposed CLF-AIAD as a new effective method for unsupervised anomaly detection that substantially surpasses the state-of-the-art baseline model by 25.82% in mean AUC score. As a future extension, we will enhance our CLF-AIAD by incorporating the inherent periodic nature of machine operation.

Acknowledgements. Research Fund KU Leuven in the context of the ReSOS project (C3/20/014) and by Ford Motor Company in the context of the Ford-KU Leuven Research Alliance project Automated S&R (KUL0134).

References

1. Bigoni, C., Hesthaven, J.S.: Simulation-based anomaly detection and damage localization: an application to structural health monitoring. Comput. Methods Appl. Mech. Eng. **363**, 112896 (2020)
2. Chen, T., Kornblith, S., Norouzi, M., Hinton, G.: A simple framework for contrastive learning of visual representations. In: International Conference on Machine Learning, pp. 1597–1607. PMLR (2020)
3. Cui, Y., Liu, Z., Lian, S.: A survey on unsupervised industrial anomaly detection algorithms. arXiv preprint: 2204.11161 (2022)
4. Defard, T., Setkov, A., Loesch, A., Audigier, R.: Padim: a patch distribution modeling framework for anomaly detection and localization. In: Pattern Recognition, pp. 475–489 (2021)
5. Dohi, K., Endo, T., Purohit, H., Tanabe, R., Kawaguchi, Y.: Flow-based self-supervised density estimation for anomalous sound detection. In: Proceedings of ICASSP, pp. 336–340 (2021)
6. Dohi, K., Imoto, K., et al.: Description and discussion on dcase 2022 challenge task 2: Unsupervised anomalous sound detection for machine condition monitoring applying domain generalization techniques. arXiv preprint: 2206.05876 (2022)
7. Eldele, E., Ragab, M., Chen, Z., Wu, M., et al.: Time-series representation learning via temporal and contextual contrasting, pp. 2352–2359, August 2021

8. Eltouny, K., Gomaa, M., Liang, X.: Unsupervised learning methods for data-driven vibration-based structural health monitoring: a review. Sensors **23**(6), 3290 (2023)

9. Gong, D., Liu, L., Le, V., Saha, B., Mansour, M.R., et al.: Memorizing normality to detect anomaly: Memory-augmented deep autoencoder for unsupervised anomaly detection. In: Proceedings of ICCV, pp. 1705–1714 (2019)

10. He, K., Fan, H., Wu, Y., Xie, S., Girshick, R.: Momentum contrast for unsupervised visual representation learning. In: Proceedings of CVPR, pp. 9729–9738 (2020)

11. He, K., Zhang, X., Ren, S., Sun, J.: Deep residual learning for image recognition. In: Proceedings of CVPR, pp. 770–778 (2016)

12. Hojjati, H., Armanfard, N.: Dasvdd: Deep autoencoding support vector data descriptor for anomaly detection. arXiv preprint: 2106.05410 (2021)

13. Hojjati, H., Armanfard, N.: Self-supervised acoustic anomaly detection via contrastive learning. In: Proceedings of ICASSP, pp. 3253–3257 (2022)

14. Jiang, A., Zhang, W.Q., et al.: Unsupervised anomaly detection and localization of machine audio: a gan-based approach. In: Proceedings of ICASSP, pp. 1–5 (2023)

15. Kawaguchi, Y., Imoto, K., et al.: Description and discussion on dcase 2021 challenge task 2: Unsupervised anomalous sound detection for machine condition monitoring under domain shifted conditions. arXiv preprint: 2106.04492 (2021)

16. Kim, M.S., Yun, J.P., Lee, S., Park, P.: Unsupervised anomaly detection of lm guide using variational autoencoder. In: Proceedings of ATEE, pp. 1–5 (2019)

17. Koizumi, Y., Saito, S., Uematsu, H., Harada, N.: Optimizing acoustic feature extractor for anomalous sound detection based on neyman-pearson lemma. In: Proceedings of EUSIPCO, pp. 698–702 (2017)

18. Koizumi, Y., Saito, S., Uematsu, H., Harada, N., Imoto, K.: Toyadmos: a dataset of miniature-machine operating sounds for anomalous sound detection. In: Proceedings of WASPAA, pp. 313–317 (2019)

19. Liu, Y., Garg, S., Nie, J., Zhang, Y., et al.: Deep anomaly detection for time-series data in industrial IoT: a communication-efficient on-device federated learning approach. IEEE Internet Things J. **8**(8), 6348–6358 (2020)

20. Liu, Z., Tang, H., Michiels, S., Joosen, W., Hughes, D.: Unsupervised acoustic anomaly detection systems based on gaussian mixture density neural network. In: Proceedings of EUSIPCO, pp. 259–263 (2022)

21. Lu, H., Fu, Y., Qin, H., Huang, S., et al.: Anomalous sounds detection using autoencoder and classification methods. Technival report, DCASE2021 Challenge (2021)

22. McLachlan, G.J.: Mahalanobis distance. Resonance **4**(6), 20–26 (1999)

23. Patcha, A.: An overview of anomaly detection techniques: existing solutions and latest technological trends. Comput. Netw. **51**(12), 3448–3470 (2007)

24. Purohit, H., Tanabe, R., et al.: Mimii dataset: sound dataset for malfunctioning industrial machine investigation and inspection. arXiv preprint: 1909.09347 (2019)

25. Ruff, L., et al.: A unifying review of deep and shallow anomaly detection. Proc. IEEE **109**(5), 756–795 (2021)

26. Sohn, K.: Improved deep metric learning with multi-class n-pair loss objective. Advances in neural information processing systems 29 (2016)

27. Suefusa, K., Nishida, T., Purohit, H., et al.: Anomalous sound detection based on interpolation deep neural network. In: Proceedings of ICASSP, pp. 271–275 (2020)

28. Taheri, H., Koester, L.W., et al.: In situ additive manufacturing process monitoring with an acoustic technique: clustering performance evaluation using k-means algorithm. J. Manuf. Sci. Eng. **141**(4) (2019)

29. Trapp, M., Chen, F.: Automotive buzz, squeak and rattle: mechanisms, analysis, evaluation and prevention. Elsevier (2011)

30. Tuli, S., Casale, G., et al.: Tranad: Deep transformer networks for anomaly detection in multivariate time series data. arXiv preprint: 2201.07284 (2022)
31. Vaswani, A., Shazeer, N., Parmar, N., Uszkoreit, J., et al.: Attention is all you need. Advances in neural information processing systems 30 (2017)
32. Yue, Z., Wang, Y., Duan, J., Yang, T., et al.: Ts2vec: Towards universal representation of time series. In: Proceedings of AAAI, vol. 36, pp. 8980–8987 (2022)
33. Yun, H., Kim, H., Jeong, Y.H., Jun, M.B.: Autoencoder-based anomaly detection of industrial robot arm using stethoscope based internal sound sensor. J. Intell. Manuf. **34**(3), 1427–1444 (2023)
34. Zhang, W., Yang, D., Peng, H., Wu, W., et al.: Deep reinforcement learning based resource management for dnn inference in industrial iot. IEEE Trans. Veh. Technol. **70**(8), 7605–7618 (2021)
35. Zhang, Z., Zhao, Z., Zhang, X., et al.: Industrial anomaly detection with domain shift: a real-world dataset and masked multi-scale reconstruction. arXiv preprint: 2304.02216 (2023)

Prediction and Analysis of Acoustic Displacement Field Using the Method of Neural Network

Xiaodong Jiao[1], Jin Tao[2], Hao Sun[1(⊠)], and Qinglin Sun[1(⊠)]

[1] College of Artificial Intelligence, Nankai University, Tianjin, China
xdjiao@mail.nankai.edu.cn, {sunh,sunql}@nankai.edu.cn
[2] Silo AI, Helsinki 00100, Finland
jin.tao@silo.ai

Abstract. Micro/nano manipulation technology holds significant value in diverse fields, such as biomedicine, precision killing, and material chemistry. Among these applications, acoustic manipulation technology stands out as a crucial approach for micro/nano manipulation. However, the determination of a control equation for acoustic manipulation proves challenging due to the complex motion characteristics exhibited by micro/nano targets. The mastery of acoustic displacement field information is essential for successful manipulation of micro targets. Addressing this challenge, this paper presents a novel method that utilizes neural networks for predicting sound wave displacement fields. Which involves collecting raw data on acoustic displacement changes under specific excitation frequencies and times utilizing the acoustic manipulation experimental platform. Using the data to train the neural network model. The network takes the initial position information of the micro target as input and accurately predicts the displacement field across the entire thin plate area, achieving a remarkable prediction accuracy of 0.5mm. Through a comprehensive analysis of the network's prediction performance using test set data, the effectiveness of the designed network in solving the acoustic displacement field problem with high accuracy is demonstrated. This research significantly contributes to the advancement of acoustic manipulation technology, enabling precise control and manipulation of micro/nano targets across various applications.

Keywords: Micro nano manipulation · Acoustic technology · Neural network · Acoustic displacement

1 Introduction

In recent years, the growing demand for micro/nano technology has necessitated an increase in the technical requirements for micro/nano manipulation. Traditional operational techniques typically rely on the direct application of mechanical external forces to the target, but this approach is only effective for

Supported by College of Artificial Intelligence, Nankai University.

© The Author(s), under exclusive license to Springer Nature Singapore Pte Ltd. 2024
B. Luo et al. (Eds.): ICONIP 2023, CCIS 1961, pp. 138–148, 2024.
https://doi.org/10.1007/978-981-99-8126-7_11

larger operational targets [1–3]. When it comes to micro and nano level particles, such methods not only fail to achieve the desired operational tasks but can also result in damage to the delicate micro targets. Consequently, several micro/nano manipulation methods have been proposed to address this challenge, including optical technology [4–6], electromagnetic manipulation technology [7–9], microfluidic manipulation technology [10–12], and acoustic manipulation technology [13–16]. Each of these methods exhibits distinct advantages and is suitable for specific scenarios. For instance, optical technology carries the risk of damaging biological molecules and chemical materials. On the other hand, electromagnetic technology is limited to manipulating particles with electromagnetic properties. In contrast, sound wave manipulation technology offers unique benefits such as high environmental compatibility, long propagation distance, and the ability to manipulate particles without restrictions based on their properties [17,18].

The focal point of this study revolves around acoustic wave manipulation technology, which is rooted in the Chladni effect initially discovered by Chladni, Ernst Florens Friedrich in 1981 [19–22]. This effect refers to the regular patterns formed by sand particles on a violin board when subjected to sound wave excitation. Building upon this principle, the control and manipulation of micro and nano particles on thin plates can be achieved by modulating and combining the frequency of sound waves. This enables a range of manipulation tasks, including single particle migration, multiple particle fusion, particle formation, and particle swarm control. These diverse manipulation functions hold significant promise in various fields, including the pharmaceutical industry, material chemistry, material molecules, and battlefield medicine.

The movement of micro and nano particles on a thin plate, driven by acoustic waves, exhibits inherent randomness and complexity. Obtaining a kinematics model of these particles through analytical algorithms proves challenging. Typically, the movement mode of particles is determined by combining force and kinematics analysis, leading to the classification of particle motion into three types on the thin plate: fixed area, sliding area, and bouncing area [23,24]. However, obtaining a comprehensive particle motion model for the entire thin plate region remains elusive. Consequently, to effectively control particle motion on thin plates, it becomes necessary to acquire the displacement field of particle motion under different acoustic frequency excitations. In light of these considerations, a data-driven approach emerges as a highly promising avenue to pursue.

Neural networks are algorithmic mathematical models that emulate the behavioral characteristics of animal neural networks, enabling distributed parallel information processing [25–28]. They can be categorized into two types: feedforward neural networks and feedback neural networks. Furthermore, neural networks can be classified based on their depth, distinguishing between shallow neural networks and deep neural networks. A deep neural network generally consists of a structure with more than two hidden layers. Moreover, in recent years, neural network technology has shone brightly in many industrial, financial, and mathematical fields [29–32].

Each neuron within a neural network multiplies the initial input value by a specific weight and then adds other input values to calculate a sum. Following neuron bias adjustment, the activation function is applied to normalize the output value. In essence, neural networks are composed of interconnected computing units, known as neurons, organized in layers. These networks possess the capability to process and classify data, yielding the desired output [33–35]. Consequently, by gathering a limited amount of acoustic displacement field data, designing the network structure, defining appropriate inputs and outputs, training the network, it becomes feasible to utilize the network for predicting the acoustic displacement field across the entire thin plate area.

This paper commences by providing an introduction to the principle of acoustic wave manipulation, primarily focusing on the Chladni effect. The control equation of this effect is analyzed, and its implementation is demonstrated through experimental means, resulting in the generation of Chladni pattern. Leveraging the established experimental platform, an acoustic displacement field dataset is collected. Subsequently, a network structure is designed, and the dataset is utilized for training purposes. This trained network is then employed to predict the complete acoustic displacement field across the thin plate. Finally, the accuracy of the network's predictions is thoroughly analyzed, and the resulting prediction outcomes are presented.

2   Related Physical Background

According to the previous introduction, a corresponding experimental platform can be built, including three modules: a metal thin plate with a fixed center, an acoustic wave signal generator, and particle image data acquisition system. The vibration control equation of a forced thin plate with a fixed center is described as

$$\rho h \frac{\partial^2 w}{\partial t^2} + D\nabla^4 w = p(x, y, t)\delta(x - x_0, y - y_0) \tag{1}$$

in which, ρ is the plate material density, h is the thickness, $p(x, y, t)$ is the external excitation, δ is the two-dimensional Dirac function, w is the thin plate deflection function, ∇^4 is the double Laplace operator, defined as $\nabla^4 = \nabla^2(\nabla^2 w)$.

Further organizing Eq. (1), and substituting the Dirac function into Eq. (1), setting the point source excitation response to the Green's function can obtain the thin plate vibration mode response function

$$\begin{aligned}
W(x, y; \omega) &= G(x, y; \omega) * \frac{\omega^2}{D} P(x, y; \omega)\delta(x - x_0, y - y_0) \\
&= G(x, y; x_0, y_0, \omega) \cdot P(x_0, y_0; \omega) \\
&= P(x_0, y_0; \omega)\frac{\omega^2}{D}\left(\frac{2}{l}\right)^2 \sum_{n_1 n_2} \frac{\cos(\frac{n_1\pi}{l}x_0)\cos(\frac{n_2\pi}{l}y_0)\cos(\frac{n_1\pi}{l}x)\cos(\frac{n_2\pi}{l}y)}{\lambda_n{}^2 - \lambda^2}
\end{aligned} \tag{2}$$

in which, P is the excitation amplitude, l is the edge length of thin plate, λ_n is the eigenvalue corresponding to the characteristic mode, n_1 and n_2 characterizing feature modes.

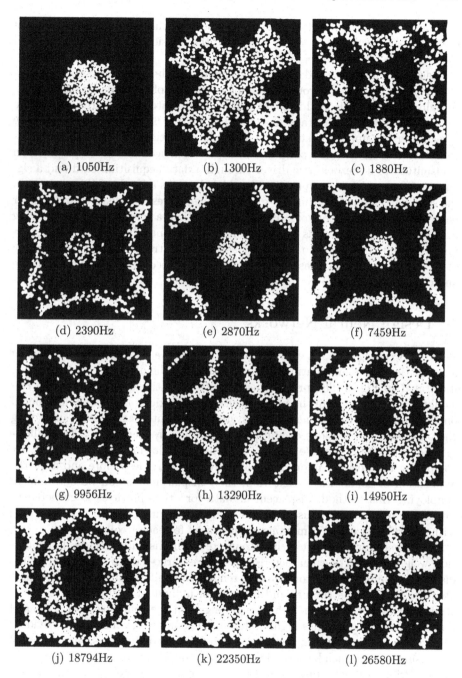

Fig. 1. Results of the Chladni experiment.

Combining Eq. (1) and Eq. (2), it can be seen that the vibration mode of the thin plate oscillator is closely related to the physical parameters and sound source signals of the thin plate system

Based on the established experimental platform, the effectiveness of the physical platform is verified through the implementation of the Chladni experiment. Initially, a square silicon thin plate with a fixed center is prepared, and white sand particles with an average size of approximately 650 microns are uniformly spread over the upper surface of the plate. An acoustic wave signal generator is constructed using Labview software. The generated acoustic wave signal is then transmitted to a piezoelectric driver through a data acquisition card and a signal amplifier, which in turn excites the thin plate. As a result of this excitation, the sand particles on the thin plate arrange themselves into distinct and regular patterns. Through careful experimental observation, a total of 12 corresponding frequencies are obtained, as illustrated in the accompanying Fig. 1. The experimental results depicted in the figure clearly demonstrate that as the frequency of the sound waves increases, the formed patterns become increasingly intricate and complex.

3   Design Neural Networks

Before proceeding with network design and training, a dataset is collected using an acoustic wave manipulation platform, following the data collection principle illustrated in Fig. 2. The process begins with the utilization of an image collection system to capture the initial positions of particles uniformly distributed on the thin plate. Subsequently, the system is activated again after subjecting the particles to continuous excitation at a specific frequency for 500ms, allowing the collection of particle position information following their movement. By comparing the position information of individual particles before and after the application of sound wave excitation, their displacements can be determined. Employing this method, displacement data for other particles is gathered to form a comprehensive dataset. In this study, approximately 130 valid particles with a particle size of 650 microns are prepared for each frequency.

Neural networks consist of interconnected neurons with adjustable connection weights, exhibiting characteristics such as large-scale parallel processing, distributed information storage, and remarkable self-organizing and learning abilities. While the structure and function of individual neurons are relatively straightforward, the system behavior arising from the collective interactions of numerous neurons becomes highly intricate. The network's flexibility is evident through the ability to configure the number of intermediate layers, the number of processing units in each layer, and the learning coefficient to suit specific requirements. Prior to functioning, neural networks must undergo a learning process based on predetermined criteria.

The neural network structure presented in this study is depicted in Fig. 3. It utilizes the acquired knowledge from the acoustic displacement field data to predict the displacement field across the whole thin plate area through the

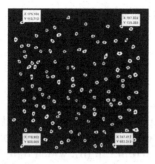 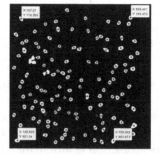

(a) Before acoustic excitation   (b) After acoustic excitation

Fig. 2. Dataset collection.

trained network. The network model, as illustrated, comprises an input layer, three hidden layers, and an output layer. The input layer consists of three neurons representing the particle position coordinates and excitation. The output layer provides information about the corresponding changes of particle displacement along the XY direction. The hidden layers consist of three layers, with 16 neurons in the first layer, 64 neurons in the second layer, and 16 neurons in the third layer, resulting in a total of 96 neurons. Additionally, in the acoustic displacement field prediction network, each layer is linearized, and then the (ReLU) function is adopted to enhance network nonlinearity. This integration of nonlinear factors within the neural network allows for effective resolution of complex acoustic displacement field predictions.

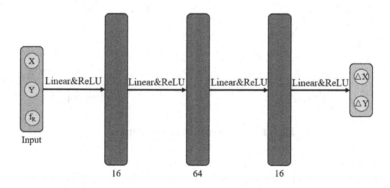

Fig. 3. Neural network structure for predicting acoustic displacement field.

The dataset consists of 130 displacement field data, with 100 sets dedicated to learning and training the acoustic displacement field network, while the remaining 30 sets are utilized for evaluating the network's prediction performance. The network model will work out the changes in particle displacement whenever frequency-related thin plate position data is sent to the network. Finally, the network output is set to predict the particle displacement motion field across the entire thin plate area.

4   Error Analysis and Results

This section examines and illustrates the particle displacement field for two sound wave frequencies. To assess the network's prediction performance, an average error function is formulated, as depicted in Eq. (3), which represents the average percentage difference between the predicted displacement values and the corresponding true values in the test set data. Through meticulous data organization and calculations, it was observed that the average error is significantly below 1%. This suggests that the network's prediction results exhibit minimal error, which is deemed acceptable and can be utilized in subsequent stages of the work.

$$E = \frac{\sum\limits_{i=1}^{N} \frac{\Delta_{pi} - \Delta_{ri}}{\Delta_{ri}}}{N} \tag{3}$$

in which, Δ_{pi} is the predictive value, Δ_{ri} is the actual value, N represents the number of test sets.

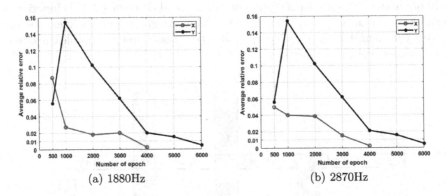

(a) 1880Hz   (b) 2870Hz

Fig. 4. Mean error evolution curves.

Figure 4 displays the average error trends in the X and Y directions, represented by the red and blue curves, respectively. As depicted in Fig. 4, it is evident that the error evolution curve gradually diminishes as the number of training iterations increases, ultimately reaching a level below 1%. Additionally, to assess the stability of the error in the test set data, the variance of the test set error was calculated using Eq. (4). When the average error falls below 1%, the corresponding variance value is less than 5%. This indicates that the error exhibits minimal volatility, and the distribution of errors between each predicted value and the corresponding true value is consistent. Consequently, the network's prediction results can be deemed reliable and acceptable.

$$\sigma_e = \sqrt{\frac{\sum\limits_{i=1}^{N} (e_i - \bar{e})}{N}} \qquad (4)$$

Compared to traditional methods of predicting acoustic displacement field, learning-based modeling prediction methods offer enhanced efficiency and accuracy. Traditional approaches often struggle with collecting extensive datasets and exhibit limited prediction accuracy. Conversely, the proposed learning-based acoustic displacement field modeling prediction method leverages the nonlinearity of the network to better approximate real-world scenarios. This enables more effective and accurate predictions, surpassing the limitations of traditionap methods.

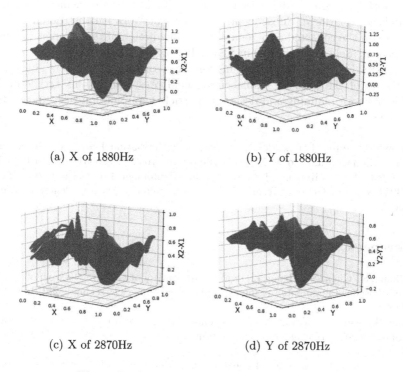

(a) X of 1880Hz   (b) Y of 1880Hz

(c) X of 2870Hz   (d) Y of 2870Hz

Fig. 5. Prediction results of displacement field.

Figure (5) illustrates the prediction results of the displacement field at frequencies 1880 Hz and 2870 Hz. Specifically, Fig. (5a) and Fig.(5b) depict the acoustic displacement field along the X and Y directions, respectively, under excitation at a frequency of 1880 Hz. Similarly, Fig.(5c) and Fig.(5d) present the corresponding results for the frequency of 2870 Hz.

5 Conclusion

Micro/nano robots are expected to hold significant application value in various research fields, including industrial production, biomedical sciences, and material chemistry. Among these, acoustic-based micro/nano robots have garnered considerable attention due to their compatibility with different operating environments and their versatility in maneuvering targets. The successful application of acoustic manipulation hinges on the precise understanding of the acoustic displacement field, which in turn facilitates the control of micro targets. Consequently, achieving high accuracy in predicting the acoustic displacement field becomes paramount. Conventional methods relying on mathematical fitting not only demand substantial human effort to collect a substantial volume of raw data but also exhibit limitations in terms of prediction accuracy and scope. In response to these requirements and existing challenges, this article presents a learning-based approach for modeling and predicting acoustic displacement fields. Initially, acoustic displacement field data is collected using an acoustic manipulation platform. Subsequently, a neural network is constructed to learn and train on the collected data, enabling the prediction of the complete acoustic displacement field across the thin plate. Through testing, this method demonstrates remarkable accuracy, with an average error rate of just 1%. The prediction results of the acoustic displacement field presented in this article offer a reliable basis for planning and designing particle manipulation tasks in practical operations.

Acknowledgment. This work was supported by the National Natural Science Foundation of China (Grant No. 61973172, 62003177, 62103204, 62003175 and 61973175), Joint Fund of the Ministry of Education for Equipment Pre research (Grant No. 8091B022133) and General Terminal IC Interdisciplinary Science Center of Nankai University.

References

1. Zhang, S., Chen, Y., Zhang, L., et al.: Study on Robot Grasping System of SSVEP-BCI based on augmented reality stimulus. Tsinghua Sci. Technol. **28**(2), 322–329 (2023)
2. Liu, S., Wang, L., Wang, X. V: Multimodal data-driven robot control for human-robot collaborative assembly. J. Manuf. Sci. Eng. Trans. ASME **2022**(5), 144 (2022)
3. Zhou, L., Zhang, L: A novel convolutional neural network for electronic component classification with diverse backgrounds. Int. J. Model. Simul. Sci. Comput. **2022**(1), 13 (2022)
4. Chen, L., Gan, W., Chen, L., et al.: Optical encryption technology based on spiral phase coherent superposition and vector beam generation system. Optik **2022**(253), 168599 (2022)
5. Rebbah, R., Messaoudene, I., Khelifi, M., et al.: Enhanced isolation of MIMO cavity antenna using substrate integrated waveguide technology. Microwave Optical Technol. Lett. **2022**(2), 64 (2022)

6. Zhang, W., Wei, X., Chen, L., et al.: Axial uniformity diagnosis of coaxial surface wave linear plasma by optical emission spectroscopy. Plasma Sci. Technol. **24**(2), 025403 (2022)
7. Brito, H.H.: Experimental status of thrusting by electromagnetic inertia manipulation. Acta Astronaut. **54**(8), 547–558 (2004)
8. Wang, J., Zhang, Z., Huang, C., et al.: Transmission-reflection-integrated quadratic phase metasurface for multifunctional electromagnetic manipulation in full space. Adv. Optical Mater. **2022**(6), 10 (2022)
9. Li, Z., Zhang, D., Liu, J., et al.: 3-D manipulation of dual-helical electromagnetic wavefronts with a noninterleaved metasurface. IEEE Trans. Antennas Propagation **70**(1), 378–388 (2022)
10. Chen, Q., Meng, Q., Liu, Y., et al.: A digital microfluidic single-cell manipulation system optimized by extending-depth-of-field device. J. Innov. Optical Health Sci. **16**(03), 2244006 (2023)
11. Wu, B., Zhou, J., Guo, Y., et al.: Preparation of HMX/TATB spherical composite explosive by droplet microfluidic technology. Defence Technol. **21**(3), 11 (2023)
12. Wang, T., Ke, M., Li, W., et al.: Particle manipulation with acoustic vortex beam induced by a brass plate with spiral shape structure. Appl. Phys. Lett. **109**(12), 2140–2143 (2016)
13. Liang, S., Liu, J., Lai, Y., et al.: Nonlinear wave propagation in acoustic metamaterials with bilinear nonlinearity. Chinese Phys. B **32**(4), 405–411 (2023)
14. Zma, C., Pfa, B.: Acoustic micro-manipulation and its biomedical applications. Engineering, 1–4 (2022)
15. Yuan, J., Meng, X., Ran, J., et al.: Manipulation of acoustic wave reflection for arbitrary reflecting surfaces based on acoustic metasurfaces. International Journal of Modern Physics, B. Condensed Matter Phys. Stat. Phys. Appl. Phys. **36**(6), 1–10 (2022)
16. Wu, Z., Pan, M., Wang, J., et al.: Acoustofluidics for cell patterning and tissue engineering. Eng. Regeneration **3**(4), 397–406 (2022)
17. Kozuka, T., Yoshimoto, T., Toyoda, M. : Two-dimensional acoustic manipulation in air using interference of a standing wave field by three sound waves. Japanese J. Appl. Phys. **61**, 1–10 (2022)
18. Qi, Y., He, H., Xiao, M.: Manipulation of acoustic vortex with topological dislocation states. Appl. Phys. Lett. **120**(21), 1–5 (2022)
19. Zha, B., Zz, A., Zl, C., et al.: Particles separation using the inverse Chladni pattern enhanced local Brazil nut effect. Extreme Mech. Lett. **2021**(49), 101466 (2021)
20. Worrell, C, L., Lynch, J, A., Jomaas, G.: Effect of smoke source and horn configuration on enhanced deposition, acoustic agglomeration, and chladni figures in smoke detectors. Fire Technol. **39**(4), 309–346 (2003)
21. Raghu, M.: A study to explore the effects of sound vibrations on consciousness. Int. J. Soc. Work Hum. Serv. Practice **6**(3), 75–88 (2018)
22. Bardell, N.S.: Chladni figures for completely free parallelogram plates: an analytical study. J. Sound Vib. **174**(5), 655–676 (1994)
23. Jiao, X., Tao, J., Sun, H., et al.: Kinematic modes identification and its intelligent control of micro-nano particle manipulated by acoustic signal. Mathematics **10**(21), 4156 (2022)
24. Aman, K., Anirvan, D.: Wave-induced dynamics of a particle on a thin circular plate. Nonlinear Dyn. **2021**(103), 293–308 (2021)
25. Lohith, M. S., Manjunath, Y., Eshwarappa, M. N. : Multimodal biometric person authentication using face, ear and periocular region based on convolution neural networks. Int. J. Image Graph. **23**(02), 2350019 (2023)

26. Chang, X. K., He, Y., Gao, Z. M.: Exponential stability of neural networks with a time-varying delay via a cubic function negative-determination lemma. Appl. Math. Comput. **2023**(438), 127602 (2023)
27. Zhang, N., Wang, Y., Zhang, X., et al.: A multi-degradation aided method for unsupervised remote sensing image super resolution with convolution neural networks. IEEE Trans. Geosci. Remote Sens. **2022**(20), 1–14 (2022)
28. Nogueira, I., Dias, R., Rebello, C.M., et al.: A novel nested loop optimization problem based on deep neural networks and feasible operation regions definition for simultaneous material screening and process optimization. Chem. Eng. Res. Des. **2022**(180), 243–253 (2022)
29. Lek, S., Delacoste, M., Baran, P., et al.: Application of neural networks to modelling nonlinear relationships in ecology. Ecol. Model. **90**(1), 39–52 (1996)
30. Kowalski, C.T., Orlowska-Kowalska, T., et al.: Neural networks application for induction motor faults diagnosis. Math. Comput. Simul. **63**(3–5), 435–448 (2003)
31. Wang, J., Fan, X., Shi, N., et al.: Convolutional neural networks of whole jujube fruits prediction model based on multi-spectral imaging method. Chin. J. Electron. **32**(3), 655–662 (2023)
32. Skrypnik, A.N., Shchelchkov, A.V., Gortyshov, Y.F., et al.: Artificial neural networks application on friction factor and heat transfer coefficients prediction in tubes with inner helical-finning. Appl. Thermal Eng. **2022**(206), 118049 (2022)
33. Atwya, M., Panoutsos, G.: Structure optimization of prior-knowledge-guided neural networks. Neurocomputing **2022**, 491 (2022)
34. Fan, Y.: A Study on Chinese-English machine translation based on migration learning and neural networks. Int. J. Artif. Intell. Tools **31**(05), 2250031 (2022)
35. Serebryanaya, L. V.: Methods for constructing artificial neural networks for data classification. Digital Transformation **28**(1), 2250031 (2022)

Graph Multi-dimensional Feature Network

Minghong Yao[1], Haizheng Yu[1(✉)], and Hong Bian[2]

[1] College of Mathematics and System Sciences, Xinjiang University, Urumqi, China
yuhaizheng@xju.edu.cn
[2] School of Mathematical Sciences, Xinjiang Normal University, Urumqi, China

Abstract. Graph Neural Networks (GNNs) have attracted extensive interest in the world because of its superior performance in the field of graph representation learning. Most GNNs have a message passing mechanism to update node representations by aggregating and transforming input from node neighbors. The current methods use the same strategy to aggregate information from each feature dimension. However, according to current papers, the model will be more practical if the feature information of each dimension can be treated differently throughout the aggregating process. In this paper, we introduces a novel Graph Neural Network-Graph Multi-Dimensional Feature Network (GMDFN). The method is accomplished by mining feature information from diverse dimensions and aggregating information using various strategies. Furthermore, a self-supervised learning module is built to keep the node feature information from being destroyed too much in the aggregation process to avoid over-smoothing. A large number of experiments on different real-world datasets have shown that the model outperforms various current GNN models and is more robust.

Keywords: Graph neural network · Multi-Dimensional node feature information · Self-Supervised learning

1 Introduction

Graph is a ubiquitous form in a variety of applications such as social networks [25], bioscience [20], knowledge graph [21] and traffic networks [1], etc. Graph Neural Networks (GNNs) are designed to learn effective representations of graphs and witnessed fabulous success in representation learning of graphs [24,29]. The main principle of GNNs is the neural message passing mechanism, where for each given node, its node representation is obtained by aggregating and transforming the representations of its neighbors [26]. Though the message passing mechanism plays a significant role in GNNs, it uses the exact same strategy to aggregate information from each feature dimension. The same strategy adopted in aggregation process leads GNNs tend to smooth the node features [7]. This aggregation operation will reduce the overall node feature difference, indicating most GNNs suffer from over-smoothing and destroying the original node feature [15,23].

© The Author(s), under exclusive license to Springer Nature Singapore Pte Ltd. 2024
B. Luo et al. (Eds.): ICONIP 2023, CCIS 1961, pp. 149–160, 2024.
https://doi.org/10.1007/978-981-99-8126-7_12

Over-smoothing can reduce the effectiveness of the learned representations and hinder the performance of downstream tasks. First, when applying multiple graph convolution layers, GNNs can suffer over-smoothing problem that the learned node embeddings become totally indistinguishable [9,11]. Second, according to the social dimension theory, the real-world social network are not homogeneous since different people are connected due to diverse reasons [18]. Based on the same extraction strategy in each dimension adopted in aggregation process, GNNs will lose the practicability in social network applications.

In this work, we aim to design a new graph neural network model that can differently use each dimension feature and preserve the original node feature. Inspired by the general framework recently introduced in [13], we propose a new aggregation scheme which enables feature dimensions to contribute differently to the aggregation process. Meanwhile, we employ a self-supervised learning strategy to predict the pairwise feature similarity from the hidden representations of given node pairs to protect the original node feature. Our proposed model, Graph Multi-Dimensional Feature Network (GMDFN), combining the above two components can enable feature dimensions to contribute differently to the aggregation process and preserve the original node features, achieving state-of-the-art performance on a wide range of benchmark datasets.

2   Related Work

GNNs generalize convolutional neural networks (CNN) to graph structure data through the message passing framework [4,16]. The design of message passing and GNN architectures are majorly motivated in spectral domain [3,8] and spatial domain [16,19]. Spectral based methods learn node representations based on graph spectral theory [17]. Bruna et al. [2] first generalize convolution operation to non-grid structures from spectral domain by using the graph Laplacian matrix. Following this work, ChebNet [3] utilizes Chebyshev polynomials to modulate the graph Fourier coefficients and simplify the convolution operation. The ChebNet is further simplified to GCN [8] by setting the order of the polynomial to 1 together with other approximations. On the other hand, spatial-based GNNs have been proposed before the spectral-based ones, which dates back to the time when deep learning was not yet popular. However, its development has stagnated since then until the emergence of GCN, which is a simplied spectral-based model and also can be treated as a spatial-based model. More spatial-based GNNs have since been developed [5,19]. The spatial-based graph convolution can be understood as propagating features across the graph, which is simple yet effective in various applications [24]. Furthermore, there are some advanced topics in GNNs such as deep graph neural networks [11], self-supervised graph neural networks [12,28] and robust graph neural networks [7,27].

Most of existing works utilize a unified way to propagate features for all feature dimensions, which may not be optimal since different dimensions of features may represent different accept of nodes and thus have different smoothness. This way can apparently lead GNNs to suffer from over-smoothing, which means most GNNs will not be applicable in some realistic applications [14]. Furthermore, over-smoothing also damage the original node feature, which makes most existing GNNs vulnerable [7]. Hence, in this paper, we try to address this issue by designing a new aggregation process to control smoothness for each dimension and employing a self-supervised task to preserve the original node feature.

3   Preliminary Study

Let $\mathcal{G} = (\mathcal{V}, \mathcal{E})$ be a graph, where $\mathcal{V} = \{v_1, v_2, \dots, v_N\}$ is the set of nodes and $\mathcal{E}$ is the set of edges. The edges are used to describe the relations between nodes, which can be also represented by an adjacency matrix $\mathbf{A} \in \mathbb{R}^{N \times N}$ where $\mathbf{A}_{ij}$ indicates existence of edge between nodes v_i and v_j. In addition, we use $\mathbf{X} = \{x_1, x_2, \dots, x_N\} \in \mathbb{R}^{N \times D}$ to indicate the node feature matrix with dimension D where x_i is the feature vector of the node v_i. Thus a graph can also be denoted as $\mathcal{G} = (\mathbf{A}, \mathbf{X})$. In the setting of semi-supervised node classification, a subset of nodes $\mathcal{V}_L \subset \mathcal{V}$ are associated with corresponding labels $\mathcal{Y}_L$. The goal of node classification is to infer the labels for the unlabeled data by learning a mapping function. Its objective can be summarized as follows,

$$\min_{\theta} \sum_{v_i \in \mathcal{V}_L} \ell\left(\hat{y}_i, y_i\right), \tag{1}$$

where $\hat{y}_i$ and y_i denote the output of the mapping function and true label for node v_i respectively, and $\ell(\cdot, \cdot)$ denotes the loss function.

The mapping function is often implemented as a graph neural network. Most graph neural networks follow a message passing scheme where the node representation is obtained by aggregating the representation of its neighbors and updating its own representation. We use $\mathbf{H}^{(l)}$ to denote the node hidden representation matrix at the l-th layer. Further, the operation in the l-th graph neural network layer can be described as,

$$\mathbf{H}_i^{(l)} = \text{Transform}\left(\text{Aggregate}\left(\mathbf{H}_j^{(l-1)} \mid v_j \in \mathcal{N}\left(v_i\right) \cup v_i\right)\right), \tag{2}$$

where $\mathbf{H}_i^{(l)}$ denotes the l-th layer hidden representation of node v_i. Then the predicted probability distribution of a k-layer graph neural network can be formulated as,

$$\hat{y}_{v_i} = \text{softmax}\left(\mathbf{H}_i^{(k)}\right). \tag{3}$$

GCN implements the message passing scheme as

$$\mathbf{H}^{(l)} = \sigma\left(\tilde{\mathbf{D}}^{-1/2}\tilde{\mathbf{A}}\tilde{\mathbf{D}}^{-1/2}\mathbf{H}^{(l-1)}\mathbf{W}^{(l)}\right), \tag{4}$$

where σ denotes the activation function such as ReLU, $\mathbf{A} + \mathbf{I}$ and $\tilde{\mathbf{D}}$ is the diagonal matrix of $\tilde{\mathbf{A}}$ with $\tilde{\mathbf{D}}_{ii} = 1 + \sum_j \mathbf{A}_{ij}$ and $\mathbf{W}$ is the parameter matrix for feature transformation.

4   The Proposed Framewoek

4.1   Multi-dimension Feature Mining Aggregation

We propose to design the GMDFN based on the graph signal denoising problem [13]. Assume that we are given a noisy graph signal $\mathbf{x} = \mathbf{f} + \eta$, where $\mathbf{f}$ is the clean graph signal and η is Gaussian noise. Note that the graph signal is single-channel, i.e., $\mathbf{x} \in \mathbb{R}^{N \times 1}$, with the i-th element corresponding to node v_i. The goal of graph signal denoising is to recover $\mathbf{f}$ from $\mathbf{x}$. Since the clean signal $\mathbf{f}$ is often assumed to be smooth with respect to the underlying graph $\mathcal{G}$, we can adopt the following objective function to recover the clean signal:

$$\min_{\mathbf{f}} g(\mathbf{f}) = \|\mathbf{f} - \mathbf{x}\|^2 + \theta \mathbf{f}^\top \mathbf{L} \mathbf{f}, \tag{5}$$

where θ denotes the smoothing coefficient controlling how smooth we want the graph signal to be, and $\mathbf{L}$ is the Laplacian matrix of the underlying graph $\mathcal{G}$. If we adopt $\mathbf{L}$, Eq. (5) can be rewritten as,

$$\min_{\mathbf{f}} g(\mathbf{f}) = \|\mathbf{f} - \mathbf{x}\|^2 + \frac{\theta}{2} \sum_{v_i \in \mathcal{V}} \sum_{v_j \in \mathcal{N}(v_i)} \left(\frac{\mathbf{f}_i}{\sqrt{d_i}} - \frac{\mathbf{f}_j}{\sqrt{d_j}} \right)^2, \tag{6}$$

where $\mathcal{N}(v_i)$ denotes the neighbors of node v_i. It is clear that the second term in Eq. (6) is small when connected nodes share similar signals, which can indicate the smoothness of $\mathbf{f}$.

In this work, we aim to give different smoothness coefficients to each node pair (v_i, v_j). Based on Eq. (6), we assign a smoothness coefficient θ_{ij} to each connected node pair (v_i, v_j) as:

$$\min_{\mathbf{f}} g(\mathbf{f}) = \|\mathbf{f} - \mathbf{x}\|^2 + \frac{1}{2} \sum_{v_i \in \mathcal{V}} \sum_{v_j \in \mathcal{N}(v_i)} \theta_{ij} \left(\frac{\mathbf{f}_i}{\sqrt{d_i}} - \frac{\mathbf{f}_j}{\sqrt{d_j}} \right)^2. \tag{7}$$

Extending Eq. (7) with multi-dimensional features, we design multi-dimension feature mining aggregation as:

$$\mathbf{H}_i^{(l)} = \sigma \left(\left(1 - \sum_{v_j \in \mathcal{N}(v_i)} \frac{\eta_{ij}}{d_i} \right) \odot \mathbf{H}_i^{(l-1)} \mathbf{W}^{(l)} + \sum_{v_j \in \mathcal{N}(v_i)} \mathbf{s}_{ij} \odot \frac{\mathbf{H}_j^{(l-1)} \mathbf{W}^{(l)}}{\sqrt{d_i d_j}} \right). \tag{8}$$

where $\mathbf{s}_{ij}$ is the smoothness vector for node pair (v_i, v_j) over all feature dimensions. We model $\mathbf{s}_{ij}$ as follows,

$$\mathbf{s}_{ij} = \lambda \cdot \text{sigmoid} \left(\left(\mathbf{H}_i^{(l-1)} \mathbf{W}^{(l)} \| \mathbf{H}_j^{(l-1)} \mathbf{W}^{(l)} \right) \mathbf{W}_s^{(l)} \right). \tag{9}$$

We can denote the output of the last layer as $\hat{\mathbf{H}}^{(L)}$, then the classification loss is shown as,

$$\mathcal{L}_{\text{class}} = \frac{1}{|\mathcal{D}_L|} \sum_{(v_i, y_i) \in \mathcal{D}_L} \ell \left(\text{softmax} \left(\hat{\mathbf{H}}^{(L)} \right), y_i \right), \tag{10}$$

where $\mathcal{D}_L$ is the set of labeled nodes, y_i is the label of node v_i and $\ell(\cdot,\cdot)$ is the loss function to measure the difference between predictions and true labels such as cross entropy.

4.2 Self-supervised Learning

Self-Supervised learning first designs a domain specific pretext task to assign constructed labels for nodes and then trains the model on the pretext task to learn better node representations. Following the joint training manner described in [6], we design a contrastive pretext task where the self-supervised component is asked to predict pairwise feature similarity. In detail, we first calculate pairwise similarity for each node pair and sample node pairs to generate self-supervised training samples. Specifically, for each node, we sample its m most similar nodes and m most dissimilar nodes. Then the self-supervised loss can be stated as,

$$\mathcal{L}_{\text{self}}(\mathbf{A}, \mathbf{X}) = \frac{1}{|\mathcal{T}|} \sum_{(v_i, v_j) \in \mathcal{T}} \left\| f_w \left(\mathbf{H}_i^{(l)} - \mathbf{H}_j^{(l)} \right) - \mathbf{S}_{ij} \right\|^2, \tag{11}$$

where $\mathcal{T}$ is the set of sampled node pairs, f_w is a linear mapping function, and $\mathbf{H}_i^{(l)}$ is the hidden representation of node v_i at l-th layer. $\mathbf{S}_{ij}$ indicates the feature similarity between node v_i and v_j:

$$\mathbf{S}_{ij} = \frac{\mathbf{x}_i^\top \mathbf{x}_j}{\|\mathbf{x}_i\| \|\mathbf{x}_j\|}. \tag{12}$$

4.3 Objective Function

The overall objective function can be stated as,

$$\min \mathcal{L} = \mathcal{L}_{\text{class}} + \lambda \mathcal{L}_{\text{self}}, \tag{13}$$

where λ is a hyper-parameter that controls the contribution of self-supervised loss.

5 Experiments

5.1 Experimental Settings

Datasets. GMDFN will test six public data sets in this experiment. These six datasets consist of three citation network datasets (Cora, Citeseer, and Cora_ML), two social network datasets (Blogcatalog and Flickers), and one web dataset (Texas). The first three datasets are assortative graph datasets and the last three are disassortative graph datasets. Table 1 displays the details of these datasets.

Table 1. Basic statistical information of the datasets.

	Assortative			Disassortative		
	Cora	Citeseer	Cora_ml	Texas	Blogcatalog	Flicker
Nodes	2485	2120	2810	183	5196	7575
Edgss	10138	7385	15692	309	171743	239738
Features	1,433	3,703	2478	1703	8189	12047
Labels	7	7	6	5	6	9

Baselines. We compare our model with the state-of-the-art GNNs and defense models to evaluate the effectiveness. The following methods including GCN [8], GAT [19], RGCN [30], SGC [22] and SimP-GCN [7] are implemented by the pytorch adversarial learning library [10].

Parameters Settings. We randomly split the dataset from each graph as: 10% training, 10%validation and 80% test. For each experiment, we record the average performance of 20 runs. The hyper-parameters of all the models are tuned based on the loss and accuracy on validation set.

5.2 Node Classification Performance

Table 2. The node classification performance on assortative graphs.

Datasets	GCN	GAT	RGCN	SGC	SimP-GCN	GMDFN
Cora	**84.14**	82.95	83.56	82.26	83.42	84.04
Cora_ml	80.98	80.52	81.56	81.17	80.87	**83.02**
Citeseer	71.95	72.09	71.83	70.86	74.28	**75.51**

Table 3. The node classification performance on disassortative graphs.

Datasets	GCN	GAT	RGCN	SGC	SimP-GCN	GMDFN
Texas	71.14	71.95	67.34	71.35	78.48	**80.09**
Blogcatalog	77.95	77.09	73.42	75.54	74.32	**83.14**
Flicker	66.98	61.52	71.98	50.25	71.73	**73.26**

To validate the GMDFN node classification performance, the GMDFN and the aforementioned baseline models are performed on the two categories datasets. The following outcomes can be achieved from the Table 2 and Table 3:

- The tables shows that GMDFN outperforms other models in terms of node classification rates on all five datasets except Cora. It should also be highlighted that GMDFN outperforms GCN on other datasets. On the Texas and Blogcatalog, for example, the model achieves in this paper outperformes GCN by 8.95% and 5.19%, respectively.
- On the disassortative graph datasets, the feature similarity between node pairs is lower than that on assortative graph datasets, so GCN, GAT and SGC learn poor representations because of the aggregation mechanism. On the other hand, GMDFN can not only learn new representations from different feature dimensions, but also preserve the feature information of nodes through self-supervised learning.
- Although the performance improvement of GMDFN is not obvious on the assortative datasets, and the performance of GMDFN on Cora is a little worse than that of GCN, it can be concluded that this method will not damage the original performance of graph neural networks, and its performance is comparable with the current great model.

5.3   Defense Performance

GNNs are highly susceptible to adversarial perturbations. This result will cause perturbations in the traditional GNN aggregation process, ultimately affecting the performance of the model. Based on this, we set up two attack methods: non-targeted adversarial attack and targeted adversarial attack, and verify the defense performance of the proposed method through the performance of each model under these two settings.

Against Non-targeted Adversarial Attack. To evaluate the node classification performance of these different methods against non-targeted adversarial attack, we use metattack and keep all the default parameter settings in the authors' original implementation. The perturbation rate is varied from 0 to 25% with a step size of 5%. All the experiments are shown in Fig. 1. The best accuracy on node classification is highlighted in bold. From the Fig. 1, several observations are derived as follows:

- From the results on the two datasets, it can be concluded that the performance of GMDFN is slightly lower than that of other models only when both datasets are not attacked by metattack. This suggests that, under typical conditions, GMDFN node classification performance is comparable to that of existing mainstream models. In other cases, the node classification performance of GMDFN is superior to other models. For example, when the perturbation rate is 5%, GMDFN outperforms other models on both datasets Moreover, this trend of leading performance will increase with the increase of perturbation rate. For example, when the perturbation rate is 25%, GMDFN has a node classification rate higher than GCN on both datasets by more than 15%. However, when the perturbation rate is 5%, GMDFN performance is only slightly higher than GCN.

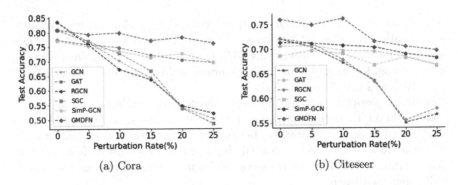

Fig. 1. Node classification performance under metattack.

– Compared with the node classification rate of SimP-GCN, regardless of the perturbation rate, the node classification rate of GMDFN is higher than that of SimP-GCN. This is because although SimP-GCN also constructs a unique aggregation mechanism, it mainly focuses on protecting node feature similarity, and is not as effective in mining information as GMDFN in exploring multi-dimensional information. In addition, the performance of GMDFN on these two datasets is also relatively stable, unlike other models where performance sharply decreases with increasing perturbation rates.

Against Targeted Adversarial Attack. In this experiment, we adopt nettack as the targeted attack method and keep its default parameter settings in the its original implementation. The perturbation number of each targeted node is set from 1 to 5 with a step size of 1. Nodes in test set whose degree larger than 10 are set as target nodes. The node classification accuracy on target nodes is shown in Fig. 2. According to the figure, we can observe that in the majority of cases, regardless of the number of perturbed nodes, GMDFN outperforms other methods. For example, on the Citeseer where each target node is affected by 1 targeted node, the node classification rate of GMDFN is improved by about 3% compared to GCN, and it is also better than other baseline models. On the Cora where each target node is affected by 5 targeted node, the node classification rate of GMDFN is improved by about 17% compared to GCN. Furthermore, GMDFN demonstrates very consistent variations in node classification rate curves under nettack, showing that GMDFN can effectively protect against this types of poisoning attack.

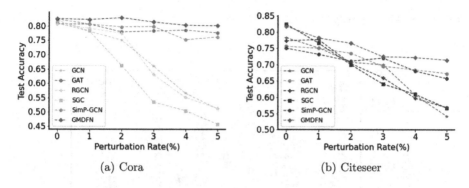

(a) Cora   (b) Citeseer

Fig. 2. Node classification performance under nettack.

Table 4. The performance of over-smoothing on Cora_ml.

Datasets	propagating steps	GCN	GAT	SGC	GMDFN
Cora_ml	2	0.81	**0.83**	0.80	**0.83**
	4	0.47	0.78	0.41	**0.82**
	6	0.33	0.67	0.28	**0.74**
	8	0.21	0.33	0.22	**0.70**
	10	0.23	0.25	0.19	**0.71**

5.4   Over-Smoothing Analysis

Most graph neural networks have to face the over-smoothing problem. When the number of steps of propagation becomes large, the node-pairs with different labels become indistinguishable and lead to over-smoothing.

Table 4 shows the node classification results of GMDFN at different propagation steps on Cora_ml. According to the table, the node classification rate of GCN, GAT and SGC decreased significantly from 0.8 to 0.2 with the increase of propagation steps, and the performance of the model was almost completely destroyed. The result is caused by over-smoothing. However, the performance of GMDFN is completely different from that of the above models, and the node classification rate of this model does not decrease much. This is because the feature information obtained from the model aggregation process in this paper comes from different dimensions of each node, instead of using the same strategy to aggregate the dimension node information as other models do. Secondly, the self-supervised learning module of GMDFN also protects the node feature information from over-smoothing by too much propagation process. Therefore, compared with the existing representative graph neural networks, GMDFN performs better and is less affected by the over-smoothing problem.

5.5 Ablation Study

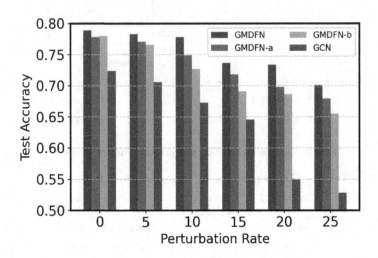

Fig. 3. Results of parameter analysis on Citeseer.

There are two variants of the model: GMDFN-a and GMDFN-b, which corresponds to multi-dimensional feature mining aggregation module without self-supervised learning module and self-supervised learning module without multi-dimensional feature mining aggregation module. The experiment is carried out on the Citeseer and Metattack is chosen as the adversarial attack. The results are shown in Fig. 3. According to Fig. 3, it can be concluded that the node classification rate of the variant models of GMDFN and GMDFN are higher than that of GCN at various perturbation rates. This further shows that the performance of GMDFN is better than GCN. It can also be observed that the node classification rate of GMDFN-a is always slightly better than that of GMDFN-b, indicating that the multi-dimensional feature mining aggregation module is better than the self-supervised learning module. From the above observation results, we can see that different modules all play a important role in the defense against poisoning attacks, and both modules are significant to GMDFN.

6 Conclusion

Graph neural networks are powerful tools in representation learning for graphs. The majority of GNN models adopt same strategy to aggregate information from each feature dimension, which not only causes the model to be over-smoothing but also causes it to lack robustness. Therefore, a novel graph neural network model called GMDFN is proposed in this paper. It consists of a module for multi-dimensional feature mining aggregation and a module for self-supervised learning. Each feature dimension information can be processed differently by the multi-dimensional feature mining aggregation module to get a better representation. The self-supervised learning module safeguards node feature information by preserving node feature information similarity. Extensive experiments demonstrate that GMDFN outperforms representative baselines on a wide range of real-world datasets.

Acknowledgements. This work is supported in part by the National Natural Science Foundation of China (12361072), Xinjiang Natural Science Foundation (2021D01C078).

References

1. Bai, L., Yao, L., Kanhere, S.S., Wang, X., Liu, W., Yang, Z.: Spatio-temporal graph convolutional and recurrent networks for citywide passenger demand prediction. In: Proceedings of the 28th ACM International Conference on Information and Knowledge Management, pp. 2293–2296 (2019)
2. Bruna, J., Zaremba, W., Szlam, A., LeCun, Y.: Spectral networks and locally connected networks on graphs. arXiv preprint arXiv:1312.6203 (2013)
3. Defferrard, M., Bresson, X., Vandergheynst, P.: Convolutional neural networks on graphs with fast localized spectral filtering. Advances in neural information processing systems 29 (2016)
4. Gilmer, J., Schoenholz, S.S., Riley, P.F., Vinyals, O., Dahl, G.E.: Neural message passing for quantum chemistry. In: International Conference on Machine Learning, pp. 1263–1272. PMLR (2017)
5. Hamilton, W., Ying, Z., Leskovec, J.: Inductive representation learning on large graphs. Advances in neural information processing systems 30 (2017)
6. Jin, W., Derr, T., Liu, H., Wang, Y., Wang, S., Liu, Z., Tang, J.: Self-supervised learning on graphs: Deep insights and new direction. arXiv preprint arXiv:2006.10141 (2020)
7. Jin, W., Derr, T., Wang, Y., Ma, Y., Liu, Z., Tang, J.: Node similarity preserving graph convolutional networks. In: Proceedings of the 14th ACM International Conference on Web Search and Data Mining, pp. 148–156 (2021)
8. Kipf, T.N., Welling, M.: Semi-supervised classification with graph convolutional networks. arXiv preprint arXiv:1609.02907 (2016)
9. Li, Q., Han, Z., Wu, X.M.: Deeper insights into graph convolutional networks for semi-supervised learning. In: Proceedings of the AAAI Conference on Artificial Intelligence, vol. 32 (2018)
10. Li, Y., Jin, W., Xu, H., Tang, J.: Deeprobust: a platform for adversarial attacks and defenses. In: Proceedings of the AAAI Conference on Artificial Intelligence, vol. 35, pp. 16078–16080 (2021)
11. Liu, M., Gao, H., Ji, S.: Towards deeper graph neural networks. In: Proceedings of the 26th ACM SIGKDD International Conference on Knowledge Discovery & Data Mining, pp. 338–348 (2020)
12. Liu, Y., et al.: Graph self-supervised learning: a survey. IEEE Trans. Knowl. Data Eng. **35**(6), 5879–5900 (2022)
13. Ma, Y., Liu, X., Zhao, T., Liu, Y., Tang, J., Shah, N.: A unified view on graph neural networks as graph signal denoising. In: Proceedings of the 30th ACM International Conference on Information & Knowledge Management, pp. 1202–1211 (2021)

14. Oono, K., Suzuki, T.: Graph neural networks exponentially lose expressive power for node classification. arXiv preprint arXiv:1905.10947 (2019)

15. Rong, Y., Huang, W., Xu, T., Huang, J.: Dropedge: towards deep graph convolutional networks on node classification. arXiv preprint arXiv:1907.10903 (2019)

16. Scarselli, F., Gori, M., Tsoi, A.C., Hagenbuchner, M., Monfardini, G.: The graph neural network model. IEEE Trans. Neural Networks **20**(1), 61–80 (2008)

17. Shuman, D.I., Narang, S.K., Frossard, P., Ortega, A., Vandergheynst, P.: The emerging field of signal processing on graphs: extending high-dimensional data analysis to networks and other irregular domains. IEEE Signal Process. Mag. **30**(3), 83–98 (2013)

18. Tang, L., Liu, H.: Scalable learning of collective behavior based on sparse social dimensions. In: Proceedings of the 18th ACM Conference on Information and Knowledge Management, pp. 1107–1116 (2009)

19. Veličković, P., Cucurull, G., Casanova, A., Romero, A., Lio, P., Bengio, Y.: Graph attention networks. arXiv preprint arXiv:1710.10903 (2017)

20. Wang, X., Flannery, S.T., Kihara, D.: Protein docking model evaluation by graph neural networks. Front. Mol. Biosci. **8**, 647915 (2021)

21. Wang, Y., Liu, Z., Fan, Z., Sun, L., Yu, P.S.: Dskreg: differentiable sampling on knowledge graph for recommendation with relational gnn. In: Proceedings of the 30th ACM International Conference on Information & Knowledge Management, pp. 3513–3517 (2021)

22. Wu, F., Souza, A., Zhang, T., Fifty, C., Yu, T., Weinberger, K.: Simplifying graph convolutional networks. In: International Conference on Machine Learning, pp. 6861–6871. PMLR (2019)

23. Wu, H., Wang, C., Tyshetskiy, Y., Docherty, A., Lu, K., Zhu, L.: Adversarial examples on graph data: Deep insights into attack and defense. arXiv preprint arXiv:1903.01610 (2019)

24. Wu, Z., Pan, S., Chen, F., Long, G., Zhang, C., Philip, S.Y.: A comprehensive survey on graph neural networks. IEEE Trans. Neural Networks Learn. Syst. **32**(1), 4–24 (2020)

25. Xiao, Y., Pei, Q., Xiao, T., Yao, L., Liu, H.: Mutualrec: joint friend and item recommendations with mutualistic attentional graph neural networks. J. Netw. Comput. Appl. **177**, 102954 (2021)

26. Xu, K., Hu, W., Leskovec, J., Jegelka, S.: How powerful are graph neural networks? arXiv preprint arXiv:1810.00826 (2018)

27. Yao, M., Yu, H., Bian, H.: Defending against adversarial attacks on graph neural networks via similarity property. AI Commun. **36**(1), 27–39 (2023)

28. You, Y., Chen, T., Wang, Z., Shen, Y.: When does self-supervision help graph convolutional networks? In: International Conference on Machine Learning, pp. 10871–10880. PMLR (2020)

29. Zhang, Z., Cui, P., Zhu, W.: Deep learning on graphs: a survey. IEEE Trans. Knowl. Data Eng. **34**(1), 249–270 (2020)

30. Zhu, D., Zhang, Z., Cui, P., Zhu, W.: Robust graph convolutional networks against adversarial attacks. In: Proceedings of the 25th ACM SIGKDD International Conference on Knowledge Discovery & Data Mining, pp. 1399–1407 (2019)

CBDN: A Chinese Short-Text Classification Model Based on Chinese BERT and Fused Deep Neural Networks

Yiyun Xing[1,2,3], Qin Lu[1,2,3]([✉]), and Kaili Zhou[1,2,3]

[1] Key Laboratory of Computing Power Network and Information Security, Ministry of Education, Shandong Computer Science Center, Qilu University of Technology (Shandong Academy of Sciences), Jinan, China
luqin@qlu.edu.cn
[2] Shandong Engineering Research Center of Big Data Applied Technology, Faculty of Computer Science and Technology, Qilu University of Technology (Shandong Academy of Sciences), Jinan, China
[3] Shandong Provincial Key Laboratory of Computer Networks, Shandong Fundamental Research Center for Computer Science, Jinan, China

Abstract. To address the common issues in Chinese short-text classification caused by the lack of contextual information, ambiguity, and sparsity of semantic features due to the short length of the text, a feature fusion-based Chinese short-text classification model CBDN is proposed. Firstly, the Chinese-BERT-wwm pre-trained model, improved by the full-word masking technique, is selected as the embedding layer to output the vector representation of the short text. Secondly, to fully extract the limited semantic features of the short text, the model employs a multi-head self-attention module and a long connected bidirectional LSTM (LC-BiLSTM) network to further learn the semantic features, and then fuses the hidden layer output vector with the feature vector further processed by these two methods. Finally, to improve the classification performance, the fused features are input into an improved "pyramid CNN" (PCNN) layer, and the short-text classification result is obtained through the classifier. The CBDN model is experimentally compared with various baseline models on the THUCNews dataset. The experimental results show that the proposed model achieves an accuracy and precision of 94.38% and 94.37%, respectively, outperforming other baseline models, indicating that the model better extracts the semantic information of short text and effectively improves the classification performance of Chinese short text.

Keywords: Text classification · Feature fusion · Chinese short text · Pre-trained model

Supported by Shandong Province Key R&D Program (Major Science and Technology Innovation Project) Project under Grants 2020CXGC010102.

© The Author(s), under exclusive license to Springer Nature Singapore Pte Ltd. 2024
B. Luo et al. (Eds.): ICONIP 2023, CCIS 1961, pp. 161–173, 2024.
https://doi.org/10.1007/978-981-99-8126-7_13

1   Introduction

With the rapid development of the Internet, a large amount of short-text data emerges every second. For example, comments on social media, microblogs and tweets, news headlines, as well as product descriptions and user reviews in the e-commerce field. The rapid and accurate classification and categorization of these short texts, which contain rich information, can greatly improve the efficiency and accuracy of information processing and decision-making. Therefore, the importance and necessity of short-text classification cannot be ignored, attracting a lot of attention and research from researchers.

While short-text classification has broad application prospects, it faces several challenges and difficulties due to the characteristics of short texts themselves.

Firstly, short texts are often short in length and do not provide sufficient contextual information, which makes text classification more difficult. For example, a comment containing only a few words may not accurately convey the user's sentiment tendency. To address this problem, Zeng et al. [1] proposed a topic memory network for short text classification, using a novel topic memory mechanism to encode potential topic representations associated with category labels. Chen et al. [2] proposed retrieving knowledge from external knowledge sources to enhance the semantic representation of short texts. Kim [3] proposed a TextCNN network model that applies convolutional neural networks to the field of natural language processing, capturing as much feature information of text as possible by using multiple convolutional kernels of different sizes. The above model uses a traditional word2vec model [4], which cannot deal well with the noise of short Chinese texts, such as abbreviations, spelling errors and lexical ambiguities.

With the development of deep learning techniques, the emergence of pre-trained models [5] has greatly addressed the noise-related issues that traditional models struggle with. Pre-trained models, trained on large-scale corpora through self-supervised learning, can capture richer language knowledge and patterns, enabling better understanding and handling of noise in Chinese short texts. Zhang et al. [6] designed a text classification method based on the BERT-Att-TextCNN model. BERT was used to obtain an accurate semantic representation of the text, and an attention mechanism was used to capture important information, calculate the weights of the word vectors to enhance the semantic representation, and finally the results were fed into a TextCNN network to improve the classification effect. Guo [7] proposed the idea based on KNN, using BERT to get the text vectors of documents, and then averaging them to get the central value of the text vector of each document for classification purpose. Ding et al. [8] proposed to use BERT to encode sliced clauses to obtain local semantic information, and then use BiLSTM to fuse the local semantic information and use attention mechanism to increase the weight of important clauses in long text to obtain text semantic features.

Although the aforementioned studies have achieved certain results in short-text classification tasks, they are designed based on the extraction of local text features and may not fully capture the global features of short texts.

This paragraph describes the improvements made in this study to address the limitations in current Chinese short-text classification tasks. A Chinese short-text classification model called CBDN (Chinese BERT and fused Deep neural Networks) is proposed. The main contributions of this study are as follows:

1) The study uses the Chinese-BERT-wwm pre-trained model, which performs better in the Chinese domain, as the text embedding model for Chinese short-text classification tasks. It replaces traditional static word embedding methods like Word2Vec and BERT-base-Chinese, which do not adhere to Chinese grammatical subword tokenization. By using a vector representation method that aligns better with Chinese linguistic features, the text features are more comprehensive.

2) Building upon the bi-directional long short-term memory (BiLSTM) network, a long connected bi-directional long short-term memory network is designed. In this network, the features outputted by all shallow layers of the deep BiLSTM network are further extracted, enhancing the model's ability to learn from sparse features.

3) The features extracted using different methods are fused and input into an improved "pyramid CNN" to strengthen the model's feature extraction capability.

2   CBDN

The CBDN Chinese short-text classification model proposed in this paper consists of three main components: the text embedding layer, the feature fusion module, and the deep semantic feature extraction module. This section will introduce the overall architecture of the model and the design details of each module. The overall structure of the model is illustrated in Fig. 1.

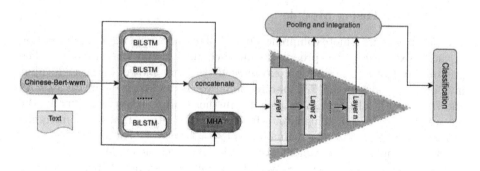

Fig. 1. CBDN model structure diagram.

2.1 Text Embedding Layer

The purpose of the text embedding layer is to represent text as computationally recognizable word vectors. In this study, we utilize a Chinese-BERT-wwm model, which is an improved version of BERT specifically designed for Chinese tasks, as the text embedding layer.

The original pre-trained BERT model was released by Google in 2018 [9]. In the task of sentiment analysis (SA), BERT achieved better performance on more complex and unfamiliar sentences compared to previous shallow, bidirectional word vector models [10]. Shortly after, Google introduced the BERT-Base-Chinese model, which was trained using a character-based tokenization approach for Chinese text. In 2021, Cui Yiming et al. [11] built the Chinese-BERT-wwm model based on the BERT-Base-Chinese model using Google's whole-word masking technique.

Under the whole-word masking approach, if a portion of a word is masked, the remaining parts of that word are also masked altogether. The handling of masking is similar to the original BERT, including replacing with the [MASK] label, keeping the original word, or randomly replacing the original word with another word. By applying whole-word masking, the model becomes more capable of capturing the boundary relationship between words.

Chinese-BERT-wwm takes Chinese Word Segmentation (CWS) into consideration, which was previously overlooked in the BERT-Base-Chinese model. During the training of the model, Harbin Institute of Technology's Language Technology Platform (LTP) was utilized as the Chinese-BERT-wwm's word segmentation tool. Chinese whole-word masking means that all corresponding Chinese characters that form an entire Chinese word will be masked. As a result, Chinese-BERT-wwm possesses more flexible and powerful text characterization capabilities for Chinese language.

2.2 Feature Fusion Module

In order to capture a more comprehensive set of features from limited information in short texts, this paper proposes a multi-channel feature fusion module. The preliminary feature information extracted by the text embedding layer is input into both the long connected bi-directional long short-term memory network and the multi-head attention module for further feature extraction. Finally, the output of the text embedding layer is fused with the features that have undergone further processing, allowing the model to extract richer global features.

By utilizing a multi-channel feature fusion module, the model is able to integrate diverse levels and perspectives of feature information, thereby capturing the global characteristics of short texts more effectively. This feature fusion approach enhances the expressive capacity and classification performance of the model.

LC-BiLSTM. The design of the Long Connected Bi-directional Long Short-Term Memory (LC-BiLSTM) layer aims to better extract global sequential information features from short texts. .

BiLSTM can encode text information in both forward and backward directions. In the process of short text classification, the short length of the text can lead to difficulties in feature extraction. By stacking single layers of LSTM in both forward and backward directions, the ability of LSTM to extract features from short texts can be enhanced to some extent. Furthermore, to deepen the BiLSTM network while ensuring that the limited features of short texts are not weakened or lost, it is proposed to directly connect the input of each BiLSTM layer with all the shallow network outputs of the current layer, thus achieving the reuse of shallow network output features. The structure of the Long Connected Bi-directional Long Short-Term Memory network is illustrated in Fig. 2.

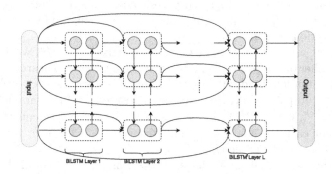

Fig. 2. LC-BiLSTM network structure.

Assuming that the network has L layers and the output dimension of each layer is h, according to the model structure, the network can be expressed as the following equation:

$$Inputlayer : x_{1:T} \tag{1}$$

$$1st layer : h_{1:T}^{(1)} = BiLSTM\left(x_{1:T}\right) \tag{2}$$

$$2nd layer : h_{1:T}^{(2)} = BiLSTM\left(\left[h_{1:T}^{(1)}, x_{1:T}\right]\right) \tag{3}$$

$$\cdots\cdots$$

$$Lth layer : h_{(1:T)}^{((L))} = BiLSTM\left(\left[h_{(1:T)}^{((L-1))}, h_{(1:T)}^{((L-2))}, \ldots, h_{(1:T)}^{((1))}, x_{(1:T)}\right]\right) \tag{4}$$

For layer L, the input $h_{1:T}^{(L-1)}, h_{1:T}^{(L-2)}, \ldots, h_{1:T}^{(1)}$ represents the forward and backward hidden states of the preceding $L-1$ layers of BiLSTM. $x_{1:T}$ represents the embedded representation of the original input sequence.

Specifically, for the forward LSTM, the input to the L-th layer is:

$$x_t^{(l, \text{ forward })} = \left[h_t^{(l-1, \text{ forward })}, h_t^{(l-2, \text{ backward })}, \ldots, h_t^{(1, \text{ forward })}, e_t \right] \quad (5)$$

For the backward LSTM, the input to the Lth layer is:

$$x_t^{(l, \text{ backward })} = \left[h_t^{(l-1, \text{ backward })}, h_t^{(l-2, \text{ forward })}, \ldots, h_t^{(1, \text{ backward })}, e_t \right] \quad (6)$$

Finally, the output of the Lth layer BiLSTM is $h_{1:T}^{(l)} = \left[h_1^{(l)}, h_2^{(l)}, \ldots, h_T^{(l)} \right]$, where $h_t^{(l)} = \left[h_t^{(l, \text{forward})}; h_t^{(l, \text{backward})} \right]$ represents the concatenated vector of the hidden states of the forward and backward LSTMs at time step t.

Multi-Head Attention. Multi-Head Attention(MHA) [12] is a technique used to enhance the focus of neural network models on input data. It divides the input data into multiple distinct subspaces, independently computes attention weights for each subspace, and then combines these computed results to obtain the final attention weights.

Specifically, assuming the input data is a matrix $X \in R^{n \times d}$, where n represents the number of samples and d represents the number of features per sample. Multi-Head Attention divides the matrix X into h subspaces, each with a size of d/h. For each subspace, Multi-Head Attention introduces three parameter matrices: $W_i^Q \in R^{d \times k}$, $W_i^K \in R^{d \times k}$, and $W_i^V \in R^{d \times k}$, where k represents the number of features for each subspace. For each subspace $i \in 1, 2, \ldots, h$, the attention weights are computed as follows:

1) For the input matrix X, linear transformations are applied using the parameter matrices W_i^Q, W_i^K, and W_i^V to obtain three matrices $Q_i = XW_i^Q$, $K_i = XW_i^K$, and $V_i = XW_i^V$, where $Q_i \in R^{n \times k}$, $K_i \in R^{n \times k}$, and $V_i \in R^{n \times k}$.

2) For matrices Q_i and K_i, the similarity matrix S_i between them is calculated using the dot product operation as $S_i = Q_i K_i^T \in R^{n \times n}$.

3) For the similarity matrix S_i, it is normalized using the $softmax$ function to obtain the attention matrix $A_i = \text{softmax}(S_i) \in R^{n \times n}$.

4) For matrices A_i and V_i, the final representation matrix $O_i = A_i V_i \in R^{n \times k}$ is calculated by taking the weighted average. The weight for each sample is determined by the corresponding row vector in the attention matrix A_i.

5) Finally, the h representation matrices O_i are concatenated along the feature dimension to obtain the final representation matrix $O \in R^{n \times hk}$, where $hk = d$.

Compared to traditional attention mechanisms, multi-head attention improves computational efficiency through parallel computation and can achieve better performance.

In the model proposed in this article, multi-head attention enhances the focus on important information in short texts by performing multi-head attention calculations on the output of the embedding layer. This improves the ability to capture important information in text sequences and increases the accuracy of short text classification.

2.3   Deep Semantic Feature Extraction Module

The reference [13], in its previous work, proposed that using a network model with a small number of hidden units in the deeper layers of a deep learning model can greatly assist in feature extraction. Based on this, we incorporated a "pyramid CNN" structure after feature fusion and designed a pooling fusion layer for feature compression, aiming to achieve improved classification performance.

The Pyramid Convolutional Neural Network (PCNN) extracts and represents features from input text at multiple scales. It consists of multiple CNN layers, with the shallowest layer having the most hidden units and the deepest layer having the fewest hidden units. To ensure stable data feature distribution and avoid overfitting caused by the deepening of the neural network, a LayerNorm layer is applied after each CNN layer to normalize the features across all samples.

Suppose the word vector matrix of the input text in the CNN layer is $X \in R^{m \times w}$, where m is the text length and w is the dimension of the word vector, and the output of the CNN in layer i is $H_i \in R^{m_i \times w}$, where m_i is the number of feature maps in the output of the CNN in layer i and w is the dimension of each feature map.

The output of each CNN layer is passed to the average pooling layer, and assuming that the output of the ith average pooling layer is P_i, we have:

$$P_i = \frac{1}{m_i} \sum_{j=1}^{m_i} H_{i,j} \qquad (7)$$

where $H_{i,j}$ denotes the jth feature map output by the ith layer CNN.

Concatenate the outputs of m pooling layers to obtain an mw-dimensional vector, denoted as:

$$P = [P_1, P_2, \dots, P_m]^{mw} \qquad (8)$$

Finally, the vector P is passed as input to the fully connected layer for classification, resulting in the classification results for short texts.

3   Experimental Settings

3.1   Dataset

In order to validate the effectiveness of the proposed model, the Chinese text classification dataset THUCNews [16] was used in the experiment. The original THUCNews dataset contains a total of 740,000 news documents, divided into 14 categories. To facilitate the validation experiment, the dataset was preprocessed. News texts from the original dataset were selected, including 10 categories: finance, real estate, stocks, education, science, society, politics, sports, gaming, and entertainment. Only the title texts of the news articles were extracted. To ensure that the news titles were short texts, titles with a character count of approximately 30 were selected as the final dataset for the experiment.

A total of 200,000 news headline data were obtained after pre-processing, with 20,000 data for each category, and a ratio of 7:1.5:1.5 was used to allocate the training set, validation set, and test set data. The details are shown in Table 1.

Table 1. THUCNews dataset settings.

Text Category	Training sets	Test Sets	Validation Sets
Finance	14,000	3,000	3,000
Realty	14,000	3,000	3,000
Stocks	14,000	3,000	3,000
Education	14,000	3,000	3,000
Science	14,000	3,000	3,000
Society	14,000	3,000	3,000
Politics	14,000	3,000	3,000
Sports	14,000	3,000	3,000
Game	14,000	3,000	3,000
Entertainment	14,000	3,000	3,000

3.2 Evaluation Metrics

Four evaluation metrics, Accuracy, Precision, Recall and F1-score, were used to comprehensively evaluate the performance of the CBDN model proposed in this paper. The evaluation indexes are calculated as shown in the following equations:

$$Acc = (TP + TN)/(TP + FP + TN + FN) \tag{9}$$

$$Pre = TP/(TP + FP) \tag{10}$$

$$Recall = TP/(TP + FN) \tag{11}$$

$$F1 = (2 * Pre * Recall)/(Pre + Recall) \tag{12}$$

3.3 Experimental Platform and Parameter Setting

The experiment was conducted using the Ubuntu 20.04 operating system. The PyTorch deep learning development framework version 1.12 was utilized, with Python as the programming language. The IDE used was PyCharm 2023.1 Professional. The GPU employed was the NVIDIA A100 SXM4 with 40GB of memory. Table 2 provides more detailed experimental parameter settings.

4 Experimental Results and Analysis

To ensure the reliability of the experimental results, three experiments were conducted for both the CBDN model and the other baseline models, each with a different random seed. The average values of the evaluation metrics for each model are reported in Table 3. All the baseline models mentioned in the reference section.

By analyzing the experimental results in Table 3, it can be observed that the CBDN model achieved the highest scores in all evaluation metrics for Chinese

Table 2. Experimental parameter settings.

Parameters	Values
Epochs	10
Batch size	64
Learning rate	1.6e-5
Pad size	32
hidden size	768
Rnn hidden	384
Convolution kernel size	(2, 3, 4)

Table 3. Experimental result.

Model	Accuracy %	Precision %	Recall %	F1-score %
BiLSTM+Att [14]	89.28	89.28	89.28	89.24
Transformer [15]	89.56	89.63	89.56	89.52
TextCNN [3]	89.74	89.79	89.74	89.75
Bert+CNN [16]	92.95	92.96	92.95	92.92
Bert+DPCNN [17]	93.06	93.11	93.06	93.07
ERNIE [18]	93.12	93.15	93.12	93.11
Bert+BiLSTM [19]	93.19	93.22	93.19	93.20
Bert [20]	93.19	93.24	93.19	93.20
Bert+RCNN [21]	93.35	93.33	93.35	93.32
CBDN	**94.38**	**94.37**	**94.38**	**94.37**

short text classification, significantly improving the effectiveness of short text classification.

In terms of accuracy, the CBDN model showed an improvement of 5.10% points compared to the traditional Word2Vec-based BiLSTM+Att model and a 4.64% point improvement compared to the TextCNN model. It can be observed that using pre-trained models for text embedding significantly enhances classification performance. This is because Word2Vec utilizes static vector representations, which may not effectively capture information such as polysemy and text sequences commonly found in Chinese. Although the BiLSTM model is capable of capturing contextual information, its effectiveness is limited. The Transformer model, despite employing a multi-layer Attention architecture to capture semantic information, also exhibited poor classification performance. This indicates that simply stacking layers in the network does not necessarily enhance classification capability. Similar conclusions can be drawn from the Bert+CNN, Bert+DPCNN, and Bert+BiLSTM models. Simply adding effective network structures after the Bert model does not necessarily yield better classification results than the Bert model itself. ERNIE, proposed by Baidu, is a pre-trained model specifically designed for Chinese natural language processing tasks and achieved similar results to Bert in our experiments.

To better demonstrate the effectiveness of CBDN in short text classification, we have generated the confusion matrix for the classification of short text titles from the THUCNews dataset in our experiments. The confusion matrix is shown in Fig. 3.

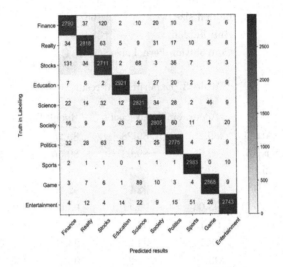

Fig. 3. Confusion matrix of classification results.

From the confusion matrix in Fig. 3, it can be observed that CBDN achieved the highest precision in classifying short texts in the education and sports domains, with accuracies of 2921/3000 and 2983/3000, respectively. The relatively poorer classification performance was mainly concentrated in the finance and stock domains. It is evident that there were many cases of mutual misclassification between finance and stock, indicating significant semantic similarity between the short text data in these domains. The presence of similar words with high feature weights in the data introduces considerable noise, resulting in suboptimal performance.

By comparing the aforementioned experimental evaluation indicators, it can be observed that the proposed CBDN model has achieved excellent results in the field of Chinese short text classification. It enhances the classification capability and effectively addresses the problem of poor classification performance caused by the sparsity of features in short text data.

5   Ablation Experiment

To verify the effectiveness of the LC-BiLSTM network module proposed in this paper for Chinese short text classification tasks and explore the influence of network depth on the experimental results, we conducted experiments by setting the network depth from 0 to 5 layers. Accuracy was used as the criterion, and the experimental comparative results are shown in Fig. 4.

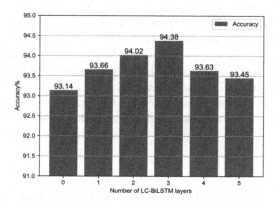

Fig. 4. Effect of LC-BiLSTM with different layers on experimental results.

It can be observed that incorporating the LC-BiLSTM into the feature fusion module effectively improves the accuracy of classification. When the network depth is set between 1 and 3 layers, the accuracy increases with the number of layers. However, if the network depth is further increased, a significant over-fitting phenomenon occurs. Therefore, our model ultimately utilizes a 3-layer LC-BiLSTM network.

6   Conclution

This paper addresses the issues of lack of contextual information, ambiguous semantic features, and sparsity in Chinese short text classification due to the length of the texts. To tackle these challenges, a Chinese short text classification model called CBDN is proposed, which leverages pre-trained models and feature fusion techniques. The core idea is to use the Chinese pre-trained model Chinese-BERT-wwm, based on the whole word masking technique, as the hidden layer. It combines innovative LC-BiLSTM networks, multi-head attention modules to further learn semantic features, and techniques such as feature fusion and pyramid CNN to effectively extract semantic information from short texts. The experimental results demonstrate that the proposed model outperforms other deep learning text classification models in terms of performance.

The main contribution of this study lies in the introduction of a novel model called CBDN specifically designed for Chinese short text classification. This model not only ensures high accuracy but also effectively utilizes the linguistic characteristics of the Chinese language to extract semantic information from the texts. Moreover, this paper proposes an innovative approach by introducing the concept of LC-BiLSTM networks. By directly connecting the input of each layer with the output of the shallow network, improved feature extraction performance is achieved. Additionally, this research incorporates various techniques for feature extraction and fusion, offering a new perspective and methodology for Chinese short text classification.

While the CBDN model demonstrates excellent performance in the experiments, there are still some issues that need to be further researched and addressed for its practical application. For instance, the Chinese-BERT-wwm pre-trained model used in this paper has a large number of parameters, resulting in high computational costs. Additionally, the noise level in the corpus also affects the stability of the model to some extent. Future work will focus on addressing these limitations and conducting further research to optimize the model accordingly.

References

1. Zeng, J., Li, J., Song, Y., Gao, C., Lyu, M.R., King, I.: Topic memory networks for short text classification. In: Proceedings of the 2018 Conference on Empirical Methods in Natural Language Processing. Brussels, Belgium: Association for Computational Linguistics, pp. 3120–3131 (2018)
2. Chen, J., Hu, Y., Liu, J., Xiao, Y., Jiang, H.: Deep short text classification with knowledge powered attention. Proc. AAAI Conf. Artif. Intell. **33**(01), 6252–6259 (2019)
3. Kim, Y.: Convolutional neural networks for sentence classification. In: Proceedings of the 2014 Conference on Empirical Methods in Natural Language Processing (EMNLP). Doha, Qatar: Association for Computational Linguistics, pp. 1746–1751 (2014)
4. Mohammed, M., Omar, N.: Question classification based on Bloom's taxonomy cognitive domain using modified TF-IDF and word2vec. PLoS ONE **15**(3), e0230442 (2020)
5. Qiu, X., Sun, T., Xu, Y., Shao, Y., Dai, N., Huang, X.: Pre-trained models for natural language processing: a survey. Sci. China Technol. Sci. **63**(10), 1872–1897 (2020)
6. Zhang, H., Shan, Y., Jiang, P., Cai, X.: A text classification method based on BERT-Att-TextCNN model. In: IEEE 5th Advanced Information Management, Communicates, Electronic and Automation Control Conference (IMCEC), Chongqing, China, pp. 1731–1735. IEEE (2022)
7. Guo, Z.: Forestry text classification based on BERT and KNN. In: 2022 International Conference on Information Technology, Communication Ecosystem and Management (ITCEM), Bangkok, Thailand, pp. 61–65. IEEE (2022)
8. Ding, W., Li, Y., Zhang, J., Shen, X.: Long text classification based on BERT. In: IEEE 5th Information Technology, Networking, Electronic and Automation Control Conference (ITNEC), Xi'an, China, pp. 1147–1151 (2021)
9. Devlin, J., Chang, M.-W., Lee, K., Toutanova, K.: BERT: pre-training of deep bidirectional transformers for language understanding. In: Proceedings of the 2019 Conference of the North American Chapter of the Association for Computational Linguistics: Human Language Technologies, Volume 1 (Long and Short Papers). Minneapolis, Minnesota: Association for Computational Linguistics, pp. 4171–4186 (2019)

10. Arora, S., May, A., Zhang, J., Rt'e, C.: Contextual embeddings: when are they worth it? In: Proceedings of the 58th Annual Meeting of the Association for Computational Linguistics. Online: Association for Computational Linguistics, pp. 2650–2663 (2020)

11. Cui, Y., Che, W., Liu, T., Qin, B., Yang, Z.: Pre-training with whole word masking for Chinese BERT. IEEE/ACM Trans. Audio, Speech Lang. Process. **29**, 3504–3514 (2021)

12. Lai, T., Cheng, L., Wang, D., Ye, H., Zhang, W.: RMAN: relational multi-head attention neural network for joint extraction of entities and relations. Appl. Intell. **52**(3), 3132–3142 (2022)

13. He, K., Zhang, X., Ren, S., Sun, J.: Deep residual learning for image recognition. In: IEEE Conference on Computer Vision and Pattern Recognition (CVPR) **2016**, 770–778 (2016)

14. Zhang, H., Shan, Y., Jiang, P., Cai, X.: A text classification method based on BERT-Att-TextCNN model. In: IEEE 5th Advanced Information Management, Communicates, Electronic and Automation Control Conference (IMCEC), Chongqing, China, pp. 1731–1735. IEEE (2022)

15. Kokab, S.T., Asghar, S., Naz, S.: Transformer-based deep learning models for the sentiment analysis of social media data. Array **14**, 100157 (2022)

16. Bello, A., Ng, S.-C., Leung, M.-F.: A BERT framework to sentiment analysis of tweets. Sensors **23**(1), 506 (2023)

17. Li, X., Ning, H.: Deep pyramid convolutional neural network integrated with self-attention mechanism and highway network for text classification. J. Phys. Conf. Ser. **1642**(1), 012008. IOP Publishing (2020)

18. Zhang, W.: Research on Chinese news text classification based on ERNIE model. In: Proceedings of the World Conference on Intelligent and 3-D Technologies (WCI3DT 2022) Methods, Algorithms and Applications, pp. 89–100. Springer (2023). https://doi.org/10.1007/978-981-19-7184-6_8

19. Li, X., Lei, Y., Ji, S.: BERT- and BiLSTM-based sentiment analysis of online Chinese buzzwords. Future Internet **14**(11), 332 (2022)

20. Pal, A., Rajanala, S., Phan, R.C.-W., Wong, K.: Self supervised BERT for legal text classification. In: ICASSP 2023–2023 IEEE International Conference on Acoustics, Speech and Signal Processing (ICASSP), pp. 1–5. IEEE (2023)

21. Nguyen, Q.T., Nguyen, T.L., Luong, N.H., Ngo, Q.H.: Fine-Tuning BERT for sentiment analysis of vietnamese reviews. In: 2020 7th NAFOSTED Conference on Information and Computer Science (NICS), Ho Chi Minh City, Vietnam, pp. 302–307. IEEE (2020)

Lead ASR Models to Generalize Better Using Approximated Bias-Variance Tradeoff

Fangyuan Wang[1]($\boxtimes$) (ID), Ming Hao[2], Yuhai Shi[2], and Bo Xu[1,3,4]

[1] Institute of Automation, Chinese Academy of Sciences, Beijing, China
{fangyuan.wang,xubo}@ia.ac.cn
[2] Academy of Broadcasting Science, National Radio and Television Administration, Beijing, China
{haoming,shiyuhai}@abs.ac.cn
[3] School of Future Technology, University of Chinese Academy of Sciences, Beijing, China
[4] School of Artificial Intelligence, University of Chinese Academy of Sciences, Beijing, China

Abstract. The conventional recipe for Automatic Speech Recognition (ASR) models is to 1) train multiple checkpoints on a training set while relying on a validation set to prevent over fitting using early stopping and 2) average several last checkpoints or that of the lowest validation losses to obtain the final model. In this paper, we rethink and update the early stopping and checkpoint averaging from the perspective of the bias-variance tradeoff. Theoretically, the *bias* and *variance* represent the fitness and variability of a model and the tradeoff of them determines the overall generalization error. But, it's impractical to evaluate them precisely. As an alternative, we take the training loss and validation loss as proxies of bias and variance and guide the early stopping and checkpoint averaging using their tradeoff, namely an Approximated Bias-Variance Tradeoff (ApproBiVT). When evaluating with advanced ASR models, our recipe provides 2.5%–3.7% and 3.1%–4.6% CER reduction on the AISHELL-1 and AISHELL-2, respectively (The code and sampled unaugmented training sets used in this paper will be public available on GitHub).

Keywords: Bias-Variance Tradeoff · Early Stopping · Checkpoint Averaging · Speech Recognition

1 Introduction

In the past ten years, neural network based End-to-End (E2E) ASR systems have achieved great progress and overwhelming success. Many advanced network

Supported by the National Innovation 2030 Major S&T Project of China under Grant 2020AAA0104202 and the Basic Research of the Academy of Broadcasting Science, NRTA, under Grant JBKY20230180.

© The Author(s), under exclusive license to Springer Nature Singapore Pte Ltd. 2024
B. Luo et al. (Eds.): ICONIP 2023, CCIS 1961, pp. 174–185, 2024.
https://doi.org/10.1007/978-981-99-8126-7_14

architectures have been proposed and applied in ASR tasks, such as Convolution Neural Networks (CNN) [1–3], Recurrent Neural Networks (RNN) [4,5], Self-Attention Networks (SAN) [6–8], and CNN/SAN hybrid networks [9–15].

Regardless of specific model architecture, most state-of-the-art ASR recipes involve two steps: 1) train multiple checkpoints on a training set while using a held-out validation set to decide when to stop training and 2) yield the final model by averaging several last checkpoints or that have the lowest losses on the held-out validation set. However, this conventional recipe has two downsides. For one, the validation loss based early stopping procedure [17–19] determines the stop point solely on the validation loss, which is more like a proxy of variance rather than the overall generalization error, without considering the bias explicitly, and may prevent the exploration of more checkpoints of lower bias-variance tradeoff scores. For another, existing checkpoint averaging procedures either prefer to pick checkpoints of lower biases (last ones in the same training trajectory, the last-k (LK) checkpoint averaging scheme) [6,20] or that of lower variances (with lower validation losses, the k-best validation loss (KBVL) checkpoint averaging scheme) [21] to average, discarding the tradeoff of bias and variance, which may limit the power of checkpoint averaging.

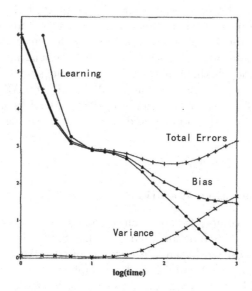

Fig. 1. Figure 15 in [16] visualizes the correlation between training loss (Learning), bias, validation loss (Total Errors), and variance. Details about this Figure are in [16].

To lead ASR models to generalize better, we find inspiration from the bias-variance tradeoff [16] and propose a reasonable and practical alternative to update the conventional recipe. 1) We argue that it's reasonable to consider the tradeoff of bias and variance. According to the definition of bias-variance decomposition [16], the generalization error consists of noise, bias, and variance. The *noise* is the response to the label error in the training set, the *bias* measures how closely the average evaluator matches the target, and the *variance* indicates how much the guess fluctuates for evaluators. Since the overall generalization error determines the final performance, any recipe only considering bias or variance cannot ensure the best generalization. Unfortunately, it's nontrivial to evaluate the bias and variance precisely using the Monte Carlo procedure [16]. 2) We propose a reasonable and practical method to approximate the bias-variance tradeoff. Figure 1 depicts the learning curves of training loss, bias, validation loss, and variance for a neural network. We observe that the training loss and bias have a positive correlation, while the trend of validation loss and variance at the end also tends to be positively correlated. This phenomenon encourages us to take the training loss and validation loss as the proxies of bias and variance and use their summary to approximate the tradeoff of bias and variance. In particular, we use a Sampled Unaugmented Training Loss (SUTL), instead of the naive training loss, as the proxy of bias to avoid using the total samples in the training set and the distortion of training loss introduced by the data augmentation [22] and regularization [23]. The core concepts are shown in Fig. 2 and Fig. 3.

We update the ASR recipe using the Approximated Bias-Variance Tradeoff from two aspects: 1) we use the ApproBiVT score to tell when the training process needs to stop, and our recipe tends to allow training longer until converged according to the ApproBiVT score; 2) we implement a k-best ApproBiVT (KBABVT) checkpoint averaging scheme to yield the final model. Experiments with various ASR models show that our recipe can outperform the conventional ones by 2.5%–3.7% on AISHELL-1 [24] and 3.1%–4.6% on AISHELL-2 [25]. To our knowledge, this is the first work that uses the training loss and validation loss together to guide the early stopping and checkpoint averaging in ASR recipes.

2 ApproBiVT Recipe

2.1 Bias-Variance Decomposition

The bias-variance decomposition of mean squared error is well understood. However, the prominent loss function used in E2E ASR is the cross-entropy. We present the decomposition for cross-entropy as in [26]. We denote the training set as D and the output distribution on a sample x of the network trained as $\hat{y} = f(x, D)$. Then let the average output of $\hat{y}$ be $\overline{y}$, that is,

$$\overline{y} = \frac{1}{Z} \exp(\mathbb{E}_D[\log \hat{y}]) \tag{1}$$

where Z is a normalization constant. According to [26], we have the following decomposition for the expected error on the sample x and $y = t(x)$ is the ground

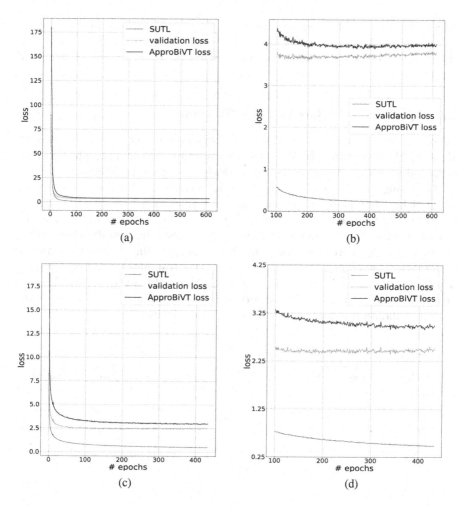

Fig. 2. The SUTL, validation loss, and ApproBiVT loss curves (overview and zoomed) of Conformer on AISHELL-1/-2 (a, b/c, d). The sampled unaugmented training loss (SUTL, shown in green) illustrates a monotonically decreasing trend. Whereas the validation loss (shown in red) and ApproBiVT loss (shown in blue) show a trend of falling first and then rising, the difference is that the ApproBiVT loss curve still decreases for many epochs after the validation loss has reached the turning point. (Color figure online)

truth label:

$$error = \mathbb{E}_{x,D}[-y \log \hat{y}]$$

$$= \mathbb{E}_{x,D}\left[-y \log y + y \log \frac{y}{\overline{y}} + y \log \frac{\overline{y}}{\hat{y}}\right]$$

$$= \mathbb{E}_x[-y \log y] + \mathbb{E}_x\left[y \log \frac{y}{\overline{y}}\right] + \mathbb{E}_D\left[\mathbb{E}_x\left[y \log \frac{\overline{y}}{\hat{y}}\right]\right] \qquad (2)$$

$$= \mathbb{E}_x[-y \log y] + D_{KL}(y, \overline{y}) + \mathbb{E}_D\left[D_{KL}(\overline{y}, \hat{y})\right]$$

$$= intrinsic\ noise + bias + variance$$

where, the *bias* is the divergence between $\bar{y}$ and y to measure how closely the average evaluator matches the target; the *variance* is the expected divergence between $\bar{y}$ and each guess $\hat{y}$ to measure how much the guess fluctuates for evaluators.

2.2   Approximated Bias-Variance Tradeoff

In general, it's nontrivial to evaluate the bias and variance precisely using the Monte Carlo procedure [16]. We need large training samples and split them into many independent sets, which is impracticable for most ASR tasks.

According to [16], see Fig. 1, each training iteration will decrease the training loss and bias but increase variance, and the validation loss starts to increase dramatically when the variance grows sharply. This encourages us to regard the validation loss as the proxy of variance rather than the overall generalization error. And also, there are some empirical evidences that networks generalize better if they are training longer [19, 27]. Since the training loss and bias decrease simultaneously in the training process, we suggest taking the training loss as a proxy of bias. Borrow the tradeoff format of Eq. 2, we take the sum of training loss and validation loss as the proxy of the generalization error.

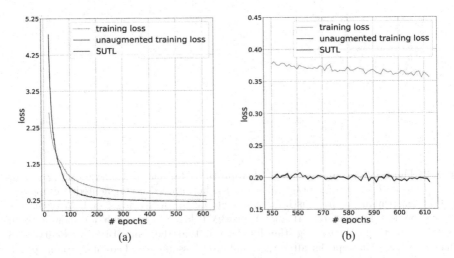

Fig. 3. The training loss, SUTL, and unaugmented training loss curves of Conformer on AISHELL-1, (a) overview (depicts from the 20th epoch), (b) zoomed. The divergence between the trend of unaugmented training loss (shown in black) and training loss (shown in green) is significant. In contrast, the trend of sampled unaugmented training loss (SUTL, shown in red) and unaugmented training loss is roughly the same. (Color figure online)

However, it may introduce significant "bias" if we take the naive training loss as the proxy of bias incurred by the difference between the training and validation

---

Algorithm 1. ApproBiVT recipe

Preparation: Construct a sampled unaugmented training set by randomly sampling from the training set.

Training:

  1: Train the ASR model iteratively, and save a checkpoint once finished an epoch.

  2: Evaluate each checkpoint on the sampled unaugmented training set and the validation set to get the SUTL and validation loss, respectively.

  3: Calculate the ApproBiVT loss according to Eq. 3.

  4: Conduct early stopping of Algorithm 2.

Inference:

  1: Conduct k-best ApproBiVT checkpoint averaging.

  2: Decode using the final model to generate the outputs.

---

processes. Let's revisit the overall training and validation processes. Typically, the conventional recipe divides the total samples into a training set, a validation set, and a test set. During training, several augmentation and regularization methods, such as speed perturbation, spectral augmentation [22], dropout [23] et al., are conducted to prevent over fitting. Whereas, no additional augmentation or regularization operations are executed in the validation process to ensure the evaluations of validation set and test set are under the same setting. To avoid the "bias" incurred by the augmentation and regularization methods, a natural solution is to take the training set as a validation set and reevaluate on each checkpoint. But it's inefficient as the size of a training set is typically much larger than that of a validation set. In our work, to be precise and efficient, we propose to use a randomly sampled training set that has the same size of validation set to conduct reevaluation on each checkpoint to get the SUTL as the proxy of bias. The loss curves of training loss, unaugmented training loss, and SUTL are illustrated in Fig. 3. Furthermore, we take the sum of SUTL and validation loss as the ApproBiVT:

$$ApproBiVT \triangleq SUTL + validation\ loss \tag{3}$$

The overall ApproBiVT recipe is shown in Algorithm 1.

2.3 ApproBiVT-Guided Early Stopping

We implement a simple ApproBiVT-guided early stopping procedure and stop the training when the ApproBiVT loss monotonically increases for S epochs, see algorithm 2. The stop point can be regarded as the inflection point of the ApproBiVT learning curve, which implies that real over fitting may occur.

2.4 k-Best ApproBiVT Checkpoint Averaging

To further release the power of checkpoint averaging, we introduce a k-best ApproBiVT checkpoint averaging procedure to average k checkpoints of the lowest ApproBiVT losses in parameter space. In contrast, the k-best validation loss based method prefers to select checkpoints of lower variances.

Algorithm 2. ApproBiVT-guided early stopping procedure

Input: $\{L_0, L_1, ...\}$: the ApproBiVT losses of checkpoints.
Parameter: S: the maximum number of epochs with monotonically increased ApproBiVT losses.
Output: the stop point E

1: Let $i = S$.
2: **while** True **do**
3:     **if** $L_i \geq L_{i-1}, L_{i+1} \geq L_i, ..., L_{i-S+1} \geq L_{i-S}$ **then**
4:         $E{=}i$, break.
5:     **else**
6:         $i{+}{+}$, continue.
7:     **end if**
8: **end while**
9: **return** E

---

3   Experiments

3.1   Dataset

We conduct most experiments on AISHELL-1 [24]. It has 150 h of the training set, 10 h of the dev set, and 5 h of the test set. To test on a large dataset, we verify our recipe on AISHELL-2 [25], which consists of 1000 h of the training set, 2 h of the dev set, and 4 h of the test set.

3.2   Experimental Setup

We implement our recipe using the WeNet 2.0 toolkit [21] and conduct with 2 NVIDIA GeForce RTX 3090 GPU cards. The inputs are 80-dim FBANK features with a 25-ms window and a 10-ms shift. SpecAug [22] with two frequency masks with maximum frequency mask (F=10) and two times masks with maximum time mask (T=50) is applied. The SpecSub [11] is used in U2++ [11] to augment data with Tmax, Tmin, and Nmax set to 30, 0, and 3, respectively. Two convolution subsampling layers with kernel size 3×3 and stride 2 are used in the front. We use 12 stacked layers of Conformer or an improved variant on the encoder side. The kernel size of the convolution layer for the Conformer and Blockformer [13] is 15, while 8 for the U2 [10] and U2++ [11] with casual convolution. On the decoder side, a CTC decoder and an Attention decoder of 6 Transformer blocks are used. The number of attention heads, attention dimension, and feed-forward dimension are set to 4, 256, and 2048, respectively. We use the Adam optimizer [28] with the Transformer schedule to train models until they converged. The value of S is 5 in Algorithm 2. For the other hyper-parameters, we just follow their default recipes.

3.3   Evaluated Models

We mainly verify our recipe on the Conformer [9] model. To test whether the gains obtained by the ApproBiVT recipe are additive with other techniques, we also evaluate our recipe with U2 [10], U2++ [11], and Blockformer [13].

3.4   Ablation Studies

Early Stop but When? First, we explore when stopping training can help to achieve better performance. We conduct the last-k checkpoint averaging using Conformer at various endpoints to detect when the training starts to show over fitting in the context of checkpoint averaging, see Fig. 4. We can observe that:

1) The over fitting seems to be postponed in the context of checkpoint averaging, especially when averaging more checkpoints.

2) Compared to the validation loss based early stopping suggested training 240 and 120 epochs, our recipe allows training more epochs until the ApproBiVT score starts to increase, see Fig. 2, thus facilitating to averaging more checkpoints. And if comparing Fig. 2 and Fig. 4, we can see that the stopping points suggested by our recipe are more reasonable than that of the validation loss based recipe in the context of checkpoint averaging.

Checkpoint Averaging but How? Second, we explore how conducting checkpoint averaging can achieve better performance. Table 1 lists the results of Conformer with different settings of early stopping and checkpoint averaging procedures. We can observe that:

1) With VL guided early stopping, models will converge with 240 epochs, while 613 with ApproBiVT loss guided procedure.

2) With the VL procedure, averaging more checkpoints can not bring additional performance gains.

3) With our early stopping, each averaging procedure can achieve better performance and averaging more checkpoints seems to hedge out the increased variance introduced by the checkpoints at the last time steps.

4) The best CER, 4.39%, is achieved by our KBABVT procedure when averaging 100 checkpoints.

5) The KBABVT checkpoint averaging can often achieve better results than the LK and KBVL methods, regardless of how many checkpoints are used to

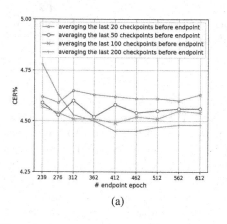
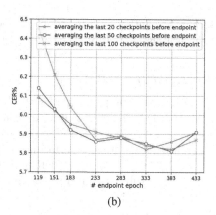

(a)   (b)

Fig. 4. The CERs of Conformer when averaging different numbers of checkpoint at various endpoints using the last-k checkpoint averaging procedure, (a) AISHELL-1, (b) AISHELL-2 (IOS test set).

average, which implies the ApproBiVT score is an effective metric for selecting the checkpoints and has verified the necessity of considering bias-variance tradeoff scores.

Table 1. Comparisons of different early stopping and checkpoint averaging procedures with Conformer on AISHELL-1 test set. The notation of VL is the abbreviation of validation loss.

Early Stopping	Avg. Procedure	Epoch Num.(#)	Avg. Num.(#)	CER (%)
VL	KBVL	240	20/50	4.56/4.64
VL	LK	240	20/50	4.62/4.64
VL	KBABVT	240	20/50	4.59/4.62
ApproBiVT	KBVL	613	20/50 100/200	4.55/4.53 4.53/4.53
ApproBiVT	LK	613	20/50/100 200/300/400	4.63/4.55/4.52 4.48/4.43/4.44
ApproBiVT	KBABVT	613	20/50 100/200	4.46/4.43 **4.39**/4.43

Table 2. Evaluation with various models on AISHELL-1.

Model Architecture	Online	Recipe	Epoch Num.(#)	Avg. Num.(#)	CER (%)
U2 [10]	Yes	VL+KBVL	180	20	5.45/5.54
		VL+LK	180	20	5.62
		Ours	584	100	**5.40**
U2++ [11]	Yes	VL+KBVL	360	30	5.05/5.12
		VL+LK	360	30	5.23
		Ours	855	200	**4.99**
Blockformer [13][a]	No	VL+KBVL	160	40	4.29/4.35
		VL+LK	160	40	4.52
		Ours	328	100	**4.23**

[a] 4.29 are the author reported results. 1 epoch in Blockformer recipe is equivalent to 3 epochs as different data preparation methods used.

3.5   AISHELL-1 Results

Table 2 lists the comparisons of several recently proposed state-of-the-art models on AISHELL-1. In general, if we use the validation loss based early stopping, the checkpoint averaging of KBVL is often superior to the LK scheme. All these models can obtain additional performance gains using our ApproBiVT recipe, with roughly 2.5%–3.7% relative CER reduction compared to their baseline counterparts. The resulting Blockformer and U2++ models, which attain 4.23% and 4.99% CERs, achieve new state-of-the-art for non-streaming and streaming E2E ASR models on AISHELL-1, respectively. These results confirm that our recipe can be used as an additive technique to enhance ASR models.

3.6 AISHELL-2 Results

To test whether the gains of AppriBiVT can scale to a large dataset, we verify our recipe with Conformer on AISHELL-2. Table 3 gives the comparison results between our ApproBiVT recipe and other recipes. Compared with the baseline recipe of WeNet, our recipe achieves 4.6%, 3.1%, and 4.4% relative CER reduction on the IOS, Android, and Mic test sets, respectively. This has proved that the ApproBiVT recipe is also an additive technique for large dataset.

Table 3. Comparisons of different recipes on AISHELL-2

Recipe	Epoch Num.(#)	Avg. Num.(#)	IOS	CER(%) Android	Mic.
VL+KBVL (WeNet)[a]	120	20	6.09	6.69	6.09
VL+LK	120	20	6.11	6.62	6.12
VL+KBABVT	120	20	6.09	6.78	6.09
ApproBiVT+KBVL	434	100	5.98	6.52	5.98
ApproBiVT+LK	434	100	5.87	6.52	5.86
Ours	434	100	**5.81**	**6.48**	**5.82**

[a](The WeNet reported results in https://github.com/wenet-e2e/wenet.)

3.7 Limitations and Scope

This work has demonstrated that the ApproBiVT is an additional technique to improve the ASR with no extra inference cost. It's no doubt that it typically costs more training time which is necessary to allow averaging more checkpoints with low bias and variance tradeoff responses.

4 Conclusion

Our results challenge the conventional recipe of early stopping and checkpoint averaging, with no model modification and extra inference computation, we are often able to produce better models using the proposed ApproBiVT recipe. Our work confirms that training longer can generalize better, gives a reasonable interpretation to indicate how long should be trained and a practical recipe to explore the power of long training. Our results rewrite the conclusion about checkpoint averaging that no performance improvement can be further squeezed out [29] and provide an effective method to yield better results.

References

1. Li, J., et al.: Jasper: an end-to-end convolutional neural acoustic model. In: Interspeech 2019–20rd Annual Conference of the International Speech Communication Association (2019)
2. Kriman, S., et al.: QuartzNet: deep automatic speech recognition with 1D time-channel separable convolutions. In: ICASSP 2020–45rd IEEE International Conference on Acoustics, Speech and Signal Processing, pp. 6124–6128. May 4–8, Barcelona, Spain (2020)
3. Han, K.J., Pan, J., Naveen Tadala, V.K., Ma, T., Povey, D.: Multistream CNN for robust acoustic modeling. In: ICASSP 2021–46rd IEEE International Conference on Acoustics, Speech and Signal Processing, pp. 6873–6877. Jun. 6–11, Toronto, Ontario, Canada (2021)
4. Chan, W., Jaitly, N., Le, Q., Vinyals, O.: Listen, attend and spell: a neural network for large vocabulary conversational speech recognition. In: ICASSP 2016–41rd IEEE International Conference on Acoustics, Speech and Signal Processing, pp. 4960–4964. Mar. 20–25, Shanghai, China (2016)
5. Rao, K., Sak, H., Prabhavalkar, R.: Exploring architectures, data and units for streaming end-to-end speech recognition with RNN-transducer. In: ASRU 2017–2017 IEEE Automatic Speech Recognition and Understanding Workshop, pp. 193–199. Dec. 16–20, Okinawa, Japan (2017)
6. Vaswani, A., Shazeer, N., Parmar, N., et al.: Attention is all you need. In: NIPS 2017–31rd Conference on Neural Information Processing Systems, pp. 5998–6008. Dec. 4–9, Long Beach, California, U.S.A. (2017)
7. Dong, L., Xu, S., Xu, B.: Speech-Transformer: a no-recurrence sequence-to-sequence model for speech recognition. In: ICASSP 2018–43rd IEEE International Conference on Acoustics, Speech and Signal Processing, pp. 5884–5888. Apr. 22–27, Seoul, South Korea (2018)
8. Moritz, N., Hori, T., Roux, J.L.: Streaming automatic speech recognition with the transformer model. In: ICASSP 2020–45rd IEEE International Conference on Acoustics, Speech and Signal Processing, pp. 6074–6078. May 4–8, Barcelona, Spain (2020)
9. Gulati, A., Qin, J., Chiu, C.C., et al.: Conformer: convolution-augmented transformer for speech recognition. In: Interspeech 2020–21rd Annual Conference of the International Speech Communication Association, pp. 5036–5040. Oct. 25–30, Shanghai, China (2020)
10. Zhang, B.B., Wu, D., Yao, Z.Y., et al.: Unified streaming and non-streaming two-pass end-to-end model for speech recognition. arXiv preprint arXiv:2012.05481 (2020)
11. Wu, D., Zhang, B.B., Yang, C., et al.: U2++: unified two-pass bidirectional end-to-end model for speech recognition. arXiv preprint arXiv:2106.05642 (2021)
12. An, K., Zheng, H., Ou, Z., Xiang, H., Ding, K., Wan, G.: CUSIDE: chunking, simulating future context and decoding for streaming ASR. arXiv preprint arXiv:2203.16758 (2022)
13. Ren, X., Zhu, H., Wei, L., Wu, M., Hao, J.: Improving mandarin speech recogntion with block-augmented transformer. arXiv preprint ArXiv:2207.11697 (2022)
14. Wang, F., Xu, B.: Shifted chunk encoder for transformer based streaming end-to-end ASR. In: ICONIP 2022–29rd International Conference on Neural Information Processing, Part V, pp. 39–51. Nov. 22–26, Virtual Event, India (2022)

15. Kim, S., Gholami, A., Eaton, A., et al.: Squeezeformer: an efficient transformer for automatic speech recognition. arXiv preprint ArXiv:2206.00888 (2022)
16. Geman, S., Bienenstock, E., Doursa, R.: Neural networks and the bias/variance dilemma. Neural Comput. **4**, 1–58 (1992)
17. Morgan, N., Bourlard, H.: Generalization and parameter estimation in feedforward netws: some experiments. In: NIPS 1990–4rd Conference on Neural Information Processing Systems (1990)
18. Reed, R.: Pruning algorithms-a survey. IEEE Trans. Neural Netw. **4**(5), 740–747 (1993)
19. Prechelt, L.: Early stopping-but when? In Neural Networks (1996)
20. Popel, M., Bojar, O.: Training tips for the transformer model. The Prague Bull. Math. Linguist. **110**, 43–70 (2018)
21. Yao, Z., Wu, D., Wang, X., et al.: WeNet: production oriented streaming and non-streaming end-to-end speech recognition toolkit. In: Interspeech 2021–22rd Annual Conference of the International Speech Communication Association, Aug. 30-Sep. 3, Brno, Czech Republic (2021)
22. Park, D.S., Chan, W., Zhang, Y., Chiu, C.C., et al.: SpecAugment: a simple data augmentation method for automatic speech recognition. In: Interspeech 2019–20rd Annual Conference of the International Speech Communication Association, pp. 2613–2617. Graz, Austria (2019)
23. Bouthillier, X., Konda, K., Vincent, P., Memisevic, R.: Dropout as data augmentation. arXiv preprint arXiv:1506.08700 (2015)
24. Bu, H., Du, J., Na, X., Wu, B., Zheng, H.: Aishell-1: an open-source mandarin speech corpus and a speech recognition baseline. In: O-COCOSDA 2017–20rd Conference of the Oriental Chapter of the International Coordinating Committee on Speech Databases and Speech I/O Systems and Assessment, pp. 1–5. Nov. 1–3, Seoul, South Korea (2015)
25. Du, J., Na, X., Liu, X., Bu, H.: AISHELL-2: transforming mandarin ASR research into industrial scale. arXiv preprint ArXiv:1808.10583 (2018)
26. Heskes, T.M.: Bias/Variance decompositions for likelihood-based estimators. Neural Comput. **10**, 1425–1433 (1998)
27. Hoffer, E., Hubara, I., Soudry, D.: Train longer, generalize better: closing the generalization gap in large batch training of neural networks. arXiv preprint ArXiv:1705.08741 (2017)
28. Jais, I.K.M., Ismail, A.R., Nisa, S.Q.: Adam optimization algorithm for wide and deep neural network. Knowl. Eng. Data Sci. **2**, 41–46 (2019)
29. Gao, Y., Herold, Y., Yang, Z., Ney, H.: Revisiting checkpoint averaging for neural machine translation. In: AACL/IJCNLP (2022)

Human-Guided Transfer Learning
for Autonomous Robot

Minako Oriyama[1]([✉]), Pitoyo Hartono[2], and Hideyuki Sawada[3]

[1] Department of Pure and Applied Physics, Graduate School of Advanced Science and
Engineering, Waseda University, Tokyo, Japan
minako.oriyama@suou.waseda.jp
[2] School of Engineering, Chukyo University, Nagoya, Japan
hartono@sist.chukyo-u.ac.jp
[3] Faculty of Science and Engineering, Waseda University, Tokyo, Japan
sawada@waseda.jp

Abstract. In recent years, neural networks have been successfully applied to many problems. However, prohibitively long learning time and vast training data are sometimes unavoidable. While the long learning time can be tolerated for many problems, it is crucial for autonomous robots learning in physical environments. One way to alleviate this problem is through transfer learning, which applies knowledge from one domain to another. In this study, we propose a method for transferring human common sense for guiding the subsequent reinforcement learning of a robot for real-time learning in a more complex environment. The efficacy of the transfer mechanism is analyzed to obtain new insights on the required prior knowledge to be transferred for training an autonomous robot. Different types of prior knowledge were analyzed and explained in this paper.

Keywords: Transfer learning · Reinforcement learning · Autonomous robots · Neural networks · Common Sense

1 Introduction

Machine learning is a widely used technique in artificial intelligence, allowing machines to identify patterns and rules in data and make predictions. In recent years, machine learning is often used to train controllers for autonomous robots. However, long training time and the requirement for a vast amount of data often prevent machine learning applications to robotics.

We aim to leverage transfer learning for real-time robot learning to address this challenge. While transfer learning is often used for classification tasks, some parts of neural networks [1, 2] that are relevant for building internal representations are transferred. However, what to transfer for some tasks remains to be discovered, including for robots' controllers. Therefore, our research focuses on identifying which knowledge should be transferred to alleviate robot learning. In this paper, we pre-train a neural network in a simple environment in an offline manner to be subsequently transferred for

© The Author(s), under exclusive license to Springer Nature Singapore Pte Ltd. 2024
B. Luo et al. (Eds.): ICONIP 2023, CCIS 1961, pp. 186–198, 2024.
https://doi.org/10.1007/978-981-99-8126-7_15

real-time learning in more complex environments. The inputs for the pre-training data are obtained by letting the robot randomly run in the physical environment, while the teacher signals are generated by human common sense. This study's primary novelty is combining human common sense and reinforcement learning [3]. This paper reports on the method to implement this idea and the contents and results of some real-time learning experiments.

Generally, transfer learning [4, 5] is a method that enhances learning accuracy and efficiency by leveraging accumulated knowledge from one task to other related tasks. Since the training process in the original domain enables a neural network to capture important features, transferring a part of that neural network to another one for learning in a similar domain is beneficial. It is obvious that since some relevant functionalities are already embedded in the new neural network, its learning burden can be significantly alleviated. Image classifications [6–8] have extensively utilized transfer learnings. In the past few years, transfer learning was also utilized in robotics. Most involve pre-training a neural network in the simulator and subsequently transfer part of the neural network for learning in physical environments (sim-to-real) [9, 10]. Our study differs from those past studies in that we do not need simulators. Here, we pre-trained a neural network in a commonsensical manner using readily available or artificially generated data, then transferred the basic knowledge for training an autonomous robot.

In this study, the task for our robot is to learn collision avoidance behavior [11–13] while randomly walking in physical space. The task is similar to [14, 15]; however, our approach does not use path planning or deep reinforcement learning. Models trained by deep reinforcement learning or path planning are generally treated as black boxes. It is difficult to interpret the details of the internal mechanisms and logic by which the models make decisions, which presents challenges in explainability. Using simple neural networks, the mechanisms and logic inside the model may be relatively easy to interpret. It is easier to make sense of the network's layers and parameters and explain the model's behavior.

A previous study [16] discusses path planning for mobile robots using deep reinforcement learning and transfer learning strategies. The authors propose transfer learning as a means to address the learning time challenge in real-time learning of autonomous robots in real-world environments, aligning with the objective of this study. However, prior studies have focused on achieving faster autonomous navigation and system stability in path planning but have yet to replicate these results in real-world scenarios successfully. Furthermore, the utilization of Deep Q-Network (DQN) [17, 18] for path planning is efficient but also increases calculation complexity.

In this study, we use a simple neural network that preserves the characteristics of the prior learning process after it is implemented for reinforcement learning in the physical environment. The task for the robot is to execute random walks while avoiding obstacles. Experiments are conducted in static and dynamic environments to show that prior knowledge helps to alleviate learning difficulty. The reinforcement learning approach used in this study is based on previous research [19]. The proposed method in the previous study is relevant to this research because it requires fewer computational resources to be embedded into the robot. However, the proposed study significantly differs from the past study in that it introduces prior learning based on the common-sensical knowledge of humans to speed up the learning process of the robot. This study links human common sense and reinforcement learning by introducing pre-training to plan the robot's movements. Here, human common sense will be used to generate teacher signals for pre-training the neural network, which is expected to make the learning process intuitive and effective.

This paper uses "transfer learning" in a slightly different context than conventional transfer learning. In conventional transfer learning, a part of a neural network trained in one domain is transferred to another neural network to be trained in a similar but different domain. In this study, we only utilized one neural network that was first trained in a supervised manner to reflect human's common-sensical decision process regarding sensory data of the physical environment and subsequently utilized in the reinforcement learning process in the real environment. Hence, here, we do not transfer a part of the neural network's structure, but we transfer prior knowledge that reflects human common sense.

This study investigates the types of prior knowledge that should be transferred to mitigate the learning difficulty. The research attempts to establish a new relationship between humans and artificial intelligence through collaborative learning. In addition to addressing the challenges of long training times and the need for extensive data in robotics, our research also contributes in the following ways, highlighting its unique novelty and significance:

- **Adaptability in Complex Environments:** We recognize the importance of reinforcement learning in addressing the challenges of complex, real-world scenarios, and thus, this research proposes pre-training for enhanced adaptability.
- **Interpretability:** Using simple neural networks enables a clearer understanding of the robot's decision-making processes, promoting interpretability.
- **Human-Guided Learning:** Incorporating human common sense into pre-training creates a more intuitive and effective learning process, fostering collaboration between humans and AI.
- **Real-World Relevance:** Our approach eliminates the need for simulators, making robot learning more applicable to complex real-world tasks.

By considering these additional aspects of our research, we address the challenges in robot learning and contribute to the development of more adaptable, interpretable, and human-informed robotic systems, thus fostering a more profound synergy between humans and artificial intelligence.

This paper is organized as follows: Sect. 2 describes the structure and algorithm of the neural network used in this study. Section 3 describes the system used in this study. Section 4 presents the experimental results and discussion of learning with the proposed method. Finally, Sect. 5 provides a summary and future work.

2 Neural Network's Structure and Learning Process

The outline of the robot's training process in this study is shown in Fig. 1, while the pre-training and the subsequent reinforcement learning are explained as follows.

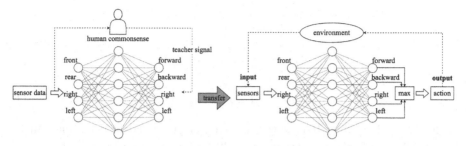

Fig. 1. Overview of the robot's training process

2.1 Human Guided Pre-training

In this study, prior knowledge is given to the robot before performing reinforcement learning. In the experimental environment, the robot performs 100 random walks, starting from the center of the environment and randomly generating actions (forward, backward, turn left, turn right). Four proximity sensor values (in the robot's front, back, left, and right) are recorded each time an action is generated.

Utilizing these sensory data, human generates a teacher signal for input at time t. Here, the teacher signals are generated based on the common sense of the human. For example, for obstacle avoidance strategy, it is common-sensical to move in the direction where the robot is the farthest from environmental obstacles. Naturally, the generated rules are subjective in that they may differ according to the experience, knowledge, or preference of the human that generates them. Here, by initializing the neural network with human knowledge, we postulate that the robot will learn better in the subsequent reinforcement learning as it does not need to acquire the required policy from scratch.

The pre-training process is illustrated on the left-hand side of Fig. 1.

After the pre-training process, the neural network is utilized for real-time reinforcement learning of the robot. The human experience is transferred to the neural network for reinforcement learning.

2.2 Real-Time Reinforcement Learning

In the reinforcement learning process, the pre-trained neural network is utilized as follows. The input layer of this neural network comprises four nodes, which receive sensory values from four proximity sensors mounted on the robot for detecting obstacles in four different directions. The output layer, which includes four nodes, is associated with an action that the robot can generate: moving forward, moving backward, turning right, and turning left.

The dynamics of the neural network are as follows. In Eq. 1, at time t, $I_j^{hid}(t)$ is the potential of the j-th hidden unit at time t, $x_i(t)$ is the i-th sensor value, w_{ij} is the weight connecting the i-th input unit and the j-th hidden unit, while $\theta_j^{hid}(t)$ is the bias of the j-th hidden unit. Similarly, $I_k^{out}(t)$ in Eq. 3 is the potential of the k-th output unit, v_{jk} is the weight connecting the j-th hidden unit and the k-th output unit, while θ_k^{out} is the k-th output unit's bias.

$$I_j^{hid}(t) = \sum_i x_i(t) w_{ij}(t) - \theta_j^{hid}(t) \tag{1}$$

$$O_j^{hid}(t) = f\left(I_j^{hid}(t)\right) \tag{2}$$

$$I_k^{out}(t) = \sum_j O_j^{hid}(t) v_{jk}(t) - \theta_k^{out}(t) \tag{3}$$

$$O_k^{out}(t) = f\left(I_k^{out}(t)\right) \tag{4}$$

In Eq. 2 and Eq. 4, $O_j^{hid}(t)$ and $O_k^{out}(t)$ are the output of the j-th hidden unit and the k-th output unit, while $f(x)$ is a sigmoid function.

At each time step, a sensor value is obtained, and the neural network calculates the output using the sensor value as input. The resulting action is $a(t)$ determined by the neuron with the highest output value, i.e., the winner neuron $w(t)$. The selected action is then executed by the robot via the $act()$ function, and the sensor value is obtained again at time $t + 1$.

$$w(t) = \arg\max_k O_k^{out}(t) \tag{5}$$

$$a(t) = act(w(t)) \tag{6}$$

Next, the robot executes reinforcement learning based on the acquired sensor values to determine the correctness of the performed action. The evaluation function $U(a(t))$ evaluates the distance between the robot and obstacles before and after the action. At time t, the two smallest sensor values among the four acquired values are denoted as $d_1(t)$ and $d_2(t)[cm]$, respectively. The difference between the mean squares of these two values before and after the action is calculated to obtain the evaluation function $U(a(t))$, shown in Eq. 8.

$$d_1(t) < d_2(t) < d_3(t) < d_4(t) \tag{7}$$

$$U(a(t)) = \sqrt{\sum_i^2 (d_i(t+1))^2} - \sqrt{\sum_i^2 (d_i(t))^2} \tag{8}$$

$U(a(t)) > 0$ indicates that the robot moved away from the obstacle. Hence the behavior is considered "good" in the reinforcement learning process, while $U(a(t)) < 0$ indicates that the robot moved towards the obstacle; hence the behavior is considered "bad".

When the action is good, the teacher signals for the winning neuron $T_w = 1$, while the other neurons receive $T_j = 0(j \neq w)$, encouraging the same behavior for similar inputs in the future. Conversely, when the action is bad $T_w = 0$, the other neurons receive $T_j = 1$. In this case, the action executed by the robot is inhibited, while other actions will encourage similar input. The ideal action is unknown; hence, this learning mechanism can be considered reinforcement learning, although its implementation mimics a supervised one.

The reinforcement learning here is masked as a supervised mechanism in this neural network by defining the loss function as follows.

$$E(t) = \tfrac{1}{2} \sum_j \left(O_j^{hid}(t) - T_j(t)\right)^2 \tag{9}$$

The rewards for the robot's action are shown as follows.

$$T_w = \begin{cases} 1 & \text{if } U(a(t)) > 0 \\ 0 & \text{if } U(a(t)) < 0 \end{cases} \tag{10}$$

$$T_j = \begin{cases} 0 & \text{if } U(a(t)) > 0 \\ 1 & \text{if } U(a(t)) < 0 \end{cases} \tag{11}$$

To minimize the loss E, the stochastic gradient descent method is employed. The partial derivative of the total error E for each neural network weight must be computed using backpropagation [20], where ε represents the learning rate of the connection weights, and α is an exponential decay factor ranging from 0 to 1.

$$\Delta w_{ij}(t) = -\varepsilon \frac{\partial E(t)}{\partial w_{ij}(t)} + \alpha \Delta w_{ij}(t-1)$$

$$\Delta v_{jk}(t) = -\varepsilon \frac{\partial E(t)}{\partial v_{jk}(t)} + \alpha \Delta v_{jk}(t-1) \tag{13}$$

In the proposed reinforcement learning algorithm, the state-action mapping is achieved using a neural network, which maps states to actions to reduce computation time, instead of using a Q-table, as in conventional Q-learning. The reinforcement learning process is illustrated with pseudocode in Algorithm 1.

Algorithm 1 Reinforcement Learning

1:	start learning
2:	acquire sensor values
3:	calculate output
4:	execute action
5:	acquire sensor values
6:	**if** evaluation is good **then**
7:	reinforce executed behavior
8:	suppress other behaviors
9:	**Else**
10:	**if** evaluation is bad **then**
11:	suppress executed behavior
12:	reinforce other behavior
13:	end if
14:	**end if**
15:	**End**

---

3 Robot Platform

In this study, we built an omnidirectional robot shown in Fig. 2. We used the "Zumo Robot for Arduino," designed for infinite orbit motion. The robot is equipped with four ultrasonic proximity sensors and a Bluetooth module. The specifications and the sensors for the robot are presented in Table 1.

The Zumo robot is a crawler robot that generates four different actions: forward, backward, right turn, and left turn. Four ultrasonic proximity sensors are installed in the

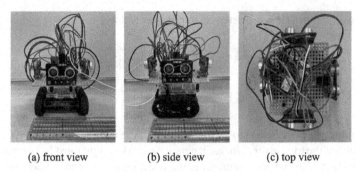

(a) front view   (b) side view   (c) top view

Fig. 2. Outline of the robot

Table 1. Robot specification

Gross weight [g]	396
Dimension [cm]	10 × 10 × 12
Sensors	Four ultrasonic proximity sensors
Communication	Bluetooth
Measurement range [cm]	2–400
Operating frequency [kHz]	40

robot to measure distances to obstacles. A 10 μs pulse is supplied as the trigger input to obtain distance measurements, which transmits 40 kHz ultrasonic waves for eight cycles to generate an echo. The distance to the object can be measured by calculating the time between the trigger signal's transmission and the echo signal's reception. The measurement range is from 2 cm to 400 cm, and deviations from this range will result in measurement errors. The specifications of the ultrasonic sensor are shown in Table 1.

A Bluetooth module is connected to the Arduino and paired with the PC to establish serial communication between them. The acquired sensor values are sent to the PC via Bluetooth via Arduino. The neural network was implemented on the PC, and training was performed using the sent sensor values as input. Upon receiving the input, the neural network calculates the output (action) the robot should perform and sends this information to Arduino via Bluetooth. After the action is executed, the PC enhances or inhibits the action based on the sensor values using an evaluation function (Eq. 8). Learning is performed by repeating this algorithm.

4 Experiments

4.1 Pre-training Experiments

The neural network is pre-trained using sensor data and human-generated teacher signals using standard backpropagation. The experimental setup is shown in Fig. 3(a). We created a controlled experimental environment measuring 150 × 150 cm by assembling 12 polypropylene panels, each with dimensions of $500 \times 500 \times 5$ mm. The training process is shown in Fig. 3(c). The graph shows an exponentially decreasing trend, indicating that the neural network has successfully learned from the human-generated data.

Subsequently, we conducted an offline experiment to assess the accuracy of the pre-training by generating actions based solely on data. Using the pre-trained neural network, we tested whether the robot could perform collision avoidance behavior in the experiment environment (Fig. 3(a)). Random walk data were fed into the neural network that predicted the action that the robot should take. Starting from the lower left corner of the environment, the robot performed 50 steps. As a result, the robot moved to the center of the environment and repeatedly traversed back and forth (Fig. 3(b)). This outcome demonstrates that the robot acquired the basic policy from human common-sensical knowledge.

However, we hypothesize that relying solely on prior learning proves inadequate for the robots to acquire the necessary skills. We tested the hypothesis by exposing the same pre-trained robot to a more complex environment. The outcomes of this experiment demonstrated that when faced with a complex environment, the robot struggled to navigate and avoid obstacles solely relying on prior learning. This outcome underscores the imperative of incorporating subsequent reinforcement learning to enhance the robot's adaptive capabilities.

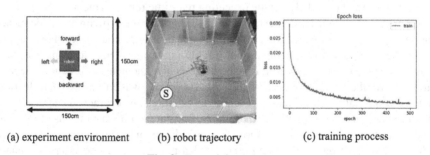

(a) experiment environment   (b) robot trajectory   (c) training process

Fig. 3. Pre-training process

4.2 Transfer Experiments

First, an experiment was conducted without prior training. Reinforcement Learning was performed in 50-time steps. After wandering around the lower left corner of the environment, the robot collided with a wall while moving. It stopped around the lower center of the environment without being able to recover (Fig. 4(a)). Figure 4(b) shows the loss function obtained from this experiment. The robot was not able to learn after 10 steps. This is due to the learning difficulty when the robot has no prior knowledge and needs to learn from scratch.

In the next experiment, reinforcement learning with prior human knowledge was executed. The learning process was executed in 50-time steps. In this case, the robot moved from the lower left corner of the environment to the upper left corner of the environment, reached the center of the environment in about 20-time steps, and then

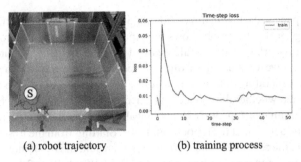

(a) robot trajectory   (b) training process

Fig. 4. Reinforcement learning without pre-training

repeatedly moved back and forth in the center until the completion of the prescribed time step. Once the robot approached the wall in the upper part of the environment, it performed collision avoidance behavior. Then it went to the environment's center without difficulty (Fig. 5(a)).

It can be observed that with correct prior information, a robot can learn quickly based on pre-training without sequentially repeating collision avoidance behavior. The loss function in Fig. 5(b) continuously decreases. Compared to Fig. 4(b), the final loss value is reduced by about 1/4, indicating that the robot efficiently learns by utilizing previously learned knowledge while executing real-time reinforcement learning.

The robot acquired a fundamental policy in the pre-training experiment, reducing hesitation (Fig. 3(b)). However, during the reinforcement learning phase, the robot struggled to find the optimal action selection initially, requiring trial and error learning over time. This led to some failures in the initial trials and gradual refinement of the action selection. Furthermore, the pre-training established a basic policy, while reinforcement learning aimed for optimal action selection, resulting in variations in accuracy and speed during the initial actions. Additionally, unlike in Fig. 3(c), the temporary increase in the loss function occurred due to initial performance degradation in learning collision avoidance behavior. However, the iterative reinforcement learning process improved performance, leading to a subsequent decrease in the loss function.

To test the efficiency of the transferred knowledge, the next step is to have the robot perform non-common sensical pre-training followed by reinforcement learning. The pre-training is based on the same random walk data but utilizes random teacher signals. In this case, the robot collided with a wall after about 15-time steps without exhibiting collision avoidance behavior. Subsequently, the robot repeatedly generated forward-only behaviors, indicating a stagnant learning process.

Figure 6(b) shows that the loss function swings immediately after the start of the experiment, indicating that the given prior information does not match the information of the environment in which sequential learning takes place. The loss function decreases after about 15-time steps, which may indicate that moving forward after a collision is considered good behavior. This experiment indicates that common-sensical pretraining is important in supporting subsequent reinforcement learning.

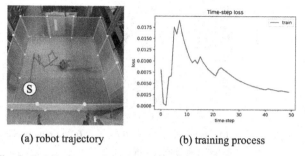

(a) robot trajectory   (b) training process

Fig. 5. Reinforcement learning with common-sensical pre-training

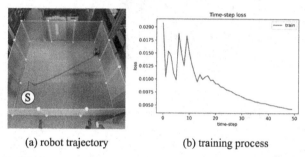

(a) robot trajectory   (b) training process

Fig. 6. Reinforcement learning with random pre-training

Next, obstacles were set up to test whether the proposed reinforcement learning can learn in a more complex environment. Here, the robot was pre-trained with common-sensical data. The robot swiftly navigated through the partitions placed in the front and rear without difficulty, reaching the center of the environment in approximately 20-time steps and repeatedly moving back and forth (Fig. 7(b) and (e)).

Despite the difference in environment between the pre-training and the reinforcement learning, the robot effectively demonstrated collision avoidance behavior by applying its prior knowledge to sequential learning. This suggests that the knowledge transfer mechanism proposed in this study is efficient.

In the last experiment, two robots generated a dynamic environment. Here, both robots are pre-trained with the same common-sense data and subsequently trained with reinforcement learning. Robot 1 started from the bottom left, while Robot 2 started from the top right. The obstacles in the environment were removed, and both robots started from a different starting point and could reach the center of the environment without cooperation. This suggests that the proposed method can be applied to cases where multiple robots work together and is an effective means of sharing knowledge in group learning and cooperative tasks. The training process of both robots is shown in Fig. 8(b) and 8(c).

The proposed method is promising when multiple robots must cooperate to accomplish a task. Pre-training on common-sense data allows each robot to share individually learned knowledge and experience. Thus, each robot could operate independently yet achieve effective cooperative behavior based on shared knowledge.

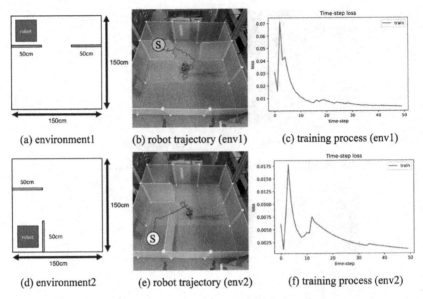

(a) environment1   (b) robot trajectory (env1)   (c) training process (env1)

(d) environment2   (e) robot trajectory (env2)   (f) training process (env2)

Fig. 7. Reinforcement learning with obstacles

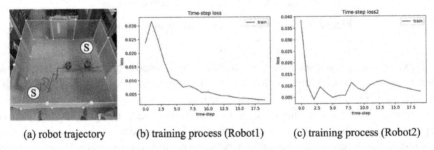

(a) robot trajectory   (b) training process (Robot1)   (c) training process (Robot2)

Fig. 8. Reinforcement learning in dynamic environment

5   Conclusion and Future Works

In this study, we proposed a method for transferring human common-sensical knowledge into a neural network. The transferred knowledge is utilized for initializing a hierarchical neural network that subsequently executes reinforcement learning. Here, the prior knowledge is generated based on human common sense. Hence, this study is a means for transferring human knowledge into neural networks. This idea is novel in that it significantly differs from conventional transfer learning, in which some parts of a neural network trained in one domain are transferred to other neural networks that learn in similar but different domains.

The proposed idea is tested in robotics experiments. The results presented in this paper demonstrate the usefulness and validity of the proposed transfer learning. While this study is still preliminary, this idea can potentially establish a new relationship between humans and artificial intelligence. In contrast to traditional machine learning approaches that focus solely on producing correct results, our approach of reflecting human decision-making characteristics in neural networks may contribute to further applications of AI in human-centric environments, like in medical, educational, and cultural fields.

Learning in highly dynamic environments is one of the most immediate future works. It is also interesting to experiment with different transfer mechanisms and build an intuitive interface that allows humans to transfer their knowledge into neural networks.

References

1. McCulloch, W.S., Pitts, W.: A logical calculus of the ideas immanent in nervous activity. Bull. Math. Biophys. **5**, 115–133 (1943). https://doi.org/10.1007/BF02478259
2. Rosenblatt, F.: The perceptron: a probabilistic model for information storage and organization in the brain. Psychol. Rev. **65**(6), 386–408 (1958). https://doi.org/10.1037/h0042519
3. Sutton, R.S., Barto, A.G.: Reinforcement learning: an introduction. IEEE Trans. Neural Netw. **9**(5), 1054 (1998)
4. Pan, S.J., Yang, Q.: A survey on transfer learning. IEEE Trans. Knowl. Data Eng. **22**(10), 1345–1359 (2010). https://doi.org/10.1109/TKDE.2009.191
5. Tan, C., Sun, F., Kong, T., Zhang, W., Yang, C., Liu, C.: A survey on deep transfer learning. In: Kůrková, V., Manolopoulos, Y., Hammer, B., Iliadis, L., Maglogiannis, I. (eds.) ICANN 2018. LNCS, vol. 11141, pp. 270–279. Springer, Cham (2018). https://doi.org/10.1007/978-3-030-01424-7_27
6. Kornblith, S., Shlens, J., Le, Q.V.: Do better imagenet models transfer better? In: 2019 IEEE/CVF Conference on Computer Vision and Pattern Recognition (CVPR), Long Beach, CA, USA, pp. 2656–2666 (2019). https://doi.org/10.1109/CVPR.2019.00277
7. Khan, N.C., et al.: Predicting systemic health features from retinal fundus images using transfer-learning-based artificial intelligence models. Diagnostics **12**(7), 1714 (2022). https://doi.org/10.3390/diagnostics12071714
8. Marmanis, D., Datcu, M., Esch, T., Stilla, U.: Deep learning earth observation classification using ImageNet pretrained networks. IEEE Geosci. Remote Sens. Lett. **13**(1), 105–109 (2016). https://doi.org/10.1109/LGRS.2015.2499239
9. Xie, Z., Clary, P., Dao, J., Morais, P., Hurst, J., Panne, M.: Learning locomotion skills for cassie: iterative design and sim-to-real. In: Proceedings of the Conference on Robot Learning, vol. 100, pp. 317–329. PMLR (2020)
10. Peng, X.B., Andrychowicz, M., Zaremba, W., Abbeel, P.: Sim-to-real transfer of robotic control with dynamics randomization. In: 2018 IEEE International Conference on Robotics and Automation (ICRA), Brisbane, QLD, Australia, pp. 3803–3810 (2018). https://doi.org/10.1109/ICRA.2018.8460528
11. Borenstein, J., Koren, Y.: Real-time obstacle avoidance for fast mobile robots. IEEE Trans. Syst. Man Cybern. **19**(5), 1179–1187 (1989). https://doi.org/10.1109/21.44033
12. Alajlan, A.M., Almasri, M.M., Elleithy, K.M.: Multi-sensor based collision avoidance algorithm for mobile robot. In: 2015 Long Island Systems, Applications and Technology, Farmingdale, NY, USA, pp. 1–6 (2015). https://doi.org/10.1109/LISAT.2015.7160181
13. Cardona, G.A., et al.: Autonomous navigation for exploration of unknown environments and collision avoidance in mobile robots using reinforcement learning. In: 2019 SoutheastCon, Huntsville, AL, USA, pp. 1–7 (2019). https://doi.org/10.1109/SoutheastCon42311.2019.9020521

14. Lin, J.-K., Ho, S.-L., Chou, K.-Y., Chen, Y.-P.: Q-learning based collision-free and optimal path planning for mobile robot in dynamic environment. In: 2022 IEEE International Conference on Consumer Electronics - Taiwan, Taipei, Taiwan, pp. 427–428 (2022). https://doi.org/10.1109/ICCE-Taiwan55306.2022.9869215

15. Feng, S., Sebastian, B., Ben-Tzvi, P.: A collision avoidance method based on deep reinforcement learning. Robotics **10**(2), 73 (2021). https://doi.org/10.3390/robotics10020073

16. Zhu, J., Yang, C., Liu, Z., Yang, C.: Path planning of mobile robot based on deep reinforcement learning with transfer learning strategy. In: 2022 37th Youth Academic Annual Conference of Chinese Association of Automation (YAC), Beijing, China, pp. 1242–1246 (2022). https://doi.org/10.1109/YAC57282.2022.10023708

17. Mnih, V., Kavukcuoglu, K., Silver, D., et al.: Human-level control through deep reinforcement learning. Nature **518**, 529–533 (2015). https://doi.org/10.1038/nature14236

18. Mnih, V., et al.: Playing Atari with deep reinforcement learning. arXiv, abs/1312.5602 (2013)

19. Hartono, P., Kakita, S.: Fast reinforcement learning for simple physical robots. Memet. Comput. **1**(4), 305–319 (2009). https://doi.org/10.1007/s12293-009-0015-x

20. Rumelhart, D., Hinton, G., Williams, R.: Learning representations by back-propagating errors. Nature **323**, 533–536 (1986). https://doi.org/10.1038/323533a0

Leveraging Two-Scale Features to Enhance Fine-Grained Object Retrieval

Yingjie Jing and Shenglin Gui[✉]

University of Electronic Science and Technology of China, Chengdu 611731, China
jingyingjie@std.uestc.edu.cn, shenglin_gui@uestc.edu.cn

Abstract. Constructing a discriminative embedding for an image based on the features extracted by a convolutional neural network (CNN) has become a common solution for fine-grained object retrieval (FGOR). However, existing methods construct the embedding based solely on features extracted by the last layer of CNN, neglecting the potential benefits of leveraging features from other layers. Based on the fact that features extracted by different layers of CNN represent different abstraction and semantic information on those levels, we believe that leveraging features from multiple layers of CNN can construct a more discriminative embedding. Upon this, we propose a simple yet efficient end-to-end model named TSF-Enhance, which leverages two-scale features extracted by the CNN to construct the discriminative embedding. Specifically, we extract features from the third and fourth layers of Resnet50 and construct an embedding based on features from these two layers respectively. When testing, we concatenate these two embeddings to get a more discriminative embedding for retrieval. Additionally, we design a Feature Enhancement Module (FEM) that consists of several common operations, such as layer normalization, to process the features. Finally, we achieve competitive results on three FGOR datasets, specifically exceeding the current state-of-the-art performance on the most challenging dataset CUB200. Furthermore, our model also demonstrates strong scalability compared to localization-based methods, achieving the best performance on two general-purpose image retrieval datasets. The source code is available at https://github.com/jingyj203/TSF-Enhance.

Keywords: Metric Learning · Fine-grained Object Retrieval · Convolutional Neural Network

1 Introduction

Fine-grained object retrieval (FGOR) is an important task in computer vision that focuses on retrieving images from multiple subcategories of a certain meta-category and returning images with the same subcategory as the given query image. Retrieving visually similar objects in practical applications remains challenging, especially when there are high intra-class variances but minimal inter-class differences. To address this issue, it is essential to learn discriminative

© The Author(s), under exclusive license to Springer Nature Singapore Pte Ltd. 2024
B. Luo et al. (Eds.): ICONIP 2023, CCIS 1961, pp. 199–211, 2024.
https://doi.org/10.1007/978-981-99-8126-7_16

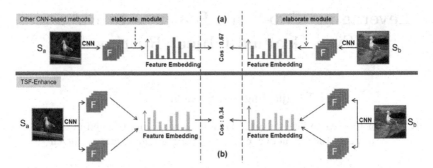

Fig. 1. Motivation of TSF-Enhance. S_a and S_b are images from different subcategories. The green features and blue features are extracted by a middle layer and the last layer of CNN respectively. The values in (a) and (b) represent the cosine similarity (Cos) between (S_a, S_b). As we can see, other CNN-based methods construct the embeddings based solely on features from the last layer of CNN, the similarity (a) between the embeddings of S_a and S_b is large. In contrast, our TSF-Enhance constructs the embeddings based on the features from a middle layer and the last layer of CNN, thus the embeddings contain more information to better distinguish subcategories. (Color figure online)

embeddings for recognizing visually similar objects in FGOR. Existing methods for FGOR can be roughly divided into two groups: metric-based and localization-based schemes.

Metric-based schemes aim to learn an embedding space where similar images are encouraged to be closer while dissimilar images are pushed away. Some methods, such as [5,6], achieve this objective by pulling images of the same class closer and pushing images of different classes farther apart. In addition, other methods, such as [16,21], use proxies to represent classes, associate images with proxies instead of other images, and encourage images to be close to proxies of the same class and far apart from those of different classes. However, since metric-based methods construct embeddings based on image-level features, they are limited by the noise in features.

To address this limitation, localization-based schemes tend to localize local parts to construct embeddings. Some methods, such as [24,29], focus on localizing the objects from images and extracting features corresponding to the objects, thereby eliminating the background noise. Other methods, such as [13,19], aim to discover and align common parts (e.g., tails and eyes) across the FGOR dataset via the guidance of additional prior knowledge (e.g., information and locations of common parts). Furthermore, recent methods, including [20,22], tend to dig into category-specific nuances that contribute to category prediction, rather than the common parts across the dataset.

Different from improving the discriminative ability of embeddings by localizing local parts, we propose a simple yet efficient end-to-end model named TSF-Enhance, which leverages two-scale image-level features extracted by CNN to construct more discriminative embeddings, as shown in Fig 1. Specifically, we extract features from the third and fourth layers of Resnet50 and construct an embedding based on features from these two layers respectively. During training, we train these two embeddings respectively. When testing, we contatenate these two embeddings to get a more discriminative embedding for retrieval. Additionally, we design a Feature Enhancement Module (FEM) to process features in order to improve performance. The FEM takes features as input and pools them through a global max pooling, which are then normalized using a layer normalization and fed into a fully connected layer. Finally, we conduct experiments on three FGOR datasets and two general-purpose image retrieval datasets that contain a wider range of object classes and exhibit greater discriminability in appearance across objects of different classes.

Main contributions of this paper are as follows:

- In FGOR, metric-based methods usually have poor performance when compared with localization-based methods due to with more noise in the embeddings. However, our method, which is essentially a metric-based method as well, achieves comparable performance with the state-of-the-art localization-based methods on three prevalent FGOR datasets.
- Localization-based methods that perform well in FGOR however are not suitable for general-purpose image retrieval due to the inherent nature of general-purpose image retrieval datasets (e.g., objects from different classes vary greatly in appearance.). In contrast, our method not only achieves competitive results on those three FGOR datasets but also demonstrates strong scalability on two general-purpose image retrieval datasets. Beyond our expectations, our method achieves the state-of-the-art results on both two general-purpose retrieval datasets.
- Unlike other methods that contain additional elaborate modules, such as the attention mechanism in [25] and the localization module in [22], our model is essentially a simple modification on Resnet50 that adds the layer normalization and the feature extraction from the third layer, and replaces the average pooling with global max pooling. Nevertheless, our model still achieves the state-of-the-art results on the most challenging dataset CUB200.

2 Related Work

Metric-based schemes aim to learn embeddings that bring similar images closer together and push dissimilar images apart from each other. Some methods, such as [5, 6, 18], achieve this by considering the pairwise relation of images. [5] exploits pairwise similarities between images in the source embedding space as the knowledge to guide target embedding model, to obtain more discriminative embeddings. [6] regularizes pairwise distances between embeddings into multiple levels and prevents the pairwise distances from deviating from its belonging level. Additionally, [16, 21] compare images with proxies instead of with one another. [16] proposes non-isotropy regularization to learn unique sample-proxy relations and eliminate semantic ambiguity. [21] proposes a proxy-based loss based on proxy-neighborhood component analysis.

An obvious shortcoming of metric-based schemes is that the embeddings are constructed using image-level features which contain a lot of noise, resulting in much non-discriminative information in these embeddings. [25] designs a discrimination-aware mechanism aiming to capture discriminative information from the embeddings. In this paper, we design a FEM that uses global max pooling to pool image-level features in order to reduce noise interference. Moreover, another limitation is that the embeddings are constructed based solely on features from the last layer of CNN, which neglects the potential benefits of leveraging features from other layers. Upon this, we explore and leverage features from multiple layers of CNN to construct more discriminative embeddings.

Localization-based schemes aim to localize local parts to construct embeddings so as to improve the discriminative ability of the embeddings. Some methods, such as [24,29], focus on localizing the objects from images by exploring activations of features. Instead of localizing objects, [13,19] tend to discover and align common parts across all subcategories via the guidance of additional prior knowledge. [13] uses the KAE-Block to output the embeddings of common parts and constructs an embedding that is aligned with the order of the common parts. [19] exploits strong neural activations on the last convolutional layer to localize the common parts. Recent approach [22] proposes two nuance regularizations to discover category-specific nuances that contribute to category prediction and align them grouped by subcategory, achieving the state-of-the-art results on three FGOR datasets.

Although [13,22] achieve impressive results in FGOR, they are not suitable for general-purpose image retrieval. The success of [13,22] on FGOR can be attributed to the fact that objects from different classes in FGOR datasets have similar appearance (e.g., the wings of birds) with only slight differences in some subtle positions, allowing [13] to distinguish subcategories by localizing the common parts across the dataset and [22] to improve retrieval performance by discovering and aligning the category-specific nuances among subcategories. However, in general-purpose image retrieval datasets, objects from different classes vary greatly in appearance and rarely share common (e.g., sofas and lamps), making it inappropriate to retrieve images based on category-specific nuances or common parts across the dataset.

3   Our Method

We aim to leverage features from multiple layers of CNN to construct more discriminative embeddings for better retrieval. To this end, we propose the TSF-Enhance model as shown in Fig 2. It introduces two components: the two branches of model and feature enhancement module.

3.1   The Two Branches of Model

Convolutional neural network (CNN) layers extract features with varying levels of abstraction and semantic information. Intuitively, leveraging features

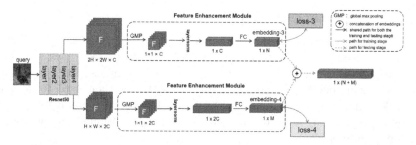

Fig. 2. TSF-Enhance Framework. 'C', 'H' and 'W' represent the channel, height and width of features, respectively. 'M' and 'N' refer to the dimensions of embedding. During training, embedding-3 and embedding-4 are trained with two losses. During testing, we concatenate the embedding-3 and embedding-4 for retrieval.

extracted by multiple layers of CNN would improve retrieval performance, but our attempts of fusing features from multiple layers to construct embeddings in various ways do not yield any improvement and even degrade the performance. We hypothesize the reason is that the feature fusion process may amplify the noise, thus reducing the discriminative ability of embeddings. In contrast to the fusion approach, we propose a novel idea, as illustrated in Fig. 2. Specifically, we extract the features from the third and fourth layers of Resnet50 and construct an embedding based on features from these two layers respectively. During training, we train these two embeddings respectively. When testing, we concatenate these two embeddings to get a more discriminative embedding for retrieval.

3.2 Feature Enhancement Module

This module is designed to process the features extracted by CNN in order to improve retrieval performance. It is mainly composed of two components: global max pooling (GMP) and layer normalization (layernorm). Global pooling is used to compress the feature points, so as to obtain a low-dimensional embedding for retrieval. There are two commonly used types of global pooling: global average pooling (GAP) and global max Pooling (GMP). Due to the presence of backgrounds and non-discriminative object parts in the images, the features extracted by the CNN contain a lot of noise. The global average pooling averages all feature points in the feature map, resulting in performance degradation. Thus, we use global max pooling (GMP) to extract the most salient feature points in features. Given a convolution feature map $M \times M$ dimension with E channels, g $\in R^{M \times M \times E}$ and a binary variable, $h_i \in \{0,1\}$. GMP is defined as:

$$GMP(g_\epsilon) = \max_h \sum_{i=1}^{M^2} h_i \cdot g_\epsilon, s.t. \sum_{i=1}^{M^2} h_i = 1, \forall \epsilon \in E \qquad (1)$$

In addition, inspired by [28] which investigates the application of layernorm without affine parameters, we add it to our model and find that this helps to improve performance in our experiments.

3.3   Loss and Temperature Scaling

Our model outputs two embeddings for each image. To ensure that both embeddings are discriminative, we train them with two proxy-based losses, loss-3 and loss-4, as shown in Fig 2. We obtain two loss values from loss-3 and loss-4, and then add them together to obtain a total loss that is then used to compute and back-propagate the gradients for updating the network parameters. These two losses are actually identical in their mathematical formulation and the loss function is as follows:

$$P_i = \frac{exp\left(-d(\frac{x_i}{||x_i||_2}, \frac{f(x_i)}{||f(x_i)||_2}) * 9\right)}{\sum_{a \in A} exp\left(-d(\frac{x_i}{||x_i||_2}, \frac{a}{||a||_2}) * 9\right)} \tag{2}$$

$$loss = -log(P_i) \tag{3}$$

A is the set of all proxies that are used to represent classes in the embedding space and x_i is the data in embedding space. These proxies are learnable and their initial positions in the embedding space are random. $f(x_i)$ is a proxy function that returns the proxy assigned to x_i, $||a||_2$ is the L2-Norm of vector a, and $d(x_i, a)$ returns the euclidean distance between vector x_i and vector a. For each x_i, the loss aims to maximize P_i. Maximizing P_i makes the x_i closer to the proxy of the same class and pushes x_i apart from other proxies. This loss function is explored by [21], and the value "9" in Eq. 2 is explored based on temperature scaling and it also helps to boost performance in our experiments.

4   Experiments

4.1   Datasets

We train and evaluate our model on three fine-grained object retrieval datasets: CUB200 [2], Cars196 [7] and FGVC Aircraft [12]. We follow the standard dataset split in [1,6,22]. CUB200 contains 200 bird subcategories with 11,788 images, we utilize the first 100 classes (5,864 images) for training and the rest 100 classes (5,924 images) for testing. Cars196 contains 196 car subcategories with 16,185 images, we utilize the first 98 classes (8,054 images) for training and the remaining 98 classes (8,131 images) for testing. FGVC Aircraft is divided into 100 classes with 10,000 images, we utilize the first 50 classes (5,000 images) for training and the rest 50 classes (5,000 images) for testing.

We also train and evaluate on two general-purpose image retrieval datasets: Stanford Online Products [14] (SOP) and In-shop Clothes Retrieval [11] (In-Shop). We follow the standard dataset split in [10,18]. SOP contains 12 super-classes which are further divided into 22,634 classes, we utilize 11,318 classes (59,551 images) for training and the rest 11,316 classes (60,502 images) for testing. In-Shop contains 7,982 classes, we utilize the first 3,997 classes (25,882 images) for training and the rest 3,985 classes (26,830 images) for testing. The test set is further partitioned into a query set with 14,218 images of 3,985 classes and a gallery set with 12,612 images of 3,985 classes.

4.2   Implementation Details

We apply the widely-used Resnet50 [3] model pre-trained on ImageNet [17] as our backbone. Before applying a crop of random size (0.08 to 1.0 of the scaled image), the original images are scaled to a random aspect ratio between 0.75 and 1.33. The cropped images are then resized to 256 × 256. We train our model using Adam optimizer without weight decay on two Tesla T4 GPUs. Before formal training, we fix the network parameters and train for five epochs to adjust the positions of the proxies in the embedding space. The other settings are presented in Table 5.

4.3   Evaluation Protocols

Our retrieval performance is measured using the Recall@K, which computes the average recall scores over all query images in the test set, strictly following the setting in [14]. Specifically, for each query, our model returns the top K most similar images. If at least one of the positive images appears in the top K results, we assign a score of 1 to that query; otherwise, we assign a score of 0.

4.4   Comparison with the State-of-the-Art Methods

A. Fine-grained Object Retrieval Results. We evaluate our TSF-Enhance against the state-of-the-art methods on three FGOR datasets (CUB200, Cars196, and FGVC Aircraft) and present the results in Table 1. The methods in Table 1 are categorized into three groups, listed from top to bottom as: (1) metric-based frameworks, (2) localization-based networks, and (3) our TSF-Enhance.

As we can see, the metric-based methods in the first group generally perform worse than the localization-based methods in the second group due to with more noise in the embeddings and all the methods in Table 1 have the lowest performance on dataset CUB200 compared to the other two datasets, suggesting dataset CUB200 is a more challenging FGOR task. The CNENet [22] achieves impressive results on datasets Cars196 and FGVC Aircraft by localizing the category-specific nuances among subcategories that contribute to retrieve cars and aircrafts with similar poses in these two datasets. Unfortunately, CNENet lowers its performance when confronting the more challenging dataset CUB200 in which birds have more variations in poses (e.g., flying and sitting) and vague features to distinguish. In contrast, our method, which is a metric-based method essentially, achieves the state-of-the-art performance on this challenging dataset and has impressive performance on the other two datasets as well.

B. General-purpose Image Retrieval Results. We also evaluate our TSF-Enhance against the state-of-the-art methods on two general-purpose image retrieval datasets (Sop and In-Shop) and present the results in Table 2 and Table 3. The localization-based methods [13, 22] for FGOR, which rely on detecting common parts across the dataset and category-specific nuances to retrieve

Table 1. Recall@k for k = 1,2,4,8 on CUB200, Cars196 and FGVC Aircraft datasets.

Method	Venue/Journal	CUB200				Cars196				FGVC Aircraft			
		1	2	4	8	1	2	4	8	1	2	4	8
CEP [1]	ECCV '20	69.2	79.2	86.9	91.6	89.3	93.9	96.6	98.4	81.3	84.3	90.1	92.3
NASA [8]	AAAI '22	70.2	80.4	88.0	–	88.5	93.3	96.1	–	–	–	–	
IBC [18]	ICML '21	70.3	80.3	87.6	92.7	88.1	93.3	96.2	98.2	–	–	–	–
PA+NIR [16]	CVPR '22	70.5	80.6	–	–	89.1	93.4	–	–	–	–	–	–
HIST [10]	CVPR '22	71.4	81.1	88.1	–	89.6	93.9	96.4	–	–	–	–	–
S2SD+PLG [15]	CVPR '22	71.4	81.1	–	–	90.2	94.4	–	–	–	–	–	–
MDR [6]	AAAI '21	71.4	81.2	88.0	92.6	90.4	94.3	96.6	98.0	–	–	–	–
ETLR [5]	CVPR '21	72.1	81.3	87.6	–	89.6	94.0	96.5	–	–	–	–	–
PNCA++ [21]	ECCV '20	72.2	82.0	89.2	93.5	90.1	94.5	97.0	98.4	–	–	–	–
DAM [25]	CVPR '21	72.3	81.2	87.8	92.7	88.9	93.4	96.0	97.7	–	–	–	–
HDCL [27]	Neurocomp '21	69.5	79.6	86.8	92.4	84.4	90.1	94.1	96.5	71.1	81.0	88.3	93.3
KAE-Net [13]	WACV '21	74.2	83.3	89.1	93.2	91.1	94.9	96.9	98.1	–	–	–	–
CNENet [22]	AAAI '22	74.5	83.1	89.2	93.8	94.2	96.9	98.2	98.8	85.6	91.5	94.8	96.8
TSF-Enhance(ours)	–	75.5	84.7	90.4	94.4	93.4	96.4	97.9	98.9	84.1	90.8	94.8	97.2

Table 2. Comparison with the state-of-the-art methods on SOP datatset. We align their models with the same 512-dimensional embedding.

Method	R@1	R@10	R@100	R@1000
CGD^{512} [4]	80.5	92.1	96.7	98.9
PCNA++^{512} [21]	80.7	92.0	96.7	98.9
PA + NIR^{512} [16]	80.7	91.5	–	–
PNP-D_q ^{512} [9]	81.1	92.2	96.8	99.0
S2SD+PLG^{512} [15]	81.3	92.3	–	–
HIST^{512} [10]	81.4	92.0	96.7	–
TSF-Enhance^{512}(ours)	81.7	92.5	96.9	99.0

Table 3. Comparison with the state-of-the-art methods on In-Shop datasets.

Method	R@1	R@10	R@20	R@30	R@40
PCNA++^{2048} [21]	90.9	98.2	98.9	99.1	99.4
Cont. w/M^{128} [23]	91.3	97.8	98.4	98.7	99.0
MDR [6]^{512}	91.3	98.2	98.8	–	99.3
CGD^{1536} [4]	91.9	98.1	98.7	99.0	99.1
HPL-PA^{512} [26]	92.5	98.0	–	–	–
IBC^{512} [18]	92.8	98.5	99.1	–	99.2
TSF-Enhance(ours)^{2048}	92.5	98.6	99.1	99.3	99.4

images, are not suitable for general-purpose image retrieval due to the significant differences in appearance among different classes in general-purpose image datasets. In contrast, our method demonstrates strong scalability, achieving the state-of-the-art results on datasets Sop and In-shop.

4.5 Ablation Experiments

We conducted experiments to demonstrate the superiority of global max pooling(GMP) over global average pooling (GAP). We replace the GMP in our model with global average pooling (GAP), then we use it as a baseline model (BL) for comparison. As shown in Table 4, GMP outperforms GAP and improves the performance by 4.3%.

Table 4. The ablative retrieval results of different global pooing methods on CUB200.

Method	R@1
BL	71.2%
BL/GMP	75.5% (4.3% ↑)

Table 5. The settings for different datasets. N and M denote dimensions of embedding-3 and embedding-4, respectively.

Dataset	BatchSize	Epoch	N	M	Total
CUB200	32	80	1024	1024	2048
Cars196	32	80	1664	384	2048
FGVC Aircraft	32	40	1792	256	2048
SOP	120	160	256	256	512
In-Shop	120	160	1280	768	2048

4.6 Discussions

The key to FGOR is to improve the discriminability of image embedding. To further illustrate the effectiveness of our proposed TSF-Enhance, which can improve the discriminative ability of embeddings by leveraging two-scale features from CNN, we evaluate the performance on three FGOR datasets using different embeddings and present the qualitative retrieval results, as shown in Table 6 and Fig. 3. We can see that using the concatenation for retrieval can distinguish subcategories better than embedding-3 and embedding-4, and outperforms embedding-3 and embedding-4 across all metrics, indicating that the concatenation constructed based on two-scale features is more discriminative than embedding-3 and embedding-4 that are constructed based on one-scale features.

Why is the concatenation more discriminative? We suspect the reasons are as follows. As demonstrated in Table 6 and Fig. 3, using embedding-3 and embedding-4 for retrieval can also achieve good performance and most of the returned images belong to the same subcategory as the query image, indicating that both embedding-3 and embedding-4 are discriminative. Therefore, both of them contain a lot of discriminative information that contributes to distinguishing subcategories. Moreover, they return different retrieval results for the same query image, indicating that their information is different. This discrepancy can be attributed to the fact that they are constructed based on different features. As a result, by concatenating these embeddings, we can obtain a more discriminative embedding, which contains more information that is discriminative than the single embedding, to better distinguish subcategories.

The more features, the better the performance? We show the retrieval performance with combinations of different layers, as shown in Table 7. We can observe that, without limiting the total dimension of embeddings, the combination of layer2 and layer4 and the combination of layer3 and layer4 can achieve

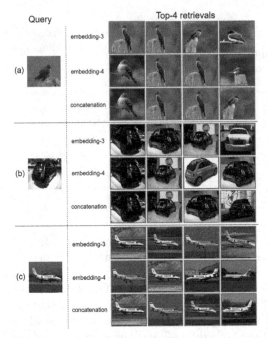

Query   Top-4 retrievals

(a) embedding-3 / embedding-4 / concatenation

(b) embedding-3 / embedding-4 / concatenation

(c) embedding-3 / embedding-4 / concatenation

Fig. 3. Qualitative retrieval results on CUB200 (a), Cars196 (b), and FGVC Aircraft (c). For each query, top-4 retrievals are presented and incorrect retrievals are indicated with a red boundary. "concatenation" refers to the concatenation of embedding-3 and embedding-4. (Color figure online)

Table 6. The performance on three FGOR datasets using different embeddings for retrieval. "concatenation" refers to the concatenation of embedding-3 and embedding-4.

Dataset	Embedding	R@1	R@2	R@4	R@8
CUB200	embedding-3	70.8	80.4	87.9	93.1
	embedding-4	70.7	81.2	88.1	93.1
	concatenation	75.5	84.7	90.4	94.4
Cars196	embedding-3	92.2	95.6	97.6	98.5
	embedding-4	86.5	92.2	95.6	97.5
	concatenation	93.3	96.3	97.9	98.9
Aircraft	embedding-3	83.4	90.3	94.7	96.9
	embedding-4	72.3	81.1	88.7	92.7
	concatenation	84.1	90.8	94.8	97.2

Table 7. The retrieval performance on CUB200 using different combinations of layers.

Combinations	Dimensions	R@1
layer4	512	69.3%
layer2 + layer4	512 512	69.9%
layer3 + layer4	512 512	73.1%
layer2 + layer3 + layer4	512 512 512	73.3%
layer4	1024	70.8%
layer2 + layer4	512 512	69.9%
layer3 + layer4	512 512	73.1%
layer2 + layer3 + layer4	341 341 342	71.8%

better performance than using layer4 alone. Furthermore, the combination of layer2, layer3 and layer4 achieves even better performance. However, leveraging features from layer2 does not improve performance as much as the leverage of features from layer3. We suspect that the low-level features extracted by layer2 contribute less to distinguishing subcategories than the higher-level features extracted by layer3. When we fix the total dimension, leveraging features from layer2 results in performance degradation. We suspect that reducing the dimensions of embedding-3 and embedding-4 leads to a reduction in discriminative information, resulting in performance degradation. And the performance improvement achieved by leveraging features from layer2 is insufficient to compensate for the performance degradation caused by reducing the dimensions of embedding-3 and embedding-4, which leads to a decrease in overall performance. Therefore, we choose to extract features from the third and fourth layers.

5   Conclusion

In this paper, we propose a simple yet efficient end-to-end model that is a simple modification on Resnet50. We aim to leverage features from multiple layers of CNN to construct more discriminative embeddings for better retrieval. Upon this, we propose a novel idea and conduct extensive experiments to explore the impact of different combinations of layers on performance. Based on the experiments, we extract features from the third and fourth layers of Resnet50 to construct an embedding based on the features from these two layers respectively. When testing, we concatenate these two embeddings to get a more discriminative embedding for retrieval. The last but the most important, our method achieves the state-of-the-art results on the most challenging dataset CUB200 and competitive results on Cars196 and FGVC Aircraft. In addition, our method demonstrates strong scalability, achieving the state-of-the-art results on two general-purpose retrieval datasets SOP and In-Shop. Compared to those localization-based methods which exploit discriminative local features among subcategories, our method could be employed in wider realistic retrieval scenarios for both general-purpose image retrieval and FGOR.

Acknowledgements. This work is supported by the Sichuan Provincial Social Science Programs Project under Grants SC22EZD065 and the Fundamental Research Funds for the Central Universities under Grants XGBDFZ04 and ZYGX2019F005. Besides, our heartfelt thanks go to the anonymous reviewers for their valuable recommendations and thoughtful feedback.

References

1. Boudiaf, M., et al.: A unifying mutual information view of metric learning: cross-entropy vs. pairwise losses. In: Vedaldi, A., Bischof, H., Brox, T., Frahm, J.-M. (eds.) ECCV 2020. LNCS, vol. 12351, pp. 548–564. Springer, Cham (2020). https://doi.org/10.1007/978-3-030-58539-6_33
2. Branson, S., Van Horn, G., Belongie, S., Perona, P.: Bird species categorization using pose normalized deep convolutional nets. arXiv preprint arXiv:1406.2952 (2014)
3. He, K., Zhang, X., Ren, S., Sun, J.: Deep residual learning for image recognition. In: Proceedings of the IEEE Conference on Computer Vision and Pattern Recognition, pp. 770–778 (2016)
4. Jun, H., Ko, B., Kim, Y., Kim, I., Kim, J.: Combination of multiple global descriptors for image retrieval. arXiv preprint arXiv:1903.10663 (2019)
5. Kim, S., Kim, D., Cho, M., Kwak, S.: Embedding transfer with label relaxation for improved metric learning. In: Proceedings of the IEEE/CVF Conference on Computer Vision and Pattern Recognition, pp. 3967–3976 (2021)
6. Kim, Y., Park, W.: Multi-level distance regularization for deep metric learning. In: Proceedings of the AAAI Conference on Artificial Intelligence, vol. 35, pp. 1827–1835 (2021)
7. Krause, J., Stark, M., Deng, J., Fei-Fei, L.: 3D object representations for fine-grained categorization. In: Proceedings of the IEEE International Conference on Cvision Workshops, pp. 554–561 (2013)

8. Li, P., Li, Y., Xie, H., Zhang, L.: Neighborhood-adaptive structure augmented metric learning. In: Proceedings of the AAAI Conference on Artificial Intelligence, vol. 36, pp. 1367–1375 (2022)
9. Li, Z., et al.: Rethinking the optimization of average precision: only penalizing negative instances before positive ones is enough. In: Proceedings of the AAAI Conference on Artificial Intelligence, vol. 36, pp. 1518–1526 (2022)
10. Lim, J., Yun, S., Park, S., Choi, J.Y.: Hypergraph-induced semantic tuplet loss for deep metric learning. In: Proceedings of the IEEE/CVF Conference on Computer Vision and Pattern Recognition, pp. 212–222 (2022)
11. Liu, Z., Luo, P., Qiu, S., Wang, X., Tang, X.: DeepFashion: powering robust clothes recognition and retrieval with rich annotations. In: Proceedings of the IEEE Conference on Computer Vision and Pattern Recognition, pp. 1096–1104 (2016)
12. Maji, S., Rahtu, E., Kannala, J., Blaschko, M., Vedaldi, A.: Fine-grained visual classification of aircraft. arXiv preprint arXiv:1306.5151 (2013)
13. Moskvyak, O., Maire, F., Dayoub, F., Baktashmotlagh, M.: Keypoint-aligned embeddings for image retrieval and re-identification. In: Proceedings of the IEEE/CVF Winter Conference on Applications of Computer Vision, pp. 676–685 (2021)
14. Oh Song, H., Xiang, Y., Jegelka, S., Savarese, S.: Deep metric learning via lifted structured feature embedding. In: Proceedings of the IEEE Conference on Computer Vision and Pattern Recognition, pp. 4004–4012 (2016)
15. Roth, K., Vinyals, O., Akata, Z.: Integrating language guidance into vision-based deep metric learning. In: Proceedings of the IEEE/CVF Conference on Computer Vision and Pattern Recognition, pp. 16177–16189 (2022)
16. Roth, K., Vinyals, O., Akata, Z.: Non-isotropy regularization for proxy-based deep metric learning. In: Proceedings of the IEEE/CVF Conference on Computer Vision and Pattern Recognition, pp. 7420–7430 (2022)
17. Russakovsky, O., et al.: Imagenet large scale visual recognition challenge. Int. J. Comput. Vis. **115**, 211–252 (2015)
18. Seidenschwarz, J.D., Elezi, I., Leal-Taixé, L.: Learning intra-batch connections for deep metric learning. In: International Conference on Machine Learning, pp. 9410–9421. PMLR (2021)
19. Shen, C., Zhou, C., Jin, Z., Chu, W., Jiang, R., Chen, Y., Hua, X.S.: Learning feature embedding with strong neural activations for fine-grained retrieval. In: Proceedings of the on Thematic Workshops of ACM Multimedia 2017, pp. 424–432 (2017)
20. Shen, Y., Sun, X., Wei, X.S., Jiang, Q.Y., Yang, J.: SEMICON: a learning-to-hash solution for large-scale fine-grained image retrieval. In: Computer Vision-ECCV 2022: 17th European Conference, Tel Aviv, Israel, October 23–27, 2022, Proceedings, Part XIV, pp. 531–548. Springer, Cham (2022). https://doi.org/10.1007/978-3-031-19781-9_31
21. Teh, E.W., DeVries, T., Taylor, G.W.: ProxyNCA++: revisiting and revitalizing proxy neighborhood component analysis. In: Vedaldi, A., Bischof, H., Brox, T., Frahm, J.-M. (eds.) ECCV 2020. LNCS, vol. 12369, pp. 448–464. Springer, Cham (2020). https://doi.org/10.1007/978-3-030-58586-0_27
22. Wang, S., Wang, Z., Li, H., Ouyang, W.: Category-specific nuance exploration network for fine-grained object retrieval. In: Proceedings of the AAAI Conference on Artificial Intelligence, vol. 36, pp. 2513–2521 (2022)
23. Wang, X., Zhang, H., Huang, W., Scott, M.R.: Cross-batch memory for embedding learning. In: Proceedings of the IEEE/CVF Conference on Computer Vision and Pattern Recognition, pp. 6388–6397 (2020)

24. Wei, X.S., Luo, J.H., Wu, J., Zhou, Z.H.: Selective convolutional descriptor aggregation for fine-grained image retrieval. IEEE Trans. Image Process. **26**(6), 2868–2881 (2017)
25. Xu, F., Wang, M., Zhang, W., Cheng, Y., Chu, W.: Discrimination-aware mechanism for fine-grained representation learning. In: Proceedings of the IEEE/CVF Conference on Computer Vision and Pattern Recognition, pp. 813–822 (2021)
26. Yang, Z., Bastan, M., Zhu, X., Gray, D., Samaras, D.: Hierarchical proxy-based loss for deep metric learning. In: Proceedings of the IEEE/CVF Winter Conference on Applications of Computer Vision, pp. 1859–1868 (2022)
27. Zeng, X., Liu, S., Wang, X., Zhang, Y., Chen, K., Li, D.: Hard decorrelated centralized loss for fine-grained image retrieval. Neurocomputing **453**, 26–37 (2021)
28. Zhai, A., Wu, H.Y.: Classification is a strong baseline for deep metric learning. arXiv preprint arXiv:1811.12649 (2018)
29. Zheng, X., Ji, R., Sun, X., Wu, Y., Huang, F., Yang, Y.: Centralized ranking loss with weakly supervised localization for fine-grained object retrieval. In: IJCAI, pp. 1226–1233 (2018)

Predefined-Time Synchronization of Complex Networks with Disturbances by Using Sliding Mode Control

Lei Zhou[1], Hui Zhao[1(⊠)] (iD), Aidi Liu[2], Sijie Niu[1], Xizhan Gao[1], and Xiju Zong[1]

[1] Shandong Provincial Key Laboratory of Network Based Intelligent Computing, School of Information Science and Engineering, University of Jinan, Jinan 250022, China
hz_paper@163.com

[2] Collaborative Innovation Center of Memristive Computing Application, Qilu Institute of Technology, Jinan 250200, China

Abstract. This paper aims to investigate the issue of predefined-time synchronization in complex networks with disturbances using sliding mode control technology. Firstly, a new predefined-time stability lemma is proposed based on the Lyapunov second method. By adjusting parameters, this new lemma can degenerate into previous ones and is, therefore, more general. Based on the proposed stability lemma and sliding mode technology, a new sliding mode surface is designed and an effective new sliding mode controller is designed to ensure that the system achieves synchronization within the predefined time. Moreover, the new effective sliding mode controller proposed in this paper has two advantages as follows: (1) The proposed sliding mode controller is robust against disturbances, which aligns more with practical application requirements. (2) The predefined time is set as the sliding mode controller parameter, avoiding overestimation of synchronization time. Finally, numerical simulations are presented to demonstrate the effectiveness of the proposed approach.

Keywords: Complex Networks · Predefined-time Synchronization · Sliding Mode Control · Disturbances

1 Introduction

Synchronization is a typical cluster behavior in complex networks, which has received much attention due to its broad applications in areas such as image encryption [1,2], information security [3,4], and public transportation [5,6]. Essentially, synchronization refers to the circumstance where two or more complex networks possessing different initial states eventually reach identical motion states under the influence of a controller over a while. Following the introduction of the concept of synchronization, different types of synchronization for complex

© The Author(s), under exclusive license to Springer Nature Singapore Pte Ltd. 2024
B. Luo et al. (Eds.): ICONIP 2023, CCIS 1961, pp. 212–225, 2024.
https://doi.org/10.1007/978-981-99-8126-7_17

networks have been proposed, including complete synchronization [7] and projective synchronization [8], along with others. In terms of convergence time, the mentioned synchronization types are divided into asymptotic synchronization [7,8] or exponential synchronization [9,10], implying that complex networks can only achieve synchronization as time tends towards infinity. Currently, numerous breakthroughs have been attained in research on asymptotic synchronization. However, in practical applications, to accelerate convergence and improve efficiency, it is often necessary for systems to achieve synchronization within the finite time. Consequently, to optimize synchronization time, the concept of finite-time synchronization has been proposed and extensively employed [11–13]. It is important to note that synchronization time is closely correlated with the initial values, but given that most practical systems possess unknown initial values, making it impossible to accurately obtain synchronization time.

To address the issue of finite-time stability mentioned above, Polyakov introduced a new concept of finite-time stability known as fixed-time stability [14], where the upper bound of the stable time is independent of the initial values. Based on this concept, research on fixed-time synchronization has attracted increasing attention from researchers [15–17]. It is worth noting that the upper bound on synchronization time depends on the parameters of the controller and the system. However, in practical applications, it is more desirable for the system to achieve synchronization within the predefined time independent of any initial value or parameter. Therefore, the concept of predefined-time stability was proposed [18]. Predefined-time stability means the system can stabilize to equilibrium within the predefined time independent of any initial values, system parameters, and controller parameters. In addition, the predefined time can also be set as a controller parameter to avoid overestimating the stability time. In recent years, predefined-time synchronization based on this concept has become a hot topic of research [19–21].

According to the concept of synchronization, it is possible to achieve synchronization in complex networks by designing suitable controllers. Currently, researchers have designed various control strategies, such as adaptive control [22,23], feedback control [24,25], and sliding mode control [26,27]. The main advantage of sliding mode control is its strong robustness to disturbances. Compared to accurately measuring the value of disturbances, it can more easily eliminate the impact of disturbances by estimating their boundaries. In practical applications, systems are inevitably influenced by disturbances and other factors. Therefore, sliding mode control has been widely used in practical applications due to its robustness against disturbances [26–30]. Although many synchronization schemes based on sliding mode control have been proposed, research on complex network synchronization based on sliding mode control technology and predefined-time stability theory has not attracted much attention due to its complexity.

Taking the above discussion as inspiration, this paper mainly investigates the predefined-time synchronization problem of complex networks with disturbances based on sliding mode control. The main contributions and features are as follows:

(1) First, a new predefined-time stability lemma is proposed based on Lyapunov second method. Compared with previous lemmas, the proposed lemma is more general and stability time has flexible operability.
(2) Accoding to the proposed lemma and sliding mode control techniques, the predefined time synchronization is achieved between complex networks with disturbances.

The remaining sections of the paper are organized as follows: Sect. 2 presents the models of complex networks, assumptions, definitions, and lemmas used in this paper, along with deriving a new predefined-time stability lemma. Section 3 provide some effective conclusions some combines based on the proposed predefined-time stability lemma and sliding mode control scheme. Section 4 validates the effectiveness of this scheme through numerical simulation experiments. Finally, Sect. 5 provides the conclusion and prospect.

2   Preliminaries

In this paper, a complex network consisting of N nodes with disturbances is described as follows:

$$\dot{x}_i(t) = f(x_i(t)) + c \sum_{p=1}^{N} a_{ip} \Gamma x_p(t) + \zeta_i(t) + u_i(t), \tag{1}$$

where $x_i(t) = (x_{i1}(t), x_{i2}(t), ..., x_{in}(t))^T$ represents the state vector of ith node, $i = 1, 2, 3, ..., N$, $f(\cdot) : R^n \rightarrow R^n$ is a continuous nonlinear function, $c>0$ is on behalf of the coupling strength, Γ denotes the inner coupling matrix. Matrix $A = (a_{ip}) \in R^{N \times N}$ is the external coupling matrix which stands for the basic topology of the complex network, if there is a connection from node j to node i, then $a_{ip} \neq 0$, otherwise, $a_{ip} = 0$, and the diagonal elements of A are defined as $a_{ii} = - \sum_{p=1, j \neq i}^{N} a_{ip}$, $\zeta_i(t) \in R^n$ denotes disturbance, $u_i(t) \in R^n$ is the control input.

The target of complex network synchronization can be described as follows:

$$\dot{s}(t) = f(s(t)), \tag{2}$$

where $s(t) = (s_1(t), s_2(t), ..., s_n(t))^T$ is the state of the target node.

Define $e_i(t) = x_i(t) - s(t)$, the error system is obtained as follows:

$$\dot{e}_i(t) = f(x_i(t)) - f(s(t)) + c \sum_{p=1}^{N} a_{ip} \Gamma e_p(t) + \zeta_i(t) + u_i(t). \tag{3}$$

Assumption 1. ([31]) Assume that the function $f(\cdot)$ satisfies the Lipschitz continuous condition, and there is a postive constant l such that

$$|f(x) - f(y)| \leq l|x - y|, \quad l \in R^+.$$

Assumption 2. Assume that the disturbance $\zeta_i(t)$ is bounded with respect to a positive number ϕ:

$$|\zeta_i(t)| \leq \phi.$$

Definition 1. *([32]) Predefined-time stability*
If the system (3) is fixed-time stability, and there exists a predefined time $T_c > 0$ which is completely independent of initial conditions, and other controller parameters such that the settling-time function $T_{e(0)} \leq T_c$. Then the system (3) is often considered to be predefined-time stability.

Lemma 1. *([33]) If $\nu_1, \nu_2, ..., \nu_m \geq 0$, $\Psi > 1$, $0 < \sigma \leq 1$, then we have*

$$\sum_{j=1}^{m} \nu_j^\sigma \geq (\sum_{j=1}^{m} \nu_j)^\sigma, \quad \sum_{j=1}^{m} \nu_j^\alpha \geq m^{1-\Psi}(\sum_{j=1}^{m} \nu_j)^\Psi.$$

Lemma 2. *For system (3), if there is a positive definite and radially unbounded Lyapunov function $V(e(t)) : C^{Nn} \to R$, T_c is a custom parameter and the function $V(e(t))$ satisfies the following two conditions:*

(1) $V(e(t)) = 0 \Leftrightarrow e(t) = 0$.
(2) For any $V(e(t)) > 0$, satisfying:

$$\dot{V}(e(t)) \leq -\frac{G_c}{T_c}(\alpha V^{1+\frac{\delta}{2}} + \beta V^{1-\frac{\delta}{2}} + cV + \rho),$$

where α, β, ρ, c, T_c, $G_c > 0$, $0 < \delta < 1$. The system (3) realized the predefined-time stability within the predefined time T_c and

$$G_c = \frac{1}{c}ln(\frac{c+\rho}{\rho}) + \frac{\pi}{\delta\beta}\sqrt{\frac{\beta}{\alpha}}.$$

Proof. For any $e(t) \in R^{Nn}$, the settling time function follows the following form:

$$T_{e(0)} \leq \frac{T_c}{G_c} \int_0^{V(e(0))} \frac{dV}{\alpha V^{1+\frac{\delta}{2}} + \beta V^{1-\frac{\delta}{2}} + cV + \rho},$$

$$\leq \frac{T_c}{G_c} \int_0^1 \frac{dV}{cV + \rho} + \frac{T_c}{G_c} \int_1^{+\infty} \frac{dV}{\alpha V^{1+\frac{\delta}{2}} + \beta V^{1-\frac{\delta}{2}}}.$$

Case 1:

$$\frac{T_c}{G_c} \int_0^1 \frac{dV}{cV + \rho} = \frac{T_c}{G_c c} ln(\frac{c+\rho}{\rho}).$$

Case 2: Let $\omega = V^{\frac{\delta}{2}}$, we have $d\omega = \frac{\delta}{2}V^{\frac{\delta}{2}-1}dV$.

$$\frac{T_c}{G_c} \int_1^{+\infty} \frac{dV}{\alpha V^{1+\frac{\delta}{2}} + \beta V^{1-\frac{\delta}{2}}} = \frac{T_c}{G_c} \int_1^{+\infty} \frac{V^{\frac{\delta}{2}-1}dV}{\alpha V^\delta + \beta},$$

$$= \frac{2}{\delta\alpha}\frac{T_c}{G_c} \int_1^{+\infty} \frac{d\omega}{\omega^2 + (\sqrt{\frac{\beta}{\alpha}})^2},$$

$$\leq \frac{T_c}{G_c}\frac{\pi}{\delta\beta}\sqrt{\frac{\beta}{\alpha}}.$$

Therefore, we have

$$T_{e(0)} \leq \frac{T_c}{G_c c} ln(\frac{c+\rho}{\rho}) + \frac{T_c}{G_c} \frac{\pi}{\delta\beta} \sqrt{\frac{\beta}{\alpha}} \leq T_c.$$

Remark 1. Predefined-time stability is a special case of fixed-time stability. The parameter G_c in Lemma 2 can be considered as the minimum upper bound of stable time in fixed-time stability, which is independent of system parameters and initial conditions. Using the Lyapunov second method in Lemma 2 provides guarantees predefined-time stability, and the predefined time T_c can be adjusted arbitrarily according to practical needs.

Remark 2. If $\rho = 0$, $c = 0$, $G_c = \frac{\pi}{\delta}$, the form of predefined-time stability in Lemma 2 will degenerate into the form of predefined-time stability in Refs. [32,34]; if $\rho = 0$, $c = 2$, $\alpha = \beta = 1$, $G_c = \frac{2}{\delta}$, the form of predefined-time stability in Lemma 2 will degenerate into the form of predefined-time stability in Refs. [35]. So it can be concluded that Lemma 2 is more generalized.

3   Main Results

The main task in this section is to design a suitable sliding surface and construct an effective controller to achieve predefined-time stability of complex networks. The synchronization time of complex networks consists of two time periods. The first period is the time required for the controller to drive the error system from the initial state to the sliding surface, which is denoted as T_{c2} later. The second period is the time required for the error system to stabilize to the origin along the sliding mode surface, which is denoted as T_{c1} later. Therefore, the total synchronization time of complex networks is $T_{c1} + T_{c2}$. The proof for the two periods will be given in the subsequent content.

3.1   Design of a New Sliding Mode Surface

Theorem 1. *If the state trajectory of the error system (3) is controlled and converged onto the designed sliding surface (4) through a subsequent designed controller (12), then the error system (3) will stabilize to the origin along the sliding surface (4) within the predefined time T_{c1}. The design of the sliding surface is as follows:*

$$s_i(t) = e_i(t) + \int_0^t [\frac{G_{c1}}{T_{c1}}(b_1 sign(e_i(\tau))|e_i(\tau)|^{1-\frac{\delta}{2}} + b_2 sign(e_i(\tau))|e_i(\tau)|^{1+\frac{\delta}{2}}$$

$$+ b_3 e_i(\tau) + b_4 sign(e_i(\tau))) + ke_i(\tau) - c\sum_{p=1}^{N} a_{ip}\Gamma e_p(\tau)]d\tau, \tag{4}$$

where b_1, b_2, b_3, $b_4 > 0$, $\alpha = b_2(Nn)^{-\frac{\delta}{4}}$, $\beta = b_1$, $c = b_3$, $\rho = b_4$, $\lambda = \lambda_{max}(A \otimes \Gamma)$, $c\lambda - k = 0$. G_{c1} is defined as follows:

$$G_{c1} = \frac{1}{c} ln(\frac{c+\rho}{\rho}) + \frac{\pi}{\delta\beta} \sqrt{\frac{\beta}{\alpha}}.$$

Proof. According to the sliding mode theory, when the error system (3) is on the sliding surface (4), we can obtain

$$s_i(t) = \dot{s}_i(t) = 0.$$

Therefore, the sliding mode dynamics can be expressed as follows:

$$
\begin{aligned}
\dot{e}_i(t) = &-[\frac{G_{c1}}{T_{c1}}(b_1 sign(e_i(t))|e_i(t)|^{1-\frac{\delta}{2}} + b_2 sign(e_i(t))|e_i(t)|^{1+\frac{\delta}{2}} \\
&+ b_3 e_i(t) + b_4 sign(e_i(t))) - c\sum_{p=1}^{N} a_{ip}\Gamma e_p(t) + ke_i(t)].
\end{aligned}
\tag{5}
$$

In order to facilitate the subsequent calculation, let $e_i = e_i(t)$, the Lyapunov function is constructed as follows:

$$V = (\sum_{i=1}^{N} e_i^T e_i)^{\frac{1}{2}}.$$

By taking its derivative, we can obtain

$$
\begin{aligned}
\dot{V} = &\frac{1}{V}\sum_{i=1}^{N} e_i^T \dot{e}_i, \\
= &-\frac{1}{V}\sum_{i=1}^{N} e_i^T[\frac{G_{c1}}{T_{c1}}(b_1 sign(e_i)|e_i|^{1-\frac{\delta}{2}} + b_2 sign(e_i)|e_i|^{1+\frac{\delta}{2}} \\
&+ b_3 e_i + b_4 sign(e_i)) - c\sum_{p=1}^{N} a_{ip}\Gamma e_p + ke_i].
\end{aligned}
\tag{6}
$$

By using Lemma 1, we have

$$
\begin{aligned}
&-\frac{1}{V}\sum_{i=1}^{N} e_i^T(\frac{G_{c1}}{T_{c1}}(b_1 sign(e_i)|e_i|^{1-\frac{\delta}{2}}) \\
= &-\frac{G_{c1}}{T_{c1}}\frac{b_1}{V}\sum_{i=1}^{N}\sum_{j=1}^{n}|e_{ij}|^{2-\frac{\delta}{2}} = -\frac{G_{c1}}{T_{c1}}\frac{b_1}{V}\sum_{i=1}^{N}\sum_{j=1}^{n}(e_{ij}^2)^{1-\frac{\delta}{4}}, \\
\leq &-\frac{G_{c1}}{T_{c1}}\frac{b_1}{V}\sum_{i=1}^{N}(\sum_{j=1}^{n}e_{ij}^2)^{1-\frac{\delta}{4}} \leq -\frac{G_{c1}}{T_{c1}}\frac{b_1}{V}(\sum_{i=1}^{N}\sum_{j=1}^{n}e_{ij}^2)^{1-\frac{\delta}{4}} = -b_1\frac{G_{c1}}{T_{c1}}V^{1-\frac{\delta}{2}}.
\end{aligned}
\tag{7}
$$

By employing the Lemma 2, we have

$$
\begin{aligned}
-\frac{1}{V}\sum_{i=1}^{N} e_i^T(\frac{G_{c1}}{T_{c1}}b_2 sign(e_i)|e_i|^{1+\frac{\delta}{2}}) = &-\frac{G_{c1}}{T_{c1}}\frac{b_2}{V}\sum_{i=1}^{N}\sum_{j=1}^{n}(e_{ij}^2)^{1+\frac{\delta}{4}}, \\
\leq &-b_2\frac{G_{c1}}{T_{c1}}(Nn)^{-\frac{\delta}{4}}V^{1+\frac{\delta}{2}}.
\end{aligned}
\tag{8}
$$

Furthermore

$$-\frac{G_{c1}}{T_{c1}}\frac{b_3}{V}\sum_{i=1}^{N}e_i^T e_i = -b_3\frac{G_{c1}}{T_{c1}}V. \tag{9}$$

$$-\frac{G_{c1}}{T_{c1}}\frac{b_4}{V}\sum_{i=1}^{N}e_i^T sign(e_i) = -\frac{G_{c1}}{T_{c1}}\frac{b_4}{V}\sum_{i=1}^{N}\sum_{j=1}^{n}|e_{ij}| \leq -\frac{G_{c1}}{T_{c1}}b_4. \tag{10}$$

Let $\lambda = \lambda_{max}(A \otimes \Gamma)$

$$-(\frac{1}{V}e_i^T\sum_{i=1}^{N}(-c\sum_{p=1}^{N}a_{ip}\Gamma e_p + ke_i)) = \frac{1}{V}(ce^T(A \otimes \Gamma)e - ke^T e),$$

$$\leq \frac{1}{V}(c\lambda e^T e - ke^T e) = 0. \tag{11}$$

Substitute (7)–(11) into (6)

$$\dot{V} \leq -b_1\frac{G_{c1}}{T_{c1}}V^{1-\frac{\delta}{2}} - b_2\frac{G_{c1}}{T_{c1}}(Nn)^{-\frac{\delta}{4}}V^{1+\frac{\delta}{2}} - b_3\frac{G_{c1}}{T_{c1}}V - \frac{G_{c1}}{T_{c1}}b_4.$$

Let $\alpha = b_2(Nn)^{-\frac{\delta}{4}}$, $\beta = b_1$, $c = b_3$, $\rho = b_4$, we have

$$\dot{V} \leq -\frac{G_{c1}}{T_{c1}}(\alpha V^{1+\frac{\delta}{2}} + \beta V^{1-\frac{\delta}{2}} + cV + \rho).$$

By combining with Lemma 2, it follows that the error system (3) will converge to zero within the predefined time T_{c1} on the sliding surface (4).

3.2 Design of a New Sliding Mode Controller

Theorem 2. *If Assumptions 1 and 2 hold, the error system (3) will converge from its initial state to the sliding surface (4) designed above, under the designed controller (12), within the predefined time T_{c2}. The design of the controller is shown as follows:*

$$u_i = -\frac{G_{c1}}{T_{c1}}(b_1 sign(e_i)|e_i|^{1-\frac{\delta}{2}} + b_2 sign(e_i)|e_i|^{1+\frac{\delta}{2}} + b_3 e_i + b_4 sign(e_i))$$

$$- me_i - \phi sign(s_i) - \frac{G_{c2}}{T_{c2}}(d_1 sign(s_i)|s_i|^{1-\frac{\delta}{2}} + d_4 sign(s_i) \tag{12}$$

$$+ d_2 sign(s_i)|s_i|^{1+\frac{\delta}{2}} + d_3 s_i),$$

where $m = k + l$, $G_{c2} = \frac{1}{c}ln(\frac{c+\rho}{\rho}) + \frac{\pi}{\delta\beta}\sqrt{\frac{\beta}{\alpha}}$, $d_1, d_2, d_3, d_4 > 0$, $\alpha = d_2(Nn)^{-\frac{\delta}{4}}$, $\beta = d_1$, $c = d_3$, $\rho = d_4$.

Proof. In order to facilitate the subsequent calculation, make $s_i = s_i(t)$, the Lyapunov function is constructed as follows:

$$V = (\sum_{i=1}^{N} s_i^T s_i)^{\frac{1}{2}}.$$

By taking its derivative, we can obtain

$$\dot{V} = \frac{1}{V} \sum_{i=1}^{N} s_i^T \dot{s}_i,$$

$$= \frac{1}{V} \sum_{i=1}^{N} s_i^T [\dot{e}_i + \frac{G_{c1}}{T_{c1}} (b_1 sign(e_i)|e_i|^{1-\frac{\delta}{2}} + b_2 sign(e_i)|e_i|^{1+\frac{\delta}{2}} + b_4 sign(e_i) \qquad (13)$$

$$+ b_3 e_i) + k e_i - c \sum_{p=1}^{N} a_{ip} \Gamma e_p],$$

$$= \frac{1}{V} \sum_{i=1}^{N} s_i^T [f(x_i) - f(s) + c \sum_{p=1}^{N} a_{ip} \Gamma e_p(t) + \zeta_i(t) + u_i(t) + \frac{G_{c1}}{T_{c1}} (b_1 sign(e_i)|e_i|^{1-\frac{\delta}{2}}$$

$$+ b_2 sign(e_i)|e_i|^{1+\frac{\delta}{2}} + b_3 e_i + b_4 sign(e_i)) + k e_i - c \sum_{p=1}^{N} a_{ip} \Gamma e_p].$$

Substitute (12) into (6)

$$\dot{V} = \frac{1}{V} \sum_{i=1}^{N} s_i^T [f(x_i) - f(s) + c \sum_{p=1}^{N} a_{ip} \Gamma e_p(t) + \zeta_i(t) - \frac{G_{c1}}{T_{c1}} (b_1 sign(e_i)|e_i|^{1-\frac{\delta}{2}} + b_4 sign(e_i)$$

$$+ b_3 e_i + b_2 sign(e_i)|e_i|^{1+\frac{\delta}{2}}) - m e_i - \phi sign(s_i) - \frac{G_{c2}}{T_{c2}} (d_4 sign(s_i) + d_1 sign(s_i)|s_i|^{1-\frac{\delta}{2}}$$

$$+ d_2 sign(s_i)|s_i|^{1+\frac{\delta}{2}} + d_3 s_i) + \frac{G_{c1}}{T_{c1}} (b_1 sign(e_i)|e_i|^{1-\frac{\delta}{2}} + b_2 sign(e_i)|e_i|^{1+\frac{\delta}{2}}$$

$$+ b_3 e_i + b_4 sign(e_i)) + k e_i - c \sum_{p=1}^{N} a_{ip} \Gamma e_p]. \qquad (14)$$

Based on Assumptions 1, 2 and condition $m = k + 1$, we have

$$\dot{V} \leq \frac{1}{V} \sum_{i=1}^{N} s_i^T [l e_i + k e_i - m e_i - \frac{G_{c2}}{T_{c2}} (d_1 sign(s_i)|s_i|^{1-\frac{\delta}{2}} + d_2 sign(s_i)|s_i|^{1+\frac{\delta}{2}}$$

$$+ d_3 s_i + d_4 sign(s_i))] + \frac{1}{V} \sum_{i=1}^{N} |s_i^T|(|\zeta_i(t)| - \phi),$$

$$\leq \frac{1}{V} \sum_{i=1}^{N} s_i^T [l e_i + k e_i - m e_i - \frac{G_{c2}}{T_{c2}} (d_1 sign(s_i)|s_i|^{1-\frac{\delta}{2}} + d_2 sign(s_i)|s_i|^{1+\frac{\delta}{2}}$$

$$+ d_3 s_i + d_4 sign(s_i))], \qquad (15)$$

$$\leq \frac{1}{V} \sum_{i=1}^{N} s_i^T [-\frac{G_{c2}}{T_{c2}} (d_1 sign(s_i)|s_i|^{1-\frac{\delta}{2}} + d_2 sign(s_i)|s_i|^{1+\frac{\delta}{2}} + d_3 s_i + d_4 sign(s_i))].$$

The following simplification process is consistent with Theorem 1. In the end, we obtain:

$$\dot{V} \le -d_1 \frac{G_{c2}}{T_{c2}} V^{1-\frac{\delta}{2}} - d_2 \frac{G_{c2}}{T_{c2}} (Nn)^{-\frac{\delta}{4}} V^{1+\frac{\delta}{2}} - d_3 \frac{G_{c2}}{T_{c2}} V - \frac{G_{c2}}{T_{c2}} d_4.$$

Let $\alpha = d_2(Nn)^{-\frac{\delta}{4}}$, $\beta = d_1$, $c = d_3$, $\rho = d_4$, we have

$$\dot{V} \le -\frac{G_{c2}}{T_{c2}} (\alpha V^{1+\frac{\delta}{2}} + \beta V^{1-\frac{\delta}{2}} + cV + \rho). \tag{16}$$

According to Lemma 2, we can conclude that the error system (3) is capable of achieving convergence to the sliding surfaces within the predefined time T_{c2}.

Remark 3. It is widely recognized that the value of the disturbance in most practical applications is typically bounded, which justifies the validity of Assumption 2 proposed in this paper. Furthermore, since $|\zeta_i(t)| \le \phi$, the term $\phi sign(s_i)$ can be employed to nullify the impact of disturbances in the design of controllers. Therefore, it is evident that Assumption 2 plays a crucial role in the formulation of effective control strategies.

Remark 4. In conclusion, the predefined-time sliding mode control scheme ensures that the error system will eventually converge to zero within the predefined time $T_{c1} + T_{c2}$, meaning that system (1) and system (2) will achieve synchronization within the predefined time $T_{c1} + T_{c2}$. This approach is particularly useful in implementing synchronization operations in various applications.

4   Numerical Simulations

In this section, the effectiveness of the proposed approach was verified through numerical simulation experiments. The selection of the Lorenz system as the target system in this paper is shown below:

$$\dot{s}(t) = f(s(t)) = Hs(t) + W(s(t)), \tag{17}$$

where

$$H = \begin{pmatrix} -10 & 10 & 0 \\ 28 & -1 & 0 \\ 0 & 0 & -8/3 \end{pmatrix}, \quad W(s(t)) = \begin{pmatrix} 0 \\ -s_1(t)s_3(t) \\ s_1(t)s_2(t) \end{pmatrix}.$$

When $s_0 = (0, 2, 9)^T$, the three-dimensional state trajectory of Lorenz system $s(t)$ is shown in Fig. 1.

Consider a network of six nodes:

$$\dot{x}_i(t) = f(x_i(t)) + c \sum_{p=1}^{6} a_{ip} \Gamma x_p(t) + \zeta_i(t) + u_i(t), \tag{18}$$

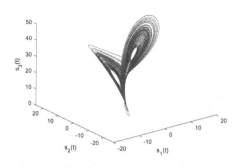

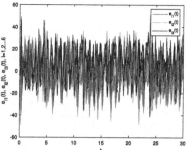

Fig. 1. The three-dimensional state trajectory of the system (17).

Fig. 2. The three-dimensional state trajectory of the error system without using controller.

where the expression of the $f(x_i(t))$ function is the same as $f(s_i(t))$, $f(x_i(t)) = Hx_i(t) + W(x_i(t))$. $c = 1$, $\Gamma = 0.4E_3$, E_3 is the third-order identity matrix,

$$A = (a_{ip}) = \begin{pmatrix} 3 & -1 & 0 & 0 & -1 & -1 \\ -1 & 1 & 0 & 0 & 0 & 0 \\ 0 & 0 & 3 & -1 & -1 & -1 \\ 0 & 0 & -1 & 3 & -1 & -1 \\ -1 & 0 & -1 & -1 & 4 & -1 \\ -1 & 0 & -1 & -1 & -1 & 4 \end{pmatrix}.$$

The initial values of the complex network are chosen as: $x_1(0) = (2, 4, 5)^T$, $x_2(0) = (1, 2.5, 0.5)^T$, $x_3(0) = (3.5, 3, 4.5)^T$, $x_4(0) = (-1, -4, -2)^T$, $x_5(0) = (-5, -3, -1.5)^T$, $x_6(0) = (-0.5, -2.5, -3.5)^T$.

Figure 2 shows the trajectory of the error system without using the controller. From Fig. 2, it is evident that without the application of a controller, the error system (3) will not converge to 0.

In the simulation experiment, it is assumed that the disturbance is $\zeta_i(t) = (2sin(t)cos(t), sin(2t), cos(t))^T$. The expression of disturbance provides information that enables us to derive that $|2sin(t)cos(t)| \leq 1$, $|sin(2t)| \leq 1$, $|cos(t)| \leq 1$, so we can choose $\phi = 1$. To satisfy Assumption 1, we assume that $l = 1$. We set the parameters of the sliding mode surface and the sliding mode controller as $d_1 = 0.6$, $d_2 = 0.8$, $d_3 = 1$, $d_4 = 1$, $b_1 = 1$, $b_2 = 2$, $b_3 = 1$, $b_4 = 1$, $\delta = 0.5$. By calculation, it can be obtained that $G_{c1} = 5.1360$, $G_{c2} = 7.4949$, $\lambda = 2.0944$, $k = 2.0944$, $m = 3.0944$. The predefined-time parameters are $T_{c1} = 0.5$, $T_{c2} = 0.1$.

Through Theorem 2, we know that the error system converges to the sliding surface (4) within the predefined time $T_{c2} = 0.1$. Figure 3 further indicates that the trajectory of the error system converges to the sliding mode surface within the predefined time $T_{c2} = 0.1$, which is consistent with theoretical analysis. According to Theorem 1 and 2, the system (17) and (18) can achieve predefined-

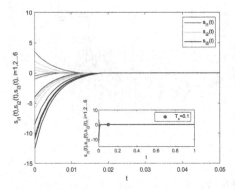
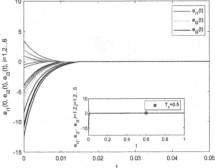

Fig. 3. The three-dimensional state trajectory of the sliding mode surface (4).

Fig. 4. The three-dimensional state trajectory of the error system with the sliding mode controller (12).

time synchronization under the action of sliding mode controller (12). Figure 4 further shows that the error system converges to the zero equilibrium point within the predefined time $T_{c1} + T_{c2}$.

In addition, we conducted a comparative experiment to verify the robustness of the sliding control scheme proposed in this paper against disturbances. The experimental group consisted of a complex network with disturbances, while the control group consisted of a complex network without disturbances. The sliding mode control scheme proposed in this paper was applied to the experimental group. In the control group, (4) was still used as the sliding mode surface, but since the network did not contain any disturbances, the form of the sliding mode controller was modified accordingly. The sliding mode controller for the control group is given by the following equation:

$$
\begin{aligned}
u_i = & -\frac{G_{c1}}{T_{c1}}\left(b_1 sign(e_i)|e_i|^{1-\frac{\delta}{2}} + b_2 sign(e_i)|e_i|^{1+\frac{\delta}{2}} + b_3 e_i \right. \\
& \left. + b_4 sign(e_i)\right) - me_i - \frac{G_{c2}}{T_{c2}}\left(d_1 sign(s_i)|s_i|^{1-\frac{\delta}{2}} \right. \\
& \left. + d_2 sign(s_i)|s_i|^{1+\frac{\delta}{2}} + d_3 s_i + d_4 sign(s_i)\right).
\end{aligned}
\tag{19}
$$

Remark 5. Compared with controller (12), there is no upper bound of disturbances in the formula of controller (19). Moreover, the control group also achieves synchronization within the predefined time, and the proof process is similar to that of the experimental group, so it is not repeated here.

The trajectory of the control group is shown in Fig. 5. By comparing Fig. 5 and Fig. 4, it was found that the trajectories of the control group and the experimental group almost overlap. This comparative experiment shows that disturbances do not affect the synchronization process, thus verifying the robustness of the sliding mode controller proposed in this paper against disturbances.

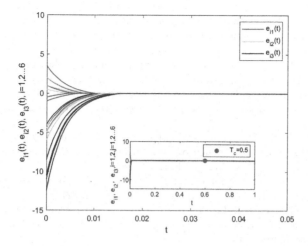

Fig. 5. The three-dimensional state trajectory of the error system without disturbances.

5   Conclusion and Prospect

This paper uses sliding mode control technology to investigate the problem of predefined-time synchronization with disturbances in complex networks. Firstly, a new lemma for predefined-time stability is proposed based on the definition of predefined-time stability. Compared with previous lemmas for predefined-time stability, the new lemma presented in this article is more generalized and can ensure that the system state variables converge to zero within the predetermined time. Secondly, a new sliding mode control method is designed based on the proposed new criterion. The technique ensures that the synchronization error system converges to zero before the predefined time, thus achieving synchronization of complex networks within the predefined time. Finally, numerical simulation results validate the effectiveness of the proposed method in this paper.

In the future, complex network models with more complex structures will be studied, and more simple and effective sliding mode controllers will be designed.

Acknowledgments. This work is supported by the National Natural Science Foundation of China (Grant Nos. 62103165, 62101213), and the Natural Science Foundation of Shandong Province (Grant No. ZR2022ZD01).

References

1. Zhou, H., Liu, Z., Chu, D., Li, W.: Sampled-data synchronization of complex network based on periodic self-triggered intermittent control and its application to image encryption. Neural Netw. **152**, 419–433 (2022)
2. Sheng, S., Zhang, X., Lu, G.: Finite-time outer-synchronization for complex networks with Markov jump topology via hybrid control and its application to image encryption. J. Franklin Inst. **355**(14), 6493–6519 (2018)

3. Zhou, L., Tan, F.: A chaotic secure communication scheme based on synchronization of double-layered and multiple complex networks. Nonlinear Dyn. **96**, 869–883 (2019)

4. Liu, D., Ye, D.: Secure synchronization against link attacks in complex networks with event-triggered coupling. Inf. Sci. **628**, 291–306 (2023)

5. Gunasekaran, N., Zhai, G., Yu, Q.: Sampled-data synchronization of delayed multi-agent networks and its application to coupled circuit. Neurocomputing **413**, 499–511 (2020)

6. Xu, C., Xu, H., Su, H., Liu, C.: Adaptive bipartite consensus of competitive linear multi-agent systems with asynchronous intermittent communication. Int. J. Robust Nonlinear Control **32**(9), 5120–5140 (2022)

7. Wu, X., Bao, H.: Finite time complete synchronization for fractional-order multiplex networks. Appl. Math. Comput. **377**, 125188 (2022)

8. Wang, S., Zheng, S., Cui, L.: Finite-time projective synchronization and parameter identification of fractional-order complex networks with unknown external disturbances. Fractal Fract. **6**(6), 298 (2022)

9. Yao, W., Wang, C., Sun, Y., Gong, S., Lin, H.: Event-triggered control for robust exponential synchronization of inertial memristive neural networks under parameter disturbance. Neural Netw. **164**, 67–80 (2023)

10. Ni, Y., Wang, Z.: Intermittent sampled-data control for exponential synchronization of chaotic delayed neural networks via an interval-dependent functional. Expert Syst. Appl. **223**, 119918 (2023)

11. Yang, D., Li, X., Song, S.: Finite-time synchronization for delayed complex dynamical networks with synchronizing or desynchronizing impulses. IEEE Trans. Neural Netw. Learn. Syst. **33**(2), 736–746 (2020)

12. Luo, Y., Yao, Y.: Finite-time synchronization of uncertain complex dynamic networks with time-varying delay. Adv. Differ. Equ. **2020**, 1–22 (2020)

13. Du, F., Lu, J.G., Zhang, Q.H.: Delay-dependent finite-time synchronization criterion of fractional-order delayed complex networks. Commun. Nonlinear Sci. Numer. Simul. **119**, 107072 (2023)

14. Polyakov, A.: Nonlinear feedback design for fixed-time stabilization of linear control systems. IEEE Trans. Autom. Control **57**(8), 2106–2110 (2011)

15. Pang, L., Hu, C., Yu, J., Wang, L., Jiang, H.: Fixed/preassigned-time synchronization for impulsive complex networks with mismatched parameters. Neurocomputing **511**, 462–476 (2022)

16. Shirkavand, M., Pourgholi, M., Yazdizadeh, A.: Robust global fixed-time synchronization of different dimensions fractional-order chaotic systems. Chaos Solitons Fractals **154**, 111616 (2022)

17. Zheng, C., Hu, C., Yu, J., Jiang, H.: Fixed-time synchronization of discontinuous competitive neural networks with time-varying delays. Neural Netw. **153**, 192–203 (2022)

18. Sánchez-Torres, J.D., Sanchez, E.N., Loukianov, A.G.: A discontinuous recurrent neural network with predefined time convergence for solution of linear programming. In: 2014 IEEE Symposium on Swarm Intelligence, pp. 1–5 (2014)

19. Assali, E.A.: Predefined-time synchronization of chaotic systems with different dimensions and applications. Chaos Solitons Fractals **147**, 110988 (2021)

20. Liu, A., et al.: A new predefined-time stability theorem and its application in the synchronization of memristive complex-valued BAM neural networks. Neural Netw. **153**, 152–163 (2022)

21. Han, J., Chen, G., Hu, J.: New results on anti-synchronization in predefined-time for a class of fuzzy inertial neural networks with mixed time delays. Neurocomputing **495**, 26–36 (2022)
22. Aghababa, M.P., Aghababa, H.P.: A general nonlinear adaptive control scheme for finite-time synchronization of chaotic systems with uncertain parameters and nonlinear inputs. Nonlinear Dyn. **69**(4), 1903–1914 (2012)
23. Li, S., Peng, X., Tang, Y., Shi, Y.: Finite-time synchronization of time-delayed neural networks with unknown parameters via adaptive control. Neurocomputing **308**, 65–74 (2018)
24. Du, H., He, Y., Cheng, Y.: Finite-time synchronization of a class of second-order nonlinear multi-agent systems using output feedback control. IEEE Trans. Circuits Syst. I Regul. Pap. **61**(6), 1778–1788 (2014)
25. Xu, D., Liu, Y., Liu, M.: Finite-time synchronization of multi-coupling stochastic fuzzy neural networks with mixed delays via feedback control. Fuzzy Sets Syst. **411**, 85–104 (2021)
26. Xiong, J.J., Zhang, G.B., Wang, J.X., Yan, T.H.: Improved sliding mode control for finite-time synchronization of nonidentical delayed recurrent neural networks. IEEE Trans. Neural Netw. Learn. Syst. **31**(6), 2209–2216 (2019)
27. Sun, J., Wang, Y., Wang, Y., Shen, Y.: Finite-time synchronization between two complex-variable chaotic systems with unknown parameters via nonsingular terminal sliding mode control. Nonlinear Dyn. **85**, 1105–1117 (2016)
28. Khanzadeh, A., Pourgholi, M.: Fixed-time sliding mode controller design for synchronization of complex dynamical networks. Nonlinear Dyn. **88**, 2637–2649 (2017)
29. Al-Mahbashi, G., Noorani, M.M.: Finite-time lag synchronization of uncertain complex dynamical networks with disturbances via sliding mode control. IEEE Access **7**, 7082–7092 (2019)
30. Hui, M., Zhang, J., Iu, H.H.C., Yao, R., Bai, L.: A novel intermittent sliding mode control approach to finite-time synchronization of complex-valued neural networks. Neurocomputing **513**, 181–193 (2022)
31. Xu, Y., Wu, X., Li, N., Liu, L., Xie, C., Li, C.: Fixed-time synchronization of complex networks with a simpler nonchattering controller. IEEE Trans. Circuits Syst. II Express Briefs **67**(4), 700–704 (2019)
32. Anguiano-Gijón, C.A., Muñoz-Vázquez, A.J., Sánchez-Torres, J.D., Romero-Galván, G., Martínez-Reyes, F.: On predefined-time synchronisation of chaotic systems. Chaos Solitons Fractals **122**, 172–178 (2019)
33. Hardy, G.H., Littlewood, J.E., Pólya, G.: Inequalities. Cambridge University Press, Cambridge Mathematical Library (1934)
34. Li, Q., Yue, C.: Predefined-time modified function projective synchronization for multiscroll chaotic systems via sliding mode control technology. Complexity **2020**, 1–11 (2020)
35. Zhang, M., Zang, H., Bai, L.: A new predefined-time sliding mode control scheme for synchronizing chaotic systems. Chaos Solitons Fractals **164**, 112745 (2022)

Reward-Dependent and Locally Modulated Hebbian Rule

Zhaofan Liu[1,2] and Da-Hui Wang[2,3(✉)]

[1] Peking University Huilongguan Clinical Medical School,
Beijing Huilongguan Hospital, Beijing 100096, China
[2] School of Systems Science, Beijing Normal University, Beijing 100875, China
wangdh@bnu.edu.cn
[3] National Key Laboratory of Cognitive Neuroscience and Learning,
Beijing Normal University, Beijing 100875, China

Abstract. In pattern classification tasks, the convolutional network has been widely used to detect the features and the error backpropagation with gradient descent (BPGD) algorithm has been used to train the network. However, the plasticity of the synapse between neurons often depends on the potential of the pre- and postsysnaptic membrane and the local concentration of neuromodulators such as dopamine. In this paper, we proposed the reward-dependent and locally modulated (RDLM) Hebbian rule to train a multi-layer network to perform the image classification tasks. We found that by introducing local modulation, the reward-dependent Hebbian rule can successfully train multi-layer networks. We have shown that the performance of our method on the MNIST and Fashion MNIST datasets can compete with the traditional BPGD algorithm. In conclusion, we proposed a biologically plausible learning rule that can compete with traditional BPGD in image classification tasks. The method can potentially be used to train the network with complex architecture for complex tasks.

Keywords: Hebbian Rule · Image Classification Tasks · Neural Network · Error-Backpropagation

1 Introduction

Recently, artificial neural networks have achieved remarkable success. They have the ability to accomplish a wide range of tasks [1,2]. The collaboration between artificial intelligence and neuroscience has yielded significant achievements, with artificial neural networks playing a crucial role in promoting neuroscience research [3,4]. Numerous studies have employed artificial neural networks to provide explanations for various phenomena in neuroscience. For instance, Yang successfully utilized artificial neural networks (ANNs) and recurrent neural networks (RNNs) to elucidate network structure and neural activity in the

Supported by NSFC under: 32171094.

© The Author(s), under exclusive license to Springer Nature Singapore Pte Ltd. 2024
B. Luo et al. (Eds.): ICONIP 2023, CCIS 1961, pp. 226–237, 2024.
https://doi.org/10.1007/978-981-99-8126-7_18

field of neuroscience [5,6]. Many investigations have also utilized deep neural networks to gain insights into sensory systems [7], and convolutional neural networks (CNNs) have been employed to explain V1 tuning properties [8].

Error backpropagation is widely and efficiently used to train the artificial neural networks, find the appropriate weights and enable the network to perform tasks [1,9]. Although error backpropagation has achieved remarkable success, its updating process is considered biologically implausible [10,11]. For instance, backpropagation relies on the symmetry of forward and feedback pathways, a characteristic rarely found in biological neural systems [12]. Backpropagation can lead to an extreme amplification of the error signal in deep neural networks [13]. Feedback signals typically modulate the activity of neural regions rather than altering connectivity between them [14,15]. In real neural systems, synapses typically undergo changes based on the activity of both pre and post-synaptic neurons [16].

At the same time, biologically inspired approaches provide alternative solutions for training neural networks. One example is the Hebbian rule. Unlike BPGD algorithm, the Hebbian rule is considered biologically plausible because it takes into account synapses that are influenced by pre- and postsynaptic activity [16], which is commonly observed in neural systems [17]. The Hebbian rule has been extensively investigated in neural systems through biophysical models [18–20]. Another example is spike-timing dependent plasticity (STDP) in Spiking Neural Networks (SNNs) which are considered the next generation of neural networks [21,22], as they are more consistent with real neurons communicating through spikes. SNNs can perform digital recognition tasks using STDP [23] and symmetric STDP [24]. Multi-layer SNNs can incorporate brain-inspired rules for classification tasks [22]. Furthermore, inspired by the Hebbian rule, many studies have implemented it to update artificial neural networks for task completion as well [25–27]. Dmitry and Hopfield modified the Hebbian rule for competing units that can detect features and added a backpropagation layer in classification tasks [28]. Zou adopted the Hebbian rule for learning continuous attractor neural networks and found that images are stored as a continuous family of stationary states of the network [29]. The Hebbian rule has also been adopted in Predictive Coding Networks, which can approximate the error of backpropagation [30]. The reward-modulated Hebbian rule has been proposed [31] and subsequently has been further developed to generate chaotic patterns in the brain [32–34], meet functional requirements [35], and establish categorical representations [36].

It is important to note that this mechanism has not yet been used for training artificial neural networks for classification tasks and could serve as a biological mechanism for readout in the brain instead of error backpropagation. Thus, the objective of this paper is to bridge the gap between neuroscience and machine learning by implementing a biologically plausible learning algorithm for classification tasks. We explore the performance of the reward-modulated Hebbian rule in classification tasks. Surprisingly, we find that in certain circumstances, the bio-inspired algorithm outperforms gradient backpropagation. This sheds light on the potential of leveraging biological mechanisms for improved machine learning performance and provides a neural mechanism for updating weights in the neural system during learning.

2   Methods

2.1   Dataset and Network Structure

We test our algorithm on the Modified National Institute of Standards and Technology(MNIST) dataset [37] and Fashion MNIST dataset [38].

MNIST dataset consists of 28 × 28 gray-scale images of hand-drawn digits. The fashion MNIST dataset is an image dataset with markers, where each image consists of 28 × 28 pixels, and each image can be viewed as a 728-dimensional vector. There are ten types of images in the dataset, which are composed of "T-shirt", "pants", "pullover", "dress", "coat", "sandals", "shirt", "Sneakers", "Bags", "Ankle boots", and each category is uniformly distributed. We randomly divide the dataset (60,000 samples) into a training set (50,000 samples) and a test set (10,000 samples).

We utilize the Hebbian learning rule to determine the weights of the neural network. One-layer neural network and two-layer of neural network are considered in our work. One layer neural network consists of a stimulus and a readout layer, and the network structure is 784-10. Two-layer neural networks have a hidden layer, and the network structure is 784-2000-10. The activation function for the nodes in the hidden layer is the ReLU function.

2.2   Reward Modulated Hebbian Rule

The traditional reward-modulated Hebbian rule can be written as follow:

$$\Delta W(t) = \eta(t)[z(t) - \overline{z}(t)][P(t) - \overline{P}(t)]r(t) \tag{1}$$

where $r(t)$ is the neural activity of hidden layer, and $z(t)$ is the neural activity of readout layer $z(t) = Wr(t) + \xi(t)$, $\xi(t)$ is the driven noise and $\overline{z}(t)$ is the mean value of $z(t)$ for a fixed interval, and $z(t) - \overline{z}(t)$ have a relaxation effect [34]. $P(t)$ is viewed as a reward signal, $P(t) = -\sum_{i=1}^{L}[z_i(t) - f_i(t)]^2$, where $f(t)$ is the label of readout layer, the subscription i is the index of readout neurons. The reward signal can be described as follow [39]:

$$M(t) = \begin{cases} 1 & \text{if } P(t) > \overline{P}(t) \\ 0 & \text{if } P(t) \leq \overline{P}(t) \end{cases} \tag{2}$$

and the update rule can be written as:

$$\Delta W = \eta(t)[z(t) - \overline{z}(t)][M(t)]r(t) \tag{3}$$

In the traditional reward-modulated Hebbian rule, the reward signal is a scalar value and cannot be used to update the feedforward neural network. We propose that for the update of the feedforward neural network, the update for each two-layer projection should be heterogeneous instead of a uniform value for all projections. Therefore, we modify the reward signal to be locally modulated for each epoch.

2.3   Reward-Dependent and Locally Modulated Hebbian Rule

For Single Layer Update. Instead of a global reward signal, we think that the reward signal should be local since that dopamine is locally released and diffused in volume. So, we develop a local reward signal and apply it to reward modulated Hebbian rule. For single-layer updates, a perturbation is given to the readout layer and results in two possible cases as shown in Fig. 1. The first case is that the classification is correct and strengthens the projection to the correct category (long-term potentiation, LTP); the other case is that the classification is incorrect and weakens the projection to the wrong category (long-term depression, LTD).

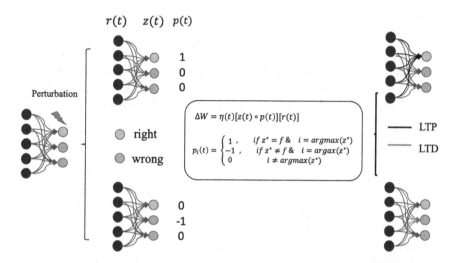

Fig. 1. Schematic diagram of reward-dependent and locally modulated Hebbian rule. The main mechanism is as follows: a perturbation is given to the readout layer and results in two cases. One case is that the classification is correct and the projection to the correct category is strengthened (long-term potentiation, LTP), such as the gray projection in the first row. The other case is that the classification is incorrect and the projection to the wrong category is weakened (long-term depression, LTD), such as the orange projection in the second row. (Color figure online)

We then formulate our ideas. We first introduce the perturbation as additive noise as follows:

$$z_i(t) = w_i r(t) + \xi_i(t) \tag{4}$$

where subscription i is the index of readout neurons. The report z^* after perturbation is:

$$z_i^* = \begin{cases} 1 & i = \mathrm{argmax}(z) \\ 0 & \text{else} \end{cases} \tag{5}$$

and the reward signal can be written as:

$$p_i = \begin{cases} 1 & z^* = f \ \& \ i = \mathrm{argmax}(z^*) \\ -1 & z^* \neq f \ \& \ i = \mathrm{argmax}(z^*) \\ 0 & i \neq \mathrm{argmax}(z^*) \end{cases} \tag{6}$$

where $z^* = f$ for correct classification and $z^* \neq f$ for incorrect classification, $i = \mathrm{argmax}(z^*)$ determine the projection to the class that is reported by the neural network. The update algorithm can be described as:

$$\Delta W(t)_{m \times n} = \eta[(z(t) - \bar{z}(t)) \circ P(t)]_{m \times k}[r(t)^T]_{k \times n} \tag{7}$$

if there is a relaxation effect on the readout layer or:

$$\Delta W(t)_{m \times n} = \eta[z(t) \circ P(t)]_{m \times k}[r(t)^T]_{k \times n} \tag{8}$$

if there is no relaxation effect on the readout layer. The $\circ$ denotes the Hadamard product. The subscription m is the batch size for each update, k is the dimension of the readout layer and n is the dimension of the hidden layer (Fig. 2).

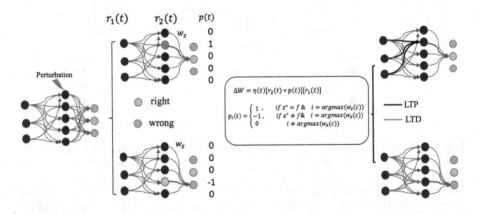

Fig. 2. Schematic diagram of a two-layer neural network updated by RDLM Hebbian rule. The main mechanism is as follows: the output layer r_2 receives a perturbation and results in two cases. One case is that the classification is correct and strengthens the neural activity of the node with the largest connection weight to the correct category, i.e., enhances the projection to this node (LTP), such as the gray projection in the first row; The other case is that the classification is wrong after the perturbation, which weakens the projection to the wrong category. The activity of the node with the largest connection weight, i.e., weakens the projection to this node (LTD), such as the green projection in the second row. (Color figure online)

For Two-Layer Update. The connection weights of the two-layer neural network can be updated using the RDLM Hebbian rule. For convenience, we denote the activity of the hidden layer as r_2 and the neural activity of the stimulus layer as r_1. The update algorithm of connection weights W_z from the hidden layer to the readout layer has been given in Eqs. (7) and (8). The multi-layer RDLM Hebbian rule updates the connection weights W from the stimulus to the hidden layer.

We add a certain perturbation to the hidden layer, thus the hidden layer neural activity can be written as $r_2(t) = \hat{r}_2(t) + \xi(t)$. The neural network reports the classification correctly or incorrectly after the perturbation. When the stimulus is classified to the correct category, we select the node j in the hidden layer with the largest weight projected to the correct category (i.e. $z^* = f$ and $j = \mathrm{argmax}(w_z(c))$) and enhance its neural activity, i.e., enhance the projection from stimulus layer to node j; when the stimulus is classified to the wrong category, we select the node j in the hidden layer with the largest weight projected to the wrong category (i.e., $z^* \neq f$ and $j = \mathrm{argmax}(w_z(c))$), weaken its neural activity, i.e., weaken the connection from stimulus layer to node j. The local reward signal can be written as follows:

$$p_i = \begin{cases} 1 & z^* = f \quad \& \quad i = \mathrm{argmax}(w_z(c)) \\ -1 & z^* \neq f \quad \& \quad i = \mathrm{argmax}(w_z(c)) \\ 0 & i \neq \mathrm{argmax}(w_z(c)) \end{cases} \tag{9}$$

where $w_z(c)$ is the weight from the hidden layer projected to the cth class, and cth class is the class reported by the neural network. The update rule of connection weights W can be written as follow:

$$\Delta W(t) = \eta(t)[r_1(t) \circ P(t)]r_2(t)^T \tag{10}$$

2.4 Measures of the Network Performance

We use the accuracy of the reports of the network and the mean square error of the reports to measure the performance of the network. The accuracy is the ratio of the correct reports to all reports, while the mean square error is $MSE = \frac{1}{n}\sum_n (y_i - \hat{y}_i)^2$, where y_i is the real label of the $i'th$ sample and $\hat{y}_i$ is the label reported by the network.

2.5 Parameters of Learning

For a single-layer network, the matrix connecting the stimulus to the readout layer is initialized using a uniform distribution between -0.01 and 0.01. The learning rate, denoted as η, is set to 0.001. During the updating process, additive noise obeys a uniform distribution with values between -1 and 1. For a two-layer network, the matrix connecting the stimulus to the hidden layer and the hidden layer to the readout layer are also initialized using a uniform distribution between -0.01 and 0.01. Other parameters are the same as in the single-layer network.

3   Results

3.1   RDLM Hebbian Rule Can Train the Network to Perform MNIST Classification Task

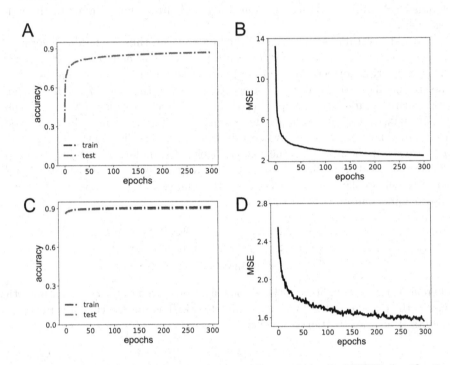

Fig. 3. RDLM Hebbian rule can train the network to perform the MNIST classification task. (A) The accuracy of a single-layer network quickly increases with the training epochs. (B) The mean square error quickly decreases during the training process. (C) The accuracy of the two-layer network. (D) The mean square error of the two-layer network during the training process.

We apply the RDLM Hebbian rule to separately train a single-layer or a two-layer network to perform the MNIST classification task. As shown in Fig. 3, the accuracy of the single-layer network performing on MNIST dataset approaches 89.1% (Fig. 3A) after 300 epochs training. The mean square error between the predicted label of the stimulus and the real label monotonically decreases during the training process (Fig. 3B). Therefore, this rule can effectively update the single-layer neural network. Similarly, we use the RDLM Hebbian rule to train a two-layer neural network to perform the MNIST classification task. The accuracy during the training process is shown in Fig. 3C, where we observe that the accuracy of the classification approach is around 90.8%. The mean squared error monotonically decreases during the training process (Fig. 3D) and is smaller than the single-layer neural network. Therefore, it can be seen that using the RDLM Hebbian rule can train the neural network to complete the MNIST classification task and the two-layer neural network outperforms the single-layer neural network. In summary, we show that the RDLM Hebbian rule can train the network to perform the MNIST classification task.

3.2   Compare the RDLM Hebbian Rule with Error-Backpropagation Algorithm

We compared RDLM Hebbian rule with the traditional BPGD algorithm. In a single-layer neural network (784-10), the RDLM Hebbian rule updates the weights based on reward signals without loss function. In contrast, the BPGD algorithm often used the cross-entropy loss function and updated the connection matrix following the derivatives of the loss function. The training and test accuracy of the network trained using RDLM Hebbian rule and error BP gradient descent algorithm are shown in Fig. 4(A) and (B), where the batch size is 100 for Fig. 4(A) and 1000 for Fig. 4(B). We can see that with a batch size of 1000, the RDLM Hebbian rule outperforms the traditional BPGD algorithm. However, when the batch size is 100, the traditional BP algorithm outperforms RDLM Hebbian rule.

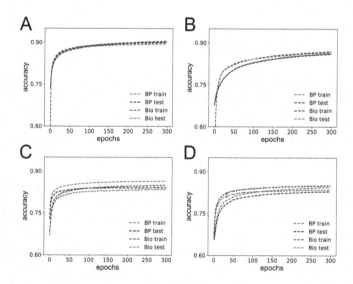

Fig. 4. Comparison of the RDLM Hebbian rule and the error backpropagation algorithm in the single-layer neural network. (A) and (B): the accuracy of the network on the MNIST classification task with batch sizes of 100 and 1000, respectively. (C) and (D): the accuracy of the Fashion-MNIST with batch sizes of 100 and 1000, respectively. Bio denotes the RDLM Hebbian rule, BP means error backpropagation algorithm.

We further train the network to perform the fashion MNIST classification task. The results are shown in Fig. 4(C) and (D). The RDLM Hebbian rule outperforms the BPGD algorithm with a batch size of 1000 during the training, while the traditional BPGD algorithm outperforms the RDLM Hebbian rule with a batch size of 100 during the training. However, the difference in the final accuracy between the two learning rules is smaller than 1%, suggesting that the RDLM Hebbian rule can almost achieve the same performance as the traditional BPGD algorithm.

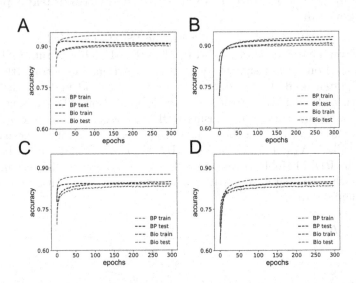

Fig. 5. Comparison of RDLM Hebbian rule and error backpropagation with gradient descent algorithms in the two-layer neural network. (A) and (B) represent the accuracy of completing the classification task in MNIST dataset classification with batch sizes of 100 and 1000, respectively. (C) and (D) represent the accuracy of completing the classification task in Fashion-MNIST with batch sizes of 100 and 1000, respectively.

We also train a two-layer neural network (784-2000-10) to perform the MNIST and fashion MNIST classification task using the RDLM Hebbian rule and the traditional BP algorithm. In the MNIST classification task, with a batch size of 100 (Fig. 5A), the accuracy of the network trained using the BP algorithm(train 94.1%, test 90.9%) is higher than that using the RDLM Hebbian rule(train 90.9%, test 90.1%). However, with training progress, the BPGD algorithm shows overfitting, while the RDLM Hebbian rule approaches the accuracy of the BP algorithm, with a difference of less than 1% in the test dataset. Likewise, with a batch size of 1000 (Fig. 5B), the performance of the RDLM Hebbian rule (train 90.7%, test 90.1%) is slightly lower than that using the BPGD algorithm (BP train 93.1%, test 90.5%), with a difference of less than 1% in the test dataset. In the fashion-MNIST dataset, with a batch size of 100 (Fig. 5C), the training accuracy of the BPGD algorithm is 87.5% and the test accuracy is 84.0%. In the RDLM Hebbian rule, the training accuracy is 84.9% and the test accuracy is 83.3%. With a batch size of 1000 (Fig. 5D), the training accuracy of the BP algorithm is 86.6%, and the test accuracy is 84.1%. In the RDLM Hebbian rule, the training accuracy is 84.6%, and the test accuracy is 83.2%.

Therefore, we can draw a conclusion: in the process of training a single-layer neural network, the accuracy of the network trained using the RDLM Hebbian rule achieves higher accuracy than that using the BPGD algorithm with larger batch size. However, given a smaller batch size, the two algorithms have similar results in performance. During the training stage, the accuracy of the network using the RDLM Hebbian rule is lower than that using the BPGD algorithm, but the difference in test set accuracy is less than 2%. Hence, the RDLM Hebbian rule is both biological plausibility and comparable accuracy to the traditional BPGD algorithm.

4   Conclusion and Discussion

By comparing the RDLM Hebbian rule and the BPGD algorithm, it can be concluded that the RDLM Hebbian rule can train the network to perform image classification tasks in different datasets. However, there are some limitations that need to be mentioned. Firstly, in the two-layer neural network, the performance is not as good as BP, with an approximate 2% performance, we will continue to improve this aspect. Secondly, the parameters used during the training process are sensitive to the model. Inappropriate parameter values, such as learning rate, can render the algorithm ineffective or result in extreme values, which is a common problem in the BPGD algorithm. Thirdly, the RDLM Hebbian rule needs to find the specific connections to update their weights, which takes more time during the training phase compared with the traditional BPGD algorithm.

Acknowledgements. The authors declare that they have no conflict of interest. Any additional information required to reanalyze the data reported in this paper is available from the lead contact upon request.

References

1. LeCun, Y., Bengio, Y., Hinton, G.: Deep learning. Nature **521**(7553), 436–444 (2015)
2. Rusk, N.: Deep learning. Nat. Methods **13**(1), 35 (2016)
3. Helmstaedter, M.: The mutual inspirations of machine learning and neuroscience. Neuron **86**(1), 25–28 (2015)
4. Hassabis, D., Kumaran, D., Summerfield, C., Botvinick, M.: Neuroscience-inspired artificial intelligence. Neuron **95**(2), 245–258 (2017)
5. Wang, P.Y., Sun, Y., Axel, R., Abbott, L.F., Yang, G.R.: Evolving the olfactory system with machine learning. Neuron **109**(23), 3879 (2021)
6. Yang, G.R., Joglekar, M.R., Song, H.F., Newsome, W.T., Wang, X.J.: Task representations in neural networks trained to perform many cognitive tasks. Nat. Neurosci. **22**(2), 297 (2019)
7. Yamins, D.L.K., DiCarlo, J.J.: Using goal-driven deep learning models to understand sensory cortex. Nat. Neurosci. **19**(3), 356–365 (2016)

8. Lindsey, S.G.S.D.J., Ocko, S.A.: A unified theory of early visual representations from retina to cortex through anatomically constrained deep CNNs. In: ICLR (2019)

9. Rumelhart, D.E., Hinton, G.E., Williams, R.J.: Learning representations by back propagating errors. Nature **323**(6088), 533–536 (1986)

10. Lillicrap, T.P., Santoro, A., Marris, L., Akerman, C.J., Hinton, G.: Backpropagation and the brain. Nat. Rev. Neurosci.

11. Whittington, J.C.R., Bogacz, R.: Theories of error back-propagation in the brain. Trends Cogn. Sci. **23**(3), 235–250 (2019)

12. Lillicrap, T.P., Cownden, D., Tweed, D.B., Akerman, C.J.: Random synaptic feedback weights support error backpropagation for deep learning. Nat. Commun. **7**(1), 13276 (2016)

13. Pascanu, R., Mikolov, T., Bengio, Y.: On the difficulty of training recurrent neural networks. JMLR.org (2012)

14. Chen, R., Wang, F., Liang, H., Li, W.: Synergistic processing of visual contours across cortical layers in v1 and v2. Neuron 1388–1402 (2017)

15. Liang, H., Gong, X., Chen, M., Yan, Y., Gilbert, C.D.: Interactions between feedback and lateral connections in the primary visual cortex. Proc. Natl. Acad. Sci. U.S.A. **114**(32) (2017)

16. Kaplan, S.J.: Organization of behavior. Yale J. Biol. Med. **23**(1), 79 (1950)

17. Caporale, N., et al.: Spike timing-dependent plasticity: a hebbian learning rule. Annu. Rev. Neurosci. **31**(1), 25–46 (2008)

18. Miller, K.D., MacKay, D.J.C.: The role of constraints in hebbian learning. Neural Comput. **6**(1), 100–126 (1994)

19. Markram, H., Gerstner, W., Sjöström, P.J.: Spike-timing-dependent plasticity: a comprehensive overview. Front. Synapt. Neurosci. **4**, 2 (2012)

20. Wang, Y., Shi, X., Si, B., Cheng, B., Chen, J.: Synchronization and oscillation behaviors of excitatory and inhibitory populations with spike-timing-dependent plasticity. Cognit. Neurodyn. 1–13 (2022)

21. Maass, W.: Networks of spiking neurons: the third generation of neural network models. Neural Netw. **10**(9), 1659–1671 (1997)

22. Yi, Z., Zhang, T., Bo, X.U.: Improving multi-layer spiking neural networks by incorporating brain-inspired rules. Sci. China Inf. Sci. **05**, 1–11 (2017)

23. Diehl, P.U., Cook, M.: Unsupervised learning of digit recognition using spike-timing-dependent plasticity. Front. Comput. Neurosci. **9**, 99 (2015)

24. Yha, B., Xh, A., Meng, D.A., Bo, X.: A biologically plausible supervised learning method for spiking neural networks using the symmetric STDP rule. Neural Netw. **121**, 387–395 (2020)

25. Sompolinsky, H.: The theory of neural networks: the Hebb rule and beyond. In: van Hemmen, J.L., Morgenstern, I. (eds.) Heidelberg Colloquium on Glassy Dynamics, pp. 485–527. Springer, Heidelberg (1987). https://doi.org/10.1007/BFb0057531

26. Dayan, P., Abbott, L.F.: Theoretical Neuroscience: Computational and Mathematical Modeling of Neural Systems. MIT Press (2005)

27. Hopfield, J.J.: Neural networks and physical systems with emergent collective computational abilities. Proc. Natl. Acad. Sci. **79**(8), 2554–2558 (1982)

28. Krotov, D., Hopfield, J.: Unsupervised learning by competing hidden units. Proc. Natl. Acad. Sci. U.S.A. (2019)

29. Zou, X., Ji, Z., Xiao, L., Mi, Y., Si, W.: Learning a continuous attractor neural network from real images. In: International Conference on Neural Information Processing (2017)

30. Whittington, J.C.R., Bogacz, R.: An approximation of the error backpropagation algorithm in a predictive coding network with local hebbian synaptic plasticity. Neural Comput. **29**(5), 1–34 (2017)
31. Loewenstein, Y., Sebastian Seung, H.: Operant matching is a generic outcome of synaptic plasticity based on the covariance between reward and neural activity. Proc. Natl. Acad. Sci. **103**(41), 15224–15229 (2006)
32. Legenstein, R., Chase, S.M., Schwartz, A.B., Maass, W.: A reward-modulated hebbian learning rule can explain experimentally observed network reorganization in a brain control task. J. Neurosci. Off. J. Soc. Neurosci. **30**(25), 8400–10 (2009)
33. Hoerzer, G.M., Robert, L., Wolfgang, M.: Emergence of complex computational structures from chaotic neural networks through reward-modulated hebbian learning. Cerebral Cortex (3), 677–690 (2012)
34. Miconi, T.: Biologically plausible learning in recurrent neural networks reproduces neural dynamics observed during cognitive tasks. Elife **6** (2017)
35. Frémaux, N., Sprekeler, H., Gerstner, W.: Functional requirements for reward-modulated spike-timing-dependent plasticity. J. Neurosci. **30**(40), 13326–13337 (2010)
36. Engel, T.A., Chaisangmongkon, W., Freedman, D.J., Wang, X.-J.: Choice-correlated activity fluctuations underlie learning of neuronal category representation. Nat. Commun. **6**(1), 1–12 (2015)
37. Lecun, Y., Bottou, L.: Gradient-based learning applied to document recognition. Proc. IEEE **86**(11), 2278–2324 (1998)
38. Xiao, H., Rasul, K., Vollgraf, R.: Fashion-mnist: a novel image dataset for benchmarking machine learning algorithms. arXiv preprint arXiv:1708.07747 (2017)
39. Legenstein, R., Chase, S.M., Schwartz, A.B., Maass, W.: A reward-modulated hebbian learning rule can explain experimentally observed network reorganization in a brain control task. J. Neurosci. **30**(25), 8400–8410 (2010)

Robust Iterative Hard Thresholding Algorithm for Fault Tolerant RBF Network

Jiajie Mai and Chi-Sing Leung[✉]

Department of Electrical Engineering, City University of Hong Kong, Hong Kong,
Hong Kong
jjmai2-c@my.cityu.edu.hk, eeleungc@cityu.edu.hk

Abstract. In the construction of a radial basis function (RBF) network, there are three crucial issues. The first one is to select RBF nodes from training samples. Two additional vital issues are addressing the realization of imperfections and mitigating the impact of outlier training samples. This paper considers that training data contain some outlier samples and that there is weight noise in the RBF weights in the implementation. We formulate the construction of an RBF network as a constrained optimization problem in which the objective function consists of two terms. The first term is designed to suppress the effect of outlier samples, while the second term handles the effect of weight noise. Our formulation has an ℓ_0-norm constraint whose role is to select the training samples for constructing RBF nodes. We then develop the robust iterative hard thresholding algorithm (R-IHT) to solve the optimization problem based on the projected gradient concept. We theoretically study the convergence properties of the R-IHT. We use several benchmark datasets to verify the effectiveness of the proposed algorithm. The performance of our algorithm is superior to a number of state-of-the-art methods.

Keywords: RBF Network · Sparse Approximation · Projected Gradient

1 Introduction

Since the radial basis function (RBF) network is a highly effective approach for modelling general nonlinear mapping with a simple network structure, it is widely used in many applications, including image processing [1] and wind turbine power modeling [2]. In the traditional training process, we need to construct RBF nodes from the training samples first. The representative selection method and training method are random selection [3], clustering, and orthogonal least squares (OLS) algorithm [4]. Recently, a fuzzy set based method is used to determine the RBF conters [5]. However, those mentioned methods cannot handle the realization imperfection.

In the realization of a neural network, there are some imperfections. For instance, when we realize a network in a digital way, there are precision errors

© The Author(s), under exclusive license to Springer Nature Singapore Pte Ltd. 2024
B. Luo et al. (Eds.): ICONIP 2023, CCIS 1961, pp. 238–251, 2024.
https://doi.org/10.1007/978-981-99-8126-7_19

in RBF weights. In analog realization, analog components usually have precision errors too. Also, the precision errors in analog components are usually specified as percentages of the analog components' nominal values. For precision errors, we can model them as multiplicative weight noise [6]. The existence of the imperfection can greatly reduce the accurate of the trained network [7,8].

In recent years, some noise tolerant algorithms [7,8] were proposed to mitigate the existence of multiplicative noise. However, the noise tolerant algorithms in [7,8] do not allow us to directly specify the number of RBF nodes in the resultant network. Also, they cannot handle the existence of outlier training samples.

This paper considers that training data contain some outlier samples, and that in the implementation there are weight noise in the RBF weights. We formulate the construction of an RBF network as a sparse optimization problem, in which the objective function consists of two terms. The first term is designed to suppress the effect of outlier samples, while the second term handles the effect of weight noise. There is an ℓ_0-norm constraint in the formulated optimization problem. The constraint allows us to select the RBF nodes from the training samples and to directly specify the number of RBF nodes in the resultant network. We then develop an algorithm, namely the robust iterative hard thresholding algorithm (R-IHT), to solve the formulated optimization problem based on the projected gradient concept. We theoretically prove that the objective value is strictly decreasing during the training iteration and that the weight vector converges. Simulations further demonstrate that the proposed algorithm outperforms many existing training algorithms.

This paper is organized as follows. Section refsec2 presents the background on RBF networks. Section 3 presents the formulation for the constrained optimization problem that includes weight noise, outlier training sample, and ℓ_0 constraint. The R-IHT is developed in Sect. 4. Also, we analyze R-IHT properties in Sect. 4. Section 5 contains the experimental results and Sect. 6 concludes the paper.

2   Background

We consider the radial basis function (RBF) concept for nonlinear regression problems. The training set is denoted as $\Lambda = \{(\boldsymbol{x}_n, y_n) : \boldsymbol{x}_n \in \Re^M,$ $y_n \in \Re, n = 1, \cdots, N\}$, where $\boldsymbol{x}_n$'s are training inputs and y_n's are training outputs. Assume that we use all training samples to construct the N RBF nodes. It should not worry the above setting because we will introduce an ℓ_0-norm constraint for selecting important RBF nodes in Sect. 3. Given an input $\boldsymbol{x}$, the network output is

$$\xi(\boldsymbol{x}) = \sum_{i=1}^{N} \beta_i \varphi_i(\boldsymbol{x}) = \sum_{i=1}^{N} \beta_i \exp\left(-\frac{1}{\omega} \|\boldsymbol{c}_i - \boldsymbol{x}\|_2^2\right), \tag{1}$$

where β_i is the output weight of the i-th node, $\varphi_i(\boldsymbol{x})$ is the output of the i-th node, $\boldsymbol{c}_i$ is the center of the i-th RBF node, and ω indicates the width parameter of RBFs. Since we use all the training samples, we have $c_i = x_i \, \forall i = 1, \cdots, N$. The conventional training objective is to minimize the training set mean square error (MSE):

$$\text{MSE} = \frac{1}{N} \sum_{n=1}^{N} (y_n - \xi(x_n))^2 = \frac{1}{N} \sum_{n=1}^{N} \left(y_n - \sum_{i=1}^{N} \beta_i \varphi_i(x_n) \right)^2 = \frac{1}{N} \| y - P\beta \|_2^2, \quad (2)$$

where $y = [y_1, \cdots, y_N]^\top$, P is an $N \times N$ matrix with $P_{n,i} = \varphi_i(x_n)$, P is symmetric because we use all the training samples to construct the RBF nodes, and $\beta = [\beta_1, \cdots, \beta_N]^\top$.

The formulation of (2) has three drawbacks. First, using all the training samples to create RBF nodes results in a network with very large size. This can also negatively impact its generalization ability. Additionally, the presence of precision errors in the output RBF weights can significantly reduce the performance of the network [7,8]. Finally, since the ℓ_2-norm concept is used in the objective function, the outlier training samples may significantly degrade the performance of the resultant network.

3   Constrained Optimization Formulation

This section presents a number of methods to resolve the mentioned drawbacks. Subsequently, we formulate the training of an RBF network as a constrained optimization problem.

Weight Noise Effect:
In the realization of a trained RBF network, there are precision errors. When precision errors exist in the RBF weights, β_n's, we can use the multiplicative noise model [6,8] to describe the effect of noise on a weight β_i, given by

$$\tilde{\beta}_i = (1 + v_i)\beta_i, \quad (3)$$

where v_i's are the normalized noise. Clearly, the noise component $v_i \beta_i$ is proportional to the weight magnitude. Assuming that the normalized noise v_i's are independent and identically distributed random variables with mean of zero and variance of σ_v^2, we can obtain the following properties:

$$\left\langle \tilde{\beta}_i \right\rangle = \beta_i, \quad \left\langle \tilde{\beta}_i^2 \right\rangle = (1 + \sigma_v^2)\beta_i^2 \quad \text{and} \quad \left\langle \tilde{\beta}_i \tilde{\beta}_j \right\rangle = \beta_i \beta_j \quad i \neq j, \quad (4)$$

where $\langle . \rangle$ indicates the expectation operator.

With weight noise, the output of a noisy network is $\xi(x) = \sum_{i=1}^{N} \tilde{\beta}_i \varphi_i(x)$, the training set error of a noisy network is

$$\text{MSE} = \frac{1}{N} \sum_{n=1}^{N} \left(y_n - \sum_{i=1}^{N} \tilde{\beta}_i \varphi_i(x_n) \right)^2. \quad (5)$$

From (4) and (5), we can show that the training error over all possible weight noise is

$$\langle\text{MSE}\rangle = \frac{1}{N}\|\boldsymbol{y}-\boldsymbol{P}\boldsymbol{\beta}\|_2^2+\frac{\sigma_v^2}{N}\boldsymbol{\beta}^\top\left(\text{diag}\left(\boldsymbol{P}^\top\boldsymbol{P}\right)\right)\boldsymbol{\beta} = \frac{1}{N}\|\boldsymbol{y}-\boldsymbol{P}\boldsymbol{\beta}\|_2^2+\sigma_v^2\boldsymbol{\beta}^\top\boldsymbol{\Theta}\boldsymbol{\beta}, \quad (6)$$

where $\text{diag}(\boldsymbol{P}^\top\boldsymbol{P})$ is a diagonal matrix, the i-th diagonal element of $\text{diag}(\boldsymbol{P}^\top\boldsymbol{P})$ is $\sum_{n=1}^{N}\varphi_i^2(\boldsymbol{x}_n)$, and the i-th diagonal element of $\boldsymbol{\Theta}$ is $\frac{1}{N}\sum_{n=1}^{N}\varphi_i^2(\boldsymbol{x}_n)$, Note that $\boldsymbol{\Theta}$ is a positive definite matrix.

With the objective function stated in (6), training RBF network is formulated as the following unconstrained optimization problem:

$$\min_{\beta}\frac{1}{N}\|\boldsymbol{y}-\boldsymbol{P}\boldsymbol{\beta}\|_2^2+\sigma_v^2\boldsymbol{\beta}^\top\boldsymbol{\Theta}\boldsymbol{\beta}. \quad (7)$$

In (7), the first term is the training set MSE, while the second term is used to suppress the effect of the multiplicative weight noise. With the second term, the resultant network has the ability to tolerate the effect of weight noise. As the objective function is a positive quadratic function, the optimal solution for (7) can be obtained easily.

Selection of RBF Nodes:
In (7), we use all the training samples to construct the RBF network. This formulation may create two disadvantages. The first one is overfitting, and the other is a waste of resources. To make the resultant network with a specified size, we can formulate the RBF training as the following constrained optimization problem, given by

$$\min_{\beta}\frac{1}{N}\|\boldsymbol{y}-\boldsymbol{P}\boldsymbol{\beta}\|_2^2+\sigma_v^2\boldsymbol{\beta}^\top\boldsymbol{\Theta}\boldsymbol{\beta}, \text{ subject to } \|\boldsymbol{\beta}\|_0\leq M, \quad (8)$$

where M is the specific number of nodes in the resultant network, which is less than N. **The formulation in (8) can handle node selection and weight noise, simultaneously. Also, we explicitly specify the number of nodes of in the resultant network.** Although the formulation in (8) is not convex, we can use the iterative hard threshold (IHT) concept [9], which is based on the projected gradient approach, to find sub-optimal solution.

Handling Outlier
In (8), the MSE term, based on the ℓ_2-norm concept, can handle Gaussian-like noise only, i.e., there are Gaussian-like noise in the training outputs. However, when the training samples contains outliers, the ℓ_2-norm is heavily affected by the magnitude of the outliers. This is because the ℓ_2-norm squares each component of the fitting error vector. Thus, outliers dominates error calculation.

To suppress the effect of outliers, we can use the ℓ_1-norm concept. The formulation stated in (8) becomes

$$\min_{\beta}\frac{1}{N}\|\boldsymbol{y}-\boldsymbol{P}\boldsymbol{\beta}\|_1+\sigma_v^2\boldsymbol{\beta}^\top\boldsymbol{\Theta}\boldsymbol{\beta}, \text{ subject to } \|\boldsymbol{\beta}\|_0\leq M. \quad (9)$$

However, the non-differentiability of the ℓ_1-norm creates difficulties to solve (9). One may argue that we can solve the non-differentiable issue by adding dummy variables with additional inequality constraints into (9). **However, this dummy variable approach creates N additional decision variables. Also, it creates $2N$ additional constraints, and thus the ℓ_0-norm constraint cannot be effectively handled.**

This paper proposes to use an approximation for the ℓ_1-norm, given by

$$\|\boldsymbol{y} - \boldsymbol{P}\boldsymbol{\beta}\|_1 \approx \frac{\sum_{n=1}^{N} \log\left(\cosh\left(a(\boldsymbol{P}_n\boldsymbol{\beta} - y_n)\right)\right)}{a}, \tag{10}$$

where a is large positive number, such as 100, and $\boldsymbol{P}_n$ is the n-th row of $\boldsymbol{P}$. One may worry that the implementation of $\log(\cosh(\cdot))$ function is a bit complication. In fact, in the optimization process, we only need to implement $\tanh(\cdot)$ function, which is a commonly used function in the neural network community. **Since $\log(\cosh(\cdot))$ is strictly convex and radially unbounded, it makes R-IHT with the convergence property.**

Robust Constrained Optimization Formulation
With the approximation, the formulation of constructing an RBF network becomes

$$\min_{\boldsymbol{\beta}} \frac{1}{N} \sum_{n=1}^{N} \frac{\log\left(\cosh\left(a(\boldsymbol{P}_n\boldsymbol{\beta} - y_n)\right)\right)}{a} + \sigma_v^2 \boldsymbol{\beta}^\top \boldsymbol{\Theta}\boldsymbol{\beta}, \quad \text{subject to } \|\boldsymbol{\beta}\|_0 \leq M. \tag{11}$$

4   R-IHT

This section presents an iterative algorithm, namely R-IHT, to solve the problem stated in (11). Afterwards, we present the convergence properties of the R-IHT.

4.1   Algorithm

Define $f(\boldsymbol{\beta}) = \frac{1}{N}\sum_{n=1}^{N} \frac{\log\left(\cosh\left(a(\boldsymbol{P}_n\boldsymbol{\beta} - y_n)\right)\right)}{a} + \sigma_v^2 \boldsymbol{\beta}^\top c\boldsymbol{\Theta}\boldsymbol{\beta}$. The gradient $\nabla f(\boldsymbol{\beta})$ of $f(\boldsymbol{\beta})$ is

$$\nabla f(\boldsymbol{\beta}) = \frac{1}{N}\boldsymbol{P}^\top \tanh\left(a(\boldsymbol{P}\boldsymbol{\beta} - \boldsymbol{y})\right) + 2\sigma_v^2 \boldsymbol{\Theta}\boldsymbol{\beta}. \tag{12}$$

With the gradient, we can use the iterative hard threshold (IHT) concept [9] to solve (11). Note that the original IHT is designed for a constrained optimization problem with a positive quadratic objective and an ℓ_0-norm constraint. In (11), **the objective function is not a quadratic objective function. Hence the existing convergence result of IHT cannot be used in our case.**

Borrowing the concept of IHT, the iterative steps of the R-IHT are given by

$$\boldsymbol{z}^{t+1} = \boldsymbol{\beta}^t - \eta\nabla f(\boldsymbol{\beta}) = \boldsymbol{\beta}^t - \eta\left(\frac{1}{N}\boldsymbol{P}^\top \tanh\left(a(\boldsymbol{P}\boldsymbol{\beta} - \boldsymbol{y})\right) + 2\sigma_v^2 \boldsymbol{\Theta}\boldsymbol{\beta}\right) \tag{13}$$

$$\boldsymbol{\beta}^{t+1} = \mathcal{P}_M\left(\boldsymbol{z}^{t+1}\right), \tag{14}$$

where η is the learning rate and $\mathcal{P}_M(\boldsymbol{z})$ is an operator. The solution of $\mathcal{P}_M(\boldsymbol{z})$ is obtained from solving an optimization problem, given by

$$\min_{\boldsymbol{\beta}} \|\boldsymbol{\beta} - \boldsymbol{z}\|_2^2 \quad \text{subject to} \quad \|\boldsymbol{\beta}\|_0 \le M. \tag{15}$$

The solution of (15) has a closed form solution, given by

$$\beta_i = \begin{cases} 0 & \text{for } |z_i| < \delta_M \\ z_i & \text{for } |z_i| \ge \delta_M \end{cases} \tag{16}$$

Constant δ_M is given as follows. If the number of nonzero elements in $\{|z_1|, \cdots, |z_N|\}$ is greater than or equal to M, then δ_M is equal to the M-th largest element of $\{|z_1|, \cdot, |z_N|\}$. Otherwise, δ_M is equal to the smallest nonzero element in $\{|z_1|, \cdots, |z_N|\}$.

In (13), we perform a gradient descent and the crucial issue is that $f(\boldsymbol{\beta})$ should be differentiable. Clearly, with our approximation $f(\boldsymbol{\beta})$ is differentiable. In (14). we project the updating vector to the feasible region and the projection operation is simple. We will investigate the convergence of the proposed R-IHT in the next section.

4.2 Convergence Behavior

The proof of convergence of the R-IHT is based on the Lipschitz continuous concept. We first prove that the objective function $f(\boldsymbol{\beta})$ is Lipschitz continuous. Afterwards, we prove that during the R-IHT iterations $f(\boldsymbol{\beta}^t)$ is strictly decreasing, i.e., $f(\boldsymbol{\beta}^t) > f(\boldsymbol{\beta}^{t+1})$ for $\boldsymbol{\beta}^t \ne \boldsymbol{\beta}^{t+1}$. Finally, we prove that $f(\boldsymbol{\beta}^t)$ and $\boldsymbol{\beta}^t$ converges.

Preliminary, several simple lemmas are first introduced.

Lemma 1: *In R-IHT, the number of nonzero elements in $\boldsymbol{\beta}^t$ is less than or equal to M.* **Proof**: This lemma is directly obtained from the definition of the projection operator stated in (15). ∎

Lemma 2: *The following inequality is held.*

$$\left\|\boldsymbol{\beta}^{t+1} - \boldsymbol{z}^{t+1}\right\|_2^2 \le \left\|\boldsymbol{\beta}^t - \boldsymbol{z}^{t+1}\right\|_2^2. \tag{17}$$

Proof: Because $\boldsymbol{\beta}^{t+1}$ is the optimal solution of

$$\min_{\boldsymbol{\beta}} \left\|\boldsymbol{\beta} - \boldsymbol{z}^{t+1}\right\|_2^2 \quad \text{subject to} \quad \|\boldsymbol{\beta}\|_0 \le M, \tag{18}$$

$\left\|\boldsymbol{\beta}^{t+1} - \boldsymbol{z}^{t+1}\right\|_2^2$ must be less than or equal to $\left\|\boldsymbol{\beta}^t - \boldsymbol{z}^{t+1}\right\|_2^2$. ∎

Lemma 3: *The objective function $f(\boldsymbol{\beta})$ is strictly convex and is L-Lipschitz continuous.*

Proof: Define $f(\beta) = g(\beta) + h(\beta)$, where $g(\beta) = \frac{1}{N} \sum_{n=1}^{N} \frac{\log\left(\cosh\left(a(P_n\beta - y_n)\right)\right)}{a}$ and $h(\beta) = \sigma_v^2 \beta^\top \Theta \beta$. Thus, we obtain

$$\nabla^2 g(\beta) = \frac{a}{N} P^\top \text{diag}\left(\text{sech}^2\left[a(P\beta - y)\right]\right) P = \frac{a}{N} P^\top D P \tag{19}$$

$$\nabla^2 h(\beta) = 2\sigma_v^2 \Theta \tag{20}$$

$$\nabla^2 f(\beta) = \nabla^2 g(\beta) + \nabla^2 h(\beta), \tag{21}$$

where D is a diagonal matrix with diagonal elements $D_{n,n}$ equal to $\text{sech}^2\left[a(P_n\beta - y_n)\right]$, where P_n is the n-th row of P.

For training inputs x_n's, of course they are bounded, i.e., $\|x_n\| \leq \infty$. Thus, the outputs of RBF nodes are in $(0, 1]$. It is because the outputs of RBF nodes for all the training samples are $\varphi_i(x_n) = \exp\left(-\omega \|c_i - x_n\|_2^2\right)$ for all $i = 1, \cdots, N$ and $n = 1, \cdots, N$, where $c_i = x_i$. Thus, the elements of P are in $(0, 1]$. Also, $\text{sech}^2\left[a(P_n\beta - y_n)\right]$ is greater than 0, and it is less than or equal to 1 for all n. Thus, $P^\top D P$ is positive definite or positive semi-definite, and the elements in $P^\top D P$ are bounded.

Similarly, in diagonal matrix Θ, the diagonal elements are bounded and greater than zero because they are equal to $\sum_{n=1}^{N} \varphi_i^2(x_n)$, where $\varphi_i(x_n) = \exp\left(-\omega \|c_i - x_n\|_2^2\right)$. Thus, Θ is positive definite and is upper bounded.

From the facts that Θ is positive definite and is upper bounded, and that $P^\top D P$ is positive definite or positive semi-definite and $P^\top D P$ are upper bounded, $\nabla^2 f(\beta)$ is positive definite and upper bounded.

As $\nabla^2 f(\beta)$ is positive definite, $f(\beta)$ is strictly convex. Since $\nabla^2 f(\beta)$ is upper bounded, we have

$$\nabla^2 f(\beta) \preceq L_g I, \tag{22}$$

where I is an identity matrix, and L_g is the largest eigenvalue of $\frac{a}{N} P^\top D P + 2\sigma_v^2 \Theta$. Note that for a strictly convex function $f(\beta)$, the condition for L-Lipschitz continuous is $\nabla^2 f(\beta) \preceq L_g I$, where L_g is positive constant. The proof is completed. ■

With Lemmas 1-3, we are ready to show the convergence behavior of R-IHT in the following theorem.

Theorem 1: *In the R-IHT, the objective function value $f(\beta^t)$ is strictly decreasing with t and converges. Also, the weight vector β^t converges.*

Proof: We first show the property of "strictly decreasing". A property of L-Lipschitz is

$$f(\alpha) \leq f(\beta) + \langle \nabla f(\beta), \alpha - \beta \rangle + \frac{L_g}{2} \|\alpha - \beta\|_2^2 \tag{23}$$

Consequently,

$$f(\beta^t) - f(\beta^{t+1}) \geq -\langle \nabla f(\beta^t), \beta^{t+1} - \beta^t \rangle - \frac{L_g}{2} \|\beta^{t+1} - \beta^t\|_2^2 \tag{24}$$

$$= \frac{1}{\eta} \langle z^{t+1} - \beta^t, \beta^{t+1} - \beta^t \rangle - \frac{L_g}{2} \|\beta^{t+1} - \beta^t\|_2^2 \tag{25}$$

$$= \frac{1}{2\eta} \left[\|z^{t+1} - \beta^t\|_2^2 + \|\beta^{t+1} - \beta^t\|_2^2 - \|z^{t+1} - \beta^{t+1}\|_2^2 \right] \tag{26}$$

$$- \frac{L_g}{2} \|\beta^{t+1} - \beta^t\|_2^2. \tag{27}$$

From Lemma 2, (27) becomes

$$f(\beta^t) - f(\beta^{t+1}) = (\frac{1}{2\eta} - \frac{L_g}{2}) \|\beta^{t+1} - \beta^t\|_2^2. \tag{28}$$

Thus, we can choose $\eta < \frac{2}{L_g}$ to make $f(\beta^t) - f(\beta^{t+1}) > 0$. From (28), the equality holds if and only if $x^t = x^{t+1}$. In other words, $f(\beta^t)$ is strictly decreasing during the R-IHT iteration. Note that η is the learning rate.

Table 1. Properties of the five datasets.

Dataset	Training set size	Testing set size	RBF width
ASN	751	752	0.1
MPG	320	72	0.5
Concrete	500	530	0.1
ABA	2000	2177	0.1
WQW	2000	2898	0.1

It is not difficult to show that $f(\beta)$ is low bounded. As $f(\beta^t)$ is strictly decreasing, $f(\beta^t)$ tends to a value for any initial β^0. Also, it not difficult to show that $f(\beta)$ is radial unbounded, i.e., $f(\beta) \to \infty$ as $\|\beta\| \to \infty$. From basic optimization theory, when $f(\beta)$ is low bounded and radial unbounded, and $f(\beta^t)$ is strictly decreasing, as $t \to \infty$ the objective function value $f(\beta^t)$ converges and β^t converges. The proof is completed. ∎

5   Simulation

Setting: Datasets
We utilize five datasets from the UCI machine learning repository [10] to test the performance of R-IHT. They are Airfoil Self Noise (ASN), Auto-MPG (MPG), Concrete, Abalone (ABA), and Wine Quality White (WQW). Detailed properties of these five datasets are listed in Table 1.

Setting: Outlier Level and Weight Noise Level
One of the key contribution in this paper is the robustness of outliers in R-IHT. To verify this property, we add outlier noise to certain proportion of training samples. The proportion is denoted as r. In this paper, $r = 0.1$ or $r = 0.3$. When a training sample is selected for adding outlier, the outlier value is given by

$$e_{Outlier} = y_{min} + \kappa\tau\left(y_{max} - y_{min}\right), \tag{29}$$

where y_{min} and y_{max} are the minimum and maximum value of training outputs y_n's, κ is the magnitude factor (1 or 2), and τ is a uniform random number between 0 to 1. In this paper, two outlier levels, $(r, \kappa) = (0.1, 1)$ and $(r, \kappa) = (0.3, 2)$, are considered.

For weight noise, we consider four weight noise levels: $\sigma_v^2 = \{0.01, 0.04, 0.09, 0.16\}$.

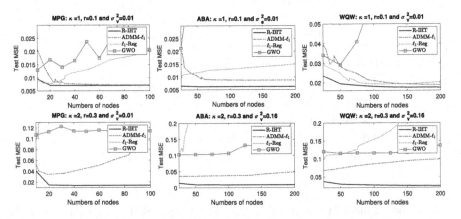

Fig. 1. Test set MSE versus number of node. First row: low outlier level and low weight noise level. Second row: high outlier level and high weight noise level.

Comparison Algorithms
We compare the proposed R-IHT with four algorithms. They are the noise-resistant ADMM-ℓ_1 [8], ℓ_1-norm regularization (ℓ_1-Reg) [11], grey wolf optimizer (GWO) [12], and the maximum mixture correntropy criterion (MMCC) [13].

Table 2. Average test MSE over ten trials with the specific number of RBF centers and outlier level $(r, \kappa) = (0.1, 1)$.

Dateset	Centers Nodes	Noise level σ_v^2	R-IHT	ADMM-ℓ_1	ℓ_1-Reg	GWO	MMCC
ASN	140	0.01	**0.0141**	0.0206	0.0154	0.0384	0.0315
		0.04	**0.0151**	0.0348	0.0192	0.0369	0.0524
		0.09	**0.0167**	0.0498	0.0255	0.0367	0.0857
		0.16	**0.0190**	0.0642	0.0343	0.0410	0.1291
Concrete	80	0.01	0.0200	0.0194	0.0169	0.0436	0.0462
		0.04	0.0209	0.0244	0.0206	0.0432	0.0785
		0.09	**0.0221**	0.0320	0.0269	0.0414	0.1287
		0.16	**0.0236**	0.0433	0.0358	0.0421	0.1910
MPG	30	0.01	**0.0097**	0.0122	0.0103	0.0167	0.1157
		0.04	**0.0104**	0.0171	0.0177	0.0173	0.1394
		0.09	**0.0112**	0.0230	0.0301	0.0188	0.1762
		0.16	**0.0121**	0.0281	0.0474	0.0197	0.2219
ABA	50	0.01	**0.0069**	0.0122	0.0112	0.0557	0.0366
		0.04	**0.0072**	0.0196	0.0185	0.0185	0.0765
		0.09	**0.0076**	0.0224	0.0309	0.0152	0.1377
		0.16	**0.0080**	0.0225	0.0481	0.0158	0.2125
WQW	125	0.01	**0.0178**	0.0209	0.0202	0.0840	0.0514
		0.04	**0.0182**	0.0252	0.0232	0.0522	0.0990
		0.09	**0.0187**	0.0352	0.0284	0.0404	0.1725
		0.16	**0.0192**	0.0407	0.0356	0.0380	0.2633

The ℓ_1-Reg and ADMM-ℓ_1 formulates the training problem as an unconstrained optimization problem with an ℓ_1-norm regularizer. The number of nodes in the resultant network is indirectly controlled by the regularization parameter. To manage the number of nodes in the resultant network, a time-consuming trial-and-error process is required to establish the regularization parameter.

The GWO algorithm can simultaneously choose the RBF centers from the training samples and train the RBF network. Since the GWO algorithm is an evolutionary based algorithm, the computational complexity of GWO is at least a few orders greater than that of our R-IHT techniques.

The MMCC algorithm [13] is originally designed to handle the outlier situation for extreme learning machines. We modify it for handle RBF networks. However, the MMCC algorithm does not have ability to select the RBF nodes.

Table 3. Average test MSE over ten trials with the specific number of RBF centers and outlier level $(r, \kappa) = (0.3, 2)$

Dateset	Centers Nodes	Noise level σ_v^2	R-IHT	ADMM-ℓ_1	ℓ_1-Reg	GWO	MMCC
ASN	100	0.01	**0.0226**	0.0737	0.1190	0.1233	0.0582
		0.04	**0.0240**	0.0616	0.1256	0.1234	0.0678
		0.09	**0.0259**	0.0610	0.1364	0.1224	0.0838
		0.16	**0.0287**	0.0672	0.1516	0.1256	0.1060
Concrete	60	0.01	**0.0628**	0.0946	0.0978	0.1420	0.0794
		0.04	**0.0643**	0.0966	0.1005	0.1448	0.1086
		0.09	**0.0571**	0.1042	0.1051	0.1396	0.1540
		0.16	**0.0630**	0.1101	0.1115	0.1444	0.2101
MPG	30	0.01	**0.0130**	0.0880	0.1111	0.1035	0.1497
		0.04	**0.0136**	0.0585	0.1465	0.1056	0.1667
		0.09	**0.0144**	0.0459	0.2057	0.1062	0.1930
		0.16	**0.0154**	0.0409	0.2885	0.1067	0.2258
ABA	50	0.01	**0.0078**	0.0589	0.1003	0.1465	0.0701
		0.04	**0.0082**	0.0389	0.1257	0.0992	0.1107
		0.09	**0.0085**	0.0333	0.1680	0.0951	0.1730
		0.16	**0.0091**	0.0334	0.2273	0.0950	0.2487
WQW	100	0.01	**0.0220**	0.1005	0.1079	0.1410	0.0898
		0.04	**0.0224**	0.0963	0.1165	0.1200	0.1431
		0.09	**0.0229**	0.0919	0.1307	0.1152	0.2254
		0.16	**0.0234**	0.0879	0.1507	0.1147	0.3263

Results: Number of RBF Nodes Versus Test Set Error

In this part, we would like to study the test set errors of the algorithms under various numbers of nodes. We consider two set of conditions: (1) $\{r = 0.1, \kappa = 1, \sigma_v^2 = 0.01\}$ and (2) $\{r = 0.3, \kappa = 2, \sigma_v^2 = 0.16\}$. Condition 1 is with low outlier level and low weight noise level, while Condition 2 is with high outlier level and high weight noise level. Note that only our R-IHT, ℓ_1-Reg, ADMM-ℓ_1 and GWO have the ability to tune the number of nodes in the resultant networks. The result is shown in Fig. 1. From the figure, we have the following observations.

- GWO: The performance of GWO remains consistently poor, regardless of whether the outlier level and weight noise level are low or high.
- ℓ_1-Reg: In some cases, such as WQW with $\{r = 0.1, \kappa = 1, \sigma_v^2 = 0.01\}$, increasing the number of nodes can improve test set error. Beyond a certain threshold, increasing the number of nodes does not result in a decrease in test set error. It should be aware that in some cases, such as MPG with

$\{r = 0.1, \kappa = 1, \sigma_v^2 = 0.01\}$, beyond a certain threshold, increasing the number of nodes results in an increases in test set error. For other cases, the performance of ℓ_1-Reg is poor.

- ADMM-ℓ_1: When the outlier level and weight noise level are low, increasing the number of nodes can improve test set error. Beyond a certain threshold, increasing the number of nodes does not result in a decrease in test set error. When the outlier level and weight noise level are high, the performance is poor.

- Our R-IHT: Regardless of whether the outlier level and weight noise level are low or high, the performance of our R-IHT is better than that of other comparison algorithm. Also, increasing the number of nodes can improve test set error. Beyond a certain threshold, increasing the number of nodes does not result in a decrease in test set error. That means, we can use a value around the threshold as the number of nodes for constructing RBF nodes.

Results: Performance Comparison Under a Specific Number of Nodes
This part compares the performance of these algorithms with respect to a specific number of RBF nodes. The specific number is determined by the R-IHT. For example, as shown in Fig. 1, in the ABA dataset, beyond 50 nodes, increasing the number of nodes does not result in a decrease in test set error. Thus, the specific number of nodes for the ABA dataset 50. We then use this specific number of nodes for other algorithms. Two outlier levels, $(r, \kappa) = (0.1, 1)$ and $(r, \kappa) = (0.3, 2)$, are considered. For weight noise, we consider four levels: $\sigma_v^2 = \{0.01, 0.04, 0.09, 0.16\}$. We run each algorithm ten times with different training and test set partitions for each setting. The performance of each algorithm under the low outlier level is presented in Table 2, while the performance under the high outlier level is presented in Table 3.

From the tables, the performance of R-IHT is superior to all the comparison in many cases. Only in a few cases in the Concrete dataset with weight noise levels of 0.01 and 0.04, the ℓ_1-Reg is better than our R-IHT.

Generally speaking, the R-HT is with the best performance and ADMM-ℓ_1 is the second best. When the weight noise level and outlier level increase, the superiority of R-IHT becomes becomes more noticeable. Using the ASN dataset as an illustration, the MSE difference between R-IHT and ADMM-ℓ_1 is 0.0065 in the case of $\{r = 0.1, \kappa = 1, \sigma_v^2 = 0.01\}$. When we increase the weight noise level to 0.16, i.e., $\{r = 0.1, \kappa = 1, \sigma_v^2 = 0.16\}$, the R-IHT technique outperforms ADMM-ℓ_1 with a test error value of 0.0452. We also provide the paired t-test result in Table 4 between the R-IHT and ADMM-ℓ_1 corresponding to Table 3.

Table 4. Paired t-test result of average MSE between the R-IHT and ADMM-ℓ_1 with outlier level $(r, \kappa) = (0.3, 2)$. When the number of trials is 10 and the confidence level is 95%, the critical t-value is 1.8331.

Dateset	Noise level σ_v^2	R-IHT	ADMM-ℓ_1	Standard Error	t value	Pct. of Improvement
ASN	0.01	**0.0226**	0.0737	0.001751485	29.15	69%
	0.04	**0.0240**	0.0616	0.001679491	22.41	61%
	0.09	**0.0259**	0.0610	0.001156789	30.36	58%
	0.16	**0.0287**	0.0672	0.001358675	28.37	57%
Concrete	0.01	**0.0628**	0.0946	0.004685609	6.79	34%
	0.04	**0.0643**	0.0967	0.004564908	7.09	33%
	0.09	**0.0571**	0.1042	0.003071183	15.33	45%
	0.16	**0.0630**	0.1101	0.003408837	13.82	43%
MPG	0.01	**0.0130**	0.0880	0.005040057	14.88	85%
	0.04	**0.0136**	0.0585	0.004412964	10.16	77%
	0.09	**0.0144**	0.0459	0.002756123	11.41	69%
	0.16	**0.0154**	0.0409	0.001950298	13.07	62%
WQW	0.01	**0.0220**	0.1005	0.000722098	108.66	78%
	0.04	**0.0224**	0.0963	0.000797445	92.63	77%
	0.09	**0.0229**	0.0919	0.000913011	75.57	75%
	0.16	**0.0234**	0.0879	0.00121256	53.14	73%
ABA	0.01	**0.0078**	0.0589	0.001509471	33.80	87%
	0.04	**0.0082**	0.0389	0.00114781	26.74	79%
	0.09	**0.0085**	0.0333	0.000572399	43.30	74%
	0.16	**0.0091**	0.0334	0.000572807	43.35	73%

6   Conclusion

This paper formulated the training process of an RBF network as a constrained optimization problem with a ℓ_0-norm. With our formulation, we can explicitly and directly control the number of RBF nodes in the resultant network. Also, the resultant network can handle weight noise and outlier training samples. We developed the R-IHT to solve the formulated constrained optimization problem. In the theoretical side, we prove that the objective function is is Lipschitz continuous. Furthermore, we prove that during the training process the objective function value converges and the estimating weight vector converges. We also use a number of datasets to demonstrate the effectiveness of the proposed algorithm.

References

1. Chen, B.H., et al.: Haze removal using radial basis function networks for visibility restoration applications. IEEE Trans. Neural Netw. Learn. Syst. **29**(8), 3828–3838 (2017)

2. Karamichailidou, D., Kaloutsa, V., Alexandridis, A.: Wind turbine power curve modeling using radial basis function neural networks and tabu search. Renew. Energy **163**, 2137–2152 (2021)
3. Haykin, S.: Neural Networks: A Comprehensive Foundation, 2nd edn. Prentice Hall PTR, Upper Saddle River, NJ, USA (1998)
4. Chen, S., Cowan, C.F.N., Grant, P.: Orthogonal least squares learning algorithm for radial basis function networks. IEEE Trans. Neural Netw. **2**(2), 302–309 (1991)
5. Giveki, D., Rastegar, H.: Designing a new radial basis function neural network by harmony search for diabetes diagnosis. Opt. Memory Neural Netw. **28**(4), 321–331 (2019)
6. Burr, J.B.: Digital neural network implementations. Neural Netw. Concepts Appli. Implement. **3**, 237–285 (1991)
7. Leung, C.S., Wan, W.Y., Feng, R.: A regularizer approach for RBF networks under the concurrent weight failure situation. IEEE Trans. Neural Netw. Learn. Syst. **28**(6), 1360–1372 (2017)
8. Wang, H., Feng, R., Han, Z.F., Leung, C.S.: ADMM-based algorithm for training fault tolerant RBF networks and selecting centers. IEEE Transactions on Neural Netw. Learn. Syst. **29**(8), 3870–3878 (2018)
9. Blumensath, T., Davies, M.E.: Iterative thresholding for sparse approximations. J. Fourier Anal. Appl. **14**, 629–654 (2008)
10. Dua, D., Graff, C.: UCI machine learning repository (2017). https://archive.ics. uci.edu/ml
11. Zhang, Q., Hu, X., Zhang, B.: Comparison of l_1-norm SVR and sparse coding algorithms for linear regression. IEEE Trans. Neural Netw. Learn. Syst. **26**(8), 1828–1833 (2015)
12. Faris, H., Mirjalili, S., Aljarah, I.: Automatic selection of hidden neurons and weights in neural networks using grey wolf optimizer based on a hybrid encoding scheme. Int. J. Mach. Learn. Cybern. **10**(10), 2901–2920 (2019)
13. Chen, B., Wang, X., Lu, N., Wang, S., Cao, J., Qin, J.: Mixture correntropy for robust learning. Pattern Recogn. **79**, 318–327 (2018)

Cross-Lingual Knowledge Distillation via Flow-Based Voice Conversion for Robust Polyglot Text-to-Speech

Dariusz Piotrowski[3][ID], Renard Korzeniowski[4][ID], Alessio Falai[4(✉)][ID],
Sebastian Cygert[2][ID], Kamil Pokora[1][ID], Georgi Tinchev[4][ID], Ziyao Zhang[4][ID],
and Kayoko Yanagisawa[4][ID]

[1] Alexa AI, Amazon, Seattle, USA
kamipoko@amazon.com
[2] Gdańsk University of Technology, Gdańsk, Poland
[3] Alexa AI, Amazon, Gdańsk, Poland
piotrod@amazon.com
[4] Alexa AI, Amazon, Cambridge, UK
{korenard,falai,gtinchev,zhaziyao,yakayoko}@amazon.com

Abstract. In this work, we introduce a framework for cross-lingual speech synthesis, which involves an upstream Voice Conversion (VC) model and a downstream Text-To-Speech (TTS) model. The proposed framework consists of 4 stages. In the first two stages, we use a VC model to convert utterances in the target locale to the voice of the target speaker. In the third stage, the converted data is combined with the linguistic features and durations from recordings in the target language, which are then used to train a single-speaker acoustic model. Finally, the last stage entails the training of a locale-independent vocoder. Our evaluations show that the proposed paradigm outperforms state-of-the-art approaches which are based on training a large multilingual TTS model. In addition, our experiments demonstrate the robustness of our approach with different model architectures, languages, speakers and amounts of data. Moreover, our solution is especially beneficial in low-resource settings.

Keywords: neural text-to-speech · multilingual synthesis · voice conversion · synthetic data · normalising flows

1 Introduction

Polyglot Text-To-Speech (TTS) systems, which rely on training data from monolingual speakers in multiple language variants (from here on referred to as *locale*), enable speakers in the training corpus to speak any language present in the same training corpus [10,14,28,29]. Such State-Of-The-Art (SOTA) models are able to achieve impressive results in cross-lingual synthesis scenarios, in terms of the high-quality naturalness and accent generated in the synthesised audios. During training, the model learns to decorrelate the embedding spaces of speaker

---

D. Piotrowski, R. Korzeniowski and A. Falai—Equal contribution.
S. Cygert—Work done while at Amazon.

© The Author(s), under exclusive license to Springer Nature Singapore Pte Ltd. 2024
B. Luo et al. (Eds.): ICONIP 2023, CCIS 1961, pp. 252–264, 2024.
https://doi.org/10.1007/978-981-99-8126-7_20

and language conditionings, which enables inference-time-synthesis of {*speaker, language*} pairs unseen during training. Such a training paradigm is the mainstream methodology of building Polyglot TTS systems, which is referred to as *Standard Polyglot* hereinafter. One major drawback of *Standard Polyglot* is that the model needs to have significant capacity to be able to disentangle speaker and language characteristics [4,29]. This means the deployment of such models on low-resource devices is extremely challenging. Moreover, existing studies [19,29] have revealed that this approach can be sensitive to data composition.

In this work, we address the issue of deploying large Polyglot models in computationally-constrained settings, by proposing a framework based on a high-capacity Voice Conversion (VC) model, working in conjunction with a low-capacity, single-speaker, monolingual acoustic model. The key point of this solution is the decoupling of speaker-language disentanglement and TTS tasks, which are realised in a single model with *Standard Polyglot* approaches. Specifically, we shift the demanding task of speaker-language disentanglement to the upstream VC model. In addition to its ability to deliver disentanglement, the robustness of our VC model makes it much less sensitive to the composition of the training corpus, thus easing the burden of scaling the framework to new speakers and locales. Extensive evaluations show that, when compared to the SOTA *Standard Polyglot* approach [14], our framework yields significantly better naturalness and accent similarity and on par speaker similarity. Moreover, the robustness of our framework is confirmed in experiments with varying model architecture, target speaker, target locale, and dataset size. To sum up, the contributions of this work are as follows:

1. We propose a new paradigm for cross-lingual TTS based on an upstream VC model and a downstream monolingual TTS model
2. We demonstrate, through extensive evaluations, that it beats SOTA and is robust with regard to data composition as well as architecture type and size
3. We demonstrate that the proposed approach is especially effective under low-resource constraints, enabling the use of lightweight architectures

2   Related Work

The idea of using synthetic data to support TTS tasks was explored before. Some works rely on VC as a data augmentation technique, for improving signal quality in low data regimes [5,21] or for style transfer tasks [2]. Most closely related to our work, authors in [4] presented a system which uses synthetic data for accent transfer tasks, but they have an upstream TTS model, while we explore VC techniques, and their downstream acoustic model and vocoder pair is more heavy-weighted than ours, as their focus is solely on reliability. Other approaches present so-called Accent Conversion (AC) techniques on disentangling speaker and accent [25,30], but they usually work in an in-lingual fashion (e.g. by converting a British English speaker to sound like an American English one); in contrast, this work focuses on more generic cross-lingual applications.

The idea of relying on cross-lingual VC to enable Polyglot TTS has already been investigated as well, but previous works [17,23] mainly present results on legacy VC and TTS models, whereas we do so using modern, SOTA architectures. Moreover, we rely on mel-spectrograms alone for both VC and TTS models, instead of using a diverse set of features, such as phonetic posteriorgrams (PPG), as was done in [23,31]. Other approaches [3] position the VC model for speaker identity conversion after the TTS model, which is not ideal for latency. In addition, some works [23,31] specifically target bilingual scenarios with various restrictions (e.g. in [23] the system is designed for one specific target speaker), whereas our approach can be scaled to any target speaker-language combination. Solutions presented in [26] can also be scaled to any speaker-language pair, but the main differences with our work are the following: (1) we show robustness under varying assumptions, thus making our experimental validation more comprehensive; (2) we specifically focus on computationally-constrained applications; (3) we show that our approach to vocoder training works well in a locale-independent fashion, further highlighting the scalability of the proposed solution.

3   Proposed Approach

3.1   Framework

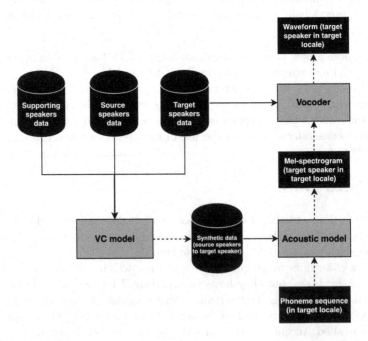

Fig. 1. High-level diagram of the proposed framework, where → means training, while --→ stands for inference.

We propose a 4-stage approach for building a Polyglot TTS system (*VC-based Polyglot*), with the goal of synthesising cross-lingual speech from a *target speaker* in a *target locale*. In addition, we define *source speakers* as the speakers whose native locale is the *target locale*; whereas the locales of *supporting speakers* are not the *target locale*. Our approach consists of the following steps.

1. Train a many-to-many VC model, using training data from *source, target* and *supporting speakers*.
2. Convert *source speakers'* identities in the original audio files to that of the *target speaker*, by using the trained VC model from 1, thus creating the synthetic dataset.
3. Use the synthetic dataset from 2 to train a single-speaker, monolingual acoustic model, to produce speech with *target speaker* identity in the *target locale*.
4. Train a speaker-specific locale-independent vocoder using original data from the *target speaker*.

When building the synthetic dataset in 2, we use original phoneme durations (from *source speakers*) instead of re-running forced alignment, as the latter leads to the acoustic model cutting off utterances and producing unnatural prosody. Then, at inference time we discard the upstream VC model and chain the trained acoustic model and vocoder, to output waveforms. Note that 2 and 3 can be repeated for generating synthetic data in multiple *target locales*, without the need to re-train the VC model. Figure 1 illustrates how different components interact with each other at training and inference times.

Note that the framework configuration described above is not the only possible option. For example, in 3, we empirically find that adding *supporting speakers* in VC model training does not improve (nor does it degrade) synthesis quality in the *target locale* for downstream models. Also, in 4, early results show that using *source speakers* alongside original data from the *target speaker* is only beneficial for some *target locales*. Moreover, an alternative solution for 4 would be to train the vocoder in a locale-specific fashion, by generating synthetic waveforms w from synthetic mel-spectrograms m using a powerful universal vocoder (such as the one from [7]), and training on (m, w) pairs. We adopt the locale-independent procedure described in 4 because doing so renders more perceptually expressive speech.

3.2   Model Description

For our upstream VC model, we initially considered normalising flows [1,12] and Variational Auto-Encoder (VAE) [8] based architectures. However, given that the latter is particularly sensitive to the dimension of the auto-encoder bottleneck [16], which clashes with our requirement for robustness, we decided to adopt the flow-based VC approach. The main reason for selecting an upstream VC model instead of a TTS one is that the former operates on speech-to-speech mappings, which are easier to learn than the more ill-defined text-to-speech ones. This, in turn, leads to more natural synthetic data than their TTS counterparts [1].

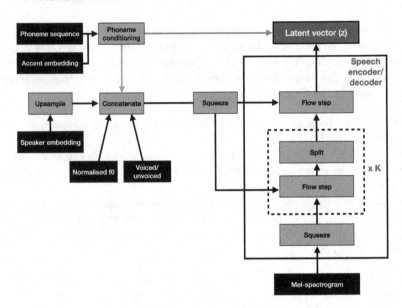

Fig. 2. The many-to-many non-parallel VC model used to generate synthetic datasets. The green line denotes phoneme conditioning, while the blue line the phoneme-based prior.

In particular, the VC model we rely on is the text-conditioned non-parallel many-to-many VC model with fixed prior described in [1] and depicted in Fig. 2. The VC model topology is based on Flow-TTS [13] and modified to accept F0 (interpolated and normalised at the utterance level) and V/UV (binary voiced/unvoiced flag) conditionings. Phoneme embeddings are also enriched with accent one-hot encodings, while speaker embeddings are extracted using a pre-trained utterance-level encoder [24]. At training time the model learns to maximise the likelihood

$$\log P_X(x|c) = \log P_Z(z|c) + \log \left| \det \frac{\partial z}{\partial x} \right| \tag{1}$$

of the prior distribution

$$c \sim N(\mu_p, \sigma_p), \mu_p = 0, \sigma_p = 1 \tag{2}$$

for all frames regardless of the speech content, where z is the encoded latent vector and x is the input mel-spectrogram. As done in [1], we rely on pre-trained phoneme alignments instead of using an attention mechanism.

As our main acoustic model we chose FastSpeech 2 (FS2) [18], which consists of feed-forward transformer-based encoder with explicit duration, energy, and F0 predictors. In order to reduce latency and model size, we remove the self-attention mechanism in the decoder, thus making it fully-convolutional. To validate the robustness of our acoustic model, we compare it with higher and lower capacity models. The former is based on gated convolutions and recurrent units (ED) [20], while the latter is LightSpeech (LS) [11], a scaled-down version of FS2 (in two variants, as described in Sect. 4.3).

With low-resource embedded devices in mind, we selected MultiBand Mel-GAN [27], which is able to generate high quality speech with only 0.95 GFLOPS, as the vocoder in our pipeline. To address audio artefacts typical to GAN-based vocoders, we standardise input features in the range of $[-1, 1]$. Empirically, we find this solution equivalent to, but more suitable than, commonly used fine-tuning techniques [9,22]. The practical reason for this is that no ground-truth waveforms corresponding to synthetic mel-spectrograms are available.

4 Experiments

4.1 Evaluation Setup

For evaluations, we use MUltiple Stimuli with Hidden Reference and Anchor (MUSHRA) [6] tests with a scale of 0 to 100, through crowd-sourcing platforms, to perceptually assess naturalness, speaker similarity and accent similarity of our syntheses, as done in [19,29]. We opted for MUSHRA tests, rather than more popular Mean Opinion Score (MOS) tests, due to the highly comparative nature of our work. In general, a MUSHRA test explicitly provides the listener with the same utterance synthesised with all of the systems to be compared, anchored by the upper and lower references. This form of testing therefore allows the listener to make direct comparisons when making their judgements, thus resulting in meaningful rank amongst systems.

For each evaluation, we use a test set of 200 utterances, unseen during training. Each of the 60 listeners evaluate a random set of 50 utterances for each evaluation. When evaluating speaker similarity, testers are given a reference system, which consists of *target speaker*'s recordings. Hence, the upper anchor is always equal to 100. Recordings from *source speakers* are instead used as the lower anchor. When evaluating naturalness and accent similarity, *source speakers*' recordings are used as the upper anchor, while the lower anchor is a system based on phoneme-mapping, whereby an input from the *target locale* is translated into the locale of the *target speaker* using hard-coded rules for mapping phones between locales (based on linguistic proximity). The phoneme-mapped input is then fed into a Tacotron 2 [22] model trained on the *target speaker*'s data.

To ensure the integrity of our evaluation results, we apply filtering methods to exclude potential cheaters, i.e., we remove submissions from listeners who scored all systems at extreme or default values for more than 5 times. Moreover, to check the statistical significance of our test results, we use a pairwise two-sided Wilcoxon signed-rank test corrected for multiple comparisons with the Holm-Bonferroni method, as done in [20].

Results in Tables 1, 2, 3, 4 and 5 are reported in terms of average MUSHRA scores[1], along with the Closing The Gap (*CTG*) percentage, which summarises

---

[1] We use boldface for the highest score per aspect if the gap between baseline and proposed system is statistically significant.

the improvement of the *VC-based Polyglot* approach with respect to (w.r.t.) *Standard Polyglot*. In particular, CTG is calculated by

$$CTG = \frac{n_v - n_s}{n_s} \times 100, \text{where}$$

$$n_i = 100 - \left(\frac{i - l}{u - l} \times 100 \right), i \in \{s, v\},$$

with s and v representing the mean MUSHRA score for the *Standard* and *VC-based* approaches, respectively, and l and u corresponding to lower and upper anchors. All CTG percentages and Difference in CTG ($DCTG$) scores presented are for $p \leq 0.05$, unless otherwise specified by [n.s.].

4.2 Data Composition

Our default training dataset D_1 consists of 124 speakers in 7 locales[2] (fr-CA, en-GB, es-MX, en-US, es-US, fr-FR and de-DE). This dataset totals to roughly 720 hours of studio-quality recordings with 24 kHz sampling rate. We extract 80-dimensional mel-spectrograms with frame length of 50 ms and frame shift of 12.5 ms, which is then used to train the VC model and the *Standard Polyglot* acoustic model. For the experiment on varying dataset size, we also test the proposed approach on a smaller dataset $D_2 \subset D_1$, which consists of data from a subset of speakers in fr-CA, es-MX and en-GB and totals to 150 hours.

4.3 Experiment Designs and Results

In this section, we describe the design and results of all experiments we run. Note that in all our evaluations, we compare our *VC-based Polyglot* approach (*VC-based*) to the *Standard Polyglot* approach (*Standard*) described in Sect. 1. In particular, Table 1 shows the evaluation results using different acoustic model architectures. Then, we further demonstrate the effectiveness of the *VC-based* approach by changing the en-US *target speaker* S_1 to en-GB speaker S_2 in Table 2, changing the *target locale* to fr-CA in Table 3, and changing the dataset to D_2 in Table 4. Finally, Table 5 shows that the *VC-based* approach renders even more superior performance when downsized. As for the training setup, we use the same hyper-parameters as described in [11,18,20]. Note that for the *VC-based Polyglot* approach, we train one single-speaker monolingual acoustic model for each *target locale*.

The goal of Experiment 1 (Table 1) is to assess the quality of the *VC-based* approach against the *Standard* baseline. To this end, we rely on dataset D_1, *target speaker* S_1, *target locale* es-MX and FS2 as model architecture. Table 1 shows the superior performance of our proposed approach in all of the evaluated aspects for the FS2 architecture. Replicating the same experiment with ED as acoustic model results in the same conclusions, except for the on-par performance in speaker similarity (more discussions on this in Sect. 4.4).

---

[2] We use the ISO 639-1 nomenclature to denote locales.

Table 1. MUSHRA and CTG scores for models evaluated in Experiment 1, for dataset D_1, speaker S_1 and locale `es-MX`.

Model	Approach	Naturalness$^\uparrow$	Speaker similarity$^\uparrow$	Accent similarity$^\uparrow$
Upper		82.60	100.00	82.25
FS2	*VC-based*	**69.60**	**66.38**	**71.06**
	Standard	64.08	64.82	66.26
	CTG	29.8%	4.4%	30.0%
ED	*VC-based*	**70.21**	67.86	**70.49**
	Standard	65.24	68.05	66.27
	CTG	28.6%	$-0.6\%^{n.s.}$	26.4%
Lower		42.62	22.37	20.70

In Experiment 2 (Table 2), we focus on evaluating the robustness of the approach when changing the *target speakers* (S_1 to S_2). Table 2 shows that when we switch to *target speaker* S_2, the *VC-based* approach is still better than the *Standard* one in all aspects other than speaker similarity.

Table 2. MUSHRA and CTG scores for models evaluated in Experiment 2, for dataset D_1, speaker S_2 and locale `es-MX`.

System	Naturalness$^\uparrow$	Speaker similarity$^\uparrow$	Accent similarity$^\uparrow$
Upper	80.24	100.00	82.48
VC-based	**70.34**	67.60	**71.85**
Standard	69.86	68.69	70.54
CTG	4.5%	$-3.5\%^{n.s.}$	10.9%
Lower	34.47	28.37	19.64

With results from Experiment 3 (Table 3), we show the robustness of our approach when the *target locale* changes. In particular, Table 3 reveals that the *VC-based* approach is still better across all aspects, when changing the *target locale* to `fr-CA`.

Table 3. MUSHRA and CTG scores for models evaluated in Experiment 3, for dataset D_1, speaker S_1 and locale `fr-CA`

System	Naturalness$^\uparrow$	Speaker similarity$^\uparrow$	Accent similarity$^\uparrow$
Upper	75.76	100.00	76.52
VC-based	**74.11**	**70.69**	**70.99**
Standard	73.67	67.00	67.60
CTG	24.6%	11.2%	38%
Lower	54.89	50.11	63.71

Furthermore, in Table 4 we show the robustness of the proposed approach when scaling down the size of the dataset. For this purpose, we repeat the experiment but switch to a smaller dataset D_2. As can be seen in Table 4, the proposed approach is preferred across the board.

Table 4. MUSHRA and CTG scores for models evaluated in Experiment 4, for dataset D_2, speaker S_1 and locale `fr-CA`.

System	Naturalness$^\uparrow$	Speaker similarity$^\uparrow$	Accent similarity$^\uparrow$
Upper	78.64	100.00	80.27
VC-based	**71.07**	**72.43**	**75.84**
Standard	66.66	68.57	73.97
CTG	36.8%	12.3%	29.9%
Lower	61.55	44.80	44.30

Finally, to demonstrate the effectiveness of our proposed approach when scaling down model capacity, in Experiment 5 we repeat the previous experiments (with both `es-MX` and `fr-CA` as *target locales*), but use a "slimmed down" version of FS2 [18], i.e. LS [11]. To further test it in a more extreme scenario, we again downsize LS, thus producing LS-S, by reducing the size of all hidden layers from 256 to 192 and removing one convolutional layer from both the encoder and the decoder. Table 5 reports the results in terms of the Difference in CTG ($DCTG$), which indicates how much more we close the gap with the smaller models w.r.t. the bigger ones. The $DCTG$ score is computed as $DCTG(CTG_s, CTG_b) = CTG_s - CTG_b$ percentage points (p.p.), where CTG_s is the CTG score for smaller models and CTG_b is the one for bigger models. Each cell of the table indicates the CTG score for a specific model in the *target locale*, representing the improvements gained from using the proposed approach. Overall, we can see that the *VC-based* approach outperforms the *Standard* baseline by a wider margin when lightweight architectures are employed.

Table 5. $(D)CTG$ scores in Experiment 5 for 4 MUSHRA tests, dataset D_1, speaker S_1 and locales `es-MX` and `fr-CA`.

Locale	Model	Naturalness$^\uparrow$	Speaker similarity$^\uparrow$	Accent similarity$^\uparrow$
`es-MX`	FS2	32.87%	−2.71%	25.42%
	LS	42.27%	−1.91%	41.24%
	$DCTG$	9.40p.p.	0.80p.p.$^{n.s.}$	15.82p.p.
`fr-CA`	FS2	28.89%	4.37%	16.42%
	LS	28.23%	4.98%	45.19%
	$DCTG$	−0.66p.p.	0.61p.p.	28.77p.p.
`es-MX`	FS2	27.55%	−0.98%	30.90%
	LS-S	38.33%	−0.23%	35.08%
	$DCTG$	10.78p.p.	0.75p.p.$^{n.s.}$	4.18p.p.
`fr-CA`	FS2	30.49%	−0.44%	25.97%
	LS-S	39.49%	−0.64%	35.54%
	$DCTG$	9.00p.p.	−0.20p.p.$^{n.s.}$	9.57p.p.

4.4 Discussion

As presented in Sect. 4.3, the proposed solution is preferred overall, while the only aspect where there is a non-statistically significant gap between *VC-based Polyglot* and *Standard Polyglot* (where the latter got more votes) is speaker similarity. We attribute this to the choice of the upstream VC model. In particular, although flow-based models are not widely-known to suffer from the speaker-accent trade-off, where speaker similarity degrades as accent similarity improves (and vice-versa), we believe that more powerful density-based models, such as those based on diffusion [15], have the potential of closing this gap in terms of speaker similarity.

We also hypothesise that the increase in naturalness scores can be attributed to the fact that in the *VC-based Polyglot* models the generated prosody follows that of the source speaker(s). In particular, many prosodic features are language-specific and inheriting them from source speakers (i.e. native speakers of the target language) shows to be beneficial. Future work will focus on assessing the impact of prosody transfer in this cross-lingual scenario.

Furthermore, we observe that the proposed approach is robust to variations in data composition: when comparing the data generated using VC models trained on vastly different datasets, there is no statistically significant preference. In comparison, previous works with the *Standard Polyglot* approach explicitly report how the mixture of training data has a significant impact on final performance [29].

5 Conclusions

In this paper, we proposed a novel paradigm for building highly robust Polyglot TTS systems. The introduction of a powerful upstream VC model successfully lifted the burden of learning disentangled speaker-language representations from the downstream acoustic model. This, in turn, allowed for effective use of high-performance architectures targeting low-powered devices. In summary, our proposed approach is able to simultaneously (1) outperform the current SOTA Polyglot TTS systems in terms of naturalness and accent similarity, while maintaining on-par performance in speaker similarity, (2) improve model robustness w.r.t. training data composition and (3) enable the use of lightweight architectures for speech synthesis on resource-constrained devices. Such results are backed by controlled experiments using different model architectures, languages, speakers and dataset sizes.

For future work, we will test how well our proposed approach scales in more diverse scenarios. In addition, we believe that by employing more advanced VC techniques, there would be further improvements in the downstream task, especially in terms of speaker similarity.

References

1. Bilinski, P., et al.: Creating new voices using normalizing flows. In: Proceedings of Interspeech 2022, pp. 2958–2962 (2022). https://doi.org/10.21437/Interspeech.2022-10195
2. Comini, G., Huybrechts, G., Ribeiro, M.S., Gabryś, A., Lorenzo-Trueba, J.: Low-data? No problem: low-resource, language-agnostic conversational text-to-speech via F0-conditioned data augmentation. In: Proceedings of Interspeech 2022, pp. 1946–1950 (2022). https://doi.org/10.21437/Interspeech.2022-10338
3. Ellinas, N., et al.: Cross-lingual text-to-speech with flow-based voice conversion for improved pronunciation (2022). https://doi.org/10.48550/ARXIV.2210.17264. https://arxiv.org/abs/2210.17264
4. Finkelstein, L., et al.: Training text-to-speech systems from synthetic data: a practical approach for accent transfer tasks. In: Proceedings of Interspeech 2022, pp. 4571–4575 (2022). https://doi.org/10.21437/Interspeech.2022-10115
5. Hwang, M.J., Yamamoto, R., Song, E., Kim, J.M.: TTS-by-TTS: TTS-driven data augmentation for fast and high-quality speech synthesis. In: ICASSP 2021-2021 IEEE International Conference on Acoustics, Speech and Signal Processing (ICASSP), pp. 6598–6602 (2021). https://doi.org/10.1109/ICASSP39728.2021.9414408
6. ITUR Recommendation: BS.1534-1. method for the subjective assessment of intermediate sound quality (MUSHRA). International Telecommunications Union, Geneva (2001)
7. Jiao, Y., Gabryś, A., Tinchev, G., Putrycz, B., Korzekwa, D., Klimkov, V.: Universal neural vocoding with parallel wavenet. In: ICASSP 2021-2021 IEEE International Conference on Acoustics, Speech and Signal Processing (ICASSP), pp. 6044–6048 (2021). https://doi.org/10.1109/ICASSP39728.2021.9414444

8. Karlapati, S., et al.: CopyCat2: a single model for multi-speaker TTS and many-to-many fine-grained prosody transfer. In: Proceedings of Interspeech 2022, pp. 3363–3367 (2022). https://doi.org/10.21437/Interspeech.2022-67

9. Kong, J., Kim, J., Bae, J.: HiFi-GAN: generative adversarial networks for efficient and high fidelity speech synthesis. In: Larochelle, H., Ranzato, M., Hadsell, R., Balcan, M., Lin, H. (eds.) Advances in Neural Information Processing Systems, vol. 33, pp. 17022–17033. Curran Associates, Inc. (2020)

10. Latorre, J., Bailleul, C., Morrill, T., Conkie, A., Stylianou, Y.: Combining speakers of multiple languages to improve quality of neural voices. In: Proceedings of 11th ISCA Speech Synthesis Workshop (SSW 2011), pp. 37–42 (2021). https://doi.org/10.21437/SSW.2021-7

11. Luo, R., et al.: Lightspeech: lightweight and fast text to speech with neural architecture search. In: ICASSP 2021, pp. 5699–5703 (2021)

12. Merritt, T., et al.: Text-free non-parallel many-to-many voice conversion using normalising flow. In: ICASSP 2022-2022 IEEE International Conference on Acoustics, Speech and Signal Processing (ICASSP), pp. 6782–6786 (2022). https://doi.org/10.1109/ICASSP43922.2022.9746368

13. Miao, C., Liang, S., Chen, M., Ma, J., Wang, S., Xiao, J.: Flow-TTS: a non-autoregressive network for text to speech based on flow. In: ICASSP 2020-2020 IEEE International Conference on Acoustics, Speech and Signal Processing (ICASSP), pp. 7209–7213 (2020). https://doi.org/10.1109/ICASSP40776.2020.9054484

14. Nachmani, E., Wolf, L.: Unsupervised polyglot text-to-speech. In: ICASSP 2019-2019 IEEE International Conference on Acoustics, Speech and Signal Processing (ICASSP), pp. 7055–7059 (2019). https://doi.org/10.1109/ICASSP.2019.8683519

15. Popov, V., Vovk, I., Gogoryan, V., Sadekova, T., Kudinov, M.S., Wei, J.: Diffusion-based voice conversion with fast maximum likelihood sampling scheme. In: International Conference on Learning Representations (2022)

16. Qian, K., Zhang, Y., Chang, S., Yang, X., Hasegawa-Johnson, M.: AutoVC: zero-shot voice style transfer with only autoencoder loss. In: Proceedings of Machine Learning Research, Long Beach, California, USA, vol. 97, pp. 5210–5219. PMLR (2019)

17. Ramani, B., Actlin Jeeva, M.P., Vijayalakshmi, P., Nagarajan, T.: Voice conversion-based multilingual to polyglot speech synthesizer for Indian languages. In: 2013 IEEE International Conference of IEEE Region 10 (TENCON 2013), pp. 1–4 (2013). https://doi.org/10.1109/TENCON.2013.6719019

18. Ren, Y., et al.: Fastspeech 2: fast and high-quality end-to-end text to speech. In: International Conference on Learning Representations (2021)

19. Sanchez, A., Falai, A., Zhang, Z., Angelini, O., Yanagisawa, K.: Unify and conquer: how phonetic feature representation affects polyglot text-to-speech (TTS). In: Proceedings of Interspeech 2022, pp. 2963–2967 (2022). https://doi.org/10.21437/Interspeech.2022-233

20. Shah, R., et al.: Non-autoregressive TTS with explicit duration modelling for low-resource highly expressive speech. In: Proceedings of 11th ISCA Speech Synthesis Workshop (SSW 2011), pp. 96–101 (2021). https://doi.org/10.21437/SSW.2021-17

21. Sharma, M., Kenter, T., Clark, R.: StrawNet: self-training WaveNet for TTS in low-data regimes. In: Proceedings of Interspeech 2020, pp. 3550–3554 (2020). https://doi.org/10.21437/Interspeech.2020-1437

22. Shen, J., et al.: Natural TTS synthesis by conditioning WaveNet on Mel spectrogram predictions. In: 2018 IEEE International Conference on Acoustics, Speech and Signal Processing (ICASSP), pp. 4779–4783 (2018). https://doi.org/10.1109/ICASSP.2018.8461368

23. Sun, L., Wang, H., Kang, S., Li, K., Meng, H.: Personalized, cross-lingual TTS using phonetic posteriorgrams. In: Interspeech 2016, pp. 322–326 (2016). https://doi.org/10.21437/Interspeech.2016-1043

24. Wan, L., Wang, Q., Papir, A., Moreno, I.L.: Generalized end-to-end loss for speaker verification. In: 2018 IEEE International Conference on Acoustics, Speech and Signal Processing (ICASSP), pp. 4879–4883 (2018). https://doi.org/10.1109/ICASSP.2018.8462665

25. Wang, Z., et al.: Accent and speaker disentanglement in many-to-many voice conversion. In: 2021 12th International Symposium on Chinese Spoken Language Processing (ISCSLP), pp. 1–5 (2021). https://doi.org/10.1109/ISCSLP49672.2021.9362120

26. Wu, J., Polyak, A., Taigman, Y., Fong, J., Agrawal, P., He, Q.: Multilingual text-to-speech training using cross language voice conversion and self-supervised learning of speech representations. In: ICASSP 2022-2022 IEEE International Conference on Acoustics, Speech and Signal Processing (ICASSP), pp. 8017–8021 (2022). https://doi.org/10.1109/ICASSP43922.2022.9746282

27. Yang, G., Yang, S., Liu, K., Fang, P., Chen, W., Xie, L.: Multi-band MelGAN: faster waveform generation for high-quality text-to-speech. In: 2021 IEEE Spoken Language Technology Workshop (SLT), pp. 492–498 (2021). https://doi.org/10.1109/SLT48900.2021.9383551

28. Zhang, Y., et al.: Learning to speak fluently in a foreign language: multilingual speech synthesis and cross-language voice cloning. In: Proceedings of Interspeech 2019, pp. 2080–2084 (2019). https://doi.org/10.21437/Interspeech.2019-2668

29. Zhang, Z., Falai, A., Sanchez, A., Angelini, O., Yanagisawa, K.: Mix and match: an empirical study on training corpus composition for polyglot text-to-speech (TTS). In: Proceedings of Interspeech 2022, pp. 2353–2357 (2022). https://doi.org/10.21437/Interspeech.2022-242

30. Zhao, G., Sonsaat, S., Levis, J., Chukharev-Hudilainen, E., Gutierrez-Osuna, R.: Accent conversion using phonetic posteriorgrams. In: 2018 IEEE International Conference on Acoustics, Speech and Signal Processing (ICASSP), pp. 5314–5318 (2018). https://doi.org/10.1109/ICASSP.2018.8462258

31. Zhao, S., Nguyen, T.H., Wang, H., Ma, B.: Towards natural bilingual and code-switched speech synthesis based on mix of monolingual recordings and cross-lingual voice conversion. In: Proceedings of Interspeech 2020, pp. 2927–2931 (2020). https://doi.org/10.21437/Interspeech.2020-1163

A Health Evaluation Algorithm for Edge Nodes Based on LSTM

Qian Sun[1], Zhengfan Wang[1], Jiarui Zhang[2(✉)], Qinglin Liu[1], and Xin Zhang[1]

[1] Network and Data Center, Northwest University, Xi'an 710127, China
sq@nwu.edu.cn
[2] The 20th Research Institute of China Electronics Technology Group Corporation,
Xi'an 710068, China
953356294@qq.com

Abstract. The health state of edge layer nodes significantly affects the reliability of the calculation and the development of related applications. Edge layer nodes health assessment is adopted to forecast node state to arrange calculation and application reasonably. The existing multidimensional data evaluation algorithms have achieved good predictive performance. However, with a small scale of training data, those algorithms could easily encounter overfitting and poor robustness. Therefore, we propose an evaluation algorithm based on the multidimensional operation data of edge layer nodes in this study. In order to solve the problem above, we propose an improved Long Short Term Memory (LSTM) model to implement the evaluation. We add feature discretization and annealing processes to the model to reduce the risk of model overfitting. Compared with typical time series prediction models, the proposed LSTM model has stronger applicability and better accuracy in the evaluation of network delay of edge layer nodes in our experiment.

Keywords: Edge layer node · Health Evaluation Algorithm · LSTM

1 Introduction

With the continuous development of the Internet of Things and edge computing, the services carried by network edge layer nodes become increasingly complex, and some network applications are affected by delays, compromising their stability. Therefore, the evaluation of network delay at edge layer nodes becomes particularly important. Past studies have explored the evaluation with regression-based algorithms, such as Autoregressive Integrated Moving Average (ARIMA) [11], exponential smoothing method [9], and classification-based models, such as the Hidden Markov Model (HMM) [1]. In the edge layer node (e.g., a large number of the edges of the access layer switches or other edge equipment at

---

Q. Sun and Z. Wang—Contribute equally to this work.

© The Author(s), under exclusive license to Springer Nature Singapore Pte Ltd. 2024
B. Luo et al. (Eds.): ICONIP 2023, CCIS 1961, pp. 265–277, 2024.
https://doi.org/10.1007/978-981-99-8126-7_21

work), the abnormal status of some key indicators (e.g., CPU utilization, memory utilization, etc.) is often associated with packet loss and calculation error. Therefore, these indicators can be used to assess the health status.

The prediction of node health state is based on time series analysis, which has been applied in many fields [2]. HMM is an important subfield of time series analysis. Saini et al. [10] present an evaluation model based on HMM. They put forward a combined model with global HMM and piecewise HMM using a genetic algorithm (GA) based framework. They also add gaussian noise to the data to verify the robustness of the model compared with the hidden Markov model. S. Wang et al. [14] propose an evaluation model for rolling bearing fault identification based on a combined global HMM with a convolutional neural network. They first use a convolutional neural network to automatically learn data features from the original vibration signal, and then construct feature visualization using the T-SNE dimension reduction algorithm. Finally, HMM is used to classify the rolling bearing state.

Lin et al. [12] designed an evaluation model based on effective variable selection and moving window, which applied it to the abnormal monitoring of boilers in the iron-making process. They proposed a variable selection method based on the coefficient of variation to select appropriate features. By combining the Markov method with moving window method, they successfully improved the accuracy of on-line fault identification by using sample dependency.

Saisai et al. [13] proposed a model to establish the component's health indicator and evaluate performance degradation that combined the Hidden Markov Model (HMM) and an improved Gated Recurrent Unit (GRU) network. The proposed model uses sensitive features extracted from life-cycle signals to establish health indicators through HMM, realizing the target of incipient degradation detection for slewing bearing. To fit the degradation curve, they used the exponential smoothing algorithm to describe the performance status and improve prediction accuracy. The degradation trend estimation and remaining life prognosis are accomplished by the GRU network, whose hyperparameters are selected through the Moth-Flame Optimization (MFO) algorithm.

Recurrent Neural Network (RNN) demonstrate effective learning performance for sequential data. Malhotra et al. [8] used a multi-layer LSTM constructed by multiple LSTM hidden layers to detect time series anomalies and faults. Their experiments show that the multi-layer LSTM network in series can achieve better results. Karim et al. [6] introduced the LSTM model into the attention mechanism and compared it with other algorithms to show that the algorithm had better performance.

The integrated models tend to perform better than single models. For example, most time series analysis models combine multiple complex classification models to improve their precision, such as the HMM model combined with the deep learning model. The experiment shows that this combination model overcomes the shortcomings of a single model and improves prediction precision. Although this approach can enhance model accuracy, it also increases the time complexity. In the case of small data samples, the combination of multiple classification models may lead to randomness in the forecasting method, thereby reducing prediction accuracy.

In order to solve this problem, an EH-LSTM (Edge Node Health Using LSTM) evaluation model is proposed in this paper, which includes the detailed design of the model structure and the algorithm of network prediction and network implementation. This model is designed to improve robustness and prediction error by using an LSTM optimization algorithm based on clustering and learning rate updating. The algorithm discretizes the continuous observation features by K-means clustering and retains the variation trend and classification criteria of the observed features. The iterative updating of the learning rate based on the annealing algorithm is helpful for the model to jump out of the local optimum. The model is applied to the actual operation data of the campus network switch and compared with various time series prediction models, such as Holt-Winters and ARIMA. The experimental results demonstrate the superior performance of the proposed LSTM prediction model and its learning rate updating algorithm in delayed state time series prediction.

2   Model Framework and Related Technology

2.1   EH-LSTM Framework

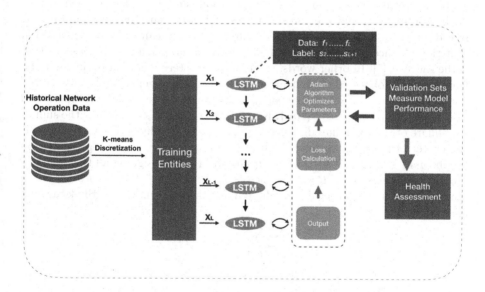

Fig. 1. The architecture of the proposed LSTM model

The overall framework of EH-LSTM is constructed in this paper, as shown in Fig. 1. The model preprocesses the original fault time series data by characteristic discrete methods, including standardization, discretization, and so on.

The discretized features have stronger classification abilities for the data labels, which will play a role in simplifying the model and reducing the risk of model overfitting. The training layer is composed of the LSTM hiding layer and the backpropagation process. The LSTM hidden layer is represented by Fig. 1, which contains T isomorphic LSTM cells constructed in chronological order. After each cycle of forward propagation and output of the neural network, the optimization algorithm mentioned in Sect. 2.3 is used to calculate the parameter gradient. The training process of deep learning is iterative, the learning rate optimization algorithm based on annealing will be described in Sect. 2.4.

2.2   Feature Discretization Method

In time series learning, the use of discrete training data can better show the time regularity of the data, and because of the reduction of the details of the data changes, it can have a better learning performance when there are fewer data samples. The running state data of edge nodes based on which network health is evaluated is a group of continuous data that presents clustering in value distance and features a large number of common values. The standardized values of all data features have clustering characteristics of different types. If the continuous data can be discretized reasonably, the LSTM training can have better performance.

In this paper, the k-means clustering method is used to quantify each feature of the training set, and the continuous normalized eigenvalues are transformed into discrete integer divisions. The feature of the K-mean clustering method is that the distance between each sample and the center point of its classification is the shortest [5]. In other words, the K-mean clustering is used to quantify the variance of the observed data included in each discrete eigenvalue. Let us call the training datasets as F, the time span as T. We show the single feature time series of F as $O = \{o_1, o_2, \ldots, o_{T-1}, o_T\}$. K-mean clustering loss function is the sum of the square of the Euclidian distance between the eigenvalue at each moment and the center of the classification. Each partition of the k-mean is represented by the integer $l \in \{1, 2, \ldots, k\}$. The coefficient matrix for assigning all the points to k classes is expressed as $R = [r_{11} \ldots r_{tl} \ldots r_{TK}]$, $(1 \leq t \leq T, r_{tl} = 0|1)$. $\overline{o_l}$ is the mean of the lth class (the center value). Then the loss function of the k-mean is calculated as.

$$J = \sum_{t=1}^{T} \sum_{l=1}^{k} r_{tl} \|o_i - \overline{o_l}\| \tag{1}$$

In this paper, k-means clustering is used to quantify the continuous time series and to estimate the initial value of the hidden state series. In order to reduce the complexity of iteration times and time, the traditional K-mean clustering was improved. The steps of the k-mean algorithm were given as follows:

Step 1: When $n = 0$, $m^{(0)} = \left\{ m_1^{(0)}, \ldots, m_l^{(0)}, \ldots, m_k^{(0)} \right\}$ is selected as the initial clustering center for the feature time series to be quantized. For each $m_l^{(0)}$, the calculation method is $m_l^{(0)} = \frac{l}{k}, 1 \leq l \leq k$;

Step 2: The partition is redone in each iteration: For clustering centers $m^{(n)} = \left\{ m_1^{(n)}, \ldots, m_l^{(n)}, \ldots, m_k^{(n)} \right\}$, the eigenvalues of each moment are squared to each class center of $m^{(n)}$. Assign the eigenvalues of each moment to the partition with the least squared difference to form a new clustering result;

Step 3: Renew the clustering centers as $m^{(n+1)} = \left\{ m_1^{(n+1)}, \ldots, m_l^{(n+1)}, \ldots, m_k^{(n+1)} \right\}$, the new class center value calculation formula is as follows.

$$m_l^{(n+1)} = \frac{\sum_{i=1}^{T} r_{tl} o_t}{\sum_{i=1}^{T} r_{tl}} \tag{2}$$

Step 4: When iterative convergence (no new cluster center is generated), the latest clustering results are output.

The k-means clustering was carried out for each feature of the training set to obtain the training set $F' = \left\{ f_1', f_2', \ldots, f_T' \right\}$, and the observed data of the rest datasets are discretized by the output clustering results.

2.3 Sequential Learning Algorithm

The running state of edge nodes changes with time-series data in a periodic and trend manner. In this section, LSTM is used to learn the time features in combination with the relevant theoretical techniques and data features in the previous section. LSTM is an algorithm based on RNN [4]. A basic RNN unit has only a single neural network layer (such as a TANH layer), and when processing long time series, backpropagation is likely to cause gradient disappearance or gradient explosion [3]. On this basis, the LSTM model improves the neural network unit and is no longer a single neural network layer. The key to the LSTM model is the cell state, represented as C_t, which is used to store the current LSTM state information and transfer it to the LSTM at the next moment. LSTM mainly consists of three different gate structures: forgetting gate, memory gate, and output gate [4]. These gates can retain and transmit the cell state and output signal. The neural network unit of LSTM is shown in Fig. 2.

The forgetting gate determines which information in cell state C_{t-1} will be forgotten. The forgetting gate contains a sigmoid neural network layer (a shallow neural network containing a hidden layer and sigmoid activation function, with parameters W_f and b_f). The mathematical form is as follows:

$$f_t = \sigma \left(W_f \cdot [s_{t-1}, o_t] + b_f \right) \tag{3}$$

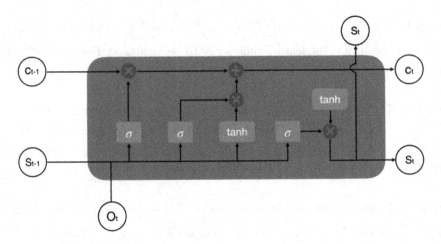

Fig. 2. The specific structure of LSTM cell

The memory gate determines which information will be retained in the newly entered information o_t and s_{t-1}. First, a sigmoid layer determines how to update the neural network parameters are W_i and b_i. Next, a tank layer creates a vector of new candidate values, $\widetilde{C_t}$, from the neural network parameters W_c and b_c. The mathematical form is as follows:

$$i_t = \sigma\left(W_i \cdot [s_{t-1}, o_t] + b_i\right) \tag{4}$$

$$\widetilde{C_t} = tanh\left(W_c \cdot [s_{t-1}, o_t] + b_c\right) \tag{5}$$

The cell state, C_t, will be update according to the cell state of forgetting gate and the output state of memory gate. The output gate determines the output of LSTM cell at time T according to the updated cell state and the input at the previous moment. The mathematical form is as follows:

$$C_t = f_t * C_{t-1} + i_t * \widetilde{C_t} \tag{6}$$

$$h_t = \sigma\left(W_h \cdot [s_{t-1}, o_t] + b_h\right) \tag{7}$$

$$s_t = h_t * tanh\left(C_t\right) \tag{8}$$

The backpropagation of the LSTM model in the training process can be roughly divided into four steps: First, the output value of LSTM cells was calculated according to the above forward calculation method. Second, calculate the error term of each LSTM cell backward, including two backward propagation directions according to time and network level. Third, calculate the gradient of each weight according to the corresponding error term. At the end, apply the gradient-based optimization algorithm to update the weight. The backpropagation of the model proposed in this paper uses the Adam optimization algorithm based on cosine annealing. Adam's algorithm can calculate adaptive learning rates for different parameters and takes up fewer storage resources. The cosine annealing algorithm is introduced into the initial learning rate setting of each small batch training, and the experiment shows that this method is better than the simple Adam algorithm.

2.4 Model Parameter Optimization Algorithm

The learning rate is an important parameter of the deep learning model. The neural network usually needs many generations of complete training. The learning rate setting is too small, so it takes too much time to converge; the learning rate setting is too large, but the oscillation around the minimum cannot converge to the minimum. The setting of the learning rate is commonly used with a fixed learning rate and learning rate decay. The mainstream learning rate setting method used by the training model is three-stage learning rate attenuation (MultiStepLR). In this section, we use a cosine annealing learning rate [7] attenuation algorithm to train the data, and the loss value of the set is calculated and verified in each step of the training. The initial learning rate was set as LR^0, the minimum learning rate was set as LR^1, the total number of training cycles was $EPOCH$, and each cycle was set as E. In this method, the learning rate decreases gradually in E of one training cycle, and is updated to LR^0 in the next cycle. The learning rate of each generation during the training process is as follows.

$$lr = \left(LR^0 - LR^1\right) cos\left(\pi\left(EPOCH\%E\right)\right) + LR^1 \qquad (9)$$

Sets the initial vector boundary to 0.01, the termination of the vector boundary is set to 10^{-6}. Compared with many sections of the vector decay, cosine vector of annealing changes as shown in the Fig. 3. Different from many sections of vector constantly decreases, the cosine annealing in vector in each cycle is smaller, and the next cycle back to the initial vector boundary. The annealing setting will help Adam optimization algorithm is better to update the network weights, jump out of local optimum.

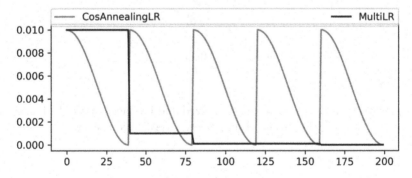

Fig. 3. Diagram of three-stage learning rate decay method and cosine annealing algorithm

3   Experiments

3.1   Experiment Design

In this section, the EH-LSTM and its parameter optimization algorithm proposed in the previous section are applied to carry out experimental verification. The experimental environment is as follows: The CPU is a 1.4 GHz quad-core Intel Core I5 with 8 GB of memory, and the operating system is macOS 10.15.4. The programming language is Python 3.7.1, and the integrated development environment is PyCharm Community Edition 2020.1.1. In the process of programming, RNN, LSTM, and GRU models are based on the Pytorch 1.4.0 program package implementation of Python. HMM is based on the HMMLearn program package in Python. The ARIMA model and the Holt-Winters model are based on the StateModels program package in Python.

The datasets used in the experiment were the operation states of a campus network switch in the actual operation environment within two months. The observed state features collected included CPU utilization rate, memory utilization rate, and network flow rate. The node health status is marked by the network manager. For a long time, the mean value of state change in a certain period can be regarded as the average state, and the average value of switch operation state in unit time is collected in 2 h. The data is divided into a training set, a verification set, and a test set. The data of the first 11 days were used as the training set for the training model, and the data of the second 2 days were used as the observation set to predict the health situation of the edge layer nodes. The node health status labeling rules are shown in Table 1.

Table 1. State partition table

Value	State
1	Lower
2	Low
3	Medium
4	High
5	Higher

About 1500 observations were normalized and Laplacian smoothed, and then the data set was divided into a training set, a verification set, and a test set by using 20% cross validation.

Contrast Models. By observing the health state time series of nodes in Fig. 4, it can be seen that there is a certain seasonality in the data changes. As can be seen from the figure below, the observed state and the time data of the node's health state are shown in the figure below. It can be seen that the health state

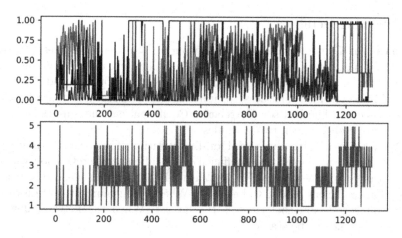

Fig. 4. Switch state time-series data

of the node changes steadily over time. In this paper, we choose not only RNN to compare with the proposed model but also four time series prediction models: GRU, HMM, Holt-Winters model, and ARIMA.

The GRU model is a simplified version of the LSTM model, but it retains the long-term memory capacity of the LSTM model. In practical application, GRU model and LSTM model have strong comparability. HMM describes the Markov process of an implicit state time series. The HMM parameter, $\lambda = [A, B, \pi]$, is the key to determining whether the hidden state time series is accurate: A is the state transition matrix, B is the observed probability matrix, and π is the initial state vector. The Holt-Winters model is a time series analysis method with trend and periodicity components. The idea is to use different characteristic components of historical data (levels, trends, and seasons) to recurse the current data. ARIMA use the number of autoregressions, difference times, and moving average terms respectively, which can be determined by observing Auto Correlation Function (ACF) and PACF, or by minimizing the AIC (Akaike Information Criterion) or BIC (Bayesian Information Criterion) values.

Evaluation. This paper evaluates all prediction models in the following two aspects: time complexity and model accuracy. For time complexity, the time consumed by each model in the construction process is counted. For model accuracy, the Root Mean Square Error (RMSE) and Cross Entropy loss Error (CEE) are selected as the measurement criteria. The RMSE is the square sum of the deviation between the predicted value and the real value at each moment of the test set, and the square root of the ratio of the time length of the test set. It measures the deviation between the predicted value and the real value, which can be used to evaluate the accuracy of the model. The CEE is used to estimate the difference between the predicted value and the real value distribution.

3.2   Training Details

Observation Data Processing. Choosing an appropriate representation of input data is an important part of the machine learning model. For continuous observation features, fitting training may lead to overfitting, and discretized features have strong robustness to abnormal data. Treat discretized data as continuous data. The Pearson Correlation Coefficient can measure the degree of linear correlation between two variables.

The K-mean algorithm was used for discretization. The variation trend of the discretized Pearson correlation coefficient with different discretization values of each feature was shown in the Fig. 5. When CPU utilization reached 4, Pearson correlation coefficient tended to be stable. When the memory utilization reaches 5, Pearson correlation coefficient is the largest. When the network flow rate reaches 6, it tends to be stable.

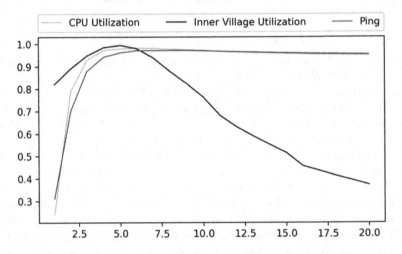

Fig. 5. Observed features discretize Pearson correlation coefficient variation trend chart (Color figure online)

The LSTM model is compared using the aforementioned definitions of discrete and continuous observations. During the iterative training process, the loss function for each step is computed using a monotonically decreasing gradient sequence. In the Fig. 6, the black curve represents the loss function of the continuous observation data after several rounds of training and parameter tuning, while the green curve represents the corresponding loss function. Under the same parameter settings, when the training data is replaced with discrete observations, the resulting loss function curves are shown in red. It is evident that the training effect of continuous observation data is subpar, quickly converging to a local optimum. On the other hand, discretizing the observation data proves beneficial in improving the accuracy of the model.

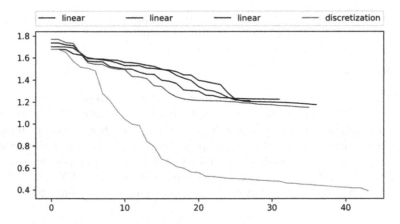

Fig. 6. Loss variation graph (black line is continuous data training, blue is network flow rate) (Color figure online)

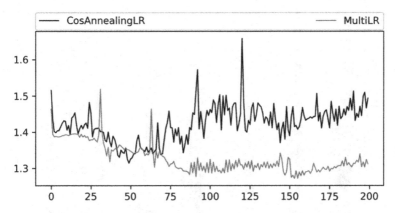

Fig. 7. The comparison of Three - step learning rate decay method and cosine annealing algorithm (Color figure online)

Parameter Optimization. In this part, we use the multi-step cosine annealing learning rate attenuation training data to calculate and verify the loss value of the set in each step of the EPOCH. A comparison is made between the vector decay algorithm and a three-stage vector decay algorithm with the same boundaries, examining the changes in the vector and the validation set loss function. The Fig. 7 illustrates that as the number of training iterations increases, the loss function of the network utilizing the vector decay algorithm shows a consistent decrease initially, reaching a local optimum at around 60 iterations. On the other hand, the loss function of the network employing the cosine annealing learning rate decay approach continues to decrease steadily with training iterations, reaching the optimal value after approximately 150 iterations. Furthermore, the loss function obtained from the cosine annealing approach is significantly lower compared to the multi-stage decay learning rate network.

3.3   Results

The CEE and RMSE are advanced measurements of model accuracy. Lower CEE and RMSE imply better prediction to fit the target value. We can see from Fig. 8 that the improved LSTM model is superior to other models in terms of RMSE and CEE, which indicates the proposed model has better performance than other models. The predicted time cost of the improved model is also lower than ARIMA, HMM, and CNN, but slightly higher than RNN and GRU. It shows that the improved LSTM model is good enough to implement in the real world.

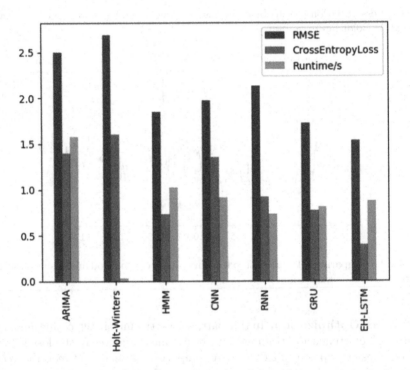

Fig. 8. Comparison of model results

4   Discussion

In this paper, we have shown that the multi-step cosine annealing learning rate and K-means discretization algorithm used in EH-LSTM can effectively improve the model's performance. Because the EH-LSTM method performed well in real-world datasets, it may be a feasible technique to predict the running state time

series of the nodes at the edge layer. Compared with the models mentioned in
Sect. 4.1.1, the EH-LSTM achieved better or comparable results, indicating that
the LSTM-based prediction model may be more robust than that based on the
classical model mentioned.

References

1. Chadza, T., Kyriakopoulos, K.G., Lambotharan, S.: Contemporary sequential network attacks prediction using hidden Markov model. In: 2019 17th International
   Conference on Privacy, Security and Trust (PST), pp. 1–3. IEEE (2019)
2. Choi, Y., Lim, H., Choi, H., Kim, I.J.: GAN-based anomaly detection and localization of multivariate time series data for power plant. In: 2020 IEEE International
   Conference on Big Data and Smart Computing (BigComp), pp. 71–74. IEEE (2020)
3. Graves, A.: Supervised sequence labelling. In: Graves, A. (ed.) Supervised Sequence
   Labelling with Recurrent Neural Networks, pp. 5–13. Springer, Heidelberg (2012).
   https://doi.org/10.1007/978-3-642-24797-2_2
4. Hochreiter, S., Schmidhuber, J., et al.: Long short-term memory. Neural Comput.
   9(8), 1735–1780 (1997)
5. Jain, A.K.: Data clustering: 50 years beyond k-means. Pattern Recogn. Lett. 31(8),
   651–666 (2010)
6. Karim, F., Majumdar, S., Darabi, H., Chen, S.: LSTM fully convolutional networks
   for time series classification. IEEE Access 6, 1662–1669 (2017)
7. Loshchilov, I., Hutter, F.: SGDR: stochastic gradient descent with warm restarts.
   arXiv preprint arXiv:1608.03983 (2016)
8. Malhotra, P., Vig, L., Shroff, G., Agarwal, P., et al.: Long short term memory
   networks for anomaly detection in time series. In: ESANN, vol. 2015, p. 89 (2015)
9. Qiao, L., Liu, D., Yuan, X., Wang, Q., Ma, Q.: Generation and prediction of
   construction and demolition waste using exponential smoothing method: a case
   study of Shandong Province, China. Sustainability 12(12), 5094 (2020)
10. Saini, R., Roy, P.P., Dogra, D.P.: A segmental hmm based trajectory classification
    using genetic algorithm. Expert Syst. Appl. 93, 169–181 (2018)
11. Schaffer, A.L., Dobbins, T.A., Pearson, S.A.: Interrupted time series analysis using
    autoregressive integrated moving average (ARIMA) models: a guide for evaluating
    large-scale health interventions. BMC Med. Res. Methodol. 21(1), 1–12 (2021)
12. Wang, L., Yang, C., Sun, Y., Zhang, H., Li, M.: Effective variable selection and
    moving window hmm-based approach for iron-making process monitoring. J. Pro-
    cess Control 68, 86–95 (2018)
13. Wang, S., Chen, J., Wang, H., Zhang, D.: Degradation evaluation of slewing bearing
    using hmm and improved GRU. Measurement 146, 385–395 (2019)
14. Wang, S., Xiang, J., Zhong, Y., Zhou, Y.: Convolutional neural network-based
    hidden Markov models for rolling element bearing fault identification. Knowl.-
    Based Syst. 144, 65–76 (2018)

A Comprehensive Review of Arabic Question Answering Datasets

Yassine Saoudi[1]([✉]) and Mohamed Mohsen Gammoudi[2]

[1] Faculty of Sciences of Tunis, University of Tunis El Manar, Tunis, Tunisia
yassine.saoudi@fst.utm.tn
[2] Higher Institute of Arts and Multimedia Manouba,
University of Manouba, Manouba, Tunisia

Abstract. The research community has shown significant interest in Question Answering (QA) due to the strong relevance of QA applications. In recent years, there has been a significant increase in the availability of publicly accessible datasets aimed at advancing research in Arabic QA systems. This survey aims to identify, summarize, and analyze current Arabic QA datasets, such as Monolingual, Multilingual, and Cross-lingual. Our research surveys the existing datasets and provides a comprehensive and multi-faceted classification. Furthermore, this study aims to guide research in Arabic QA by providing the latest updates about the state-of-the-art in this field and identifying shortcomings in the current datasets to develop more substantial and improved collections. Finally, we discuss the existing challenges in Arabic QA datasets and highlight their potential benefits for future research.

Keywords: Arabic QA datasets · Arabic QA Corpus · Survey

1 Introduction

Question Answering (QA) is a crucial and prominent research area in Natural Language Processing (NLP). The main goal of a QA system is to deliver precise answers to questions expressed in natural language. The recent advancements of Large Language Models (LLMs) such as BERT (Bidirectional Encoder Representations from Transformers) [13] and its variants, as well as the family of unsupervised transformer-based generative language models recently like GPT-4 [25] (Generative Pre-trained Transformer), and those previous variants (GPT-1, GPT-2 and GPT-3) have played a pivotal role in driving substantial advancements across various NLP tasks, including QA. Given the rapid advancements in NLP techniques, including the development of powerful language models and deep learning algorithms, evaluating the current state of Arabic QA and exploring the existing datasets available in Arabic QA is imperative. By conducting a comprehensive survey, we can provide researchers and practitioners with a valuable overview of the Arabic QA datasets and explain their limitations and areas

© The Author(s), under exclusive license to Springer Nature Singapore Pte Ltd. 2024
B. Luo et al. (Eds.): ICONIP 2023, CCIS 1961, pp. 278–289, 2024.
https://doi.org/10.1007/978-981-99-8126-7_22

for improvement. Our survey will contribute to advancing Arabic QA research and inspire the creation of enhanced and high-quality Arabic datasets in the future.

Several studies have been conducted to survey Arabic QA research, with the primary objective of providing readers with a comprehensive overview of the current works in Arabic QA systems. Notable recent surveys include Chandra et al. [11] and Alwaneen et al. [2]. Chandra et al. presented a comprehensive survey focusing on non-English QA datasets. However, due to the inclusion of multiple languages (French, German, Japanese, Chinese, Arabic, and Russian), they only explored a limited number of Arabic QA datasets. Alwaneen et al. conducted a comprehensive survey on Arabic QA systems, covering recent advancements, language-related challenges, QA tools and reviewing various Arabic QA systems and evaluation measures. However, this survey needed an in-depth discussion of critical insights and future directions regarding Arabic QA datasets.

Our paper presents a comprehensive survey of Arabic QA datasets. This comprehensive compilation includes datasets for various problem domains such as monolingual, multilingual, cross-lingual, and QA datasets based on machine reading comprehension. Our objective is to identify any shortcomings in the existing datasets, enabling the development of a more substantial and improved collection by leveraging the available resources.

The rest of this survey is organized as follows. Section 2 presents the review methodology employed in this study. Next, Sect. 3 extensively reviews monolingual QA datasets, while Sect. 4 explores multilingual and Cross-lingual QA datasets. Finally, in Sect. 5 we give an in-depth discussion, offering detailed recommendations and guidelines for future work on Arabic QA dataset development.

2   Methodology

While previous surveys on Arabic QA systems have been conducted, our survey aims to comprehensively analyze all existing Arabic QA datasets. In this research endeavor, we delve deep into examining the current state of research concerning the construction of Arabic QA datasets. Our primary focus is to address key research questions such as: What is the contemporary situation of building Arabic QA datasets? What experiments have been assembled, and what level of success have they attained? Identifying the existing gaps is another crucial aspect we aim to explore. Additionally, we strive to propose ways to enhance the current Arabic QA datasets, thus contributing to the advancement of this field.

In order to address these questions, we studied, reviewed, and explored pertinent research papers. The research papers were collected from the initial proposal of the first Arabic QA dataset from 2001 to 2023. This comprehensive review of available research papers is crucial to gain insights into Arabic QA systems' current state and identify potential improvement avenues. The literature collection encompassed many reputable publishers, including Springer, Elsevier, IEEE, ACM, Wiley, and Taylor & Francis. Additionally, we conducted comprehensive searches in prominent databases such as Google Scholar, Scopus, Web of

Science, and DBLP. The papers were primarily gathered using specific keywords related to Arabic QA, such as "Arabic question answering," "question answering systems," "Arabic question answering dataset," "Arabic question answering resource," "Multilingual and Cross-lingual Arabic question answering dataset,'" and "Arabic question answering corpus." Three hundred thirty-six papers were initially collected, of which only 33 were selected. We exclusively selected papers for inclusion in the Arabic QA dataset papers that contained the keyword "Arabic" along with the keywords "dataset," "corpus," or "resource" in either their title, abstract, or keyword section.

In the following section, we categorize the Arabic datasets to understand their development comprehensively. First, we classify them into monolingual and multilingual datasets. Then, each dataset is further categorized based on the types of questions and answers it contains. Finally, we provide information for each dataset regarding the publication year, size, format, source/domain, online accessibility (URL link), formulation, and metrics used. Our study identified fifteen monolingual QA datasets and five multilingual datasets, including Arabic, covering the period from 2001 to 2023.

3   Monolingual Datasets for Arabic QA Systems

The research community has shown significant interest in investigating Arabic question-answering datasets. However, due to the challenges of the Arabic language, Arabic datasets are less developed for QA than English and other Latin languages.

TREC: stands for Text Retrieval Conference. The annual conference series and evaluation forum focuses on information retrieval (IR) research and evaluation. TREC has organized various tracks and tasks related to information retrieval and QA, including those specific to Arabic language processing. These tasks aim to encourage the advancement of effective Arabic QA systems and facilitate research in the field. [15] is the first cross-language test set for Arabic questions consisting of 75 questions with no answers.

ArabiQA is a Factoid QA system that includes a dataset of deux cents questions and 11k documents from the Arabic version of Wikipedia. The dataset was introduced in a paper by Benajiba et al. [10]. They employed the ANERsys named entity recognition system to identify and categorize named entities in the obtained passages as part of their methodology.

DefArabicQA: is a QA dataset introduced in 2010 [31]. It consists of 50 questions and answers in Arabic definition questions about a person or an organization ('What is', 'Who is'). Their primary emphasis was on definition questions, and they employed an answer ranking module that relied on the frequency of words.

QArab-Pro is an Arabic question-answering dataset introduced in 2011 [1]. It comprises 75 reading comprehension tests and 335 questions from Arabic

Wikipedia. The dataset includes various question types, such as 45 "why" questions, 53 "who" questions, 47 "where" questions, 34 "how many/much"' questions, 93 "what/which" questions, and 63 "when" questions. QArab-Pro was developed using rules specific to each question type and incorporates several NLP components, including question classification, question reconstruction, and stemming.

QA4MRE 2012 (QA for Machine Reading Evaluation) [32] is the first Machine Reading Comprehension (MRC) Arabic dataset designed explicitly for the evaluation of QA systems. This dataset combines factoid and non-factoid questions, making it a hybrid dataset. The dataset comprises 140 multiple-choice questions from a single document and a background collection. The questions cover five different categories: factoid, purpose, causal, method, and determining truth. Each test document is associated with ten questions, further divided into three difficulty levels: difficult, intermediate, and simple. The simple questions are based on a single sentence fact, where the answer can be directly inferred from the given sentence. Intermediate questions require the interpretation of multiple sentences, and the answers may require synthesizing information from different parts of the document. Difficult questions, on the other hand, require utilizing external world knowledge inferred from the provided background collection.

QA4MRE 2013: [27] is an Arabic QA dataset that was introduced as part of the CLEF (Conference and Labs of the Evaluation Forum) evaluation campaign in 2013. It extends the QA4MRE-2012 dataset and aims to assess system-level inference capabilities in QA. The dataset consists of 240 questions, an increase from the previous year. However, besides factoid and non-factoid questions, this dataset also includes questions with no correct answers. Specifically, six of the 15 questions provided per document do not have a correct answer.

AQA-WebCorp: (Arabic QA Web Corpus) [9] is a factoid Arabic QA dataset consisting of 250 question-texts pairs. In addition, the dataset comprised 100 frequently asked questions (FAQs); 100 questions were gathered from forums, 25 were translated from the TREQ, and 25 were translated from the CLEF.

LEMAZA: [8] is a non-factoid QA dataset. It focuses explicitly on why questions and is designed with the application of rhetorical structure theory (RST) in mind. The dataset comprises 100 why questions. Also, the authors developed a method based on RST for the answer extraction module of their LEMAZA system.

TALAA-AFAQ: introduced by Aouicha et al. [4], is a collection of 2002 question-answer pairs for Arabic factoid question answering. The dataset encompasses classes such as Date, Name, Quantity, Date, Time, and Location. The authors followed a series of steps to create this dataset. Initially, they collected questions from various sources and categorized them based on their types. Next, they extracted answer patterns used for training the QA System. The vocalization of the questions aided the authors in the annotation and chunking process. Lastly, both semantic and syntactic features were extracted from the dataset.

AskFM[1]: is an Islamic dataset of 98,000 questions and their answers.

DAWQAS: (Dataset for Arabic Why QA System) [17] team created a non-factoid dataset specifically for answering "why" questions in Arabic. They generated the dataset through a semi-automatic process consisting of several steps. They retrieved "why" titles from the web in the first step. Then, they cleaned the collected data in the second step to ensure data quality. The third step involved preprocessing the text to prepare it for further analysis. Moving on to the fourth step, the team reclassified the responses by merging or replacing classes to enhance the dataset's coherence. In the fifth step, they assigned a probability score to each sentence in the dataset, indicating the presence of a rhetorical relation in the response text. They finally assigned the highest-probability rhetorical relation label to each response. In the sixth step, a native Arabic speaker manually verified and determined the exact position of the answer within the response. Lastly, in the last step, the team generated the dataset file.The DAWQAS dataset covers various topics, such as politics, economy, technology, nature, and animals. The team collected the dataset from diverse sources, including Ibelieveinsci.com, Mawdoo3.com, Albayan Ae, Arabic.rt.com, and Limaza.com.

ARCD: (Arabic Reading Comprehension Dataset) [24] is a factoid dataset that consists of 1,395 questions. These questions were posed by crowd workers and are based on Wikipedia articles. Additionally, ARCD includes a machine translation of the Stanford QA Dataset (Arabic-SQuAD). Each tuple in ARCD contains passages, questions, and answers (p, q, a).

Arabic-SQuAD: [24] is an MRC dataset designed explicitly for the Arabic language. It is the translated version of the well-known English SQuAD (v1.1.) [28]. The English-SQuAD, also known as Stanford QA Dataset, serves as a widely used benchmark for evaluating the performance of QA models in English. It consists of over 100,000 QA pairs derived from 442 English Wikipedia articles. In the case of Arabic-SQuAD, the dataset comprises 48,344 questions translated from the English-SQuAD dataset. It encompasses 10,364 paragraphs sourced from 231 articles, providing a rich and diverse set of questions for evaluation purposes.

AQAD: (Arabic Question-Answer Dataset) Atef et al. [7] have created an Arabic QA dataset based on the SQuAD (Stanford QA Dataset). The authors retrieve the passages in AQAD from the corresponding Arabic Wikipedia pages corresponding to the SQuAD passages in Chinese and English. In the first step, the authors filter the paragraphs in Arabic Wikipedia that semantically match those in English Wikipedia. Once they identify the matching passages, they translate the questions and answers related to those passages from English to Arabic. As a result, the AQAD dataset consists of over 17,000 tuples of paragraphs, questions, and answers.

AR-ASAG: also known as the Arabic dataset for automatic short answer grading evaluation [26], focuses on the cybercrime course. It comprises 2133 pairs

---

[1] https://omarito.me/arabic-askfm-dataset/.

of model and student answers, corresponding to 48 questions. The authors utilized the COALS stand for Correlated Occurrence Analogue to Lexical Semantic algorithm to establish the semantic space for word distribution. They employed the summation vector model, term weighting, and familiar words to assess the similarity between a teacher's model and a student's answer.

AyaTEC [23] is a test collection designed for verse-based QA focused explicitly on the Holy Qur'an. It consists of 207 questions that cover various topics within the Qur'an. These questions include both factoid and non-factoid types. Additionally, AyaTEC provides 1,762 corresponding answers for these questions, allowing researchers and developers to evaluate their QA systems in the context of Quranic verses. Table 1 summarizes the datasets discussed in this section. Furthermore, the table compares the listed datasets' domain coverage, size, source, and online accessibility.

Table 1. Datasets for Arabic QA systems

Datset	Year	Size	Format	Source/Domain	URL link
TREC (test-set)	2001	75 Q	Txt	Arabic newswire	https://shorturl.at/DLY12
ArabiQA [10]	2007	200 Q	N/A	Wikipedia	Not Available
Def-Arabic QA [31]	2010	50 Q	N/A	Wikipedia and Google search	Not Available
QArab-Pro [1]	2011	335 Q	N/A	Wikipedia	Not Available
QA4MRE [32]	2012	140 Q	xml	Climate change Aids, and Music	https://shorturl.at/oyCGS
QA4MRE [27]	2013	240 Q	xml	Climate change Aids, and Music	https://shorturl.at/rNQVX
AQA-WebCorp [9]	2016	250 Q	N/A	health, medicine, culture, Islam	Not available
LEMAZA [8]	2017	100 Q	N/A	Arabic Text	Not available
TALAA-AFAQ [4]	2017	2,002 QA	Txt, xml	web, set of QA4MRE@CLEF and TREC Arabic questions)	https://shorturl.at/jIJW0
DAWQAS [17]	2018	3,205	xlsx	General Websites	https://shorturl.at/klxJR
Arabic-SQuAD [24]	2019	48,344 Q	Json	Translation from SQuAD [28]	https://shorturl.at/kPUXZ
ARCD [24]	2019	1,395 Q	Json	Wikipedia articles MT of the SQuAD	https://shorturl.at/kPUXZ
AQAD [7]	2020	1,7000 QA	Json	Translated from SQuAD	https://shorturl.at/cfghJ
AR-ASAG [26]	2020	2,133	Txt, xml, and .db	Short answering for grading evaluation	https://shorturl.at/eswRX
AyaTEC [23]	2020	207 Q, 1762 A	Txt, xml	Holy Qur'an	https://shorturl.at/CGQ29

4 Available Multilingual and Cross-Lingual Datasets for the QA System Include Arabic

Multilingual QA is a subfield at the intersection of QA, Information Retrieval (IR), and NLP. In recent years, several multilingual Arabic QA systems have been developed to address the challenges posed by monolingual QA and the complexities of accessing and retrieving information in multiple languages. The primary approach in multilingual QA involves translating the pertinent sections of a question using machine translation systems. These translations are then utilized to access relevant information, enabling the retrieval of answers in different languages. This section aims to recognize, summarize, and analyze the multilingual and cross-lingual QA datasets, including those that involve the Arabic language. Table 2 compares the domain coverage, size, source, and online availability of Arabic multilingual and cross-lingual datasets.

XQuAD: [5], which stands for Cross-lingual QA Dataset, is a multilingual dataset created to evaluate cross-lingual QA. The dataset comprises 240 paragraphs and 1,190 question-answer pairs that have been professionally translated from SQuADv1.1 into 10 different languages, including Arabic.

MLQA: (Multilingual QA) [20] is a multilingual QA dataset designed to evaluate the performance of models across multiple languages. It includes QA pairs in seven languages, including Arabic. The dataset contains 12k English instances and 5k in each other language.

TyDi QA: (Typologically Diverse QA) [12] is a comprehensive multilingual dataset that showcases topological diversity. It encompasses eleven languages, including Arabic. Skilled annotators collected the dataset with a concise prompt sourced from Wikipedia. Based on the prompt, they were then tasked with formulating "information-seeking" questions in various typologically diverse languages.

XOR-TYDI-QA: (Cross-lingual Open-Retrieval QA) [6] is a substantial dataset constructed using questions from TyDi QA that do not have corresponding answers in the same language. This dataset combines information-seeking questions, multilingual QA, and open-retrieval QA to create a comprehensive multilingual open-retrieval QA dataset that facilitates cross-lingual answer retrieval. It comprises 40,000 information-seeking questions, including those in the Arabic language.

MKQA: (Multilingual Knowledge Questions and Answers) [22] is an extensive collection of multilingual open-domain QA datasets. It serves as an evaluation benchmark for assessing the quality of QA across 26 typologically diverse languages, including Arabic. The dataset was created by sampling and translating 10k queries from the Google Natural Questions dataset [19].

Table 2. Available Multilingual and Cross-lingual datasets for the QA system.

Datset	Year	Size	Format	Source	URL link
XQuAD [5]	2019	240 para. 1190 Q/A pairs	Json	Professional Translator from SQuAD v1.1 [28]	https://shorturl.at/szLV7
MLQA [20]	2020	> 12k INST EN. 5k other lang.	Json	Context: General Wikipedia	https://shorturl.at/fyCQU
TyDi QA [12]	2020	204,000 QA pairs	Json	Context: General Wikipedia	https://shorturl.at/cwENQ
MKQA [21]	2021	260k QA pairs	Jsonl	Context: General	https://shorturl.at/oNPTY
XORQA [6]	2021	40k Q	Jsonl	Human Translator from TyDi QA [12]	https://shorturl.at/acllJ

For more details, the following Table 3 shows the properties of various advanced Multilingual and Cross-lingual datasets, including the Arabic language. The Arabic question-answering datasets were divided into three categories of questions: factoid, non-factoid, hybrid, and MRC QA. Table 4 above provides more details on the available question-answering datasets in Arabic. (Concerning the formulation column, p represents a paragraph, q represents the question, and a represents the answer). Note: The performance in the tables' columns is associated with the authors who proposed each dataset.

Table 3. Properties of Multilingual and Cross-lingual datasets.

Datset	Cross-lingual	Topological diversite	Language	Total example
TyDi QA [12]	✗	✓	11	204K
MKQA [21]	✗	✓	26	260k
MLQA [20]	✓	✓	7	26k
XQuAD [5]	✗	✓	11	13k
XORQA [6]	✓	✓	7	40k

Table 4. Available QA datasets in Arabic comparison.

Datset	Questions type	Formulation	Performance
TREC	non-Factoid	q	–
ArabiQA [10]	Factoid	q,a	P: 63.21 R: 49.04 F1: 55.23MRR
DefArabicQA [31]	"Who is", "What is"	q,a with doc	expl: 0.70, MRR expo2: 0.81
QArabPro [1]	'Who is'	q,a	P: 0.93, R: 0.86, F1: 0.89
QA4MRE-2012 [32]	Factoid and non-factoid	doc.,q and mult. answers	Overall c@1 0.19
QA4MRE-2013 [27]	Factoid and non-factoid	doc.,q and multi. answers	Average c@1: 0.24
AQA-WebCorp [9]	factoid	q,a	c@1: 0.89
LEMAZA [8]	nonFactoid	q,a	R: 72.7, P: 79.2, c@1: 78.7
TALAA-AFAQ [4]	Factoid	q,a	–
AskFM [14]	Factoid	q,a	–
DAWQAS [17]	Why	q,a	F1: 71.2
ARCD [24]	Factoid	p,q,a	F1: 61.3
Arabic-SQuAD [24]	Factoid	p,q,a	F1: 27.6
AQAD [7]	Factoid and non-Factoid	p,q,a	F1: 37, EM: 33
AR-ASAG [26]	Factoid	q,a	Pearson: 0.72
AyaTEC [23]	Factoid and non-Factoid	q,a	F1(R): 0.61
XQuAD [5]	Factoid and non-Factoid	p,q,a	For Ar: F1: 0.66
MLQA [20]	Factoid and non-Factoid	p,q,a	ar. F1: 54.8, EM: 36.3
TyDi QA [12]	Factoid and non-Factoid	p,q,a	ar. Passage Answer: F1 62.5
XOR-TYDI-QA [6]	Factoid and non-Factoid	p,q,a	For Arabic (Macro avg.): F1: 18.7, EM: 12.1, BLEU: 16.8
MKQA [22]	Factoid and non-Factoid	p,q,a	F1: 32.42

5   Discussion

In Sect. 3 and 4, we analyzed the available literature on building an Arabic dataset for QA tasks. Our examination revealed several challenges and difficulties that persist in current research:

Low Resources 1: Many of the mentioned datasets are derived from translations of other languages, often utilizing datasets such as English-SQuAD, TREC, or CLEF. However, it has been observed in [3] that Arabic-SQuAD and ARCD datasets contain text elements in languages other than Arabic (including unknown sub-words and characters). This problem arises due to the inadequate quality of the training samples translated from the English SQuAD dataset. In addition, as mentioned in [2], it is worth noting that utilizing translated datasets can introduce language-related complications.

Low Resources 2: Most Arabic QA datasets heavily depend on limited-sized mentioned in Table 1 (column: size). Therefore, it becomes impractical to compare these systems as each system utilizes its unique dataset. Additionally, there is a general reluctance among authors to share their working code. However, there are a few exceptions, such as Arabic-squad and TyDi-QA datasets that incorporate the Arabic language.

Complex Challenges: Notably, alongside the general challenges encountered in constructing large-sized and good-quality Arabic datasets is the difficult challenge in the language itself, that Arabic pronouns also vary based on gender and sentence position. Additionally, there are differences in the rules, sentence structure, and Arabic diacritical marks, which signify vowel sounds and pronunciation, as well as cultural background, between Arabic and English.

Lack of Benchmark References and Human Performance Benchmarks: The limited accessibility of specific QA datasets hinders researchers from effectively evaluating their methods and understanding their limitations. Furthermore, while publicly available datasets are crucial resources, challenges arise when comparing different approaches due to variations in data partitioning. Therefore, establishing standardized and consistent evaluation procedures is of utmost importance. The comparison between human performance and deep learning models is crucial to assess the disparity between the two.

5.1   Guidelines for Arabic QA Dataset

Taking into consideration our extensive analysis of the literature, we propose the following guidelines for future research on Arabic QA datasets:

First, A notable observation highlights that a considerable amount of time in the machine-learning process is dedicated to data preparation. According to [16,33], it is crucial to emphasize that, regardless of the quality of the machine learning algorithms, optimal performance can only be achieved with high-quality data. Moreover, data collection holds significance as recent deep-learning approaches rely less on feature engineering and more on extensive data availability. However, data processing becomes a crucial step to ensure data quality, which includes essential steps such as data validation, cleaning, and integration techniques.

Second, data augmentation techniques serve as a convenient manner to artificially augment the size of a given dataset. Therefore, these techniques can enhance the interpretation and performance of Deep Learning (DL) models. Previous studies on non-Arabic QA datasets have demonstrated notable enhancements through data augmentation techniques. For instance, Roy et al. [29] presented a data augmentation technique where they fine-tuned the T5 model to generate new questions using customer reviews. Their findings revealed that using the Amazon review dataset, our model with the augmentation technique surpasses the state-of-the-art models. Furthermore, Yang et al. [33] introduced a data augmentation technique utilizing distant supervision, which leverages both positive and negative examples. Their research demonstrated significant improvements in effectiveness compared to previous approaches when applied to English QA datasets.

Lastly, another perspective is developing interactive Arabic QA systems that enable users to engage with the system and offer feedback on the accuracy and relevance of the system's outputs. This interactive functionality can enhance user experience and facilitate the system in learning from valuable user feedback.

In conclusion, investigation of Arabic QA for some specific domains: There is a need to investigate and build an Arabic QA dataset for several other domains, such as Judicial and Legal domains (criminal law, intellectual property, tax law). Furthermore, creating datasets in specific fields is paramount to cater to the increasing demand for specialized Arabic QA systems. These datasets will foster advancements in research and benefit various industries and applications that rely heavily on efficient QA capabilities. Additionally, we propose the exploration of dialectical content found on social media platforms as a means to construct comprehensive Arabic question-answering datasets.

6 Conclusion

This paper extensively reviewed the significant contributions to developing Arabic QA datasets. Our findings indicate a growing interest in Arabic QA within the NLP research community. However, we concluded that there is a limited large-sized and good-quality number of Arabic QA datasets. The scarcity of Arabic QA datasets has resulted in a limited number of studies focusing on Arabic language QA systems, in contrast to the extensive research conducted in English. The currently available datasets often need help with issues like the large size and low quality, primarily due to inadequate pre-processing or incorrect annotation. Furthermore, most of the listed datasets are translations from other languages, including the well-known Arabic-SQuAD, a translation of the original SQuAD dataset.

Experiments demonstrate that the quality of the available QA resources heavily influences the performance of QA systems. Due to the language challenge,

Arabic QA resources could be better in many domains than non-Arabic datasets. Moreover, recently, researchers have exploited the potential of chatGPT, such as Siu et al. [30] and Khoshafah [18] mentioned that GPT models have limited capabilities for low-resource languages such as Arabic. We hope this survey offers researchers a comprehensive overview of Arabic QA, encompassing available resources and highlighting open challenges. We intend to contribute towards advancing the state-of-the-art in this domain and encourage further progress.

References

1. Akour, M., Abufardeh, S., Magel, K., Al-Radaideh, Q.: Qarabpro: A rule based question answering system for reading comprehension tests in arabic. Am. J. Appl. Sci. **8**(6), 652–661 (2011)
2. Alwaneen, T.H., Azmi, A.M., Aboalsamh, H.A., Cambria, E., Hussain, A.: Arabic question answering system: a survey. Artifi. Intell. Rev., 1–47 (2022)
3. Antoun, W., Baly, F., Hajj, H.: Araelectra: pre-training text discriminators for arabic language understanding. arXiv preprint arXiv:2012.15516 (2020)
4. Aouichat, A., Guessoum, A.: Building TALAA-AFAQ, a corpus of Arabic FActoid question-answers for a question answering system. In: Frasincar, F., Ittoo, A., Nguyen, L.M., Métais, E. (eds.) NLDB 2017. LNCS, vol. 10260, pp. 380–386. Springer, Cham (2017). https://doi.org/10.1007/978-3-319-59569-6_46
5. Artetxe, M., Ruder, S., Yogatama, D.: On the cross-lingual transferability of monolingual representations. arXiv preprint arXiv:1910.11856 (2019)
6. Asai, A., et al.: Xor qa: cross-lingual open-retrieval question answering. arXiv preprint arXiv:2010.11856 (2020)
7. Atef, A., Mattar, B., Sherif, S., Elrefai, E., Torki, M.: Aqad: 17,000+ arabic questions for machine comprehension of text. In: 2020 IEEE/ACS 17th International Conference on Computer Systems and Applications (AICCSA), pp. 1–6. IEEE (2020)
8. Azmi, A.M., Alshenaifi, N.A.: Lemaza: an arabic why-question answering system. Nat. Lang. Eng. **23**(6), 877–903 (2017)
9. Bakari, W., Bellot, P., Neji, M.: Aqa-webcorp: web-based factual questions for Arabic. Proc. Comput. Sci. **96**, 275–284 (2016)
10. Benajiba, Y., Rosso, P., Lyhyaoui, A.: Implementation of the arabiqa question answering system's components. In: Proceedings of Workshop on Arabic Natural Language Processing, 2nd Information Communication Technologies International Symposium, ICTIS-2007, Fez, Morroco, April. pp. 3–5. Citeseer (2007)
11. Chandra, A., Fahrizain, A., Laufried, S.W., et al.: A survey on non-english question answering dataset. arXiv preprint arXiv:2112.13634 (2021)
12. Clark, J.H., et al.: Tydi qa: a benchmark for information-seeking question answering in typologically diverse languages. Trans. Assoc. Comput. Ling. **8**, 454–470 (2020)
13. Devlin, J., Chang, M.W., Lee, K., Toutanova, K.: Bert: pre-training of deep bidirectional transformers for language understanding. arXiv preprint arXiv:1810.04805 (2018)
14. Foong, Y.J., Oussalah, M.: Cyberbullying system detection and analysis. In: 2017 European Intelligence and Security Informatics Conference (EISIC), pp. 40–46. IEEE (2017)

15. Gey, F.C., Oard, D.W.: The trec-2001 cross-language information retrieval track: searching arabic using english, french or arabic queries. In: TREC, vol. 2001 (2001)
16. de Hond, A.A., et al.: Guidelines and quality criteria for artificial intelligence-based prediction models in healthcare: a scoping review. NPJ Digital Med. **5**(1), 2 (2022)
17. Ismail, W.S., Homsi, M.N.: Dawqas: a dataset for Arabic why question answering system. Proc. Comput. Sci. **142**, 123–131 (2018)
18. Khoshafah, F.: Chatgpt for arabic-english translation: Evaluating the accuracy. Preprint, it has not been peer-reviewed by a journal (2023)
19. Kwiatkowski, T., et al.: Natural questions: a benchmark for question answering research. Trans. Assoc. Comput. Ling. **7**, 453–466 (2019)
20. Lewis, P., Oğuz, B., Rinott, R., Riedel, S., Schwenk, H.: Mlqa: evaluating cross-lingual extractive question answering. arXiv preprint arXiv:1910.07475 (2019)
21. Longpre, S., Lu, Y., Daiber, J.: Mkqa: a linguistically diverse benchmark for multilingual open domain question answering. arXiv preprint arXiv:2007.15207 (2020)
22. Longpre, S., Lu, Y., Daiber, J.: Mkqa: a linguistically diverse benchmark for multilingual open domain question answering. Trans. Assoc. Comput. Ling. **9**, 1389–1406 (2021)
23. Malhas, R., Elsayed, T.: Ayatec: building a reusable verse-based test collection for Arabic question answering on the holy Qur'an. ACM Trans. Asian Low-Res. Lang. Inform. Process. (TALLIP) **19**(6), 1–21 (2020)
24. Mozannar, H., Hajal, K.E., Maamary, E., Hajj, H.: Neural arabic question answering. arXiv preprint arXiv:1906.05394 (2019)
25. OpenAI: Gpt-4 technical report (2023)
26. Ouahrani, L., Bennouar, D.: Ar-asag an arabic dataset for automatic short answer grading evaluation. In: Proceedings of the 12th Language Resources and Evaluation Conference, pp. 2634–2643 (2020)
27. Peñas, A., Hovy, E., Forner, P., Rodrigo, Á., Sutcliffe, R., Morante, R.: QA4MRE 2011-2013: overview of question answering for machine reading evaluation. In: Forner, P., Müller, H., Paredes, R., Rosso, P., Stein, B. (eds.) CLEF 2013. LNCS, vol. 8138, pp. 303–320. Springer, Heidelberg (2013). https://doi.org/10.1007/978-3-642-40802-1_29
28. Rajpurkar, P., Zhang, J., Lopyrev, K., Liang, P.: Squad: 100,000+ questions for machine comprehension of text. arXiv preprint arXiv:1606.05250 (2016)
29. Roy, K., Goel, A., Goyal, P.: Effectiveness of data augmentation to identify relevant reviews for product question answering. In: Companion Proceedings of the Web Conference 2022, pp. 298–301 (2022)
30. Siu, S.C.: Chatgpt and gpt-4 for professional translators: Exploring the potential of large language models in translation. Available at SSRN 4448091 (2023)
31. Trigui, O., Belguith, L.H., Rosso, P.: Defarabicqa: Arabic definition question answering system. In: Workshop on Language Resources and Human Language Technologies for Semitic Languages, 7th LREC, Valletta, Malta, pp. 40–45 (2010)
32. Trigui, O., Belguith, L.H., Rosso, P., Amor, H.B., Gafsaoui, B.: Arabic QA4MRE at CLEF 2012: Arabic question answering for machine reading evaluation. In: Forner, P., Karlgren, J., Womser-Hacker, C. (eds.) CLEF 2012 Evaluation Labs and Workshop, Online Working Notes, Rome, Italy, 17–20 September 2012. CEUR Workshop Proceedings, vol. 1178. CEUR-WS.org (2012). http://ceur-ws.org/Vol-1178/CLEF2012wn-QA4MRE-TriguiEt2012.pdf
33. Yang, W., Xie, Y., Tan, L., Xiong, K., Li, M., Lin, J.: Data augmentation for bert fine-tuning in open-domain question answering. arXiv preprint arXiv:1904.06652 (2019)

Solving Localized Wave Solutions of the Nonlinear PDEs Using Physics-Constraint Deep Learning Method

Yanan Guo[1,2](✉) ⓘ, Xiaoqun Cao[2] ⓘ, Mengge Zhou[2], Kecheng Peng[3], and Wenlong Tian[3]

[1] Naval Aviation University, Huludao, Liaoning, China
[2] College of Meteorology and Oceanography, National University of Defense Technology, Changsha, Hunan, China
guoyn18@163.com
[3] College of Computer, National University of Defense Technology, Changsha, Hunan, China

Abstract. In the field of nonlinear science, localized waves hold significant research value, and their theories have found applications across various domains. Partial Differential Equations (PDEs) serve as crucial tools for studying localized waves in nonlinear systems, and numerical methods for PDEs have been widely employed in the numerical simulation of localized waves. However, the complexity of solving partial differential equations has impeded progress in the study of nonlinear localized waves. In recent years, with the rapid advancement of deep learning, Physics-Informed Neural Networks (PINNs) have emerged as powerful tools for solving PDEs and simulating multiphysical phenomena, attracting significant attention from researchers. In this study, we apply an improved PINNs approach to solve PDEs governing localized waves. The enhanced PINNs not only incorporates the constraints of the PDEs but also introduces gradient information constraints, further enriching the physical constraints within the neural network model. Additionally, we employ an adaptive learning method to update the weight coefficients of the loss function and dynamically adjust the relative importance of each constraint term in the entire loss function to expedite the training process. In the experimental section, we selected Boussinesq equation and nonlinear Schrödinger equation (NLSE) for the study and evaluated the accuracy of localized wave simulation results through error analysis. The experimental results indicate that the improved PINNs are significantly better than traditional PINNs, with shorter training time and more accurate prediction results.

Keywords: Physics constraint · Deep learning · Self-adaptive loss function · Partial differential equations · Localized wave

© The Author(s), under exclusive license to Springer Nature Singapore Pte Ltd. 2024
B. Luo et al. (Eds.): ICONIP 2023, CCIS 1961, pp. 290–302, 2024.
https://doi.org/10.1007/978-981-99-8126-7_23

1   Introduction

In recent years, the emergence of big data and significant advancements in computing power have propelled the rapid development of deep learning technology, leading to remarkable achievements across various domains [1–4]. Large, diverse datasets and the ability to process them efficiently have enabled deep learning algorithms to excel in complex tasks. For example, in the field of image classification, accuracy and robustness have improved dramatically. Machines can now recognize objects and patterns in images with unprecedented accuracy. In the field of natural language processing (NLP), deep learning models have brought about a revolution. They have enabled machines to understand and generate human language, driving advances in machine translation, sentiment analysis, chatbots, and more. These technologies are reshaping the way we communicate and interact with computers. Time series forecasting is critical in a variety of industries, including finance, weather forecasting, and healthcare, and has benefited greatly from deep learning. Deep neural networks have demonstrated their ability to capture intricate temporal patterns and make accurate predictions that improve decision-making processes and resource allocation. As deep learning technology continues to evolve, it has the potential to fundamentally change the way we tackle complex challenges in different fields.

Furthermore, the application of deep learning goes far beyond the scope of computer science and is penetrating all corners of the real world, playing an increasingly critical role in different fields. In the field of natural science research, traditional scientific studies usually rely on laboratory experiments and theoretical models. However, researchers are increasingly recognizing that deep learning techniques can provide them with novel insights and solutions to complex problems. Today, traditional methods based on experiments and theories are merging with data-driven deep learning techniques. Deep learning has emerged as a powerful research paradigm that is rapidly reshaping several scientific fields, including physics, chemistry, materials science, and biology. For example, in physics, deep learning is being used to analyze large data sets generated by large particle accelerator, facilitating the discovery of new phenomena in particle physics. In chemistry, deep learning has accelerated the process of materials discovery, leading to the identification of novel catalysts and materials. In biology, deep learning is being used to analyze biomedical images to aid in disease diagnosis and drug development. Studies have shown that deep learning is not only excellent at processing large-scale data, but also at discovering underlying patterns in the data. This versatility makes it a promising tool for a wide range of applications in natural science research. Scientists are actively exploring ways to combine deep learning techniques with traditional scientific methods to accelerate discovery and innovation and push the frontiers of scientific exploration. As mentioned above, deep learning has become an integral part of contemporary scientific research, leading us into a frontier era of data-driven science.

Remarkable advancements have been achieved in this burgeoning interdisciplinary research domain. One particularly noteworthy breakthrough is the utilization of deep learning in weather and climate prediction, where it has substantially enhanced the accuracy of forecasting severe weather events like super hurricanes and rainstorms [5]. This application has had a profound impact on safeguarding lives and property, mitigating the effects of natural disasters, and improving our understanding of climate patterns. Another prominent area of research in this field involves harnessing deep neural networks to tackle the formidable challenge of solving partial differential equations (PDEs) [6,7]. Building upon earlier work, Raissi et al. introduced a groundbreaking concept known as Physics-Informed Neural Networks (PINNs) that incorporates physical constraints into the neural network architecture, thereby offering a more robust approach to solving PDEs [8]. The introduction of PINNs sparked intensive research efforts, with numerous scholars dedicating themselves to their development and refinement in recent years. Consequently, a multitude of innovative PINN variants have emerged and found application across diverse fields, including fluid mechanics, solid mechanics, thermodynamics, and materials science [9]. The evolution of PINNs has not only revolutionized the way we approach complex physical problems but has also opened up exciting possibilities for simulating and understanding intricate physical phenomena. Their versatility extends their reach to a myriad of scientific and engineering disciplines, holding the potential to reshape how we model and solve complex systems across the sciences. As researchers continue to delve into the possibilities offered by deep learning and PINNs, we can anticipate even more groundbreaking discoveries and practical applications in the years to come.

Localized waves are a significant research topic in the field of nonlinear science [10], playing a pivotal role in enhancing our understanding of the physical principles governing nonlinear systems. In recent years, researchers have employed various numerical techniques to investigate a wide range of localized wave phenomena. These methods encompass finite difference method, spectral method, and finite element methods, among others. The results of these simulations have not only deepened our comprehension of complex system behaviors but have also unveiled new categories of localized waves that were previously undiscovered. However, conventional numerical computational methods face several challenges, including limited computational efficiency and truncation errors. These challenges have prompted researchers to explore innovative numerical simulation approaches. In response to these difficulties, this study delves deeply into the application of deep learning techniques for simulating localized wave phenomena. Specifically, we introduce a novel architecture and loss function designed for simulating localized wave phenomena, building upon the foundation of traditional physics-informed neural networks. These innovations are rigorously validated through a series of numerical experiments. This research represents a significant advancement in our ability to accurately model and simulate localized wave phenomena.

2   Method

Physics-Informed Neural Networks (PINNs) have become increasingly popular in recent years as a new way to solve PDEs in various scientific fields. However, traditional PINNs have some limitations, including the inability to accurately model localized wave phenomena, such as solitary waves and rogue waves. In order to improve the accuracy and efficiency of the simulation of localized wave phenomenon, this section focuses on the improvement of PINNs using the gradient information of physical variables. At the same time, the weight coefficients of the loss function terms are optimized using an adaptive approach. For the following partial differential equations:

$$
\begin{aligned}
u_t + \mathcal{N}_x[u] &= 0, \quad x \in \Omega, t \in [T_0, T] \\
u(x, T_0) &= h(x), \quad x \in \Omega \\
u(x, t) &= g(x, t), \quad x \in \partial\Omega, t \in [T_0, T]
\end{aligned}
\tag{1}
$$

Next, a neural network $\widehat{u}(x, t; \theta)$ is created to approximate the solution of the PDEs, where θ represents the parameters of the neural network. The prediction results generated by the trained neural network $\widehat{u}(x, t; \theta)$ should comply with the specified initial and boundary conditions, and must also obey the physical constraints of the PDEs. The differential constraints of the partial differential equation are embedded in a neural network using automatic differentiation techniques and a residual network is built to fulfill the second requirement. The mathematical definition of this residual network is shown in (2).

$$
f(x, t; \theta) := \frac{\partial}{\partial t}\widehat{u}(x, t; \theta) + \mathcal{N}_x[\widehat{u}(x, t; \theta)]
\tag{2}
$$

In the original physics-informed neural networks, only PDEs residuals $f(\mathbf{x})$ are forced to be zero. However, further analysis shows that the derivative of $f(\mathbf{x})$ is also zero due to the fact that $f(\mathbf{x})$ is zero for any $\mathbf{x}$. Therefore, we can assume that the exact solution of PDEs is smooth enough so that the gradient of PDEs residuals $\nabla f(\mathbf{x})$ exists, and propose gradient-enhanced PINNs so that the derivative of PDEs residuals is also zero [11], which can be expressed in mathematical form by the following formula:

$$
\nabla f(\mathbf{x}) = \left(\frac{\partial f}{\partial x_1}, \frac{\partial f}{\partial x_2}, \ldots, \frac{\partial f}{\partial x_d} \right) = \mathbf{0}, \quad \mathbf{x} \in \Omega
\tag{3}
$$

Based on the definition provided earlier, it is logical to derive the loss function for the gradient-enhanced PINNs as follows:

$$
\begin{aligned}
L(\theta; N) =&\, w_o L_o\left(\theta; N_o\right) + w_f L_f\left(\theta; N_f\right) + w_b L_b\left(\theta; N_b\right) \\
&+ w_i L_i\left(\theta; N_i\right) + \sum_{i=1}^{d} w_{g_i} L_{g_i}\left(\boldsymbol{\theta}; \mathcal{T}_{g_i}\right)
\end{aligned}
\tag{4}
$$

where the loss of the derivative with respect to x_i is:

$$
L_{g_i}\left(\boldsymbol{\theta}; \mathcal{T}_{g_i}\right) = \frac{1}{|\mathcal{T}_{g_i}|} \sum_{\mathbf{x} \in \mathcal{T}_{g_i}} \left| \frac{\partial f}{\partial x_i} \right|^2
\tag{5}
$$

where $\mathcal{T}_{g_i}$ is the set of residual points for the derivative $\frac{\partial f}{\partial x_i}$. It should be noted that $\mathcal{T}_{gi}(i = 1, 2, \cdots, d)$ may be different for different equations. Numerous studies have emphasized the significance of selecting appropriate weight coefficients for the loss function when employing neural networks to solve partial differential equations. Consequently, the effective selection of optimal weight coefficients for the loss function is a critical consideration in PINNs design [12]. To enhance the accuracy and robustness of localized wave simulations, this paper conducts a comprehensive investigation into the PINNs loss function and its associated weight coefficients. Subsequently, we introduce an adaptive learning method for the selection of these weight coefficients. To identify the optimal weights for the loss function, we formulate the following hypothesis:

Hypothesis. *A loss function term has been satisfied when its variance has been optimized to zero.*

Based on this assumption, it is concluded that loss terms with constant values should not be further optimized. However, classical PINNs cannot automatically adjust the optimization process. Because the coefficients of each loss term of classical PINNs are generally treated as fixed constants, the optimization of the relevant terms cannot be adaptively adjusted during the training process. Therefore, we have designed adaptive weight updating methods for gradient-enhanced PINNs. Note that sometimes loss terms with large magnitudes may also have high absolute variance, although the variance of the loss terms is also relatively small. Therefore, the reference variance measure alone is not sufficient. Therefore, a coefficient c_L of the variance of the loss term is introduced [13], which is defined as follows:

$$c_L = \frac{\sigma_L}{\mu_L} \tag{6}$$

where L denotes the loss function, while σ_L and μ_L are the standard deviation and mean of loss L, respectively. The coefficient of variation, also referred to as the relative standard deviation (RSD), remains unaffected by the scale of the sample values. Consequently, it disassociates the magnitude scale of the loss term from its weighting. For instance, a loss term with a smaller magnitude that exhibits frequent variations is deemed more significant and warrants a higher weight. Conversely, a loss term with a larger magnitude that remains relatively constant during training is regarded as less significant and is thus assigned a lower weight. Based on the above considerations, the loss ratio l_i is defined:

$$l_i^t = \frac{L_i^t}{\mu_{L_i^{t-1}}} \tag{7}$$

where L_i^t denotes the ith term of the loss function defined in equation (4) and t is the number of iterations. $\mu_{L_i^{t-1}}$ is the average value of the loss term L_i for iterations up to $t-1$. At the tth iteration, the coefficient of variation of the loss ratio is used to calculate the weight value.

$$w_{l_i^t} = \frac{1}{z_t} c_{l_i^t} = \frac{1}{z_t} \frac{\sigma_{l_i^t}}{\mu_{l_i^t}} \tag{8}$$

where z_t is a normalized parameter that is independent of moment i. The mathematical definition of z_t is given below as follows:

$$z_t = \sum_{i=1} c_{l_i^t} \tag{9}$$

The above equation constructs the constraint $\sum_{i=1} w_{l_i^t} = 1$. Next, Welford's algorithm [14] is used to calculate the coefficient of variation and the loss ratio.

$$\mu_{L_i^t} = \left(1 - \frac{1}{t}\right)\mu_{L_i^{t-1}} + \frac{1}{t}L_i^t \tag{10}$$

$$\mu_{l_i^t} = \left(1 - \frac{1}{t}\right)\mu_{l_i^{t-1}} + \frac{1}{t}l_i^t \tag{11}$$

$$M_{l_i^t} = \left(1 - \frac{1}{t}\right)M_{l_i^{t-1}} + \frac{1}{t}\left(l_i^t - \mu_{l_i^{t-1}}\right)\left(l_i^t - \mu_{l_i^t}\right) \tag{12}$$

$$\sigma_{l_i^t} = \sqrt{M_{l_i^t}}. \tag{13}$$

The online training process, guided by the algorithm mentioned earlier, exhibits the capability to dynamically adjust weights. Specifically, it responds to a decrease in the loss rate, meaning when the loss falls below the average loss, by increasing the loss weights. This adaptive mechanism prevents the overlooking of losses that contribute significantly to optimization while mitigating the impact of extreme outliers on the loss values. Concurrently, if the standard deviation of the loss rate increases during training, it necessitates a corresponding increase in the loss weights. Essentially, as the loss rate displays greater variability throughout the training process, it signifies a need for increased learning, leading to higher weights assigned to the relevant terms. In summary, by adaptively learning the optimal weight coefficients of each loss function term through the above algorithm, it can effectively accelerate the training process of the neural network model and achieve neural network parameter optimization faster. For a convergent loss function, after a sufficient number of training iterations, the online training mean and standard deviation will ultimately converge towards their true counterparts, representing a refinement of the training process.

3   Experiments and Results

In this section, we approximated the localized wave solutions of the Boussinesq equation and the nonlinear Schrödinger equation (NLSE) using the improved PINNs and analyzed the convergence speed and accuracy of the proposed method. The accuracy of the approximate solution obtained by the neural network model was measured by the relative L2 error, which is defined as follows:

$$\text{L2 error} = \frac{\sqrt{\sum_{i=1}^{N}|\hat{u}(x_i, t_i) - u(x_i, t_i)|^2}}{\sqrt{\sum_{i=1}^{N}|u(x_i, t_i)|^2}} \tag{14}$$

where $u(x_i, t_i)$ represents the exact solution and $\hat{u}(x_i, t_i)$ represents the approximate solution.

3.1   The Two-Soliton Solution of Boussinesq Equation

In this subsection, we investigate the colliding-soliton solution of the Boussinesq equation using the previously described modified PINNs approach. The Boussinesq equation belongs to a category of geophysical models extensively applied in meteorology, oceanography, and various other domains. Numerical computation methods for this equation are crucial topics within the field of fluid dynamics. The mathematical formulation of the Boussinesq equation is provided below:

$$u_{tt} - u_{xx} - u_{xxxx} - 3\left(u^2\right)_{xx} = 0 \tag{15}$$

In the experiments, we selected a finite region for the numerical simulation of solitary waves, where the spatial domain range is $[-15, 15]$ and the time range is $[-5, 5]$. We considered the initial value problem for the Boussinesq equation with Dirichlet boundary conditions, and the initial conditions set in this experiment is given as follows:

$$
\begin{aligned}
u(x, -5) =& 2\frac{1.21e^{1.1x-8.1765} + 1.21e^{1.1x+8.1765} + 4.84e^{2.2x-0.9717}}{1 + e^{1.1x-8.1765} + e^{1.1x+8.1765} + e^{2.2x-0.9717}} \\
& - 2\frac{(1.1e^{1.1x-8.1765} + 1.1e^{1.1x+8.1765} + 2.2e^{2.2x-0.9717})^2}{(1 + e^{1.1x-8.1765} + e^{1.1x+8.1765} + e^{2.2x-0.9717})^2}
\end{aligned}
\tag{16}
$$

Under the above conditions, the solitary wave solution of the Boussinesq equation is given as follows:

$$
\begin{aligned}
& u(x, t) \\
& = 2\frac{k_1^2 e^{k_1 x + \omega_1 t + \delta_1} + k_2^2 e^{k_2 x + \omega_2 t + \delta_2} + (k_1 + k_2)^2 e^{(k_1+k_2)x+(\omega_1+\omega_2)t+\delta_1+\delta_2+\delta_0}}{1 + e^{k_1 x + \omega_1 t + \delta_1} + e^{k_2 x + \omega_2 t + \delta_2} + e^{(k_1+k_2)x+(\omega_1+\omega_2)t+\delta_1+\delta_2+\delta_0}} \\
& \quad - 2\frac{\left(k_1 e^{k_1 x + \omega_1 t + \delta_1} + k_2 e^{k_2 x + \omega_2 t + \delta_2} + (k_1 + k_2) e^{(k_1+k_2)x+(\omega_1+\omega_2)t+\delta_1+\delta_2+\delta_0}\right)^2}{\left(1 + e^{k_1 x + \omega_1 t + \delta_1} + e^{k_2 x + \omega_2 t + \delta_2} + e^{(k_1+k_2)x+(\omega_1+\omega_2)t+\delta_1+\delta_2+\delta_0}\right)^2} \\
& e^{\delta_0} = -\frac{(\omega_1 - \omega_2)^2 - (k_1 - k_2)^2 - (k_1 - k_2)^4}{(\omega_1 + \omega_2)^2 - (k_1 + k_2)^2 - (k_1 + k_2)^4},
\end{aligned}
\tag{17}
$$

where $k_1 = k_2 = 1.1$, $\delta_1 = \delta_2 = 0$ and $\omega_1 = -\omega_2 = \sqrt{k_1^2 + k_1^4}$. To accurately simulate the solitary wave solution of the Boussinesq equation, we developed a neural network with six hidden layers, each containing 64 neurons. We set the initial weight coefficients of each loss function term to 0.2 and refined them throughout the training process using an adaptive learning method. The training data are obtained by random sampling, the number of sample points for the initial and boundary conditions is 300, and 25 000 sample points are randomly selected in the spatio-temporal domain to construct the physical constraints. During the training process, we also used a small number of observations, with a data volume of 400. We set the learning rate to 0.001 during training and employed different gradient descent algorithms to optimize training, with L-BFGS used for the first 5000 epoch iterations and Adam used thereafter until convergence was achieved. After training, we used the trained model to approximate the solitary wave solution of the Boussinesq equation. Figure 1 illustrates the exact solution and prediction results obtained using the improved PINNs. The Fig. 1 shows that the improved PINNs successfully simulate the solitary wave solution of the Boussinesq equation. Additionally, Fig. 2 compares the true and approximate solutions at different times, revealing that the predicted solutions are very close to the true solutions. We conducted several experiments to measure the error between the approximate and exact solutions and found that the relative L2 norm error between the predicted and true solutions is approximately $8.33 * 10^{-4}$. In comparison, the relative L2 norm error when using the original PINNs with the same training data and parameter settings is approximately $1.02 * 10^{-2}$. Figure 3 shows the error statistics of the improved and original PINNs. Finally, we analyzed the training time of the neural network models before and after the improvement and found that the average training time of the improved PINNs is approximately 51.5% of the training time of the original PINNs. Our experiments show that the improved PINNs offer more accurate predictions, better training efficiency, and shorter training times than the original PINNs when simulating the solitary wave solution of the Boussinesq equation.

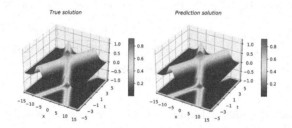

Fig. 1. A comparison between the exact soliton solution (left) and learned soliton solution (right) of the Boussinesq equation.

3.2 The Second-Order Rogue Wave Solution of Nonlinear Schrödinger Equation

In this subsection, we studied the second-order rogue waves of the nonlinear Schrödinger equation using the improved PINNs method described above. The mathematical form of the nonlinear Schrödinger equation is defined as follows:

$$iq_t + \alpha q_{xx} + \beta |q|^2 q = 0 \tag{18}$$

where $i = \sqrt{-1}$, α, β are parameters chosen according to different cases, and q is the complex-valued solution of the NLSE with respect to x and t. In this experiment, the values of α and β are taken as 0.5 and 1, respectively.

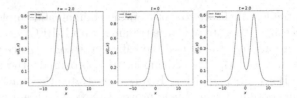

Fig. 2. Comparison of the exact soliton solution of the Boussinesq equation with the approximate solution of the improved PINNs at different times, with the blue solid line representing the exact solution and the red dashed line representing the neural network-predicted solution. (Color figure online)

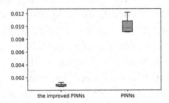

Fig. 3. For the prediction problem of soliton solution of the Boussinesq equation, visualize the error statistics results of the improved PINNs and the original PINNs.

In the experiments, the following initial condition was considered:

$$q(x, -6) = \left[1 - \frac{-192x^4 - 41760x^2 - 1282428}{64x^6 + 6960x^4 + 231372x^2 + 3652521} \right.$$
$$\left. - \frac{i(2304x^4 + 161232x^2 + 3068196.0)}{64x^6 + 6960x^4 + 231372x^2 + 3652521} \right] e^{-6i} \tag{19}$$

Similarly, for the numerical simulation of rogue waves, we chose a finite region with a spatial domain range of $[-6, 6]$ and a time range of $[-6, 6]$. Under these conditions, the second-order rogue wave solution of the nonlinear Schrödinger equation can be expressed as follows:

$$q(x, t) = \left(1 - \frac{G + iK}{D} \right) e^{it} \tag{20}$$

$$G(t, x) = 12[3 - 16x^4 - 24x^2 \left(4t^2 + 1 \right) - 80t^4 - 72t^2 + 100t] \tag{21}$$

$$K(t, x) = 24[(15 - 16x^4 + 24x^2)t - 8(4x^2 + 1)t^3$$
$$- 16t^5 + 25(2t^2 - 2x^2 - \frac{1}{2})] \tag{22}$$

$$D(t, x) = 64x^6 + 48x^4(4t^2 + 1) + 12x^2(3 - 4t^2)^2$$
$$+ 25[4t(12x^2 - 4t^2 - 9) + 24] + 9 \tag{23}$$
$$+ 64t^6 + 432t^4 + 396t^2$$

In the context of predicting the second-order rogue wave solution for the nonlinear Schrödinger equation, we designed a neural network comprising seven hidden layers, each housing 64 neurons. Initially, we assigned an initial weight coefficient of 0.2 to each term within the loss function. These coefficients underwent dynamic optimization using adaptive learning methods throughout the training process. For the construction of our training dataset, we considered the initial condition as a Dirichlet boundary condition within the spatiotemporal domain. To compile the dataset, we employed random sampling, selecting 400 training data points encompassing both initial and boundary conditions. In addition to this, we incorporated 25,000 residual points to comprehensively capture the system's behavior. Furthermore, we bolstered the training process by including 500 observational data points. Regarding hyperparameters, we configured the learning rate for this experiment to 0.001. Different optimization algorithms were employed at distinct stages of the training process. Initially, we utilized the L-BFGS optimizer for the initial 5,000 epoch iterations, transitioning to the Adam optimizer afterward, which continued until achieving convergence. The model was then utilized to approximate the second-order rogue wave solution of the nonlinear Schrödinger equation to test the algorithm's effectiveness, with the exact solution of the Schrödinger equation and the prediction results of the improved PINNs presented in Fig. 4. The Fig. 4 shows that the improved PINNs successfully simulate the second-order rogue wave solutions of the nonlinear Schrödinger equation. The predicted solutions at different moments were compared to the true solutions, and the comparison results are presented in Fig. 5. The predicted solutions at different moments are very close to the true solutions. Additionally, the errors between the approximate and exact solutions were statistically measured through several experiments. Rigorous statistics indicated that the relative L2 norm error between the predicted result and the true solution is about $6.85 * 10^{-4}$. Compared to the relative L2 norm error of the results obtained from the original PINNs with the same training data and parameter settings, which is about $9.46 * 10^{-3}$, this demonstrates the accuracy improvements of the improved PINNs. Figure 6 displays the error statistics of the improved PINNs and the original PINNs. Finally, training efficiency analysis was conducted to compare the improved neural network model before and after the improvement. The statistical analysis revealed that the average time to complete the training task of the improved PINNs was about 46.2% of the training time of the original PINNs. This experiment shows that for the rogue wave simulation of the nonlinear Schrödinger equation, the improved PINNs possess higher training efficiency than the original PINNs, and the prediction results of the improved PINNs are more accurate.

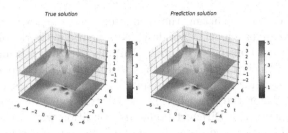

Fig. 4. A comparison between the exact second-order rogue wave solution (left) and learned second-order rogue wave solution (right) of the nonlinear Schrödinger equation

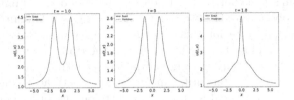

Fig. 5. Comparison of the exact second-order rogue wave solution of the nonlinear Schrödinger equation with the approximate solution of the improved PINNs at different times, with the blue solid line representing the exact solution and the red dashed line representing the neural network-predicted solution (Color figure online)

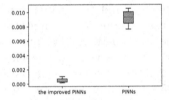

Fig. 6. For the prediction problem of second-order rogue wave solutions of the nonlinear Schrödinger equation, visualize the error statistics results of the improved PINNs and the original PINNs.

4   Conclusion

In summary, this paper presents an innovative approach for simulating localized wave phenomena using an improved physics-informed neural network (PINNs). PINNs are a type of deep learning method that can incorporate physical laws and constraints into the neural networks to enhance the accuracy of the simulation of the localized wave propagation. However, traditional PINNs often suffer from slow convergence and poor generalization due to their dependence on a pre-determined weighting of the loss function constraints. To address these challenges, we propose an improved PINNs model that incorporates gradient information and adaptive learning to dynamically adjust the weights of each constraint during the training process. In our experiments, we chose the Boussinesq equation and the nonlinear Schrödinger equation for the study and analyzed the errors of the simulation results to evaluate the accuracy of the model. The results of the experiments show that the improved PINNs model achieves significantly faster training speeds and higher prediction accuracy than traditional PINNs. Given the successful simulation of localized wave phenomena by enhanced PINNs, our future endeavors will involve their expansion and application to various complex system simulation challenges. These include simulating oceanic mesoscale eddies, modeling ocean waves, and simulating atmospheric pollutant dispersion processes. Furthermore, our ongoing research will delve into several intriguing aspects of PINNs. This includes addressing the challenge of effectively handling noisy data and exploring efficient integration with traditional numerical methods.

References

1. Li, S., Song, W., Fang, L., Chen, Y., Ghamisi, P., Benediktsson, J.A.: Deep learning for hyperspectral image classification: an overview. IEEE Trans. Geosci. Remote Sens. **57**(9), 6690–6709 (2019)
2. Guo, Y., Cao, X., Liu, B., Gao, M.: Cloud detection for satellite imagery using attention-based u-net convolutional neural network. Symmetry **12**(6), 1056 (2020)
3. Otter, D.W., Medina, J.R., Kalita, J.K.: A survey of the usages of deep learning for natural language processing. IEEE Trans. Neural Netw. Learn. Syst. **32**(2), 604–624 (2020)
4. Guo, Y., Cao, X., Liu, B., Peng, K.: El niño index prediction using deep learning with ensemble empirical mode decomposition. Symmetry **12**(6), 893 (2020)
5. Yin, J., Gao, Z., Han, W.: Application of a radar echo extrapolation-based deep learning method in strong convection nowcasting. Earth Space Sci. **8**(8), e2020EA001621 (2021)
6. Guo, Y., Cao, X., Liu, B., Gao, M.: Solving partial differential equations using deep learning and physical constraints. Appl. Sci. **10**(17), 5917 (2020)
7. Guo, Y., Cao, X., Peng, K.: Application of improved physics-informed deep learning based on activation function for solving nonlinear soliton equation. In: 2023 International Joint Conference on Neural Networks (IJCNN), pp. 1–10. IEEE (2023)
8. Raissi, M., Perdikaris, P., Karniadakis, G.E.: Physics-informed neural networks: a deep learning framework for solving forward and inverse problems involving nonlinear partial differential equations. J. Comput. Phys. **378**, 686–707 (2019)
9. Cai, S., Mao, Z., Wang, Z., Yin, M., Karniadakis, G.E.: Physics-informed neural networks (PINNs) for fluid mechanics: a review. Acta. Mech. Sin. **37**(12), 1727–1738 (2021). https://doi.org/10.1007/s10409-021-01148-1
10. Manafian, J., Lakestani, M.: N-lump and interaction solutions of localized waves to the (2+ 1)-dimensional variable-coefficient caudrey-dodd-gibbon-kotera-sawada equation. J. Geom. Phys. **150**, 103598 (2020)
11. Yu, J., Lu, L., Meng, X., Karniadakis, G.E.: Gradient-enhanced physics-informed neural networks for forward and inverse PDE problems. Comput. Methods Appl. Mech. Eng. **393**, 114823 (2022)
12. Xiang, Z., Peng, W., Liu, X., Yao, W.: Self-adaptive loss balanced physics-informed neural networks. Neurocomputing **496**, 11–34 (2022)

13. Groenendijk, R., Karaoglu, S., Gevers, T., Mensink, T.: Multi-loss weighting with coefficient of variations. In: Proceedings of the IEEE/CVF Winter Conference on Applications of Computer Vision, pp. 1469–1478 (2021)
14. Welford, B.: Note on a method for calculating corrected sums of squares and products. Technometrics 4(3), 419–420 (1962)

Graph Reinforcement Learning
for Securing Critical Loads by E-Mobility

Borui Zhang[1], Chaojie Li[1(✉)], Boyang Hu[1], Xiangyu Li[1], Rui Wang[1],
and Zhaoyang Dong[2]

[1] School of Electrical Engineering and Telecommunications, University of New South
Wales, Sydney, Australia
{borui.zhang,chaojie.li,boyang.hu1,xiangyu.li1,rui.wang20}@unsw.edu.au
[2] School of Electrical and Electronics Engineering, Nanyang Technological
University, Tengah, Singapore
zy.dong@ieee.org

Abstract. Inefficient scheduling of electric vehicles (EVs) is detrimental to not only the profitability of charging stations but also the experience of EV users and the stable operation of the grid. Regulating the charging market by dynamic pricing is a feasible choice for EV coordinated scheduling. Power outages caused by natural disasters have always been a serious threat to critical loads such as hospitals and data centers. With the development of vehicle-to-grid (V2G) technology, the potential to attract EV users to the stations near the critical loads through dynamic pricing and to aggregate EVs into a flexible emergency supply to maintain the critical load is being explored. However, determining charging prices in real-time that are both attractive to users and profitable to stations is a challenging task, which is further complicated by the relationships and interactions between multiple stations. Therefore, this paper proposes the graph reinforcement learning (GRL) approach to seek the optimal pricing strategy to address the above problems. The experiment results show that the proposed method can effectively achieve profit maximization and EV scheduling for critical load maintenance.

Keywords: Dynamic pricing · charging station · electric vehicle · critical load · graph reinforcement learning

1 Introduction

Transportation electrification is considered a sustainable solution to alleviate the climate issues and the dependence of conventional transportation on fossil fuels [1]. According to the market study [2], the electric vehicles (EVs) in Australia [3] are expected to reach 49% and 100% of the annual new vehicle sales by 2030 and 2040 respectively. However, the rapid growth in EV penetration and the heterogeneity of EV charging behaviors complicate the coordinated scheduling of EVs. Inefficient scheduling may result in congestion at some charging stations and idleness at others [4], which not only wastes charging resources but also affects the experience of EV users. Moreover, frequent connections and disconnections

© The Author(s), under exclusive license to Springer Nature Singapore Pte Ltd. 2024
B. Luo et al. (Eds.): ICONIP 2023, CCIS 1961, pp. 303–314, 2024.
https://doi.org/10.1007/978-981-99-8126-7_24

of EVs exacerbate the system fluctuations, and the distribution network may also suffer from overload risks caused by the surges in charging demand during peak periods [5].

As a promising option to address the above challenges, charging station dynamic pricing has been studied to coordinate and schedule EVs to be charged [6–10]. Specifically, the charging stations are able to regulate the charging market and guide the charging behaviors of EV users by adjusting charging prices to achieve the scheduling targets such as profit maximization on the station side [6,7], EV user satisfaction maximization on the user side [8], peak load reduction on the grid side [9] or multi-objective optimization on multiple sides [10]. With the development of vehicle-to-grid (V2G) technology, the potential of EVs in power system operation is gradually being explored relying on its strengths such as fast response and bi-directional regulation [11]. Based on V2G, active power can be interacted between EVs and the grid to achieve frequency control [12], peak shaving and congestion management [13]. In addition to active power delivery, the reactive power compensation can also be applied via V2G for voltage regulation in distribution networks [14].

Power outages caused by natural disasters have always been a serious threat to critical loads such as hospitals and data centers. If EVs can be scheduled by charging station dynamic pricing and aggregated as a flexible emergency supply to support the critical loads, the economic and technical burden on the distribution network would be relieved significantly during the outages. However, the fast attraction of sufficient EVs to maintain the critical load faces challenges in terms of the unpredictability of outages and the uncertainty of user behaviors, which requires real-time pricing decisions that trades off the station profit, the user willingness and the critical load demand.

In recent years, reinforcement learning (RL) has become popular in making real-time optimal decisions. In particular, multi-agent RL (MARL) considering the interactions of multiple agents is expected to address these challenges in the multi-station charging market. Due to the spatiotemporal randomness in critical load demand, the relationships between stations is highly uncertain and dynamic, which limits the station cooperation and hence the performance of traditional MARL. In order to solve this dilemma, we construct the station relationships as graphs and propose the graph RL (GRL) equipped with the graph attention network (GAT) to handle the communication and cooperation of multiple stations, and seek the optimal pricing strategies to maximize the station profits and schedule EVs to maintain the critical load.

2   Problem Formulation

2.1   Preliminary

In this paper, we consider a region including N charging stations, where each station determines the charging price dynamically according to the charging supply and demand. To formulate the problem, we start with the following definitions:

Charging Request. At each time slot t, there are n EVs with charging requests $Q_t = \{q_1, q_2, ..., q_n\}$ in the region. Each EV user will compare the charging prices and distances of the charging stations and submit the request q_n including arrival time $t_a^{q_n}$, expected departure time $t_d^{q_n}$, energy required e^{q_n} and its location to the selected station.

Wait and Leave Mechanism. There are two situations where EV users will wait, one is actively waiting as they are not satisfied with all the stations, and the other is passively waiting as their selected station is full. In most cases, EV users get impatient by waiting and will leave this region if they wait for more than a certain period of time.

Critical Load Demand. We simulate a power outage in the distribution network caused by a natural disaster, which interrupts the power supply to the node where a hospital is located and takes several hours to restore. We consider some charging stations to be located near the hospital. To maintain critical load in the hospital during this period, dynamic pricing is used to schedule EVs willing to provide emergency power supply through V2G technology to the target charging stations.

Figure 1 shows the framework of GRL-based charging station dynamic pricing considering the critical load demand.

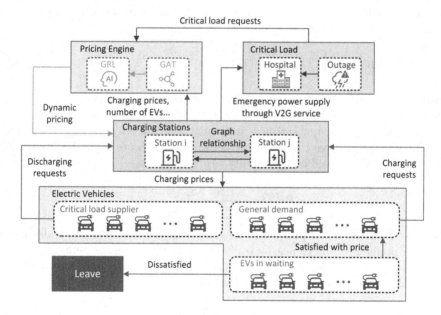

Fig. 1. The framework of GRL-based charging station dynamic pricing considering the critical load demand.

2.2   System Model

In order to ensure the profit of charging stations and the charging willingness of EV users, the charging price $p_t \in [p_{t,min}, p_{t,max}]$ is constrained by the lower and upper price boundaries. The charging willingness of EV users in terms of charging price is described as follows:

$$w_{t,p} = \frac{p_{t,max} - p_t}{p_{t,max} - p_{t,min}} \tag{1}$$

In addition to the willingness directly related to the charging price, user willingness is also influenced by market prices $p_{t,m}$ which is estimated based on the electricity price, the market supply and demand in real-time. The charging willingness $w_{t,m}$ of EV users in terms of market price is calculated by:

$$w_{t,m} = \frac{1 - (p_t - p_{t,m})/m}{2} \tag{2}$$

where the price differences $p_t - p_{t,m} \in [-m, +m]$ is limited by the threshold m, which describes the user sensitivity of market prices. Distance also plays a decisive role in the decisions of EV users, the willingness in terms of distance is estimated by $w_{t,d} = 1 - d_n/d_{max}$, where d_n and d_{max} denote the distance between EV and station and the distance boundary respectively.

To better regulate market behavior through pricing, we assume that the impact of price on EV users includes station selection and demand response. At the station selection stage, user n will evaluate each station according to $w_{t,p}$, $w_{t,m}$ and $w_{t,d}$, and then give a score $S_n^{CS_i}$:

$$S_n^{CS_i} = \begin{cases} \beta_p w_{t,p}^{CS_i} + \beta_m w_{t,m}^{CS_i} + \beta_d w_{t,d}^{CS_i}, & S_n^{CS_i} \geq S_{min} \\ 0, & S_n^{CS_i} < S_{min} \end{cases} \tag{3}$$

where β_p, β_m and β_d denote the weight of willingness in terms of $w_{t,p}$, $w_{t,m}$ and $w_{t,d}$ respectively. Users will not consider stations with scores below the threshold S_{min}. If all prices are not satisfied, users will wait for price updates until they get impatient and leave the region. According to the score $S_n^{CS_i}$, EV users will select station CS_i based on probability (4):

$$Pr_n^{CS_i} = \frac{\alpha^{CS_i} S_n^{CS_i}}{\sum_{i=1}^{N} \alpha^{CS_i} S_n^{CS_i}} \tag{4}$$

where α^{CS_i} denotes the preference weight of the station CS_i, which describes the user preference due to nearby traffic flow, charging demand, etc.

At demand response stage, the required energy e^{q_n} in request q_n is assumed to be elastic according to the demand willingness $w_{t,e}$:

$$w_{t,e} = 1 - \frac{p_t - p_{t,d}}{p_{t,max} - p_{t,min}} \tag{5}$$

where $p_{t,d} \in [p_{t,min}, p_{t,max}]$ is the user-expected price used to adjust the range of demand variation with the price. Once a charging station has completed a charging request q_n, it will receive a profit $profit^{q_n}$ based on the final demand $e_t^{q_n} w_{t,e}$, the electricity price $p_{t,e}$ and the charging price p_t, otherwise there will be no profit.

$$profit^{q_n} = \begin{cases} (p_t - p_{t,e}) \cdot e_t^{q_n} \cdot w_{t,e}, & \text{Completed} \\ 0, & \text{Incompleted} \end{cases} \tag{6}$$

In general cases, the target of the station CS_i is to maximize the station profits $profit_t^{CS_i}$ during $t = 1, 2, ..., T$:

$$\max \sum_{t=1}^{T} profit_t^{CS_i} \tag{7}$$

If the outage event occurs, the hospital will request sufficient power capacity from nearby charging stations to maintain its critical load. $C_t \in [0, 3]$ denotes a critical load request. The larger the value of C_t, the higher the demand, while $C_t = 0$ represents there is no demand. Note that the supply to the charging stations is also interrupted by the outage event, which means that all charging requests are suspended and will be continued when the supply is restored. We assume during this period, the stations will only accept EVs providing emergency supplies. After the supply restoration, the hospital will cover all the expenses of the requested stations.

3 Methodology

3.1 Markov Decision Process

In order to address the proposed problem as an RL task, the problem needs to be formulated in the form of a Markov Decision Process (MDP) as follows.

State. We define the charging station pricing engines as the RL agents and the information observed by each agent as the states s_t in MDP which specifically include the market charging prices $p_{t,m}$, the number of charged EVs $n_t^{CS_i}$, and the critical load request $C_t^{CS_i}$ in each station.

Action. After observing the information from the charging market, the stations and critical load, each pricing engine will determine the charging price $p_t^{CS_i}$ in real-time, which is defined as the action a_t.

Reward. The reward r_t of each agent is defined as the station profit $profit_t^{CS_i}$ of each station CS_i at time slot t if there is no critical load request. Otherwise, the joint target of nearby stations is to supply the critical load based on request C_t. In this case, the reward function can be rewritten as:

$$r_t = C_t \cdot (P_{c,t} - P_{c,t-1}) \cdot \epsilon \tag{8}$$

where $P_{c,t}$ is the overall power capacity to the critical load and ϵ is a constant to scale the reward.

The target of the RL problem is to seek the optimal pricing strategy π^* that maximizes the expected cumulative reward Q_π in (9) by constantly interacting with the environment.

$$Q_\pi(s_t, a_t) = \mathbb{E}_\pi \left[\sum_{t=0}^\infty \gamma^t r_t(s_t, a_t) \mid s_t = s, a_t = a \right] \tag{9}$$

where γ denotes the discount factor.

3.2 Algorithm

To describe the relationship between the agents, the graph $\mathcal{G} = (\mathcal{V}, \mathcal{E})$ is constructed, where $\mathcal{V}$ and $\mathcal{E}$ denote the set of nodes and edges respectively. If there is a relationship between two nodes, the corresponding edges exist, which is represented as 1 in the adjacency matrix $\mathcal{A}$ and 0 otherwise. Each node represents an agent and the node feature is its limited local observation. All the node features form the feature matrix $\mathcal{F}$. Typically, the agents are not related to each other, but in case of a random outage event in time and space, the agents around the critical load will become cooperative and share their observations until the power is restored. These uncertain and dynamic relationships between agents are hard to be recognized by traditional RL approaches.

In order to extract the properties of the graph, we first encode the local states $s_{t,i}$ of each node i through multi-layer perceptron (MLP) to facilitate the complicated observation communication. Then, the obtained feature vectors h_i are integrated by the GAT layer, which employs the multi-head mechanism to expand the horizon for better feature capture. Each feature vector h_i is converted to query $W_Q h_i$, key $W_K h_i$ and value $W_V h_i$ representation through each attention head to calculate the importance factor $\alpha_{i,j}$ of node j to node i scaled by function δ:

$$\alpha_{i,j} = \frac{\exp\left[\delta\left(W_Q h_i \cdot W_K h_j\right)\right]}{\sum_{k \in N_i} \exp\left[\delta\left(W_Q h_i \cdot W_K h_k\right)\right]} \tag{10}$$

Then the feature values of each attention head k are weighted as input to the activation function σ and concatenated as the new features h_i' that aggregates the information from neighboring nodes.

$$h_i' = \left\|_{k=1}^{K} \sigma\left(\sum_{j \in N_i} \alpha_{i,j} W_V h_i\right)\right. \tag{11}$$

The updated features of each agents from the graph are then input to the Q-network to approximate the Q-function $\hat{Q}_w := (s, a; w)$. The agents will execute the action randomly with probability ϵ, otherwise the action with the largest value of the Q-functions. To find the optimal policy, the Q-function is iterated to an approximation of the Bellman Optimization Eq. (12) by minimizing the loss function (13):

$$Q_\pi^*(s_t, a_t) = r_t(s_t, a_t) + \gamma \mathbb{E}_{s_{t+1}} \left[\max_{a_{t+1}} Q_\pi^*(s_{t+1}, a_{t+1}) \right] \tag{12}$$

$$\mathcal{L}(w) = \left(r_t + \gamma \max_a Q_{(s_{t+1}, a_{t+1}; \hat{w})} - Q(s_t, a_t; w) \right)^2 \tag{13}$$

Algorithm 1: GRL for dynamic pricing problem.

Initialize the parameters and replay buffer D;
Initialize the adjacency matrix $\mathcal{A}$ and feature matrix $\mathcal{F}$;
for *Episode* $e = 1, 2, 3 \cdots E$ **do**
    Reset the episode start time slot and randomly choose an outage start time slot and duration;
    for $t = 1, 2, 3 \cdots T$ **do**
        if $C_t = 1$ **then**
            Update the adjacency matrix A, the feature matrix F and the reward function r_t;
        end
        Each agent observes the local state s_t;
        Execute pricing action a_t randomly with probability ϵ, otherwise $a_t = \text{argmax}_a Q(s_t, a_t)$;
        Obtain profit r_t and update state to s_{t+1};
        Store $(s_t, a_t, r_t, s_{t+1}, \mathcal{A})$ in D;
        Randomly select a mini-batch of $(s_t, a_t, r_t, s_{t+1}$ from D;
        Update the policy by minimizing (13);
    end
end

---

4   Case Study

4.1   Configuration

As shown in Fig. 2, a region in Sydney, Australia including five charging stations is chosen for the simulation, where each station has 25 chargers with rated power of 7kW. EVs will consider the nearby stations as potential options and the average charging energy required per EV is 25kWh. When a critical load demand exists, the surrounding stations CS_1, CS_2 and CS_3 will cooperate to supply the critical load. The lengths of the training episode and testing episode are 24 h and 48 h respectively, where the time interval is 15 min. The start time slot of the power outage is randomly selected and the duration is from 3 to 5 h.

In the environment, the willingness weights β_p, β_m and β_d have identical value of 1/3. The preference weights α^{CS_1}, α^{CS_2}, α^{CS_3}, α^{CS_4} and α^{CS_5} are 1.00,

0.95, 1.05, 0.90 and 1.10 respectively. The reward scaling factor ϵ is 0.5. For hyper parameter setting, the hidden units and attention heads of GAT are 64 and 4. The discount factor and buffer size are 0.99 and 1,000,000. ReLU activation function and Adam optimizer are employed with a learning rate of 0.001. The algorithm is implemented in Python based on PyTorch.

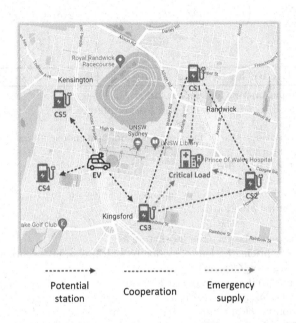

Fig. 2. Relationship between the stations, EVs, and critical load.

4.2   Performance Analysis

General Case. We first consider the general case without critical load demand. To compare pricing strategies, all stations are first assumed to be based on TOU pricing, i.e. TOU electricity price plus fixed service fee. Figure 3 shows the charging power of each station under TOU pricing. Then the stations changes the pricing strategy to dynamic pricing as shown in Fig. 4.

Table 1 shows the profit per station and the overall profit based on TOU pricing and dynamic pricing respectively. Compared to TOU pricing, the profits of all the stations based on dynamic pricing can be improved. Specifically, the profits of CS_1, CS_2, CS_3, CS_4 and CS_5 are 48.03%, 14.25%, 24.27%, 18.99% and 27.60% higher respectively. In addition, the 26.51% growth in the overall profits through dynamic pricing reflects improved user satisfaction with price and increased charging demand.

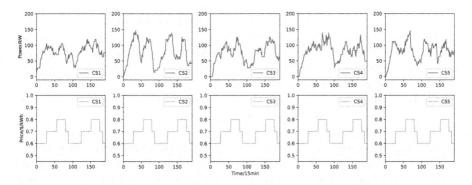

Fig. 3. Charging power of each station under TOU pricing.

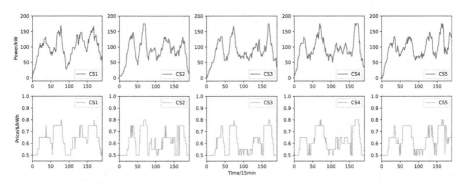

Fig. 4. Charging power of each station under dynamic pricing.

Table 1. Charging profits

Station	TOU Pricing	Dynamic Pricing
CS_1	$ 218.03	$ 322.75
CS_2	$ 228.67	$ 261.25
CS_3	$ 230.70	$ 286.70
CS_4	$ 217.05	$ 258.27
CS_5	$ 239.05	$ 305.02
Overall Profit	$ 1133.50	$ 1433.99

Special Case with Critical Load Demand. In the special case with critical load demand, there is a power outage in the hospital caused by a natural disaster. The stations CS_1, CS_2 and CS_3 temporarily become cooperative during the outage by sharing their local observations and are thus expanded to a broader view for collaborative decision making. Other stations keep the original MDP, relationships and pricing strategies.

Figure 5 shows the station response to critical load demand at various levels. As shown in the red area, the power outage occurs at the time slot 40 and interrupts the power supply to the node where these stations and the hospital are located until the supply is restored after 4 h. Normally the number of EVs in the stations is decreasing during this low-demand period. However, under the GRL-based pricing strategy, when these stations receive the critical load requests, it responds rapidly to constantly attract EVs to provide emergency supply from the stations to the hospital through V2G technology. When $C_t = 0$, the stations do not respond since there is no demand from the critical load. As C_t increases, i.e., the critical load demand increases, the stations begin to respond to the request and attract EVs to supply the demand until power is restored. Each enabled station is able to reach its full capacity at 175 kW to supply the highest demand at $C_t = 3$.

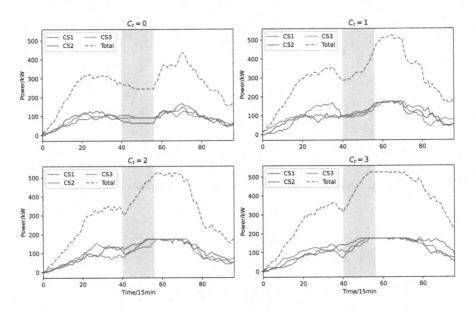

Fig. 5. GRL-based station response to critical load demand at various levels.

To demonstrate the superiority of our proposed GRL algorithm, the MARL without GAT under the same parameter configuration is trained for comparison. Table 2 presents the start capacity, the end capacity and the capacity variation of GRL and MARL under various critical load demand during the 4-hour outage. It can be seen that the growth in supply capacity of the stations under GRL is faster and higher than MARL, which indicates that GAT is able to capture node relationships and interactions and share observations to facilitate cooperation between agents. It is worth noting that the insignificant difference of capacity variation between GRL and MARL at $C_t = 3$ does not mean that GRL is underperforming in this case. This is because the supply capacity has reached the maximum capacity 3×175=525kW by the fourth hour.

Table 2. Performance comparison between GRL and MARL

Algorithm	Demand Level	Start Capacity	End Capacity	Capacity Variation
GRL	$C_t = 0$	273 kW	245 kW	−28 kW
GRL	$C_t = 1$	287 kW	455 kW	+ 168 kW
GRL	$C_t = 2$	315 kW	511 kW	+ 196 kW
GRL	$C_t = 3$	315 kW	525 kW	+ 210 kW
MARL	$C_t = 0$	273 kW	217 kW	−56 kW
MARL	$C_t = 1$	301 kW	434 kW	+ 133 kW
MARL	$C_t = 2$	301 kW	469 kW	+ 168 kW
MARL	$C_t = 3$	308 kW	511 kW	+ 203 kW

5   Conclusion

This paper studies the charging station dynamic pricing problem considering the critical load demand. An environment including multiple charging stations, EVs and a hospital as the critical load is modeled, which considers the interaction of charging stations, EV charging behaviors, and critical load demand that occurs randomly. The GRL method is applied to address the problem formulated in the form of a graph-based RL task. From the simulation results, the GRL method is able to maximize the profit of stations and schedule EVs to maintain the critical load. In the future, we will deploy a stronger algorithm to address the problem with more stations in a larger region and explore more potentials of dynamic pricing in V2G services.

References

1. Hao, H., et al.: Impact of transport electrification on critical metal sustainability with a focus on the heavy-duty segment. Nat. Commun. 10(1), 5398 (2019)
2. Energeia: Australian electric vehicle market study (2018). https://arena.gov.au/knowledge-bank/australian-electric-vehicle-market-study/
3. Li, C., Dong, Z., Chen, G., Zhou, B., Zhang, J., Yu, X.: Data-driven planning of electric vehicle charging infrastructure: a case study of Sydney, Australia. IEEE Trans. Smart Grid 12(4), 3289–3304 (2021). https://doi.org/10.1109/TSG.2021.3054763
4. Li, X., Li, C., Luo, F., Chen, G., Dong, Z.Y., Huang, T.: Electric vehicles charging dispatch and optimal bidding for frequency regulation based on intuitionistic fuzzy decision making. IEEE Trans. Fuzzy Syst. 31(2), 596–608 (2023). https://doi.org/10.1109/TFUZZ.2022.3220964
5. Yang, B., Li, J., Han, Q., He, T., Chen, C., Guan, X.: Distributed control for charging multiple electric vehicles with overload limitation. IEEE Trans. Parallel Distrib. Syst. 27(12), 3441–3454 (2016). https://doi.org/10.1109/TPDS.2016.2533614
6. Fang, C., Lu, H., Hong, Y., Liu, S., Chang, J.: Dynamic pricing for electric vehicle extreme fast charging. IEEE Trans. Intell. Transp. Syst. 22(1), 531–541 (2021). https://doi.org/10.1109/TITS.2020.2983385

7. Kim, Y., Kwak, J., Chong, S.: Dynamic pricing, scheduling, and energy management for profit maximization in PHEV charging stations. IEEE Trans. Veh. Technol. **66**(2), 1011–1026 (2017). https://doi.org/10.1109/TVT.2016.2567066
8. Zhang, Q., Hu, Y., Tan, W., Li, C., Ding, Z.: Dynamic time-of-use pricing strategy for electric vehicle charging considering user satisfaction degree. Appl. Sci. **10**(9), 3247 (2020)
9. Moghaddam, Z., Ahmad, I., Habibi, D., Masoum, M.A.S.: A coordinated dynamic pricing model for electric vehicle charging stations. IEEE Trans. Transp. Electrification **5**(1), 226–238 (2019). https://doi.org/10.1109/TTE.2019.2897087
10. Luo, C., Huang, Y.F., Gupta, V.: Stochastic dynamic pricing for EV charging stations with renewable integration and energy storage. IEEE Trans. Smart Grid **9**(2), 1494–1505 (2018). https://doi.org/10.1109/TSG.2017.2696493
11. Peng, C., Zou, J., Lian, L.: Dispatching strategies of electric vehicles participating in frequency regulation on power grid: a review. Renew. Sustain. Energy Rev. **68**, 147–152 (2017)
12. Rao, Y., Yang, J., Xiao, J., Xu, B., Liu, W., Li, Y.: A frequency control strategy for multimicrogrids with V2G based on the improved robust model predictive control. Energy **222**, 119963 (2021)
13. Prakash, K., et al.: Bi-level planning and scheduling of electric vehicle charging stations for peak shaving and congestion management in low voltage distribution networks. Comput. Electr. Eng. **102**, 108235 (2022)
14. Hu, J., Ye, C., Ding, Y., Tang, J., Liu, S.: A distributed MPC to exploit reactive power V2G for real-time voltage regulation in distribution networks. IEEE Trans. Smart Grid **13**(1), 576–588 (2022). https://doi.org/10.1109/TSG.2021.3109453

Human-Object Interaction Detection with Channel Aware Attention

Zhenhua Wang[1(✉)], Jiajun Meng[2], Yaoxin Yue[1], and Zhiyi Zhang[1]

[1] Intelligent Media Processing Group, College of Information Engineering,
Northwest A&F University, Yangling 712100, Shaanxi, China
zhenhuawang@nwsuaf.edu.cn, {seyaoxin,zhangzhiyi}@nwafu.edu.cn
[2] School of Computer Science and Technology, Zhejiang University of Technology,
Hangzhou 310023, China
1112012021@zjut.edu.cn

Abstract. Human-object interaction detection (HOI) is a fundamental task in computer vision, which requires locating instances and predicting their interactions. To tackle HOI, we attempt to capture the global context information in HOI scenes by explicitly encoding the global features using our novel channel aware attention mechanism. Our observation is that the context of an image, including people, objects and background plays important roles in HOI prediction. To leverage such information, we propose a channel aware attention, which applies global average pooling on the features to learn their channel-wise inter-dependency. Based on the channel aware attention, we develop a channel aware module and a channel aware encoder. Handling features in channel dimensions makes it convenient to encode the global features as well as to learn semantic features. Empirically, our model outstrips the strong baseline by 3.2 points on V-COCO and 0.79 points on HICO-DET respectively. The visual analysis demonstrates that our method is able to capture abundant interaction-related features by attending to relevant regions.

Keywords: Human Object Interaction Detection · Channel Attention Mechanism · Context Information

1 Introduction

Human Object Interaction Detection (HOI) [1,2] aims to locate both the human and object that are interacting, and to classify their interactive relation. These two sub-tasks require the predictive model capturing rich global context in the feature extraction phase. Recently, methods exploiting transformer [3,4] are adopted to tackle the HOI problem, achieving significant improvement. Thanks to the attention mechanism [5,6], these models are able to attend to extra regions of images instead of only focusing on RoIs (*e.g.,* bounding boxes). However, the dense attention mechanism only computes weights for each pair of tokens and lacks an explicit attention over the global scene. This high density probably makes it blind in terms of extracting effective semantic features from the entire

© The Author(s), under exclusive license to Springer Nature Singapore Pte Ltd. 2024
B. Luo et al. (Eds.): ICONIP 2023, CCIS 1961, pp. 315–327, 2024.
https://doi.org/10.1007/978-981-99-8126-7_25

image. As illustrated by the attention maps in the second column of Fig. 1, conventional attention module [3] only focuses on the object (such as the computer and the ball in the second column), which can be inadequate to HOI when surrounding objects, *e.g.*, the hand, room and ground, are also important clues (at least supplementary cues) for HOI detection.

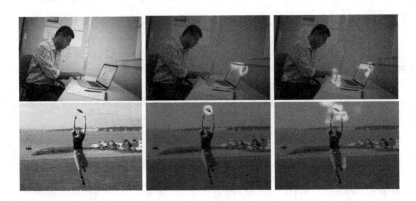

Fig. 1. Here we visualize the attention maps learned by QPIC [3] and our model. The three columns from left to right, respectively list the original images, the attention maps by QPIC and the attention maps produced by our model. Our method is able to capture extra useful contextual information for HOI.

In this work, we attempt to decrease the learning difficulties of transformer and leverage the global context information for HOI. In particular, we propose a channel aware attention (CAA). Different from the dot product attention which learns weights for each pair of tokens, the CAA mechanism learns a normalized weight for each channel based on the global information as well as the correlation of different channels, and multiplies the weight back to each channel, which enables the selection of important channels. Moreover, our proposed CAA explicitly encodes the global features of the entire image, which provides extra contextual features for HOI tasks.

Based on CAA, we propose channel aware module (CAM) and channel aware encoder (CAE). The CAM incorporates global features into the base features extracted by the backbone. The CAE, which contains CAA in each layer, is used to replace the vanilla encoder of Transformer. HOI detection requires both local representation and global context, in order to locate human and object instances and distinguish their interactive relations. We argue that the proposed channel aware attention can better capture the contextual information.

In summary, our contributions are of three folds: 1) We introduce a channel-aware attention mechanism which is able to capture the global context information by learning inter-dependency between channels. To our best knowledge, we are the first to incorporate global features in a channel-wise manner on HOI tasks. 2) We present a Transformer-based HOI detector based on CAA, which is able to enhance the contextual representation in different learning phases. 3)

The proposed method achieves promising results on two challenging benchmarks including VCOCO [1] and HICO-DET [2], which validates the effectiveness of the proposed channel-aware attention mechanism.

2 Related Work

2.1 Human-Object Interaction Detection

HOI task was initially proposed by Gupta *et al.* [1], which requires estimating the bounding boxes of human and objects, and meanwhile predicting their interactive relations. To tackle this, mainstream methods follow either two stage or one stage paradigms.

Two Stage Methods. Two stage methods [7–11] first leverage off-the-shelf object detectors to generate bounding boxes of human and objects. Afterwards, the representations of targets are cropped (typically using RoIAlign and the bounding boxes) from the original feature map of the image, and are combined to generate representations for each *human-object* pair. Based on the pairwise representation, on the second stage, a feed-forward network (FFN) is taken to predict their interactive category. Recently, in order to capture global context of HOI, graph neural networks are leveraged [12,13] to bridge the communication of features of different targets.

One Stage Methods. One stage HOI detection methods locate positions of human and object and predict the relation of human-object pairs with a unique model. Some models, such as PPDM [14] and IPNet [15], formulate HOI tasks as a problem of key-point detection and grouping. UnionDet [16] directly locates the region of interaction, which boosts the HOI performance significantly. ASNet [17] shapes HOI as a set prediction, and predicts all elements (each element is a combination of the pair of targets and their interaction) in the set simultaneously. Some works [3,4] introduce the detection paradigm of DETR [6] to HOI, building transformer-based HOI detectors. In this paper, we also adopt an architecture similar to DETR [6], but pay particular attention to enhance the contextual representation via the proposed channel-aware attention mechanism.

2.2 Attention Mechanism

Research on attention mechanism has seen a great growth in recent years. Transformer [5], initially proposed for machine translation, uses scaled dot product attention to implement self attention and cross attention. Graph attention network [18] represents each entity as a vertex, and the relation between entities being the attention weight between corresponding vertexes. SENet [19] introduces channel-wise attention mechanism to refine the convolutional features.

Due to its flexibility and effectiveness, attention has been widely adopted in many works. iCAN [8] proposes an instance-centric attention module, which is able to enhance the appearance features of instances. DCA [9] presents a context attention framework to capture the contextual feature. Recently, DETR-like [6] architectures have been proposed [3,4,20], which leverage learnable query embeddings and predictive heads to tackle the HOI task. Chan [21] adopted compound attention to enhance region proposal for visual tracking. Bai [22] incorporated attention mechanism with adversarial training into instance-level image retrieval.

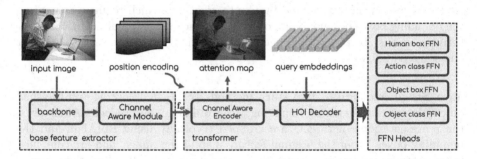

Fig. 2. Overview of our method. The framework consists of three components following the convention of DETR [6]. The first component is a base feature extractor, which we choose ResNet50. We apply our channel aware module to the base features for enhancement. The second component is a Transformer with an encoder and a decoder, where the original encoder is replaced by our channel aware encoder. The third component is a group of FFNs which predict the final results (boxes of human and objects, object classes and interaction classes).

Our method could be viewed as an extension of Transformer by incorporating the channel-wise attention mechanism. Specifically, we propose an attention mechanism which allows to capture HOI context by learning the interdependence among channels. Based on this, we separately propose a channel-aware module and channel-aware encoder, which enable the learning of contextual features in different granularity. The difference between the aforementioned attention-based methods and our approach is that we learn an embedding representing the context from the perspective of channel, which enhances the contextual representation and helps the encoder capture task-specific information in the scene.

3   Methodology

3.1   Overview

The pipeline of our CAA framework is illustrated in Fig. 2, where we concentrate on capturing features using our CAA mechanism. To this end, we first propose a CAA to learn the channel-wise inter-dependency and global interaction context. Based on the CAA, we construct a CAM and CAE. The former is used to enhance the features from backbone, and the latter cooperates with Transformer decoder to learn the image-scale contextual information.

3.2 Channel-Aware Attention

To enhance the learning of HOI context, we propose the CAA, which is sketched by Fig. 3. Given an input $\mathbf{x} \in \mathbb{R}^{C \times H \times W}$, we first calculate the mean over the spatial axes using average pooling (AvgPool):

$$\mathbf{u} = \frac{1}{HW} \sum_{i=0}^{H-1} \sum_{j=0}^{W-1} \mathbf{x}_{:,i,j}, \tag{1}$$

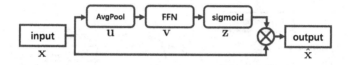

Fig. 3. The proposed Channel-Aware Attention.

where the resulting vector $\mathbf{u} \in \mathbb{R}^C$. The AvgPool operation compresses the global features into a vector representation efficiently. The vector $\mathbf{u}$ is then passed through a FFN, which is taken to learn the channel-wise inter-dependency, using

$$\mathbf{v} = \sigma(W_1\mathbf{u} + b_1)W_2 + b_2, \tag{2}$$

where W_1, W_2, b_1, b_2 are learnable parameters, σ is ReLU activation function, and $\mathbf{v} \in \mathbb{R}^C$.

To obtain the final activation, we apply Sigmoid function to acquire the vector $\mathbf{z} = \text{sigmoid}(\mathbf{v})$, where $\mathbf{z} \in \mathbb{R}^C$. The Sigmoid operation finally normalizes the output of the activation to ensure that each entry belongs to the range of $(0, 1)$. Therefore, the activation $\mathbf{z} \in (0, 1)^C$. Afterwards, we multiply the original features $\mathbf{x}$ by the activation $\mathbf{z}$ and acquire $\hat{\mathbf{x}} = \mathbf{x} \otimes \mathbf{z}$, where $\otimes$ denotes channel-wise multiplication, and the activated features $\hat{\mathbf{x}} \in \mathbb{R}^{C \times H \times W}$.

Our observation is that the activation $\mathbf{z}$ indicates the importance of each channel, and multiplying it to the channel highlights important channels, which makes it easier to learn meaningful representations. Moreover, the CAA reserves the shape of the input, which makes it flexible to subsequent usage. Here we adopt the AvgPool operation because it is simple and efficient.

3.3 The Channel-Aware Module

The typical pipeline of HOI detector contains a CNN backbone, taking images as input and producing features as output. Conventional methods directly receive these features as base inputs for downstream HOI tasks.

We apply our CAA on the features from CNN, reweighing features of each channel. We believe that CAA is able to select important features from the raw features, which benefits the downstream HOI task. Since this operation does not change the shape of features, the reweighed features are still in $\mathbb{R}^{C \times H \times W}$ space.

The CAA learns the channel-wise dependency and global information, and compresses this information to get a single value. We think this compressed value indicates the importance of current channel, which can be used to re-calibrate the features of the current channel. Such operation explicitly injects global information into features, which provides an effective enhancement of task-specific representations. As shown in Fig. 2, the CAM is essentially a CAA following the backbone, which fine-tunes its representation by incorporating channel and contextual information.

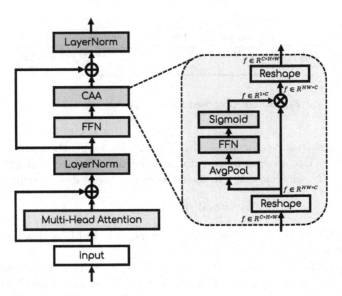

Fig. 4. Illustration of the CAE module. The left-side shows the architecture of CAE which follows the design of encoder of Transformer. Different from the encoder of Transformer, our CAE contains a CAA module, which is illustrates by the diagram on the right-side.

3.4   The Channel-Aware Encoder

The feature map $\mathbf{f} \in \mathbb{R}^{C \times H \times W}$ from CAM is transformed into $\mathbf{f_e} \in \mathbb{R}^{HW \times C}$, where HW is the number of tokens (each vector is termed as token) and C is the feature length of each token. We construct channel aware encoder (CAE) based on the observation that the encoder of Transformer preserves the tokens in each encoder layer with shape and order unchanged. We can recover the spatial property of the tokens by changing the shape of feature map from $\mathbb{R}^{HW \times C}$ to $\mathbb{R}^{C \times H \times W}$. Therefore, we are able to apply our CAA on the intermediate features in the encoder. Figure 4 illustrates the detailed architecture of the CAE. Each encoder layer contains two components including a FFN and a multi-head attention block. The tokens are first passed through the multi-head attention and the FFN module. Afterwards, we apply the CAA to the tokens with necessary transformations of shape, forming the CAE layer. Similar to the original encoder layer, the CAE layer is also stackable, and we stack the layers in the same pattern as the vanilla encoder.

4 Experiment and Result

In this section, we perform experiments on two popular benchmarks VCOCO [1] and HICO [2], and ablate our models to verify the effectiveness of its building blocks.

4.1 Datasets and Evaluation Metrics

Datasets. We evaluate our methods on two widly-adopted HOI detection benchmarks: HICO-DET and V-COCO. HICO-DET consists of 38,118 and 9,658 images for training and testing respectively, which contains 80 objects and 117 action categories. V-COCO consists of 5,400 images in the tarinval subset and 4,946 images in the test subset, which contains 80 objects and 29 action categories.

Table 1. Comparison of HOI detection performance (mAP, %) on HICO-DET [2] and V-COCO [1] test sets. The first section of the table presents the results of two stage methods, and the second section gives the results of one stage approaches. The results with the highest scores in each section are highlighted in bold.

| Method | Backbone | HICO-DET | | | | | | V-COCO | |
| | | Default | | | Known Objects | | | | |
		Full	Rare	Non-rare	Full	Rare	Non-rare	$Role_1$	$Role_2$
InteractNet[7]	ResNet-50-FPN	9.94	7.16	10.77	-	-	-	40.0	-
TIN[23]	ResNet-50	17.03	13.42	18.11	19.17	15.51	20.26	47.8	54.2
Gupta[1]	ResNet-152	17.18	12.17	18.68	-	-	-	-	-
VCL[11]	ResNet-50	23.63	17.21	25.55	25.98	19.12	28.03	48.3	-
DRG[12]	ResNet-50-FPN	24.53	19.47	26.04	27.98	23.11	29.43	51.0	-
IDN[10]	ResNet-50	24.58	20.33	25.86	27.89	23.64	29.16	53.3	60.3
FCL[24]	ResNet-50	25.27	20.57	26.67	27.71	22.34	28.93	52.4	-
SCG[13]	ResNet-50-FPN	**29.26**	**24.61**	**30.65**	**32.87**	**27.89**	**34.35**	**54.2**	**60.9**
PPDM[14]	Hourglass-104	21.94	13.97	24.32	24.81	17.09	27.12	-	-
HOTR[20]	ResNet-50	25.10	17.34	27.42	-	-	-	55.2	64.4
HOI-Trans[4]	ResNet-101	26.61	19.15	28.84	29.13	20.98	31.57	52.9	-
AS-Net[17]	ResNet-50	28.87	**24.25**	30.25	31.74	**27.07**	33.14	53.9	-
QPIC[3]	ResNet-50	29.07	21.85	31.23	31.68	24.14	33.93	58.8	61.0
QPIC[3]	ResNet-101	29.90	23.92	31.69	32.38	26.06	34.27	58.3	60.7
Ours	ResNet-50	29.86	23.58	31.74	32.19	25.49	34.20	62.4	64.5
Ours	ResNet-101	**30.32**	24.21	**32.14**	**32.68**	26.66	**34.48**	**64.6**	**66.7**

Evaluation Metrics. Following previous works [3], we employ mean average precision (mAP) as the evaluation metric of detection. One prediction has to satisfy two conditions to be true positive: 1) IOU between predicted human and object bounding boxes and ground-truth boxes is greater than 0.5; 2) Predicted object and action categories are correct.

Here, we adopt QPIC [3] as the baseline model, and construct our model by inserting CAM and customizing the encoder of transformer with CAE (see Fig. 2).

4.2   Implementation Details

We use both Resnet50 [25] and Resnet101 [25] as backbones to extract image features, followed by a CAA to enhance the condensed features. Afterwards, we enhance the vanilla transformer encoder by incorporating CAA into the encoder. The transformer decoder remains the same as QPIC [3]. We set the transformer dimension to 256, and the number of queries to 100.

We initialize the network with the parameters of DETR pretrained on MS-COCO [26]. The network is trained in 150 epochs with a batch size of 16. We optimize the model using Adam optimizer by setting the initial learning rates to 10^{-4} and 10^{-5}, for the backbone and other modules respectively. The weight decay is set to 10^{-4}. For Hungarians costs and loss weights, we adopt the identical hyper-parameters as QPIC. During inference, following the protocol in [3], we choose top 100 predictions with the highest scores as the final results.

Table 2. Comparison of HOI detection performance (mAP, %100) on the HICO-DET and V-COCO test sets. Bold values are best results.

| Method | Backbone | CAM | CAE | HICO-DET | | | | | | V-COCO | |
| | | | | Default | | | Known Objects | | | | |
				Full	Rare	Non-rare	Full	Rare	Non-rare	$Role_1$	$Role_2$
Baseline	ResNet-50			29.07	21.85	31.23	31.68	24.14	33.93	58.8	61.0
Baseline	ResNet-101			29.90	23.92	31.69	32.38	26.06	34.27	58.3	60.7
Ours	ResNet-50	✓		29.67	23.17	31.62	32.00	25.88	33.82	62.0	64.2
Ours	ResNet-50	✓	✓	29.86	23.58	31.74	32.19	25.49	34.20	62.4	64.5
Ours	ResNet-101	✓	✓	**30.32**	**24.21**	**32.14**	**32.68**	**26.66**	**34.48**	**64.6**	**66.7**

4.3   Performance Analysis

Table 1 presents the results of both two stage and one stage model, with the top section of the table listing the results of two stage methods, and the bottom section providing the results of one stage approaches. The results of our approaches (under different backbones) are highlighted in gray. On both HICO-DET and V-COCO, our method outperforms listed one stage methods by clear margins. While our approach obtains only slight improvement on HICO-DET compared with QPIC (a strong baseline), it surpasses QPIC significantly on V-COCO (3.6 and 3.5 points in terms of $Role_1$ and $Role_2$, using ResNet-50 as the backbone), showing the effectiveness of our method.

The Effectiveness of CAA. Recall that the CAM and CAE modules are based on our proposed channel aware attention. Results in Table 2 demonstrate that the incorporation of CAM and CAE is able to improve the performance on both datasets. Specifically, the inclusion of CAM only outperforms the Baseline (QPIC) by 0.6 points on HICO-DET and 3.2 points on V-COCO. The incorporation of CAE further improves the performance, gaining 0.79 points on HICO-DET and 3.6 points on V-COCO in total. Note that the improvement on VCOCO is more significant than HICO (3.6 vs. 0.79). The reason is probably that the model weights, which are taken to initialize our model, are pre-trained on COCO dataset [26], from which the VCOCO dataset originates. To summarize, these results in Table 2 demonstrate that the channel aware attention is an effective mechanism in terms of promoting the discrimination ability of representations of Transformer encoder.

Table 3. The effect of CAA positions in the encoder. Here HICO-DET test is used. All results are obtained under the default settings of this dataset.

Position	Full	Rare	Non-rare
Pre	29.35	22.25	31.48
Post	29.11	23.05	30.92
Indentity	29.55	23.05	31.49
Standard	**29.86**	**23.58**	**31.74**

The Position of CAE. In order to figure out the best configuration of CAE, we test four possible positions of CAA within CAE. Figure 5 depicts the four configurations of CAA. Table 3 presents the results of the four configurations under the default setting of HICO-DET. The standard configuration gives best results. This is probably because that the CAA in the standard configuration directly follows FFN, enabling the channel-wise feature refinement. Compared with the rest configurations, the standard version tends to gather global features benefiting the human-object pair detection, as well as to ease training. Therefore, we adopt the standard configuration in constructing the encoder of Transformer.

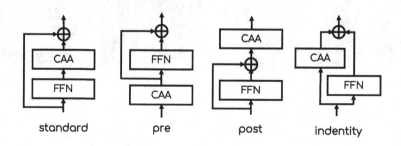

Fig. 5. Four possible positions of CAA in an encoder layer.

Stronger Backbone. The last two rows of Table 2 show the results of our full model (CAM+CAE) using ResNet-50 and ResNet-101 respectively. With stronger backbone (ResNet-101 [25]), our method notably outperforms QPIC [3] by clear margins (6.3 points on V-COCO and 0.42 points on HICO-DET). Note that the performance of QPIC with stronger backbone on VCOCO suffers from the issue of over-fitting mainly because of the shortage of data [3]. In comparison, our method with stronger backbone is still able to improve the performance. This is probably because that the proposed approach is able to capture more essential features for HOI, reduing the amount of noise within the learned representation.

4.4 Visual Analysis

Semantic Visualization. Due to the semantic nature of HOI, each action can be associated with several objects and vice versa. As the Fig. 6 shows, we group visualized samples according to object and action types. The visual results demonstrate that our method performs well under both scenarios, being able to handle the semantic roles of both actions and objects.

Attention Map Visualization. To investigate the regions that our proposed CAA focuses on, we visualize the attention maps produced by the encoder of QPIC and our approach. Figure 7 presents ground truth, detection results using our model as well as the attention maps generated by both models. In comparison with QPIC, which adopts vanilla Transformer encoder, our method visually

  (a) motorcycle   (b) kite   (c) cut   (d) ride

Fig. 6. Results on HICO-DET. Each of the first two columns (from left to right) presents several different actions corresponding to the same object category. Each of the last two columns shows various objects corresponding to the identical action class. Red and green boxes denote human and objects of each interaction respectively. The values are the confidence scores of actions and objects. (Color figure online)

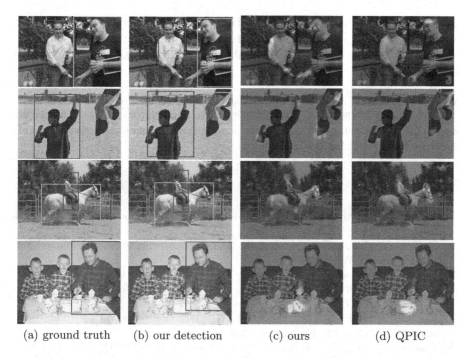

(a) ground truth   (b) our detection   (c) ours   (d) QPIC

Fig. 7. Visualization of detections and attention maps. The first two columns show the ground truth boxes and detected boxes, with red boxes denoting human and green boxes denoting objects. The last two columns visualize the attention maps of our method and QPIC respectively. (Color figure online)

attends to the regions of informative human body parts such as hands and feet, which are highly representative in terms of the recognition of the corresponding interactions. We believe that the features extracted from such regions are beneficial to the task of HOI, contributing to the improvement of our quantitative results.

5   Conclusion

We have presented a channel-aware attention mechanism, in order to learn more effective contextual representation for HOI. Based on the proposed attention, a channel-aware module and a channel-aware encoder are proposed to enhance the contextual representations at different levels. In particular, the channel-aware module is applied to the features extracted by the CNN backbone, which is a low-level enhancement to local features in different channels. In comparison, the channel-aware encoder replaces vanilla Transformer encoder, which enables the refinement of global representations within different channels. Experimental results demonstrate that our method achieves leading performance on two challenging benchmarks of HOI compared with SOTA approaches. Moreover, both ablation study and the visualization of the learned attention maps indicate that the proposed attention mechanism is able to capture richer interaction-specific information, leading to a better generalization ability to both small and large scale datasets.

Acknowledgments. This work was supported by Zhejiang Provincial Natural Science Foundation of China (No. LY21F020024), National Natural Science Foundation of China (No. 62272395), and Qin Chuangyuan Innovation and Entrepreneurship Talent Project (No. QCYRCXM-2022-359).

References

1. Gupta, S., Malik, J.: Visual semantic role labeling. arXiv preprint arXiv:1505.04474 (2015)
2. Chao, Y.-W., Liu, Y., Liu, M., Zeng, H., Deng, J.: Learning to detect human-object interactions. In: WACV (2018)
3. Tamura, M., Ohashi, H., Yoshinaga, T.: QPIC: query-based pairwise human-object interaction detection with image-wide contextual information. In: CVPR, pp. 10405–10414 (2021)
4. Zou, C., et al.: End-to-end human object interaction detection with hoi transformer. In: CVPR, pp. 11825–11834 (2021)
5. Vaswani, A.,et al.: Attention is all you need. In: NeurIPS (2017)
6. Carion, N., Massa, F., Synnaeve, G., Usunier, N., Kirillov, A., Zagoruyko, S.: End-to-end object detection with transformers. In: ECCV, pp. 213–229 (2020)
7. Gkioxari, G., Girshick, R., Dollár, P., He, K.: Detecting and recognizing human-object interactions. In: CVPR, pp. 8359–8367 (2018)
8. Gao, C., Zou, Y., Huang, J.-B.: iCAN: instance-centric attention network for human-object interaction detection. arXiv preprint arXiv:1808.10437 (2018)
9. Wang, T., et al.;, Deep contextual attention for human-object interaction detection. In: ICCV, pp. 5694–5702 (2019)
10. Li, Y.-L., Liu, X., Wu, X., Li, Y., Lu, C.: Hoi analysis: integrating and decomposing human-object interaction. In: NeurIPS, pp. 5011–5022 (2020)
11. Hou, Z., Peng, X., Qiao, Y., Tao, D.: Visual compositional learning for human-object interaction detection. In: ECCV, pp. 584–600 (2020)
12. Gao, C., Xu, J., Zou, Y., Huang, J.-B.: DRG: dual relation graph for human-object interaction detection. In: ECCV, pp. 696–712 (2020)
13. Zhang, F.Z., Campbell, D., Gould, S.: Spatially conditioned graphs for detecting human-object interactions. In: ICCV, pp. 13 319–13 327 (2021)
14. Liao, Y., Liu, S., Wang, F., Chen, Y., Qian, C., Feng, J.: PPDM: parallel point detection and matching for real-time human-object interaction detection. In: CVPR, pp. 482–490 (2020)
15. Wang, T., Yang, T., Danelljan, M., Khan, F.S., Zhang, X., Sun, J.: Learning human-object interaction detection using interaction points. In: CVPR, pp. 4116–4125 (2020)
16. Kim, B., Choi, T., Kang, J., Kim, H.J.: UnionDet: union-level detector towards real-time human-object interaction detection. In: ECCV, pp. 498–514 (2020)
17. Chen, M., Liao, Y., Liu, S., Chen, Z., Wang, F., Qian, C.: Reformulating HOI detection as adaptive set prediction. In: 2021 IEEE/CVF Conference on Computer Vision and Pattern Recognition (CVPR) (2021)

18. Veličković, P., Cucurull, G., Casanova, A., Romero, A., Lio, P., Bengio, Y.: Graph attention networks. arXiv preprint arXiv:1710.10903 (2017)
19. Hu, J., Shen, L., Sun, G.: Squeeze-and-excitation networks. In: CVPR, pp. 7132–7141 (2018)
20. Kim, B., Lee, J., Kang, J., Kim, E.-S., Kim, H.J.: HOTR: end-to-end human-object interaction detection with transformers. In: CVPR, pp. 74–83 (2021)
21. Chan, S., Tao, J., Zhou, X., Bai, C., Zhang, X.: Siamese implicit region proposal network with compound attention for visual tracking. IEEE Trans. Image Process. **31**, 1882–1894 (2022)
22. Bai, C., Li, H., Zhang, J., Huang, L., Zhang, L.: Unsupervised adversarial instance-level image retrieval. IEEE Trans. Multimedia **23**, 2199–2207 (2021)
23. Li, Y.-L., et al.: Transferable interactiveness knowledge for human-object interaction detection. In: CVPR (2019)
24. Hou, Z., Yu, B., Qiao, Y., Peng, X., Tao, D.: Detecting human-object interaction via fabricated compositional learning. In: CVPR, pp. 14646–14655 (2021)
25. He, K., Zhang, X., Ren, S., Sun, J.: Deep residual learning for image recognition. In: CVPR, pp. 770–778 (2016)
26. Lin, T.-Y. et al.: Microsoft COCO: common objects in context. In: ECCV, pp. 740–755 (2014)

AAKD-Net: Attention-Based Adversarial Knowledge Distillation Network for Image Classification

Fukang Zheng, Lin Zuo$^{(\boxtimes)}$, Feng Guo, Wenwei Luo, and Yuguo Hu

University of Electronic Science and Technology of China, Chengdu, China
linzuo@uestc.edu.cn

Abstract. Deep neural networks have achieved remarkable success in various research fields, but they face limitations. Firstly, complex models are often required to handle challenging scenarios. Secondly, limited storage and processing power on mobile devices hinder model training and deployment. To address these challenges, we propose a novel approach using a compact and efficient student model to learn from a cumbersome teacher model. To enhance feature map information extraction, we introduce an attention structure that leverages the rich features in the teacher model's feature maps. Adversarial training is incorporated by treating the student model as a generator and employing a discriminator to differentiate between teacher and student feature maps. Through an iterative process, the student model's feature map gradually approximates that of the teacher while improving the discriminator's discrimination abilities. By leveraging the knowledge of the teacher model and incorporating attention mechanisms and adversarial training, our approach provides a compelling solution to the challenges of complex model architectures and limited hardware resources. It achieves impressive performance enhancements with the student model.

Keywords: Knowledge Distillation · Feature map information extraction · Adversarial Training

1 Introduction

In recent years, deep neural networks have demonstrated remarkable performance in various tasks, prompting investigations into the impact of network depth on their success [1]. However, deeper networks often suffer from increased latency and sequential processing, limiting their suitability for time-sensitive applications and posing challenges for parallelization. This issue becomes more pronounced when deploying algorithms on resource-constrained edge devices. To address these limitations, knowledge distillation (KD) has emerged as a common approach [2]. In KD, a sophisticated teacher network produces soft labels for each training instance [3], and a student model is trained to predict these

© The Author(s), under exclusive license to Springer Nature Singapore Pte Ltd. 2024
B. Luo et al. (Eds.): ICONIP 2023, CCIS 1961, pp. 328–339, 2024.
https://doi.org/10.1007/978-981-99-8126-7_26

labels using a distillation loss. However, the conventional KD method primarily relies on logit information [4], which may underutilize the valuable output information from the teacher's feature maps during model training. To overcome this limitation, Some approaches propose enhancing the student model by incorporating information from the teacher model's feature maps [5,6]. By leveraging the rich feature information contained in these maps, we aim to improve the student model's performance. The utilization of feature maps provides additional guidance and enhances the learning process. Moreover, inspired by the selective perception ability of the human visual system, we integrate attention mechanism into the information transfer process from the feature maps. This allows the student model to focus on relevant and informative aspects of an image while suppressing irrelevant features [7]. Incorporating attention mechanisms enhances the student model's ability to prioritize essential information and improves overall performance.

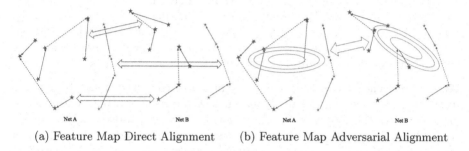

(a) Feature Map Direct Alignment   (b) Feature Map Adversarial Alignment

Fig. 1. Feature maps are represented by dots, with different colors indicating different feature maps. The thin arrows depict the changes in feature map data points as the iteration progresses, while the thick arrows demonstrate how the feature maps are compared between different networks. (a) The network is trained to minimize the distance between points of the same color. (b) The distribution between different networks is mimicked through adversarial loss.

We further leverage the concept of Generative Adversarial Networks (GAN) to align the data between the teacher and student models [8]. Our enhanced KD model employs a GAN architecture, where the student model acts as a generator and the discriminator distinguishes between the teacher's output and the content generated by the student. Through adversarial training, our model progressively aligns the student model's output with that of the teacher, facilitating knowledge transfer and enhancing performance. This integration of GAN dynamics into the knowledge distillation process improves data alignment and promotes effective knowledge transfer. Figure 1 embodies the idea of this method.

In summary, our contributions are as follows:

- Attention-based feature map transfer: We introduce an effective attention structure for transferring feature map knowledge, enhancing learning and performance.

- Integration of a trainable discriminator: We propose incorporating a trainable discriminator in the teacher-student architecture, allowing flexible and effective learning.

- Attention-based Adversarial Knowledge Distillation Network (AAKD-Net): Our AAKD-Net model serves as a bridge between the teacher and student models. Through alternating training with the discriminator, our method narrows the gap and improves performance of the student model architecture.

2   Related Work

This section surveys previous research and studies, summarizing existing knowledge and approaches.

Knowledge Distillation: Knowledge distillation, proposed by Hinton et al. [2], transfers knowledge using soft label information from teacher networks. Researchers have explored distilling feature information instead of solely focusing on output distillation. For example, FitNet [9] learns hidden information from middle layers, while attention-based methods leverage spatial attention maps. However, these methods still rely on hand-designed loss functions.

Attention: The attention mechanism efficiently completes tasks by guiding weight updates of feature maps based on task results. Popular attention models like SANet [10] and VSG-Net [11] have been widely applied in deep learning tasks such as natural language processing, image recognition, and speech sound recognition. These models represent a core technology that requires further exploration in deep learning.

Generative Adversarial Networks: GAN [8], derived from the idea of generating realistic images using random noise, consists of a generator and a discriminator. The discriminator distinguishes real and fake images, while the generator aims to deceive the discriminator by synthesizing realistic images. Training GAN involves a competitive process, and researchers have developed different variants to improve its effectiveness. CycleGAN [12] is used for domain adaptation, and BigGAN [13] generates high-fidelity images with low variety disparity.

The combination of GAN and KD can be explored in various ways: (1) GAN-generated Images for KD: Using GAN to generate synthetic images for KD [14]. (2) Model Compression for GAN: Applying KD to compress GAN models, reducing their complexity and resource requirements [15–17]. (3) GAN Discriminators for Training Classifiers: Utilizing GAN discriminators to train classifiers (student models) [18,19]. KDGAN learns real data distributions, while Chung exploits teacher feature map information through adversarial methods. Our proposed methods leverages adversarial training to improve the student model's extraction of feature map information from the teacher model, enhancing precision.

3   Methods

In this section, we present our proposed AAKD-Net, a new structure aimed at optimizing the KD process. We explore the integration of attention mechanism to enhance the interaction between the teacher and student networks. By incorporating these optimizations, we can enhance the effectiveness and efficiency of KD in compressing and transferring knowledge between networks.

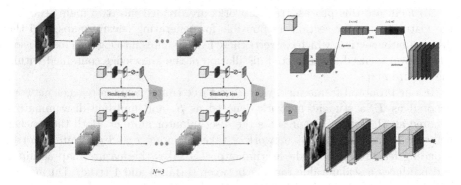

Fig. 2. Proposed Structure: Input data is simultaneously fed into both the teacher and student networks, generating intermediate and final outputs. These outputs are utilized for the computation of Similarity Loss and are further utilized in the adversarial training process against multiple discriminator networks. Additionally, attention mechanisms are employed to extract and emphasize important feature information.

We begin by describing the process of transferring knowledge from the teacher network to the student network. This involves leveraging the expertise of the teacher to guide the learning of the student, ensuring that valuable information is effectively conveyed. In addition, to achieve more flexible and effective learning, we introduce a GAN-based adversarial loss method. Unlike traditional approaches that impose strict constraints through a specific distance function, our method allows for more adaptable learning, avoiding unnecessary limitations and promoting better performance. Through iterative training, we enable the exchange of knowledge between the teacher and student networks. This iterative process facilitates the transfer of valuable insights and leads to improved accuracy and performance in the student network architecture. To ensure accurate and effective information exchange, we incorporate attention mechanisms. These mechanisms enable the models to focus on the most relevant and informative aspects, enhancing learning and overall performance. By integrating attention mechanisms and employing iterative training with GAN-based adversarial loss, our proposed AAKD-Net offers a comprehensive approach to knowledge distillation. This approach not only improves the transfer of knowledge between networks but also enhances the learning capabilities and performance of the student network. Our overall approach is shown in Fig. 2.

3.1   Knowledge Distillation with Adversarial Training

In our AAKD-Net, we leverage adversarial training as a means to transfer knowledge between networks. This approach is based on the concept of a dynamic zero-sum game, as depicted in Eq. (1). The adversarial feature graph distillation is applied to both the teacher and student networks.

$$\min_c \max_d V(c, d) = \mathbb{E}_{\boldsymbol{y} \sim p_u} [\log p_d^g(\boldsymbol{x}, \boldsymbol{y})] + \mathbb{E}_{\boldsymbol{y} \sim p_c} [\log (1 - p_d^e(\boldsymbol{x}, \boldsymbol{y}))]. \quad (1)$$

To facilitate this process, the networks are divided into two main sections: the feature extractor section, responsible for generating feature maps, and the discriminator section, which converts these feature maps into logits. This division allows for a focused and targeted distillation of the knowledge contained within the feature maps.

In our proposed framework, we utilize three components: a teacher network denoted as T, a student network denoted as S, and multiple discriminators denoted as D_k, where k represents the discriminator number. Both the teacher network T and the student network S take an input x and generate a corresponding feature map. Each discriminator D_k takes the feature map as input and produces a scalar value ranging between 0 (false) and 1 (true). During the training process, the desired output for the discriminators is 1 when the feature map comes from the more capable network, which is the teacher network T. Conversely, if the feature map comes from the weaker network, which is the student network S, the desired output is 0. The objective of each discriminator D_k is to minimize the discriminator loss term L_{D_k} by effectively distinguishing between the feature map distributions of the teacher network and the student network. On the other hand, the teacher network T aims to minimize the loss term L_{D_k} by spoofing the discriminator D_k, tricking it into classifying the feature map from the teacher network as true and producing an output of 1. Similarly, the goal of the student network S is to minimize the loss term L_{D_k} as well. By minimizing this loss term, the student network improves its parameters and learns to simulate the feature map distribution of the teacher network T.

The structure of the discriminator is composed of a series of Conv2d-BatchNorm-LeakyReLU layers followed by a Conv2d-Sigmoid layer. This architecture allows the discriminator to learn the discriminative features necessary for distinguishing between the feature maps. The structure of one of the discriminators is shown in the Fig. 3.

Overall, the adversarial loss for the discriminator(as shown in Eq. (2)) and student net (as shown in Eq. (3)) can be expressed as follows:

$$\mathcal{L}_{D_k} = [1 - D_k (T(x))]^2 + [D_k (S(x))]^2, \quad (2)$$

$$\mathcal{L}_S = [1 - D_k (S(x))]^2. \quad (3)$$

In this loss term, the discriminators aim to minimize the squared difference between their outputs when evaluating the feature maps from the teacher network T and the student network S. By minimizing the discriminator loss L_{D_k}

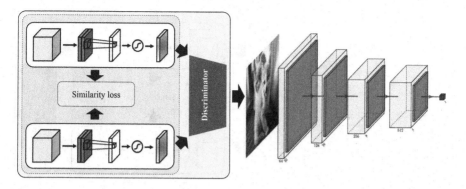

Fig. 3. One of the discriminative network architecture in the AAKD-Net module

for both the teacher and student networks, the student network can simulate the feature map distribution of the teacher network, thereby achieving effective knowledge transfer between the networks. In our experiments, we incorporated three feature extraction locations within our architecture. These locations are designed to reduce the spatial size and number of channels in the feature map through convolution operations, ultimately leading to the production of a single scalar value at the end. To normalize this scalar value within the range of 0 and 1, we apply a Sigmoid function. To distinguish between the different feature map distributions, we utilize the corresponding discriminators. These discriminators are responsible for evaluating and discerning the characteristics of the feature maps generated by the teacher and student networks. In addition, we introduce a transfer layer denoted as A_T, which leverages an attention-based structure to extract key information from the feature map. This transfer layer enhances the discriminative power and importance of certain features, facilitating more effective knowledge transfer (as shown in Eqs. (4) and (5)).

$$\mathcal{L}_{D_k} = [1 - D_k (A_T(T(x)))]^2 + [D_k (A_T(S(x)))]^2, \tag{4}$$

$$\mathcal{L}_s = [1 - D_k (A_T(S(x)))]^2. \tag{5}$$

3.2 Knowledge Distillation with Attention

In our proposed AAKD-Net architecture, we draw inspiration from the SE-Net (Squeeze-and-Excitation Network) block [20]. In the SE-Net block, each feature map undergoes compression through global average pooling, resulting in a real number (as shown in Eq. (6)) (Fig. 4). This real number encapsulates the global information of the feature map. Additionally, the compressed representation of each feature map is combined into a vector and utilized as weights for their respective groups. Based on this, we incorporated attention mechanism in the intermediate layer between the teacher and student networks, which allowed us to selectively emphasize important features and compute the corresponding loss function. By applying attention, we were able to focus on specific

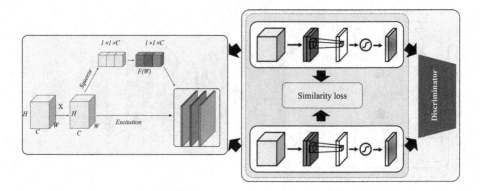

Fig. 4. The attention architecture in the AAKD-Net module

regions or channels in the feature maps, effectively enhancing the learning process and guiding the student network to better mimic the teacher's knowledge. The attention-based calculations in the intermediate layer played a crucial role in determining the loss function and facilitating knowledge transfer between the two networks.

$$z_c = \frac{1}{H \times W} \sum_{i=1}^{H} \sum_{j=1}^{W} u_c(i, j), \tag{6}$$

The height and width of the feature map are denoted as H and W respectively. The variable u represents the convolution result, while z represents the global attention information specific to the corresponding feature map. Passing this vector through two fully connected layers with an activation function allows us to adjust the weights based on their importance for the recognition task. This adjustment involves scaling up the weights of more important feature maps and scaling down the weights of less important ones.

Specifically, the ReLU activation function (σ) and sigmoid activation function (δ) are utilized, along with two different fully connected operations denoted as W_1 and W_2.

$$s = \sigma(W_2 \delta(W_1 z)), \tag{7}$$

The resulting vector s represents the degree of importance assigned to each feature map. By employing Eq. (8),

$$x_c = s_c u_c, \tag{8}$$

the vector s effectively stimulates the original feature map, guiding its continuous update in a direction favorable to the recognition task.

By incorporating these mechanisms, we can effectively capture and leverage the most informative features from the feature maps, allowing for enhanced discriminative power and performance in recognition tasks. Our process is shown in Algorithm 1.

Algorithm 1: Attention-based Adversarial Knowledge Distillation Network

---

input : A pretrained teacher model $T(x, \theta^t)$
output: A comparable student model $S(x, \theta^s)$

1 Randomly initialize a student model $S(x, \theta^s)$ and three discriminators $D_k(z, \theta^{d_k})$;

2 **for** *each training epochs* **do**

3   **1.Discriminator Optimisation Stage**;

4   **for** *each step k* **do**

5   Generate Feature x_t from $T(x, \theta^t)$;

6   Generate Feature x_s from $S(x, \theta^s)$;

7   Using Eqs. (5)–(8) get $At(x_t)$ and $At(x_s)$;

8   Update θ^{d_k} to **minimize discrepancy** with $Loss\_D_i(At(x_t), At(x_s))$;

9   **end**

10   **2.Student Model Optimisation Stage**

11   Update θ^s to **maximize discrepancy** with $Loss\_S(At(x_t), At(x_s))$;

12 **end**

---

4   Experiments

We evaluated the efficiency of our distillation method in the area of image classification. The effectiveness of KD depends on the construction of the teacher model and the student model, as well as the way knowledge is transferred between them. For a fair comparison, we reproduced the algorithm on code from other methods. Our experiments are based on the PyTorch [21] framework. We a description of the dataset situation in Sect. 4.1, details of the experiments and the underlying experimental setup in Sect. 4.2, and a comparison of the results of our experiments with other models is given in Sect. 4.3. Finally, we present the ablation experiment in Sect. 4.4.

4.1   Dataset

In our experiments, we used CIFAR100 [22], all of which have an image size of 32*32. There are 100 classes, each containing 600 images. Each class has 500 training images and 100 test images. The final result was determined by calculating the average of the 5 results obtained from the CIFAR-100 runs.

4.2   Experiments Settings

Since our method uses two loss terms, logit-based loss and feature map-based loss, we use different learning details for each loss term. For the overall process, we used 200 epoches of training to make a fair comparison of the results. For the logit-based loss, the initial learning rate was initially set to 0.1, multiplied by 0.1 at the 100th and 150th epochs. We optimised the logit-based loss using

SGD with a maximum batch size of 128, a momentum of 0.9 and a weight decay of 5e-4, for feature map-based loss, the discriminator learning rate was used at 2e-4.

4.3 Performance Comparison

In our experiments, we extensively utilized the WideResnet [23] method as the baseline for our research. To evaluate the effectiveness of our proposed AAKD-Net model for image classification tasks, we conducted a comprehensive comparison with other existing models. The results of this comparison are presented in Table 1.

Table 1. The performance of the various knowledge distillation methods on the CIFAR-100. The classification is measured by the error rate (%). Where baseline of the model without distillation training.

Teacher Net		WideResnet28-4 (21.09)		
Student Net	WideResnet16-4	WideResnet28-2	WideResnet16-2	Resnet56
Baseline	22.72	24.88	27.32	27.68
KD [2]	21.69	23.43	26.47	26.76
FitNets [24]	21.85	23.94	26.30	26.35
AT [9]	22.07	23.80	26.56	26.66
FT [25]	21.72	23.41	25.91	26.20
AB [5]	21.36	23.19	26.01	26.04
Proposed	**21.25**	**22.79**	**25.58**	**25.87**

The purpose of this comparison was to assess the performance achieved by distilling the knowledge from our proposed method. We explored four different types of distillation models and analyzed their impact. The following results are compared to baseline. Firstly, we examined the effect of changing the model depth alone, which resulted in a notable improvement of 1.47% compared to the untrained model. Secondly, we focused on altering the model width, and this modification led to an even more significant improvement of 2.09%. Additionally, we experimented with changing both the model depth and width simultaneously, resulting in a performance enhancement of 1.86%. Lastly, we investigated the impact of altering the relationship between the teacher and student networks, which yielded a performance improvement of 1.81%. Based on our extensive experiments, we found that the most significant performance gains were achieved when solely modifying the model's width.

The superiority of our AAKD-Net model over other models demonstrates the effectiveness of our proposed method in enhancing knowledge extraction. By leveraging the KD technique, we successfully transferred valuable knowledge from the teacher network to the student network, leading to improved accuracy and performance. Our methods highlight the potential of KD in the context of model architecture optimization. By exploring different aspects of the model, such as the depth or width of the teacher-student model, we can further enhance the performance of deep learning models for various tasks.

4.4 Ablation Study

We performed ablation experiments to systematically evaluate the impact of different components on the overall performance. The results are summarized in Table 2. We used the common WideResNet 16-4 as the baseline model and gradually introduced ablation components to measure their effects. Among the ablation components, incorporating attention to facilitate the transfer of teacher features yielded the most significant improvement. This attention mechanism allowed the student network to selectively focus on informative features, enabling accurate and efficient exchange of valuable information between the teacher and student. Additionally, we integrated trainable discriminators to guide the knowledge transfer process. By avoiding strict adherence to a specific distance function for teacher-student imitation, we achieved improved performance. This flexible approach allowed the student network to leverage the discriminators and adaptively enhance its learning from the teacher. Overall, the combination of attention-based feature transfer and trainable discriminators resulted in a substantial improvement in performance. These components synergistically worked together, ensuring effective information exchange and leveraging the strengths of each method.

Table 2. Ablation study of proposed method. The results are presented in the form of error rate (%).

	Baseline	Baseline+Attention	Baseline+Loss	Baseline+Attention+Loss
Error	24.88	23.70	24.17	22.79
Diff	–	−1.18	−0.71	−2.09

5   Conclusion

In our study, we proposed AAKD-Net, a novel model architecture that addresses the limitations of lacking comprehensive knowledge in the output layer. We incorporate an attention mechanism to focus on relevant features and diminish the importance of irrelevant ones. Additionally, we explore adversarial training as a loss function to accurately model the teacher's knowledge distribution. Extensive experiments show significant improvements in the performance of the lightweight student model for image classification tasks. Overall, our AAKD-Net model effectively addresses the limitation by leveraging intermediate layer distillation, attention mechanism, and adversarial training, leading to improved accuracy and effectiveness in image classification.

Acknowledgement. This work was supported by the National Natural Science Foundation of China under Grant No. 61877009, No. 62276054 and the Sichuan Science and Technology Program under contract number 2023YFG0156.

References

1. Goyal, A., Bochkovskiy, A., Deng, J., Koltun, V.: Non-deep networks. In: Koyejo, S., Mohamed, S., Agarwal, A., Belgrave, D., Cho, K., Oh, A. (eds.) Advances in Neural Information Processing Systems, vol. 35, pp. 6789–6801. Curran Associates Inc. (2022)
2. Hinton, G., Vinyals, O., Dean, J.: Distilling the knowledge in a neural network. arXiv:1503.02531 (2015)
3. Lopez-Paz, D., Bottou, L., Schölkopf, B., Vapnik, V.: Unifying distillation and privileged information. arXiv preprint arXiv:1511.03643 (2015)
4. Ba, J., Caruana, R.: Do deep nets really need to be deep? Adv. Neural Inf. Process. Syst. **27** (2014)
5. Heo, B., Lee, M., Yun, S., Choi, J.Y.: Knowledge transfer via distillation of activation boundaries formed by hidden neurons. In: Proceedings of the AAAI Conference on Artificial Intelligence, vol. 33, pp. 3779–3787 (2019)
6. Zhou, C., Neubig, G., Gu, J.: Understanding knowledge distillation in non-autoregressive machine translation. arXiv preprint arXiv:1911.02727 (2019)
7. Guo, M.-H., et al.: Attention mechanisms in computer vision: a survey. Comput. Vis. Media **8**(3), 331–368 (2022)
8. Goodfellow, I., et al.: Generative adversarial networks. Commun. ACM **63**(11), 139–144 (2020)
9. Zagoruyko, S., Komodakis, N.: Paying more attention to attention: improving the performance of convolutional neural networks via attention transfer. arXiv preprint arXiv:1612.03928 (2016)
10. Zhong, Z., et al.: Squeeze-and-attention networks for semantic segmentation. In: Proceedings of the IEEE/CVF Conference on Computer Vision and Pattern Recognition, pp. 13065–13074 (2020)
11. Ulutan, O., Iftekhar, A., Manjunath, B.S.: Vsgnet: spatial attention network for detecting human object interactions using graph convolutions, in: Proceedings of the IEEE/CVF Conference on Computer Vision and Pattern Recognition, pp. 13617–13626 (2020)
12. Zhu, J.-Y., Park, T., Isola, P., Efros, A.A.: Unpaired image-to-image translation using cycle-consistent adversarial networks. In: Proceedings of the IEEE International Conference on Computer Vision, pp. 2223–2232 (2017)
13. Brock, A., Donahue, J., Simonyan, K.: Large scale gan training for high fidelity natural image synthesis. arXiv preprint arXiv:1809.11096 (2018)
14. Ding, X., Wang, Y., Xu, Z., Wang, Z.J., Welch, W.J.: Distilling and transferring knowledge via cgan-generated samples for image classification and regression. Expert Syst. Appl. **213**, 119060 (2023)
15. Fang, G., Song, J., Shen, C., Wang, X., Chen, D., Song, M.: Data-free adversarial distillation. arXiv preprint arXiv:1912.11006 (2019)

16. Chen, H., et al.: Data-free learning of student networks. In: Proceedings of the IEEE/CVF International Conference on Computer Vision, pp. 3514–3522 (2019)
17. Chen, H., et al.: Distilling portable generative adversarial networks for image translation. In: Proceedings of the AAAI Conference on Artificial Intelligence, vol. 34, pp. 3585–3592 (2020)
18. Wang, X., Zhang, R., Sun, Y., Qi, J.: Kdgan: knowledge distillation with generative adversarial networks. Adv. Neural Inf. Process. Syst. **31** (2018)
19. Chung, I., Park, S., Kim, J., Kwak, N.: Feature-map-level online adversarial knowledge distillation. In: International Conference on Machine Learning, pp. 2006–2015. PMLR (2020)
20. Hu, J., Shen, L., Sun, G.: Squeeze-and-excitation networks. In: Proceedings of the IEEE Conference on Computer Vision and Pattern Recognition (CVPR) (2018)
21. Paszke, A., Gross, S., Massa, F., Lerer, A., Bradbury, J.P.: An imperative style, high-performance deep learning library. Adv. Neural Inf. Process. Syst. **32**
22. Krizhevsky, A., et al.: Learning multiple layers of features from tiny images (2009)
23. Zagoruyko, S., Komodakis, N.: Wide residual networks. arXiv preprint arXiv:1605.07146 (2016)
24. Romero, A., Ballas, N., Kahou, S.E., Chassang, A., Gatta, C., Bengio, Y.: Fitnets: hints for thin deep nets. arXiv preprint arXiv:1412.6550 (2014)
25. Kim, J., Park, S., Kwak, N.: Paraphrasing complex network: network compression via factor transfer. Adv. Neural Inf. Process. Syst. **31** (2018)

A High-Performance Tensorial Evolutionary Computation for Solving Spatial Optimization Problems

Si-Chao Lei[1], Hong-Shu Guo[1], Xiao-Lin Xiao[2], Yue-Jiao Gong[1]($\boxtimes$),
and Jun Zhang[3]

[1] South China University of Technology, Guangzhou, Guangdong, China
gongyuejiao@gmail.com
[2] South China Normal University, Guangzhou, Guangdong, China
[3] Hanyang University, Ansan, South Korea

Abstract. As a newly emerged evolutionary algorithm, tensorial evolution (TE) has shown promising performance in solving spatial optimization problems owing to its tensorial representation and tensorial evolutionary patterns. TE algorithm sequentially performed different tensorial evolutionary operations on a single individual or pairs of individuals in a population during iterations. Since tensor algebra considers all dimensions of data simultaneously, TE was explicitly parallel in dimension level. However, it was burdened with intensive tensor calculations especially when encountering large-scale problems. How to extend TE to efficiently solve large-scale problems is one of the most pressing issues currently. Toward this goal, we first devise an efficient TE (ETE) algorithm which expresses all the evolutionary processes in a unified tensorial computational model. Compared to TE, the tensorial evolutionary operations are directly executed on a population rather than sole individuals, enabling ETE to achieve explicit parallel in both dimension and individual levels. To further enhance the computational efficiency of ETE, we leverage the compute unified device architecture (CUDA), which provides access to computational resources on graphics processing units (GPUs). A CUDA-based implementation of ETE (Cu-ETE) is then presented that utilizes GPU to accelerate tensorial evolutionary computation. Notably, Cu-ETE is the first implementation of tensorial evolution on GPU. Experimental results demonstrate the enhanced computational efficiency of both ETE (CPU) and Cu-ETE (GPU) over TE (CPU). By harnessing the power of tensorial algebra and GPU acceleration, Cu-ETE opens up new possibilities for efficient problem-solving in more complex and large-scale problems across various fields of knowledge.

Keywords: Evolutionary computation · Tensor algebra · Graphics Processing Unit (GPU) · Compute Unified Device Architecture (CUDA)

© The Author(s), under exclusive license to Springer Nature Singapore Pte Ltd. 2024
B. Luo et al. (Eds.): ICONIP 2023, CCIS 1961, pp. 340–351, 2024.
https://doi.org/10.1007/978-981-99-8126-7_27

1 Introduction

Tensorial evolution (TE) is a newly developed evolutionary algorithm (EA) designed to tackle general-purpose spatial optimization problems [11]. While TE shares similarities with other population-based EAs that aims to search for optimal solutions through a series of evolutionary operators, it stands out due to two distinctive features. Firstly, TE employs a multidimensional tensor structure to represent solutions, in contrast to traditional EAs that flatten solutions into one-dimensional (1D) chromosomes. The tensorial representation enables TE to capture the interdependencies and interactions among variables more effectively and concisely. Secondly, TE develops tensorial variation operators (crossover and mutation), that evolve individuals in the tensor space for spatial optimization. These operators leverage tensor algebra to evolve different dimensions of individuals simultaneously. The above two features enable TE to effectively explore the search space while eliminating the operational inefficiencies associated with other 1D EAs when dealing with high-dimensional spatial optimization.

However, TE encountered efficiency challenges in solving large-scale spatial optimization problems. TE is an iteration-based algorithm that performs evolution generation by generation. During each generation, the individuals are evolved one by one using different tensorial variation operators. When dealing with larger instances, the intensive tensor operations would greatly hinder the efficiency of problem-solving in evolutionary computation (EC). Tensor algebra allows for considering all dimensions of a tensor at once. Under this view, when we superpose individuals to a population tensor, the TE could evolve all individuals simultaneously, along with all their dimensions. Obviously, the number of tensor operations implemented in each generation would be directly reduced.

Inspired by the above-mentioned issues, in the paper, we present an efficient TE (ETE) algorithm based on our previous work on TE [11] to fully exploits the inherent heavy parallelism in tensor algebra. Besides, previous studies have shown the significant speed-up potential in the parallel implementations of EAs using graphics processing units (GPUs) [2,10]. The Compute Unified Device Architecture (CUDA) provides access to computational resources to GPUs [1]. Therefore, a CUDA-based implementation of ETE (Cu-ETE) is then presented that distributes the evolutionary computation into multiple processing units. Incorporating GPU acceleration into ETE further leads to substantial time savings. Overall, the combination of tensor algebra and GPU acceleration contributes to a high-performance tensorial evolutionary computation, and ultimately boosts the performance of problem-solving for large-scale spatial optimization problems. The main contributions can be summarized as follows: (1) develop an ETE algorithm that expresses all the evolutionary operators in a unified tensorial evolutionary computation model, enabling the evolution process highly parallelizable; (2) implement the Cu-ETE on GPU to enhance the efficiency strength of ETE on large-scale spatial problem optimization; (3) compare both ETE and Cu-ETE with TE [11]. The experimental results reveal the remarkable performance of ETE and Cu-ETE in large-scale spatial optimization.

The remainder of the paper is organized as follows: Section 2 introduces the background of tensor algebra and CUDA-based GPU computing. Section 3 presents the detailed design of ETE and Cu-ETE. The experimental results are reported in Sect. 4. Finally, conclusions are made in Sect. 5.

2 Background and Preliminaries

2.1 Tensor and Tensor Algebra

Tensor is a mathematical abstraction of scalars, vectors (1D), and matrices (2D) to a general multidimensional data with an arbitrary number of dimensions [7]. A tensor is typically denoted as $T_{I \times J \times K \times \cdots}$ where $I \times J \times K \times \cdots$ is the size of a tensor. The dimension of a tensor is often called the order. The tensor provides a natural and concise mathematical representation for formulating multilinear data structures. Various tensor operations also offer an interface that implements different transformations on the data for solving problems in multidimensional environments. Here, we give the definition of the t-product operation of two third-order tensors based on tensor slices, formulated as

$$A * B = bvfold(bcirc(A) \times bvec(B)) \tag{1}$$

where $A_{n_1 \times n_2 \times n_3}$ and $B_{n_2 \times n_4 \times n_3}$ are two third-order tensors. The $bcirc$ is to unfold A along the third direction where $A^{(a)}$ for $a \in [1, n_3]$ is a frontal slice of A, and similarly for $bvec$.

In recent decades, the advent of deep learning has sparked significant advancements in tensor libraries, which are the basis of deep neural networks. Prominent tensor libraries such as TensorFlow, PyTorch, NumPy, and, JAX, have emerged that provide extensive support for accessing tensor operations, and effectively leverage computational resources from GPUs to accelerate data processing. One notable advantage of using GPUs for tensor computations is their exceptional parallelization capabilities. Breaking tensor operations into massive independent calculations offers high scalability of computations and shows better efficiency over CPU [9].

2.2 CUDA-Based GPU Computation

CUDA is a parallel computing architecture that runs on modern NVIDIA GPUs [1]. It groups hundreds of streaming processors (SPs) into several streaming multiprocessors (SMs). All SMs share the global memory with high latency, and each SM contains a number of SPs that share the on-chip control logic units, low latency shared memory, registers, and more. CUDA enables the full utilization of both the CPU (host) and GPU (device) in a heterogeneous computing system. The CPU is used to manage I/O operations and coordinate the overall execution processes of a CUDA program. The parallel processing capabilities of the GPU are utilized to speed up calculations through kernel functions. In a CUDA program, kernel functions are typically executed on the GPU in a Single

Instruction Multiple Data (SIMD) fashion, where a number of threads are executed simultaneously. These threads are organized into blocks, and the blocks are further organized into grids. All threads in a block share a low latency memory which can only be accessed inside the block, while threads across blocks interact using global memory with high latency and larger space. The maximal number of threads in a block depends on the compute capability of the GPU.

3   Tensorial Evolution

Traditional population-based EAs distribute different populations or individuals across hardware computational resources to accelerate the evolution process [4,5,12]. These distributed EAs are generally parallel at the population level or individual level. However, such parallel implementation is difficult or impossible to separate the dimensions of individuals into hardware resources, hindering practical applications in high-dimensional environments [13]. Our previous work in [11] has proposed a TE algorithm that is capable of executing different dimensions of individuals simultaneously based on tensor algebra while not relying on hardware. However, TE is sub-optimal in efficiency since it performs evolution on individuals one by one. Therefore, we propose an ETE algorithm to fully exploit the parallel routine of tensor algebra.

3.1   Overview of ETE

ETE starts from the initialization of a population. The population is defined as $\mathcal{P}_{M \times N \times NP}$ where $M \times N$ is the individual size and NP is the population size. Each frontal slice of $\mathcal{P}$ corresponds to a candidate solution. Several common tensorial evolutionary operators in ETE are elaborated below.

Selection. We express two classical selection operators separately: roulette wheel selection (RWS) and tournament selection (TS). For RWS, the individuals are selected proportional to their fitness values. Given $\mathcal{P}$ and the corresponding fitness vector $F_{NP \times 1}$, the selection tensor is obtained as $\mathcal{T}_s = ((1 \times rw^T) < (vsel^T \times 1^T)) \times 1$ where $1_{NP \times 1}$ is an all 1 s matrix. The $rw_{NP \times 1}$ is a roulette wheel selection matrix computed by $rw = 1^L \times Prob$ where $1^L_{NP \times NP}$ is a lower triangular 0-1 matrix, and $Prob_{NP \times 1}$ stores the selection probabilities calculated by $F[i]/\sum F$. The random vector $vsel_{1 \times NP}$ is generated in the range of $[0, 1]$ to simulate the probability selection process of individuals. Then, the selection is implemented by a slice operation as

$$\mathcal{P} = \mathcal{P}\,[M|N|\mathcal{T}_s] \tag{2}$$

where $M = \{1, \cdots, M\}$, $N = \{1, \cdots, N\}$, $\mathcal{T}_s \in \{1, \cdots, NP\}$, and $[\cdot|\cdot|\cdot]$ denotes an indexing operation.

For TS, several tournaments are performed among a few individuals chosen from the population at random. The individual with better fitness is selected within each tournament. Considering the tournament of size 2 as an example, the selection tensor is constructed as $T_s = vsel_1(F[vsel_1] > F[vsel_2]) + vsel_2(F[vsel_1] < F[vsel_2])$ where $vsel_1, vsel_2 \in \mathbb{Z}^{NP \times 1}$ are two vectors randomly sampled from $\{1, \cdots, NP\}$. The corresponding two individuals indicated by $vsel_1[i]$ and $vsel_2[i]$ constitute a tournament. Then, the selection is implemented by Eq. (2).

Crossover. The crossover recombines individuals in the population by exchanging local spatial regions. A random number $\alpha \in [0,1]$ is generated to control the crossover direction: If $\alpha \leq 0.5$, the crossover tensor is generated as $T_c[i|i|t_1] = 1, T_c[i|i|t_2] = 0$ for all $i \in [1, vc_1]$, and $T_c[i|i|t_1] = 0, T_c[i|i|t_2] = 1$ for all $i \in [1, vc_2]$ where t_1, t_2 are two random integers sampled from $\{1, \cdots, NP\}$. The size of T_c is $M \times M \times NP$. The $vc_1 \in \mathbb{Z}^{rl \times 1}$ is a vector containing all the numbers in $\langle rs \rangle_M, \cdots, \langle rs + rl \rangle_M$ where $\langle \cdot \rangle_M$ is a modular function over M, and $rs, rl \in [1, M]$ are two random integers for locating the starting position of crossover and the region of crossover. The vc_2 contains all the remaining numbers in $\{1, \cdots, M\}$ but not in vc_1. The crossover is then implemented as

$$\mathcal{P}[M|N|NP_c] = (T_c * \mathcal{P})[M|N|NP_c] \tag{3}$$

where $NP_c = \{1, \cdots, NP_c\}$ and $NP_c = NP \times p_c$ given the crossover probability p_c. While for the case $\alpha > 0.5$, the crossover tensor is generated similarly but changing all the above M values to N. In this case, the size of T_c is adapted to $N \times N \times NP$. Then, the crossover is implemented as

$$\mathcal{P}[M|N|NP_c] = (\mathcal{P} * T_c)[M|N|NP_c] \tag{4}$$

Local Mutation. The local mutation varies genes on individuals based on its local spatial topology given a kernel defined as $Ker_{K \times K}$. To handle the genes located on the boundaries, the population tensor is padded before mutation as $\mathcal{P}_{lm} = padding(\mathcal{P}, pad)$ where pad denotes the padding size. The local mutant tensors store the sampling locations of the mutated genes, and the corresponding mutation weights, which are formulated as $T_{lm1}[i|i-1, i, i+1|t] = Ker[K|1]$, $T_{lm2}[i|i-1, i, i+1|t] = Ker[K|2]$, and $T_{lm3}[i|i-1, i, i+1|t] = Ker[K|K]$ where $i \in [2, M+1], t \in [1, NP_m]$ are random integers, and $K = \{1, \cdots, K\}$. The size of each local mutant tensor is $(M + pad) \times (M + pad) \times NP$. Then, the local mutation is implemented by t-product based convolution as shown in Fig. 1, which is formulated as

$$\mathcal{P}[M|N|NP_m] = ((T_{lm1} * \mathcal{P}_{lm})[M_1|N|NP_m] + (T_{lm2} * \mathcal{P}_{lm})[M_1|N_1|NP_m] \\ + (T_{lm3} * \mathcal{P}_{lm})[M_1|N_2|NP_m])[M|N|NP_m] \tag{5}$$

where $M_1 = \{2, \cdots, M+1\}$, $N = \{1, \cdots, N\}$, $N_1 = \{2, \cdots, N+1\}$, $N_2 = \{3, \cdots, N+2\}$, and $NP_m = \{1, \cdots, NP_m\}$.

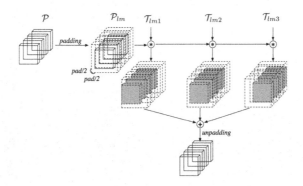

Fig. 1. Local mutation ($Ker = 3, pad = 2$). All the intermediate results in marked by dotted lines.

Global Mutation. The global mutation performs random Gaussian distribution on the population $\mathcal{P}$. Given the global mutation probability p_{gm}, the global mutant tensor is defined as $\mathcal{T}_{gm} = (tRnd < p_{gm}) \times (tGRnd \times (ub - lb))$. The size of $\mathcal{T}_{gm}$ is consistent with $\mathcal{P}$. The $tRnd \in \mathbb{R}^{M \times N \times NP}$ is a random tensor in $[0, 1]$ to select genes to mutate and $tGRnd \in \mathbb{R}^{M \times N \times NP}$ is a random Gaussian tensor to control the degree of Gaussian distribution. The ub and lb are the upper bound and lower bound for the search space of a given problem. The global mutation is implemented as

$$\mathcal{P} = \mathcal{T}_{gm} + \mathcal{P}. \tag{6}$$

3.2 CUDA-Based ETE Implementation

ETE is highly parallelizable due to its multidimensional data-parallel structure, which motivates us to develop a CUDA-based ETE algorithm (Cu-ETE). Cu-ETE encapsulates the major tensorial evolutionary operators into different kernel functions, and these kernel functions are communicated via global memory. Figure 2 illustrates the flowchart of Cu-ETE. It's worth highlighting that both ETE and Cu-ETE are highly versatile tensorial evolutionary optimizers, which may conflict with achieving maximum efficiency in GPU implementation. This is the first implementation of Cu-ETE. For this reason, certain aspects such as the maximization of the utilization of memory or device related to GPU optimization, are not considered in Cu-ETE. It is anticipated to further improve the implementation efficiency of Cu-ETE for practical applications by incorporating these aspects.

Pre-load and Memory Allocation. The task-oriented data of the problem and the required parameters of the algorithm frequently accessed by kernel functions is pre-loaded onto the device for better caching. We need to allocate memory space on the device to store the population tensors, fitness vector, and different evolutionary tensors. To avoid frequent memory transfer, the population

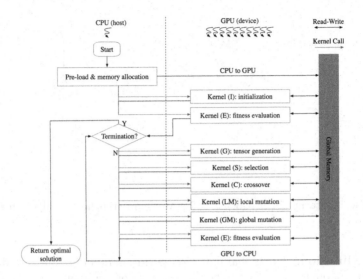

Fig. 2. The flow chart of Cu-ETE.

always resides in the global memory of the GPU and is transferred to the CPU only at the end of Cu-ETE.

Kernel Functions. After pre-loading, Cu-ETE implements ETE iterations by launching several kernel functions successively, as shown in Fig. 2. Specifically, the kernels (I) and (E) are firstly invoked in sequence by the host. Then, the host iteratively invokes a sequence of kernels ((G), (S), (C), (LM), (GM), (E)) to evolve population until the stopping condition is satisfied.

Kernel function (I) initializes $\mathcal{P}$ containing NP randomly generated and constrained individuals with NP parallel threads. The $\mathcal{P}$ is written into global memory since the size of block memory and the number of threads in a block may not be enough to handle the population. This kernel produces random numbers using a GPU-based random number generator in the random number generation library cuRAND of CUDA [1]. In kernel (I), we set the number of threads within a block to 32 to better utilize the concurrency of GPU, and the total number of threads equals NP.

Kernel function (E) evaluates the fitness of individuals in global memory. In this kernel, each thread handles one individual in the population. Specifically, an individual in the population is loaded from global memory to shared memory for low-latency access. Then, the fitness of the corresponding individual is calculated by a single thread. Finally, the obtained fitness is written into F in global memory accordingly. To avoid bank conflicts, we use zero padding that pads zeros to shared memory for extra space of an individual within a bank, ensuring there is only one individual in each bank. In kernel (E), the number of threads in a block is set to 32 to avoid losing the concurrency of the GPU. We do not use a large setting to enable each thread in a block to have sufficient shared memory. The parallel fitness evaluation can bring the prominent speed up in Cu-ETE since it is the most computationally expensive component in EC.

Kernel function (G) generates T_s and T_{gm}, then stores them into global memory. We need NP parallel threads are used to generate the two tensors with each thread handling one frontal slice. Each frontal slice of T_s and T_{gm} is associated to one individual in $\mathcal{P}$. In this kernel, both the uniform and Gaussian random number generators from cuRAND are used to generate random numbers.

Kernel function (S) selects individuals in $\mathcal{P}$. Each individual and its fitness value are directly associated with a thread. The selection is implemented by NP parallel threads based on T_s generated in kernel (G). Both T_s and $\mathcal{P}$ are loaded to shared memory for fast access, and then each selected individual along with its fitness is written into global memory accordingly.

Kernel function (C) performs crossover on $\mathcal{P}$. The most time-consuming step in the crossover is the t-product operation of T_c and $\mathcal{P}$. The t-product operation can be broken into massive independent multiplication and addition calculations between elements in T_c and $\mathcal{P}$. Therefore, we distribute the t-product operation into multiple parallel threads with each thread handling several elements in $\mathcal{P}$. If the problem scale is relatively small, we can set each thread to handle one element of $\mathcal{P}$. For example, $4,000$ threads are needed in total when a P_c of size $4 \times 5 \times 200$ is given. However, such element-wise parallelism might not be satisfied for higher-dimensional problems since the number of threads on a GPU is restricted to guarantee sufficient shared memory within each thread block. In this work, we set the total number of threads to NP with each thread handling $M \times N$ elements to ensure the scalability of Cu-ETE. Since each thread is directly associated with one individual, only the top NP_c threads would write the results into $\mathcal{P}$ in global memory accordingly for the next evolution. In kernel (C), the number of threads within a block is set to 32 for better utilization of the concurrency of GPU.

Kernel function (LM) performs local mutation on $\mathcal{P}$ with NP parallel threads. Since the local mutation requires to pad $\mathcal{P}$, each thread would first pad the boundaries of each frontal slice of $\mathcal{P}$ and stored it into $\mathcal{P}_{lm}$ accordingly in global memory. All the $T_{lm1,lm2,lm3}$ and $\mathcal{P}_{lm}$ are loaded from global memory to shared memory. Then, the t-product and addition calculations between $T_{lm1,lm2,lm3}$ and $\mathcal{P}_{lm}$ are distributed to NP parallel threads with each thread handling several elements of $\mathcal{P}_{lm}$. Finally, only top NP_m threads would write the results into global memory.

Kernel function (GM) performs global mutation on $\mathcal{P}$ with NP parallel threads. Specifically, for each thread, the two corresponding frontal slices in T_{gm} and $\mathcal{P}$ are loaded from global memory to shared memory for fast access. The two frontal slices are added together, and the result is written into global memory accordingly. Notably if more threads are available, we can use $M \times N \times NP$ parallel threads with each thread handling one element of $\mathcal{P}$. However, to ensure the scalability of Cu-ETE to high-dimensional problems, the number of threads in a block is set to 32 similar to other kernels.

4 Experiment

In this section, we compare the performance of TE (CPU), ETE (CPU), and Cu-ETE (GPU) with respect to the execution time and solution quality on two classical spatial optimization problems.

4.1 Experimental Setup

All the comparative algorithms are implemented in Python whose ecosystem gives strong support for tensor and tensor computations. Furthermore, we deploy our Cu-ETE on GPU with pyCUDA [8] that enables powerful Python access to NVIDIA's CUDA parallel computing. The experiment is run on Windows 10 with Python 3.7 and CUDA toolkit 11.4. We use an Intel(R) Core(TM) i5-7400 CPU with 3 GHz and 8G memory, and an NVIDIA GeForce GT 710 GPU supporting compute capability 3.5. The parameters are recommended as [11].

4.2 Case Study 1: Facility Layout Problem

The Facility Layout Problem (FLP) requires uniquely assigning a specific number of facilities to a 2D plane so that the assignment cost is minimized [3]. The fitness function is defined as

$$f_1(X) = \sum_i \sum_j FI_{ij} C_{ij} D_{ij} \tag{7}$$

where FI_{ij} is the flow intensity between facility i and j, C_{ij} is the assignment cost, and D_{ij} is the rectangular distance between facility i and j.

Figure 3 shows the fitness values along the optimization. The quality of solutions obtained by TE and ETE is very close, while a better solution is produced by Cu-ETE. Different random states across threads bring better randomness during iterations, which is an auxiliary for Cu-ETE to explore the search space to some extent. Figure 4 shows the comparison of TE, ETE, and Cu-ETE in terms of execution time. Firstly, we fix the generation to 1000, while the population size varies from 100 to 1000. The result is shown in the left figure. Less time is consumed by Cu-ETE than both the TE and ETE implemented on the CPU, and the execution efficiency of Cu-ETE is more obvious as the population size increases. Secondly, we fix the population size to 200, and the number of generations varies from 100 to 5000. The result is shown in the right of Fig. 4. The execution time needed for each generation of Cu-ETE is significantly less than that of TE and ETE. As the generation goes on, the execution times for all TE, ETE, and Cu-ETE experience growth, while the degree of growth in

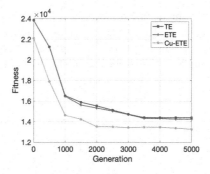

Fig. 3. Fitness of TE, ETE, and Cu-ETE with different generations.

TE and ETE is more significant. The speedup values of ETE over TE, Cu-ETE over ETE, and Cu-ETE over TE is varying from 1.2–1.7, 18.2–26.5, 34.3–48.5, respectively. The most time-consuming operator in EC is the fitness evaluation. Only the Cu-ETE is able to calculate the fitness of each individual in parallel, leading to a significant time saving compared to TE and ETE. The experimental results imply the potential computational efficiency of Cu-ETE to solve more complex problems where a larger population size and more fitness evaluations are required.

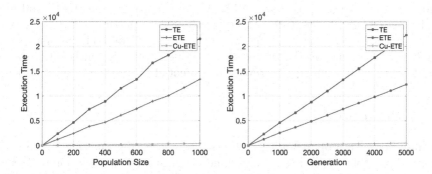

Fig. 4. Execution times of TE, ETE, and Cu-ETE with different population sizes and different generations.

4.3   Case Study 2: Image Steganography Problem

The Image Steganography Problem (ISP) intends to find an optimal 2D permutation of image blocks to minimize the image distortion after block-based image mapping between two images [6]. The cost is evaluated by the peak-signal-to-noise ratio as

$$f_2(X) = \sum_i \sum_j 10 \log_{10}(255^2 \times \frac{TP}{||p_{ij}^{(1)} - p_{ij}^{(2)}||^2}) \tag{8}$$

where $p_{ij}^{(1)}$ is the pixels of the original image, $p_{ij}^{(2)}$ denotes the pixels of the image obtained after mapping, and TP is the total number of pixels. The problem dimension of ISP has varying sizes, depending on the selected size of image blocks. The problem dimension is generally set from $4 \times 4D$ to $128 \times 128D$.

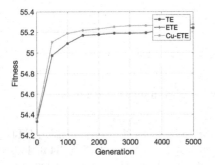

Fig. 5. Fitness on ISP.

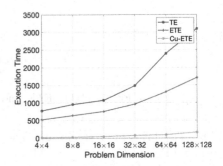

Fig. 6. Execution time on ISP.

The convergence curve on ISP is shown in Fig. 5, and the execution times are shown in Fig. 6. The Cu-ETE consumes much less execution time than TE and ETE. The speedup ratios of ETE over TE, Cu-ETE over ETE, and Cu-ETE over TE are 1.3–1.8, 19.1–24.3, and 30.5–46.4, respectively. As the problem dimension increases, the speedup of Cu-ETE over TE or ETE becomes more significant than that on the ISP of smaller scales, which is consistent with the conclusion on FLP and further reveal the potential of Cu-ETE.

5   Conclusion

In this paper, we proposed a high-performance tensorial evolutionary algorithm to speed up problem-solving for high-dimensional spatial optimization problems. Firstly, a general ETE algorithm was presented that manipulates all the evolutionary operations within a unified tensor computation model. Then, we implemented a Cu-ETE algorithm based on ETE using CUDA that distributed the tensorial evolutionary computations into parallel GPU threads, which further greatly reduced the execution time of the tensorial evolutionary process. Overall, the high-performance tensorial evolutionary computation showed promising performance for improving the efficiency and scalability of TE algorithms for solving spatial optimization problems. It's worth noting that TE and Cu-ETE are both general evolutionary optimizers capable of solving various spatial optimization problems when a proper fitness function is given. In this view, different variants of TE and Cu-ETE are expected to be developed to solve more practical applications. Moving forward, ETE and Cu-ETE have the potential to advance the field of spatial optimization and the efficiency of evolutionary computation.

Acknowledgements. This work was supported in part by the National Natural Science Foundation of China under Grant 62276100, in part by the Guangdong Natural Science Funds for Distinguished Young Scholars under Grant 2022B1515020049, in part by the Guangdong Regional Joint Funds for Basic and Applied Research under Grant 2021B1515120078, and in part by the TCL Young Scholars Program.

References

1. Toolkit documentation NVIDIA CUDA. Nvidia (2014)
2. Cheng, J.R., Gen, M.: Accelerating genetic algorithms with GPU computing: a selective overview. Comput. Ind. Eng. **128**, 514–525 (2019)
3. Drira, A., Pierreval, H., Hajri-Gabouj, S.: Facility layout problems: a survey. Annu. Rev. Control. **31**(2), 255–267 (2007)
4. Gong, Y.J., et al.: Distributed evolutionary algorithms and their models: a survey of the state-of-the-art. Appl. Soft Comput. **34**, 286–300 (2015)
5. Huang, Y., Feng, L., Qin, A.K., Chen, M., Tan, K.C.: Toward large-scale evolutionary multitasking: a GPU-based paradigm. IEEE Trans. Evol. Comput. **26**(3), 585–598 (2021)
6. Hussain, M., Wahab, A.W.A., Idris, Y.I.B., Ho, A.T., Jung, K.H.: Image steganography in spatial domain: a survey. Sign. Process. Image Commun. **65**, 46–66 (2018)
7. Itskov, M.: Tensor Algebra and Tensor Analysis for Engineers. ME, Springer, Cham (2015). https://doi.org/10.1007/978-3-319-16342-0
8. Klöckner, A., Pinto, N., Lee, Y., Catanzaro, B., Ivanov, P., Fasih, A.: PyCUDA and PyOpenCL: a scripting-based approach to GPU run-time code generation. Parallel Comput. **38**(3), 157–174 (2012)
9. Kłosko, J., Benecki, M., Wcisło, G., Dajda, J., Turek, W.: High performance evolutionary computation with tensor-based acceleration. In: Proceedings of the Genetic and Evolutionary Computation Conference, pp. 805–813 (2022)
10. Krömer, P., Platoš, J., Snášel, V.: Nature-inspired meta-heuristics on modern GPUs: state of the art and brief survey of selected algorithms. Int. J. Parallel Prog. **42**, 681–709 (2014)
11. Lei, S.C., Xiao, X., Gong, Y.J., Li, Y., Zhang, J.: Tensorial evolutionary computation for spatial optimization problems. IEEE Trans. Artif. Intell. (2022)
12. Wong, T.H., Qin, A.K., Wang, S., Shi, Y.: cuSaDE: a CUDA-based parallel self-adaptive differential evolution algorithm. In: Handa, H., Ishibuchi, H., Ong, Y.-S., Tan, K.-C. (eds.) Proceedings of the 18th Asia Pacific Symposium on Intelligent and Evolutionary Systems - Volume 2. PALO, vol. 2, pp. 375–388. Springer, Cham (2015). https://doi.org/10.1007/978-3-319-13356-0_30
13. Zhan, Z.H., et al.: Matrix-based evolutionary computation. IEEE Trans. Emerg. Top. Comput. Intell. **6**(2), 315–328 (2021)

Towards Better Evaluations of Class Activation Mapping and Interpretability of CNNs

Xinru Xiao, Yi Shi[✉], and Jiesheng Chen

Xi'an Jiaotong University, Xi'an 710049, China
shiyi@mail.xjtu.edu.cn

Abstract. As deep learning has been widely used in real life, there is an increasing demand for its transparency and its interpretability has received much attention from all walks of life. Current efforts in this field includes post-hoc visualization technique and intrinsically interpretable framework. However, there are still shortcomings in both of these techniques. In the post-hoc visualization techniques, the metric evaluating CAM method suffers from the ambiguity of evaluation object. We proposed a pair of quantitative evaluation metrics based on threshold cropping. The explanation maps were obtained by threshold cropping in order to minimize the variation of input images, thus making the evaluation object more focused on the CAM method itself. Experimental results show that these metrics can evaluate the accuracy and intensity of various CAM methods comprehensively, and get the conclusion that the gradient-based CAM methods are more stable, and Score-CAM is susceptible to model. Meanwhile, most of the existing intrinsically interpretable frameworks tend to enhance the interpretability of models by mapping their intermediate results to concepts that can be understood by humans. However, the intermediate results of the model often contain a variety of information that hinders its correspondence with single concept, which is manifested as a many-to-many relationship between filters and classes in convolutional classification networks. To address this situation, we proposed an interpretable training framework based on mutual information neural maximization to alleviate filter-class entanglement. MIS metric, classification confusion matrix and adversarial attack experiments all confirmed the validity of this method.

Keyword: Interpretability · Mutual Information Maximization · Class Activation Mapping · Deep Learning

1 Introduction

As deep learning has been successfully used in a range of tasks, especially in image recognition, the number of studies on deep learning interpretability is increasing year by year, because of its good contribution to improve models. Secondly, there is an increasing demand for model transparency in daily life [1]. For example, in the field of automatic driving to clarify the party responsible for a vehicle accident [2] and in the medical field for regulatory purposes. The deep learning interpretability will provide a solid foundation for the safe application of AI on a large scale. Currently, the interpretability methods

© The Author(s), under exclusive license to Springer Nature Singapore Pte Ltd. 2024
B. Luo et al. (Eds.): ICONIP 2023, CCIS 1961, pp. 352–369, 2024.
https://doi.org/10.1007/978-981-99-8126-7_28

of deep learning can be divided into post-hoc visualization technique and intrinsically interpretable framework. In post-hoc technique, perturbation analysis and class activation mapping (CAM) [3] are used to generate saliency maps to reflect the influence of different pixels in an image on inferred results. There are many varieties of CAM methods, such as Grad-CAM [4], XGrad-CAM [5], Grad-CAM++ [6], Smooth Grad-CAM++ [7], Score-CAM [8], etc. There is an urgent need for a quantitative metric to evaluate different CAM methods. However, the existing metrics often evaluate both the CAM methods and the trained deep vision models themselves, such as confidence drop metric [9–12], which obtains the explanation map in a way that obscures the object of evaluation of the confidence drop metric and thus cannot objectively and accurately assess the merits of CAM methods. Secondly, in the existing works on intrinsically interpretable framework, in order to enhance the explicable ability of the model, it is necessary to correspond the intermediate results (e.g., feature maps) of the model to concepts that can be understood by humans [13]. However, the intermediate results of the model often contain multiple information at the same time, and the response of the filter to semantic concepts may not be unique. The many-to-many relationship between filters and classes makes the decision-making process of the model difficult to understand.

Aiming at the above shortcomings, this paper proposed the following improvement methods:

- 1) To fill up the deficiencies in the evaluation metrics of CAM, this paper proposed a pair of evaluation metrics of CAM based on threshold cropping: the Average Cropping Confidence Drop (ACCD) and the Average Cropping Ratio (ACR), which introduce as little gradient information as possible that does not exist in the original input image, so that the evaluation metrics focus more on the CAM method itself. At the same time, the ratio of explanation map to original input image is also included in the evaluation of the CAM methods. Experiment results showed that the above metrics can effectively evaluate the advantages and disadvantages of the four kinds of CAM methods, and that the gradient-based CAM methods are more stable, and the Score-CAM method is susceptible to model.
- 2) The fact lies in the many-to-many characteristic between filters and classes. We use mutual information to measure the degree of dependence between filters and different classes. An interpretable method of filters based on the maximization of mutual information (MIM) is proposed to correspond each filter to the information of a single class. The structure and hyper-parameters selection of mutual information estimator based on neural network were analyzed experimentally, followed by the feasibility of mutual information maximization. The results showed that the proposed method can transform the many-to-many relationship between filters and classes into many-to-one relationship without compromising the model performance.

The rest of this paper is organized as follows. Section 2 provides an overview of the related work. Section 3 presents two metrics for evaluating CAM and introduces the filter interpretable method based on mutual information maximization. The experimental results and conclusion will be presented in Sect. 4 and Sect. 5 respectively.

2 Related Work

For the definition of deep learning interpretability, Doshi-Veleze et al. [14] defined machine learning interpretability as the ability to express comprehensible terms to humans. In fact, the interpretability of deep learning means that people can obtain extra information from the model except the predicted results, so that they can infer the thought and process of the predicting. Using the time of interpretation, the interpretability research can be divided into two types: post-hoc visualization technique and intrinsically interpretable framework.

Post-hoc methods focus on the means of interpretable analysis of the completed training model. Ribeiro et al. The method based on CAM [3] is to locate the salient region of a class on the input image by weighting and summing of the feature maps of the last convolutional layer, but CAM requires inserting GAP (Global Average Pooling) into the model and retraining the model, which makes the method less applicable. Selvaraju et al. [4] improved CAM and proposed Grad-CAM, a gradient information-based class-activated map, which obtained the gradient of the score for a class with respect to feature map activations of a convolutional layer, computing the weight by backpropagating, overstepped the structural obstacles and greatly expanded the applicability of the CAM methods. Chattopadhay et al. [6] proposed Grad-CAM++, which focuses on considering high degree information to compute the weight. Other similar methods include XGrad-CAM [5], Smooth Grad-CAM++ [7] et al. However, the gradient information in neural networks may be noisy, which leads to a larger amount of noise in the explanation map. Wang et al. [8] proposed a confidence-based method, Score-CAM, which gets rid of the reliance on gradient information, by taking the activation map of filter as a mask and dot the input image, using its prediction probability to take a weighted sum over the activation map. It differs from RISE only in the way it gets the mask. This method can also be classified as the method of perturbing the input detection model, so it also inherits the disadvantage of such methods (i.e., instability).

The intrinsically interpretable framework refers to mining the model for its own interpretability prior to training. By limiting the complexity of the model or adding additional training objectives, it provides an inherent decision logic and makes the decision criteria have a simple correspondence with a human-understandable concept. Chen et al. [15] proposed prototypical part network, and the network dissects the image by finding prototypical parts, and combines evidence from the prototypes to make a final classification, where the feature expression is obtained from the prototype extraction module and then the similarity score is calculated with the fixed conceptual prototype features. This score is multiplied by the weight of the full connection layer to get the classification results. Wickramanayake et al. [16] proposed training an understandable convolutional neural network through concept guidance. It is optimized by using the cosine similarity of convolutional features to adjective-noun pairs as part of the loss, such that the feature expressions correspond to specific concepts. The quantity and quality of adjective-noun pairs have a crucial effect on the performance of the model. Liang et al. [17] proposed class-specific gate matrices to guide the training of convolutional neural networks, in an attempt to differentiate the many-to-many correspondence between filters and classes. However, the differentiation effect of filters will be affected by the class-specific gate matrix, and the class-specific gate path tend to converge prematurely during the training process, making the filters unable to differentiate further. Literature [6] takes Average Drop (lower is better) and Increase in Confidence (higher is better) as quantitative evaluation metrics of CAM. The difference between the confidence of the explanation map and that of the original image is calculated and averaged over the entire dataset. Poppi et al. [18] proposed a quantitative metric ADCC to evaluate the performance of different CAM methods, which calculates coherency, complexity, and confidence drop for each input image separately, and compute the harmonic mean of them to get the ADCC score of an image, to average over the entire dataset. Literature [3] proposed an evaluation metric for weak supervised object positioning, i.e., the class activation map obtained by CAM method is binarized by setting a certain threshold. Then the largest connected domain is framed by the smallest external rectangle. If the intersection ratio between the rectangular frame and the real rectangular frame is greater than a certain threshold, the positioning is considered successful.

3 Methods

3.1 Average Cropping Confidence Drop (ACCD)

Chattopadhay et al. [6] proposed "the Average drop%" metric to evaluate the accuracy of the CAM method. The smaller the indicator, the better the CAM, which is calculated by the following formula:

$$Confidence\,Drop = \frac{max(0, o_c - e_c)}{o_c} * 100, \qquad (1)$$

where, the original image is judged to be class c with the confidence of o_c, explanation map is e_c. The method of explanation map is obtained as follows:

$$E_c = Up(CAM_c) \odot I. \qquad (2)$$

$Up(\cdot)$ represents up-sampling, which is generally bilinear interpolation. The operation symbol $\odot$ represents the multiplication of the corresponding elements of two matrices, and I is the original input image. The Average drop% suffers from the problem of ambiguity in the evaluation object. Since the explanation map looks "dark" compared to the original image, as shown in Fig. 1, the model will be less able to generalize well to images. Therefore, the greater drop in average confidence was not necessarily caused by the inaccurate identification position of the CAM method, but also by the fact that the model did not learn the features well enough to allow us to control variables to compare the advantages and disadvantages of various CAM methods. As shown in Fig. 2, the three explanation maps from left to right were obtained by Grad-CAM, Grad-CAM++ and Score-CAM using Eq. (2), and their confidence drops were 0.52, 0.61 and 0.22, respectively. However, the difference between the three explanation maps is not actually large, so it is further proved that the confidence drop score is influenced by both the deep vision model and the CAM method.

In this paper, an average cropping confidence drop metric is proposed. The basic calculation formula is still the same as Eq. (1), and the difference lies in the way of obtaining the explanation map. Specifically, the up-sampled class activation map S_{ij}^c

(a) Input image   (b) CAM   (c) Explanation map

Fig. 1. Examples of input image, class activation map and "dark" explanation map.

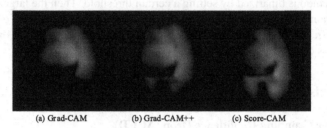

(a) Grad-CAM   (b) Grad-CAM++   (c) Score-CAM

Fig. 2. Different explanation maps from three CAM methods.

is binarized with a fixed threshold, to get the resulting feature map $M_c(x)$, which is then dot multiplied with the original input image to obtain the explanation map E_c. The calculation formula is as follows:

$$M_c(x) = \left\{ m_{ij} | m_{ij} = \begin{cases} 1 \; if \; S_{ij}^c \geq threshold \\ 0 \; if \; S_{ij}^c < threshold \end{cases} \right\}, \tag{3}$$

$$E_c = M_c(x) \odot I, \tag{4}$$

here, m_{ij} is the pixel on class activation map. This method evaluates the quality of the CAM method based on the salient region. The explanation map obtained by threshold cropping do not look darker, and are fed into the model as part of the original input image for prediction, thus introducing as little noise gradient information as possible. The evaluation of this paper is more concerned with the CAM method itself. Generally speaking, the lower the attention of the area identified by the CAM, the smaller the contribution made by this area in predicting the class, and vice versa. The explanation maps obtained by different CAM methods using Eq. (3) and (4) under the same cropping threshold can be considered as the areas that different CAM methods should focus on under the same contribution. The smaller the confidence drop, the more accurately the CAM method identifies the area of interest, while the method has stronger target localization ability.

Unfortunately, the above methods still have defects. One of them is that the evaluation effect relies heavily on the threshold. If most pixels of the original input image are higher than the threshold, the explanation map is basically the same as the original one, there is no sense of computing this metric. Thus, an average cropping ratio is proposed to solve this problem.

3.2 Average Cropping Ratio (ACR)

We defined the Cropping Ratio of a single explanation map by calculating the ratio of $M_c(x)$ to the total number of pixels, indicating the proportion of the pixel information contained in the explanation map to the original input image, as shown in Eq. (5):

$$Cropping\ Ratio = \frac{\sum_i \sum_j m_{ij}}{i \times j}. \tag{5}$$

When different CAM methods have the same cropping threshold and similar average cropping confidence drop, the smaller the ACR, the fewer pixels identified by this method, and the higher its relative contribution in confidence, i.e., the CAM identifies more accurate areas of interest. In Sect. 3, ACR and ACCD will be used to evaluate and analyze the CAM methods of Grad-CAM, Grad-CAM++, XGrad-CAM and Score-CAM through experiments.

3.3 Interpretable Framework Based on Mutual Information Maximization (MIM)

Studies in [13, 19] have shown that filters of different layers in convolutional neural networks have the ability to extract different semantic concepts from the input images. It is worth noting, however, that the response of a filter to a semantic concept may not be unique. Olah et al. [12] pointed out the filter that responds to multiple semantic concepts simultaneously exists. As shown in Fig. 3, when an image of class "mountain" is input into the convolutional neural network, multiple filters response to it. The coupling of filters with multiple semantic concepts, the many-to-many characteristic between filters and classes mentioned above, greatly limits one's ability to reason neural networks and makes the decision-making process of neural networks difficult to understand.

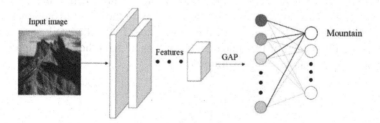

Fig. 3. Multi-semantic response properties of filters.

The proposed interpretable training framework based on mutual information maximization can transform the many-to-many characteristic of filters to many-to-one. In the training process, the mutual information estimator constructed by neural network is used to estimate the mutual information between the activation value of feature and the class label, and the mutual information is maximized by back-propagation of neural network. The random variable X consisting of a one-dimensional feature of GAP (to be the same as the dimension of class label) and the random variable Y consisting of a binarized class tag (set to 1 if it belongs to a certain class and set to 0 if it does not) are used as the object of mutual information maximization.

Based on the mutual information estimator proposed by Belghazi et al. [20], we build the interpretable framework. The mutual information is calculated by parameterizing the family of functions F with the neural network parameter θ ($\theta \in \Theta$). The neural network for estimating the mutual information between two random variables is obtained by continuously updating the neural network through negative gradients (gradient ascent) and then finding a tighter lower bound on the mutual information. The specific calculation process is as follows:

The mutual information of two continuous random variables X and Y is calculated as shown in Eq. (6):

$$I(X; Y) = \iint p_{X,Y}(x, y) \log \frac{p_{X,Y(x,y)}}{p_X(x) \cdot p_Y(y)} dx dy, \tag{6}$$

the calculation of mutual information can be further written as relative entropy (KL divergence):

$$I(X, Y) = D_{KL}(P_{XY} \| P_X \otimes P_Y), \tag{7}$$

the relative entropy can be reformulated into the Donsker-Varadhan [21] (DV) form:

$$D_{KL}(U \| V) = \sup_{T:\Omega \to R} \mathbb{E}_U[T] - \log(\mathbb{E}_V[e^T]), \tag{8}$$

let F be any class of functions $T : \Omega \to R$ that T satisfies the integrability constraints of the theorem. The lower bound on the mutual information is as shown in Eq. (9):

$$D_{KL}(U \| V) \geq \sup_{T \in F} \mathbb{E}_U[T] - \log(\mathbb{E}_V[e^T]). \tag{9}$$

The proposed network architecture based on mutual information maximization is shown in Fig. 4. The activation value of the k-th filter in the last convolutional layer of the n-th input image is denoted as a_k^n, the value after GAP is denoted as G_k^n, and the corresponding class label of the image is denoted as L^n, and the class of interest is c. The formula for calculating L^n is as follows:

$$L^n = \begin{cases} 1 \ if \ L^n = c \\ 0 \ if \ L^n \neq c \end{cases}, \tag{10}$$

the superscript $r.v.$ represents a random variable corresponding to a distribution. Each neural network mutual information estimator can only estimate the mutual information between $G_k^{r.v.}$ and $L_c^{r.v.}$, and its mutual information is written as $MI_{k,c}$. n estimators are required to maximize the mutual information for n filters. We use cross entropy loss L_{CE} as the classification loss and combine it with $MI_{k,c}$ to get the total loss. When mutual information is maximized for a single filter k and a single class c, the total loss is calculated as shown in Eq. (11):

$$L_{total} = L_{CE} - mi_{rate} * MI_{k,c},$$

where, mi_{rate} is the weight in mutual information maximization. Since the gradient descent is used to maximize the mutual information, a minus sign needs to be added in front of the mutual information. It can be further extended to multi-filters and a single-class c for mutual information maximization, and the total loss is calculated as shown in Eq. (12):

$$L_{total} = L_{CE} - mi_{rate} * \sum_{k \in K_c} MI_{k,c},$$ (12)

K_c represents the set of filters that maximize mutual information with the target class c. Similarly, when mutual information is maximized for multi-filters and multiple object classes (C), the total loss is:

$$L_{total} = L_{CE} - mi_{rate} * \sum_{c \in C} \sum_{k \in K_c} MI_{k,c}.$$ (13)

See Sect. 4 for relevant experimental procedures and results.

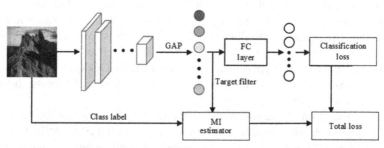

Fig. 4. Network architecture for mutual information maximization. Using MI estimator to maximize the mutual information for target filter k and class c, the total loss is calculated as shown in Eq. (11).

4   Experiment

We performed an extensive set of experiments to assess the effectiveness of our CAM evaluation metric in several CAM methods, data sources, and model architectures, in order to compare the advantages and disadvantages of different CAM methods. In addition, the structure and hyper-parameters of the mutual information estimator are compared and analyzed by several experiments, and the many-to-one and many-to-many (i.e., the relationship between filter to class) mutual information maximization experiments are carried out respectively.

4.1   Metrics Validity Verification Experiment Details

Trained and validated three visual deep learning model (ReNnet18, ResNet34 and ResNet50 [22]) on Cifar10 and BCCD[1] (Blood Cell Count and Detection) respectively.

---

[1] https://www.kaggle.com/datasets/paultimothymooney/blood-cells, the BCCD dataset of 9957 images, by the 410 original images enhancement, depending on the white cell types are

The experimental comparison and analysis of Grad-CAM, Grad-CAM++, XGrad-CAM and Score-CAM were conducted on three well-trained ResNet models, using the evaluation metrics of CAM proposed in Sect. 3.1, 3.2. We trained all sets of weights using stochastic gradient descent with momentum. The input images were resized to 224 × 224 to meet the requirements of the ResNet model, then the image normalization was added, but the data augmentation was not. The same operation was performed on dataset BCCD. The cropping threshold was specified manually. In order to comprehensively compare the ability of CAM methods to extract the salient regions, 9 cropping thresholds starting from 0.1, with intervals of 0.1 and ending at 0.9 were set, and ACCD and ACR of different CAM methods with different thresholds were calculated.

4.2 Results of Metrics Validity Verification Experiment

It can be clearly seen from Fig. 5 that four CAM methods' ACCD and ACR results vary with different threshold. It can be seen from the experimental results that compared to both GradCAM and XGradCAM, GradCAM++ has a lower ACCD under different cropping thresholds, yet it also has slightly higher ACR of its explanation map. This indicates that GradCAM++ identifies larger areas of interest with the same cropping threshold, and the more marked areas of interest indeed reduce the value of ACCD, which proves that the extra areas of interest identified by GradCAM++ are effective and accurate. Although this effect may fluctuate across models and datasets, the overall performance is consistent.

It can be seen from the results based on the ResNet18 model that the Score-CAM method is able to obtain a smaller ACCD while still ensuring a low ACR. It indicates that the gradient information-based CAM method identifies a possible redundancy in the area of interest of the model, which side-steps the argument of the literature [8] that there is noise in the gradient information of the model.

From the results based on ResNet34 model, it can be seen that although the Score-CAM method has a low ACR, its ACCD value is very high. The stability of ACCD curve indicates that the significance area marked by Score-CAM is actually inconsistent with the area more concerned by the model. Therefore, the ACCD of Score-CAM method is generally high, and the results on ResNet50 are roughly similar. In summary, Score-CAM weights the activation maps in a way that generally keeps the ACR low, but sometimes at the cost of ignoring some information that is important to the model. In principle, the Softmax operation will stretch the difference between each input element, which may cause the activation information of some filters underestimated. In addition, the generating mode of the explanation map in the Score-CAM method is derived from the original CAM method, which also introduces the problem of ambiguity in the evaluation objects (see Sect. 3.1 for details).

Then, in order to further analyze the Grad-CAM and XGrad-CAM where the curves nearly overlap, the cropping confidence drop (CCD, i.e., about per image not average) difference and cropping ratio (CR, i.e., about per image not average) difference of the two CAM explanation maps were calculated respectively under different cropping thresholds.

---

divided into four types of images. They are Eosinophil, Lymphocyte, Monocyte and Neutrophil respectively.

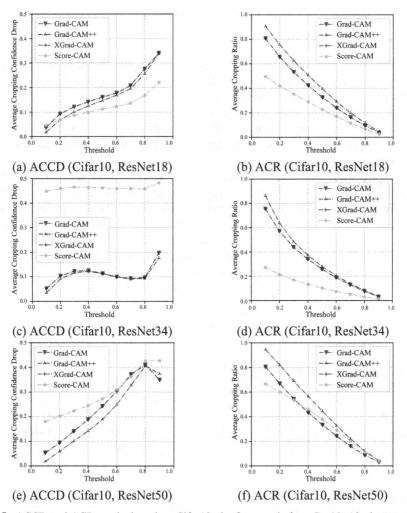

(a) ACCD (Cifar10, ResNet18)

(b) ACR (Cifar10, ResNet18)

(c) ACCD (Cifar10, ResNet34)

(d) ACR (Cifar10, ResNet34)

(e) ACCD (Cifar10, ResNet50)

(f) ACR (Cifar10, ResNet50)

Fig. 5. ACCD and ACR results based on Cifar10, the first row is from ResNet18, the second is ResNet34, the third is ResNet50.

In Table 1 and Table 2, it can be seen that CCD and CR obtained by Grad-CAM are the same as the ones obtained by XGrad-CAM in most cases. It is found that the labels (indicating which input image) set of the explanation maps whose CR difference is equal to 0 are the proper subset of the labels set of the explanation maps whose CCD difference is equal to 0, i.e., the explanation maps with the same CR must have the same CCD. This indicates that with the interval of cropping threshold set to 0.1, the focus area and the intensity of focus labeled by Grad-CAM and XGrad-CAM are basically identical. Even if there are differences in the intensity of focus between XGrad-CAM and Grad-CAM, the difference does not exceed 0.1 and the overall shape of the class activation map is similar. The experimental results on the BCCD dataset are similar to the pattern of Cifar10 above, i.e., Grad-CAM++ still identifies the model's focus area more accurately in general. In contrast, the Score-CAM method although having the minimum ACR, the ACCD on the BCCD dataset is mostly inferior to the CAM methods based on gradient information.

Overall, the CAM method based on gradient information performs more consistently. However, there may be an increase in ACR due to noise contained in the gradient.

Among the gradient-based methods, the model areas of interest identified by Grad-CAM++ generally have better accuracy, and the Grad-CAM and XGrad-CAM perform similarly and slightly worse than Grad-Cam++. Although Score-CAM performs well on ACR, its performance on ACCD is inconsistent, indicating that the method is more susceptible to the model.

It is worth stating that it is not recommended to harmonize average ACCD and ACR into one indicator, as their different range of variation and meaning make it difficult to determine the respective weights of these two indicators when computing harmonic mean. Of course, it is possible to assign artificially larger weight to the more important indicator, but such subjective assignment can make the analysis of experimental results difficult and also cause the experimental conclusions to be strongly subjective. Therefore, in this paper, instead of harmonized averaging the proposed indicators, we choose to evaluate different CAM methods by means of joint indicators.

Table 1. The comparison of CCD by two methods for training ResNet18 on Cifar10 with several thresholds.

CCD (Grad-CAM)-CCD (XGrad-CAM)	Cropping Threshold								
	0.1	0.2	0.3	0.4	0.5	0.6	0.7	0.8	0.9
>0/count	8	9	7	10	16	16	21	29	16
=0/count	49989	49982	49986	49968	49963	49964	49965	49944	49960
<0/count	3	9	7	22	21	20	14	27	24

Table 2. The comparison of CR by two methods for training ResNet18 on Cifar10 with several thresholds.

CR (Grad-CAM)-CR (XGrad-CAM)	Cropping Threshold								
	0.1	0.2	0.3	0.4	0.5	0.6	0.7	0.8	0.9
>0/count	9	11	9	15	21	19	11	28	25
=0/count	49985	49981	49986	49965	49962	49963	49965	49944	49960
<0/count	3	9	7	22	21	20	14	27	24

4.3 Constructing the Mutual Information Estimator

Structure. $G_0^{r,v.}$ and $L_7^{r,v.}$ were used for experimental data X and Y (see Sect. 3.3), and ResNet20 was used for data generation model. We first set the estimator with the structure of activation followed by addition, as shown in Fig. 6(a). It is found that the all the outputs of the models converge to 0, i.e., the estimator considers X and Y to be independent and the mutual information to be 0, which is contrary to the fact. Then the experiment is repeated using different activation functions, and the mutual information results are all 0. This shows that it prevents the model correctly estimating the mutual information between random variables correctly when the linear layer is added by the activation function. Therefore, it is reasonable to assume that the first activation function in the model must be set after X and Y "sees or corresponds" each other, i.e., the activation function is set after the addition of the linear layers, as shown in Fig. 6(b). The results showed that the outputs of each model (i.e., mutual information estimation value) converge to similar locations and only fluctuate within a certain upper and lower range. It can be inferred that the nonlinear transformation of the linear layer containing X or Y information alone will change the information it originally carries, and the information that X and Y can "sees" each other will be changed, resulting in the failure of the neural network to estimate mutual information. However, it is feasible to carry out the nonlinear transformation of the linear layer containing both X and Y information.

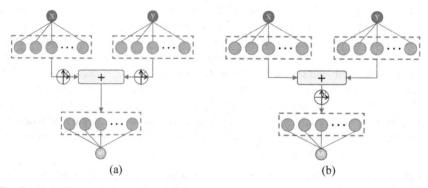

(a)   (b)

Fig. 6. (a) Activation followed by addition model. (b) Addition followed by activation model.

Number of Hidden Layers and Hidden Layer Units. We tested multiple number of hidden layers, and found that the increase in the number of hidden layers, led to downward spiky fluctuations in the output (mutual information estimation)-iteration curve during the training process, as shown in Fig. 7, which required dozens of iterations to converge back to the original position. In contrast, the model structure with only one linear layer before activation function rises smoothly to convergence with different numbers of hidden layer units, and does not produce large downward spiky fluctuations after convergence. At the same time, the more units in the hidden layer, the faster the model convergence. In order to balance the number of model parameters and the convergence speed, the number of linear layers before the activation function is set as 1, the number of neurons in each linear layer is 64, and the activation function is set as Leaky ReLU (to avoid neuron death). The learning rate is 0.1 with decay rate of 0.1, the batch size is 128, and the epoch is 128.

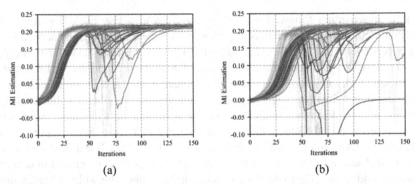

Fig. 7. (a) Convergence of the models that have two linear layers before activation function (the number of hidden layer units in each layer is 64). (b) Convergence of models that have three linear layers before activation function (the number of hidden layer units in each layer is 32).

4.4 Training Strategy

Estimator Training. In the training strategy based on generative adversarial network [23], the training of the estimator can be divided into two stages. In the first stage, the neural MI estimator between $G_k^{r.v.}$ and $L_c^{r.v.}$ is trained by using gradient descending, which is denoted as $MINE_{k,c}$. In the second stage, the parameters of the mutual information estimator are fixed and the main network parameters are updated according to the total loss. After many iterations, the distribution of $G_k^{r.v.}$ and $L_c^{r.v.}$ may change, and the estimated value of $MINE_{k,c}$ may not be close enough to the true value. So, after every few iterations, it is necessary to update $MINE_{k,c}$ from the second stage back to the first stage to ensure that the gradient information returned to the convolutional neural network through the mutual information estimator can indeed improve the mutual information between $G_k^{r.v.}$ and $L_c^{r.v.}$. And so forth until the network converges.

Complete Training. Firstly, ResNet20 model was trained on Cifar10 using the method described in [17] (CSG) until convergence. Then, the proposed interpretable framework was used for further training of the above model. The calculation formula of loss was shown in Eq. (12), (13). The same training strategy, hyper-parameters and optimizer were used in the two training stages. We train the model using Stochastic Gradient Descent (SGD) with mini-batching and the momentum updating rule. Mini-batches of size 128 are used with a fixed momentum of 0.9, and initial learning rate is 0.1 with decay rate 0.1, mi_{rate} is 0.05.

4.5 Results of Mutual Information Maximization

Firstly, the mutual information maximization experiment of single-class to multi-filters was carried out. Accuracy and MIS [17] metrics were used to measure recognition performance and correspondence between filters and classes. The formula of MIS are as follows:

$$MIS = \frac{1}{|K_c|} \sum_{k \in K_c} MI_{k,MMI_k}, \tag{14}$$

$$MMI_k = \max_{c \in C} MI_{k,c}, \tag{15}$$

$$K_c = \{k | MMI_k = c, k \in K\}, \tag{16}$$

where MMI_k represents class, $MMI_k = c$ means the k-th filter's GAP information is most closely associated with identifying class-c. K_c is a set. Higher MIS indicates higher class-specificity and lower filter-class entanglement. Let $MMI_k = 0$ (class "0" represents airplane), as seen in Table 3, our framework had higher MIS value and the same Accuracy comparing with CSG (Class-Specific Gate) model. we also visualized MI matrix which demonstrates that our framework (CSG+MIM) yields a sparse matrix where each filter is only related to one or few classes. There is a clearer comparison between our framework and CSG model about class-0 showed by Fig. 8. The filters in our model responds more to information about class-0 and less to information about other classes.

Table 3. Comparison of model performance (Accuracy and MIS) before and after mutual information maximization.

Model	Accuracy	MIS
CSG	0.9175	0.1655
Single-class to multi-filters MIM	0.9197	0.1757
Multi-group single-class to multi-filters MIM	0.9181	0.1856
Multi-group single-class to multi-filters MIM from scratch	0.9193	0.1858

The activation values of filter-8 (after GAP) were counted and their frequency distribution histograms were shown in Fig. 9. Where, the green represents the features related to class-0, while the red represents the features related to other class. Before the information maximization, the overlap area between the red and green regions is larger. After the mutual information maximization, the green part is shifted to the right and

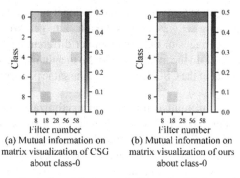

(a) Mutual information on matrix visualization of CSG about class-0

(b) Mutual information on matrix visualization of ours about class-0

Fig. 8. Comparison about class-0 before and after mutual information maximization.

the overlap area is reduced, which makes the filter and some single class have stronger correlation.

To further examine the relationship between the filters and the classes after maximizing the mutual information, we masked the filters that maximize the mutual information with class-0. The output showed that CNN model fails to predict "airplane" correctly, which indicates that these filters do respond to information related to the class-0 and are indispensable in discriminating this class. The classification confusion matrices are shown in the Fig. 10.

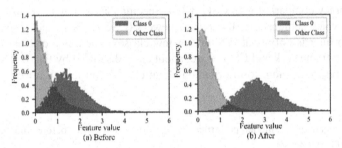

Fig. 9. Frequency distribution histogram of the activation value of the filter-8 (after GAP) before and after mutual information maximization.

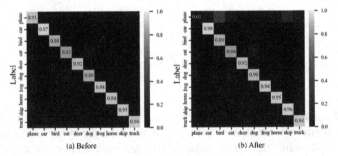

Fig. 10. Comparison of model recognition confusion matrices before and after masking filters highly related to the class-0.

Meanwhile, we conducted adversarial attack experiments on models before and after MIM. Cifar10 training set was used as the attack target. We recorded the number of successfully attacked images of the model under different perturbation ranges. The results are shown in Fig. 11. In the model with MIM, the number of successfully attacked images is smaller than that of the CSG model under the same perturbation range, as shown in Table 4 for details.

In addition, multi-group mutual information maximizations were carried out for single-class to multi-filters, (i.e., 10 classes together, Eq. (13)). As shown in Table 3, the MIS value is higher than the MIS that only maximizes single class. This shows that the mutual information maximization interpretable framework can not only be applied to the local maximization of single-class to multi-filters, but also to the global maximization of each filter and its respective classes.

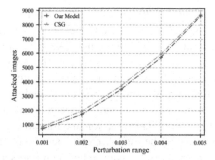

Fig. 11. Comparison of model performance under adversarial attack.

Table 4. Comparison of the number of successfully attacked images in different perturbations range.

Perturbations range	Number of images attacked	
	CSG	Our Model
0.001	859	701
0.002	1941	1714
0.003	3716	3469
0.004	5949	5717
0.005	8757	8640

Finally, we used MIM interpretable framework combined with CSG to train the model from scratch. The results are shown in Table 3, and our approach can be used not only to improve the post-hoc interpretability of the models, but also to play a supervisory role in the process of model training.

To sum up, after the original network is trained with only the accuracy metric, various information may be scattered among different filters. However, by means of MIM supervision, similar information can be integrated into some filters without compromising the model performance.

5 Conclusion

In this work, two metrics are proposed to evaluate the class activation mapping methods: the average cropping confidence drop and the average cropping ratio, in order to ameliorate the ambiguity problem of the existing evaluation metrics. When getting an explanation map by CAM, the threshold cropping binarization is used instead of the direct point-wise multiplication to minimize the introduction of noise gradient information in the original image, so that the confidence drop evaluation metric is set to pay more attention to CAM methods and excludes other factors such as model generalization ability. Meanwhile, the average cropping ratio of class-discriminative saliency map under a fixed threshold is used as a joint metric, to evaluate at different granularity whether the activation area is consistent with the area the model actually focuses on. Experimental results show that the gradient-based CAM methods, such as Grad-CAM, XGrad-CAM and Grad-CAM++, have good stability and accuracy, but the area of interest are generally large. However, the area identified by confidence-based Score-CAM is generally small, but its accuracy can fluctuate greatly among different models.

Aiming at the many-to-many characteristic between the filters and classes, an interpretable framework based on mutual information maximization is proposed in this paper to enhance the interpretability of the model. Experiments show that the proposed approach can further transform the correspondence between filters and classes to many-to-one without compromising the model performance. Moreover, the above framework is easy to migrate, it can maximize the mutual information between any two objects with correlation relationship. Future work involves designing more metrics for model interpretability evaluation, and interpretability of deep learning in specialized domain scenarios (e.g., medical imaging domain), while how to combine interpretability technologies with multimodal learning will be one of the hot research topics in the future.

References

1. Preece, A., Harborne, D., Braines, D., Tomsett, R., Chakraborty, S.: Stakeholders in explainable AI. In: Proceedings of the AAAI Fall Symposium on Artificial Intelligence in Government and Public Sector, Arlington, Virginia, USA, pp. 1–6 (2018)
2. Cysneiros, L.M., Raffi, M., Do Prado Leite, J.C.S.: Software transparency as a key requirement for self-driving cars. In: 2018 IEEE 26th International Requirements Engineering Conference (RE), Banff, Alberta, Canada, pp. 382–387 (2018)
3. Zhou, B., Khosla, A., Lapedriza, A., Oliva, A., Torralba, A.: Learning deep features for discriminative localization. In: Proceedings of the IEEE Conference on Computer Vision and Pattern Recognition, Las Vegas, NV, USA, pp. 2921–2929 (2016)
4. Selvaraju, RR., Cogswell, M., Das, A., Vedantam, R., Parikh, D., Batra, D.: Grad-CAM: visual explanations from deep networks via gradient-based localization. In: Proceedings of the IEEE International Conference on Computer Vision, Venice, Italy, vol. 128, pp. 336–359 (2017)
5. Fu, R., Hu, Q., Dong, X., Guo, Y., Gao, Y., Li, B.: Axiom-based Grad-CAM: towards accurate visualization and explanation of CNNs. In: 31st British Machine Vision Conference, Online, pp. 1–13 (2020)
6. Chattopadhay, A., Sarkar, A., Howlader, P., Balasubramanian, VN.: Grad-CAM++: generalized gradient-based visual explanations for deep convolutional network. In: 2018 IEEE Winter Conference on Applications of Computer Vision (WACV), Lake Tahoe, NV, USA, pp. 839–847 (2018)
7. Omeiza, D., Speakman, S., Cintas, C., Weldermariam, K.: Smooth Grad-CAM++: an enhanced inference level visualization technique for deep convolutional neural network models. arXiv:1908.01224 (2019)
8. Wang, H., et al.: Score-CAM: score-weighted visual explanations for convolutional neural networks. In: Proceedings of the IEEE/CVF Conference on Computer Vision and Pattern Recognition Workshops, Seattle, WA, USA, pp. 111–119 (2020)
9. Wang, H., Naidu, R., Michael, J.: SS-CAM: smoothed Score-CAM for sharper visual feature localization. arXiv:2006.14255 (2020)

10. Jung, H., Oh, Y.: Towards better explanations of class activation mapping. In: Proceedings of the IEEE/CVF International Conference on Computer Vision, Montreal, BC, Canada, pp. 1336–1344 (2021)
11. Krizhevsky, A., Hinton, G.: Learning multiple layers of features from tiny images (2009). https://www.cs.toronto.edu/~kriz/learning-features-2009-TR.pdf
12. Olah, C., Cammarata, N., Schubert, L., Goh, G., Petrov, M., Carter, S.: Zoom in: an introduction to circuits. Distill **5**(3), e24–001 (2020). https://doi.org/10.23915/distill.000 24.001
13. Bau, D., Zhou, B., Khosla, A., Oliva, A., Torralba, A.: Network dissection: quantifying interpretability of deep visual representation. In: Proceedings of the IEEE Conference on Computer Vision and Pattern Recognition, Honolulu, HI, USA, pp. 6541–6549 (2017)
14. Doshi-Velez, F., Kim, B.: Towards a rigorous science of interpretable machine learning. arXiv: 1702.08608 (2017)
15. Chen, C.F., Li, O., Tao, C.F., Barnett, A.J., Su, J., Rudin, C.: This looks like that: deep learning for interpretable image recognition. In: Advances in Neural Information Processing Systems, Vancouver, BC, Canada, vol. 32, pp. 8928–8939 (2019)
16. Wickramanayake, S., Hsu, W., Lee, M.L.: Comprehensible convolutional neural networks via guided concept learning. In: 2021 International Joint Conference on Neural Networks (IJCNN), Shenzhen, China, pp. 1–8 (2021)
17. Liang, H., et al.: Training interpretable convolutional neural networks by differentiating class-specific filters. In: Vedaldi, A., Bischof, H., Brox, T., Frahm, J.M. (eds.) ECCV 2020. LNCS, vol. 12347, pp. 622–638. Springer, Cham (2020). https://doi.org/10.1007/978-3-030-58536-5_37
18. Poppi, S., Cornia, M., Baraldi, L., Cucchiara, R.: Revisiting the evaluation of class activation mapping for explainability: a novel metric and experimental analysis. In: Proceedings of the IEEE/CVF Conference on Computer Vision and Pattern Recognition, Online, pp. 2299–2304. IEEE/CVF, Electrical Network (2021)
19. Zeiler, M.D., Fergus, R.: Visualizing and understanding convolutional networks. In: Fleet, D., Pajdla, T., Schiele, B., Tuytelaars, T. (eds.) ECCV 2014. LNCS, vol. 8689, pp. 818–833. Springer, Cham (2014). https://doi.org/10.1007/978-3-319-10590-1_53
20. Belghazi, M.I., et al.: Mutual information neural estimation. In: International Conference on Machine Learning, Stockholm, Sweden, vol. 80, pp. 531–540 (2018)
21. Donsker, M.D., Varadhan, S.S.: Asymptotic evaluation of certain Markov process expectations for large time. IV. Commun. Pure Appl. Math. **36**(2), 183–212 (1983)
22. He, K., Zhang, X., Ren, S., Sun, J.: Deep residual learning for image recognition. In: 2016 IEEE Conference on Computer Vision and Pattern Recognition (CVPR), Las Vegas, NV, USA, pp. 770–778 (2016)
23. Goodfellow, I.J., et al. Generative adversarial nets. In: Proceedings of the 27th International Conference on Neural Information Processing Systems, vol. 2, pp. 2672–2680. MIT Press, Cambridge (2014)

Contrastive Learning-Based Music Recommendation Model

Minghua Nuo[1,2,3]($\boxtimes$), Xuanhe Han[1], and Yuan Zhang[1]

[1] College of Computer Science, Inner Mongolia University, Hohhot 010021, Inner Mongolia, China
nuominghua@163.com
[2] National and Local Joint Engineering Research Center of Intelligent Information Processing Technology for Mongolian, Hohhot 010021, Inner Mongolia, China
[3] Inner Mongolia Key Laboratory of Mongolian Information Processing Technology, Hohhot 010021, Inner Mongolia, China

Abstract. In the rapidly evolving era of digital multimedia, the overwhelming rate of music publication poses a challenge for users seeking efficient access to their preferred songs. Music recommendation systems aim to address this issue but still encounter problems such as overfitting, the cold start problem for new users, and result bias. To tackle these challenges, we propose an optimized music recommendation model called Contrastive Learning for Music Recommendation (CLMR), leveraging contrastive learning techniques. CLMR leverages the bipartite graph information between users and songs and introduces a contrastive learning framework to enhance the representation of sparse data, thereby improving recommendation accuracy and mitigating data sparsity issues. To combat sampling bias, a comparative learning approach is employed within CLMR, utilizing Gaussian noise to construct more effective positive samples. This method enhances the model's learning capability and robustness in challenging environments. Experimental comparisons with traditional recommendation models based on content filtering, collaborative filtering, and supervised learning demonstrate that the proposed CLMR model outperforms them, achieving superior performance in terms of NDCG and Recall metrics.

Keywords: Recommender System · Contrastive Learning · Graph Convolutional Neural Network

1 Introduction

Digital music, propelled by the Internet, has revolutionized listening habits [1]. The profusion of online music resources requires expedited discovery of personally preferred songs, reducing search time. Music recommendation services act as a vital link between users and music resources.

Deep learning technology has showcased remarkable achievements in speech recognition, computer vision, and machine translation [2]. Its application in recommender systems presents new opportunities [3]. Through deep neural networks, intricate user

© The Author(s), under exclusive license to Springer Nature Singapore Pte Ltd. 2024
B. Luo et al. (Eds.): ICONIP 2023, CCIS 1961, pp. 370–382, 2024.
https://doi.org/10.1007/978-981-99-8126-7_29

feedback patterns can be extracted, enabling the derivation of preferences from raw data and generating abstract high-level feature representations [4]. This facilitates efficient identification of similar data, resulting in accurate reflections of user preferences and personalized content recommendations [5]. Deep learning enhances recommendation effectiveness and information acquisition efficiency across various domains [6].

Music recommendation algorithms face challenges due to expanding databases and diverse user needs [3]. Overcoming the cold start problem and updating existing algorithms are urgent requirements [7]. Additionally, scarce feedback data for less popular items and non-uniform item representation necessitate the development of novel algorithms [8]. These advancements aim to better understand user-item relationships and improve recommendation system performance. This study delves into audio features, item characteristics, and user preferences, employing contrastive learning and optimizing music data feature extraction and user interest models. Novel ideas and directions for future research are explored [9].

In this context, we propose CLMR, a Contrastive Learning model for Music Recommendation, addressing data sparsity in collaborative filtering-based music recommendation. To address sampling bias in the contrastive learning auxiliary task, we introduce a comparative learning method within CLMR. By incorporating Gaussian noise, we enhance feature representation, facilitating the construction of more effective and smoother positive samples. This adjustment improves representation distribution uniformity, enhancing the model's learning capacity in challenging environments while ensuring overall effectiveness and robustness.

2 Related Works

2.1 Music Recommender System

The field of music recommendation has gained significant attention and become an important branch of multimedia retrieval, driven by the increasing dependence and demand for recommendation systems [10]. Deep learning has been extensively applied in various recommendation scenarios. Wang et al. developed a music recommendation model combining a deep belief network with a probabilistic graph structure, incorporating collaborative filtering interaction matrix and audio content information [11]. Lee et al. introduced a deep music content-user feature vector model that combines user feature-based collaborative filtering and audio feature-based content filtering for hybrid recommendation [12]. Margon et al. incorporated arousal, valence, and depth factors into a collaborative filtering model to enhance music recommendation with content awareness and alleviate the cold start problem [13].

To address the challenges in music recommendation, it is crucial for researchers to delve deeper into these issues and propose innovative methods and algorithms. Specifically, explicit solutions are needed to tackle the cold start problem, mitigate data sparsity, and minimize the impact of bias. Advancements in these areas will not only contribute to the field of music recommendation but also enhance the overall user experience by providing more relevant and diverse recommendations.

2.2 Contrastive Learning

Contrastive Learning (CL) utilizes data augmentation to bring similar instances closer together in representation space while pushing dissimilar instances further apart, effectively leveraging the data itself for model training supervision [14]. CL serves as a method to address the issue of data sparsity in recommendation systems and can enhance the recommendation performance for long-tail items [15]. Some studies employ data augmentation techniques to enhance model recommendation performance by masking highly correlated features [16]. Additionally, some researchers reshape the distribution of sequence representation through regularization comparisons, alleviating representation degradation and enhancing recommendation performance to some extent [17]. For example, Huang et al. apply two different sets of dropout masks on a transformer-based backbone model to retain comprehensive semantic information [18]. Moreover, in the context of structure-level comparison recommendation, Zhang et al. model intra-session and inter-session structural information using a hypergraph encoder and apply CL to the session recommendation task to distinguish between positive and negative sample pairs [19]. These studies offer novel ideas and methodologies for combining recommender systems with CL.

3 Contrastive Learning Model for Music Recommendation Literature Review

3.1 Supervised Music Recommendation Model

Construct the Collaborative Bipartite Graph. Firstly, in the LGMR model of this paper, the collaborative bipartite graph is used to represent the interaction information between each user-item. User $\mathcal{U}$ and project I a node set respectively, and define the observed interaction information for $O^+ = \{y_{ui} | u \in \mathcal{U}, i \in I\}$, including y_{ui} indicates whether the user u had interaction with the project i. During this process, if user u has listened to song i before, y_{ui} is set to 1; If user u has not heard song i before, then y_{ui} is set to 0. Next, a bipartite graph $\mathcal{G} = (\mathcal{V}, \mathcal{E})$ is constructed, in which the node set $\mathcal{V} = \mathcal{U} \cup I$ contains all users and items, and the edge set $\mathcal{E} = O^+$ represents the observed interactions. The vertex set $\mathcal{V}$ can be divided into two independent subsets, and the two vertices connected by each edge belong to two non-adjacent subsets. Therefore, the user-project collaboration bipartite graph is constructed in this way, and the specific process is shown in Fig. 1.

LGMR Model Recommendation Process. In this paper, we use a lightweight and efficient GCN network (LightGCN) [20], which only contains the neighbor aggregation process, uses linear propagation to learn the potential features of user- item, and then generates the final node representation by iterative weighted summation layer by layer. The node aggregation diagram is shown below (Fig. 2).

The model propagates and aggregates the node information in the bipartite graph through lightweight convolutional layers and layer combinations, so as to better capture user preferences and song item information to generate a new node, which integrates neighbor information and is more helpful to obtain accurate recommendation results.

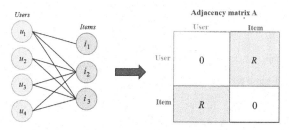

Fig. 1. Schematic diagram of building the collaborative bipartite graph

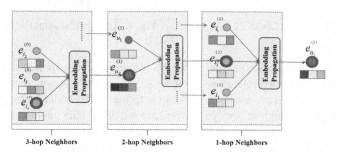

Fig. 2. Schematic diagram of the node propagation aggregation process

3.2 Unsupervised Contrastive Learning Music Recommendation Model

In this section, we tackle the issue of user interest neglect in collaborative filtering recommendation algorithms when faced with sparse data and encountering long-tail phenomena. To address this problem, we enhance the existing LGMR model by introducing Contrastive Learning as an auxiliary task for joint training. Consequently, we propose a novel recommendation model for Music Recommendation called CLMR, which effectively integrates user preferences into the recommendation process. By incorporating Contrastive Learning, CLMR aims to overcome the limitations of collaborative filtering algorithms and provide more accurate and personalized recommendations in the context of music recommendation.

CLMR Model Architecture. Contrastive learning task in CLMR model may face the problem of song information sampling bias caused by randomly extracting training data. To solve this problem, we add noise perturbation to the graph node embedding representation to enhance positive samples and ensure the uniformity of the representation space, so that the model can learn in difficult environments. In addition, a fully connected Projection Head layer is added after the encoder to enhance the robustness and generalization ability of the model, so as to improve the recommendation performance of the model. The CLMR model framework is shown in Fig. 3.

Graph Data Augmentation. The uniform representation of graph augmentation operators is given in Eq. (1):

$$Z_1{}^{(l)} = H\left(Z_1{}^{(l-1)}, s_1(\mathcal{G})\right), Z_2{}^{(l)} = H\left(Z_2{}^{(l-1)}, s_2(\mathcal{G})\right), s_1, s_2 \sim S \qquad (1)$$

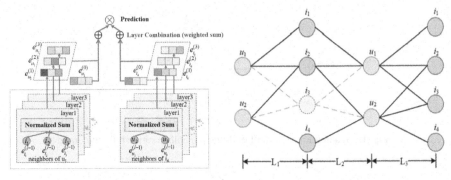

Fig. 3. The architecture of CLMR model   **Fig. 4.** Schematic diagram of Edge Dropout

In the above formulation, $s1$ and $s2$ are two random choices applied independently on the graph to build two related node views $Z1^{(l)}$ and $Z2^{(l)}$. Augmenting the data is a key step in contrastive learning, which can help the model better learn positive and negative samples. Therefore, this paper introduces a novel data augmentation method suitable for recommendation algorithms: Edge Dropout (ED).

This method randomly removes edges in the graph with probability to generate a new graph structure. The ED diagram is shown in Fig. 4 and this independent process can be expressed as Eq. (2):

$$s_1(\mathcal{G}) = (\mathcal{V}, M_1 \odot \mathcal{E}), s_2(\mathcal{G}) = (\mathcal{V}, M_2 \odot \mathcal{E}) \tag{2}$$

where are two masking vectors on the edge set, which are used to preserve edges that are partially connected between neighbors, thus forming two subgraphs. Coupling these two subgraphs together is able to effectively capture useful information in the local node structure and further enhance the robustness of the model through representation learning.

Noise Disturbance. This section mainly introduces how to add Gaussian noise to the multi-view node embedding in a targeted way, and use LightGCN for node construction and aggregation of perturbed subgraphs. Through the process, the model can more effectively aggregate information such as user preferences and item labels, and finally generate new nodes. The difficult nodes containing disturbance information help the model to find positive sample pairs, improve the benefit of contrastive learning, and further improve the accuracy of the recommendation algorithm. The above process is divided into two steps. In the first step, noise is added to the node embedding. In the second step, the LightGCN model is used to aggregate the neighborhood information to generate new perturbed nodes.

Noise Embedding. Since it is difficult and time-consuming to construct a graph structure for uniform representation, we shift our focus to the embedding space in this paper. Inspired by adding small perturbations to the input picture to build adversarial examples, we add random noise to the graph structure representation for effective augmentation. Formally, given node i and its embedded representation e_i on the two- dimensional spatial graph structure, the representation-level enhancement is implemented as follows,

$$e_i' = e_i + \Delta_i', e_i'' = e_i + \Delta_i'' \tag{3}$$

where the added noise vectors and the constraints that need to be satisfied are given in Eqs. (4) and (5),

$$||\Delta||_2 = \epsilon \tag{4}$$

$$\Delta = \tilde{\Delta} \odot sign(e_i), \tilde{\Delta} \in \mathbb{R}^d \sim U(0, 1) \tag{5}$$

From the above formula, we can see that the first constraint is used to control the size of the noise and is numerically equivalent to a point on the hypersphere with radius. The second constraint requires, and to be in the same vector Angle to ensure that large deviations do not occur after adding noise, while not reducing the number of effective positive samples.

Noisy Node Representation. Firstly, the user-song interaction matrix is denoted as, where and represent the number of users and the number of songs respectively. If the user has listened to the song, let be 1. If the user hasn't heard the song, it will be 0. The adjacency matrix of user-song items is given in Eq. (6),

$$A = \begin{pmatrix} 0 & R \\ R^T & 0 \end{pmatrix} \tag{6}$$

Secondly, we set the feature representation of the input layer (i.e., layer 0) to be a matrix (the size of feature representation is), and the matrix form of LightGCN graph convolution is shown in Eq. (7),

$$E^{(k+1)} = \left(D^{-\frac{1}{2}}AD^{-\frac{1}{2}}\right)E^{(k)} \tag{7}$$

where, is the diagonal matrix of, and the element represents the number of nonzero vectors in the row of the adjacency matrix A (that is, the degree matrix).

Finally, the feature representation matrix for model prediction is obtained, as shown in Eq. (8),

$$\begin{aligned} E &= \alpha_0 E^{(0)} + \alpha_1 E^{(1)} + \alpha_2 E^{(2)} + ... + \alpha_k E^{(k)} \\ &= \alpha_0 E^{(0)} + \alpha_1 \tilde{A} E^{(0)} + \alpha_2 \tilde{A}^2 E^{(0)} + ... + \alpha_k \tilde{A}^k E^{(0)} \end{aligned} \tag{8}$$

In the above equation, is the symmetric normalized matrix. In addition, because random noise of different scales is applied to the current node embedding at each layer, the final representation of the perturbed nodes is generated as shown in Eq. (9),

$$\begin{aligned} E' = \frac{1}{k}\left(\left(\tilde{A}E^{(0)} + \Delta^{(1)}\right) + \left(\tilde{A}\left(\tilde{A}E^{(0)} + \Delta^{(1)}\right) + \Delta^{(2)}\right)\right) \\ + ... + \left(\tilde{A}^K E^{(0)} + \tilde{A}^{K-1}\Delta^{(1)} + ... + \tilde{A}\Delta^{(K-1)} + \Delta^{(K)}\right) \end{aligned} \tag{9}$$

3.3 Contrastive Loss Function

In this section, the self-supervised contrastive learning auxiliary task and the main supervised recommendation task (LGMR) are jointly trained, and the optimal encoder and projection layer are learned by minimizing the joint loss to generate high-quality recommendation results. The loss function for each partial training objective is presented separately in the following.

Main Task Loss Function. CLMR takes LGMR recommendation model as the main supervision task, and the Pairwise Bayesian Personalized Ranking (BPR) loss function as the optimization objective of the main task. The purpose of this loss function is to select the music items that fit the user's preferences from their implicit feedback, and the calculation is given in Eq. (10),

$$L_{main} = \sum_{(u,i,j)\in O} -\log\left(\hat{y}_{ui} - \hat{y}_{uj}\right) \tag{10}$$

where, $O = \left\{(u,i,j)|(u,i) \in O^+, (u,j) \in O^-\right\}$ is the training dataset, O^+ represents the observed interactions and O^- epresents the unobserved interactions. By minimizing the contras loss function, LGMR can improve the prediction value of observed items and reduce the prediction value of unobserved items to achieve better recommendation effect.

The Contrastive Loss Function. After creating enhanced views, in this paper, the different perturbed views of the same node are regarded as positive sample pairs $\left(\left\{(z_u', z_u'')|u \in U\right\}\right)$, and the perturbed views of different nodes are regarded as negative sample pairs $\left(\left\{(z_u', z_v'')|u, v \in U, u \neq v\right\}\right)$. Next, the positive sample pairs are used to supervise the consistency between different views of the same node, while the negative sample pairs are used to enhance the difference between different nodes, so that the views obtained by the enhancement of different nodes are more different. Specifically, the contrastive learning task in this paper is optimized by maximizing the consistency of positive pairs and minimizing the consistency of negative pairs, which is formulated as shown in Eq. (11),

$$L_{cl}^{user} = \sum_{u\in U} - \log\frac{exp(s(z_u', z_u'')/\tau)}{\sum_{v\in U} exp(s(z_u', z_v'')/\tau)} \tag{11}$$

In the above formula, $s(\cdot)$ is used to measure the similarity between two vectors, and cosine similarity is used as the measurement method, and the hyperparameter τ represents the temperature coefficient in softmax. Similarly, the contrastive loss function L_{cl}^{item} of item can be obtained, and the total objective function of the contrastive learning task can be obtained by combining these two contrastive learning tasks as shown in (12),

$$L_{cl} = L_{cl}^{user} + L_{cl}^{item} \tag{12}$$

The Joint Loss Function. In order to improve the performance of the overall recommendation task, this paper takes the contrastive learning self-supervised task proposed in Sect. 3.3 as an auxiliary task, and uses multi-task training to jointly optimize the LGMR recommendation task in Sect. 3.2, so as to strengthen the feature representation in the main task and auxiliary task. The loss function for joint training is given in Eq. (13),

$$L = L_{main} + \lambda_1 L_{cl} + \lambda_2 ||\Theta||_2^2 \tag{13}$$

In the above equation, the parameter Θ is the parameter set of the main task model L_{main}. The contrastive learning loss L_{cl} does not introduce additional parameters; therefore, λ_1 and λ_2 are hyperparameters that control the contrastive loss and L_2 the regularization strength, respectively.

4 Experiments

4.1 Datasets

This paper conducts experiments to evaluate the effectiveness of the CLMR model on a public dataset: Last.FM. The datasets are detailed as follows: **Last.FM:** This dataset comes from the Last.FM online music system and takes music tracks played by users as items. To ensure data quality, data from January 2015 to June 2015 were selected and users and items with at least 10 interactions were selected as the filtering criteria. The detailed statistical information is shown in Table 1 below:

Table 1. Statistics for the datasets Last.FM

Datasets	Number of users	Number of items	Interactive information	Sparsity
Last.FM	23566	48123	3034796	99.72%

4.2 Training Settings

In the setting of experimental hyperparameters in this chapter, the hierarchical structure of LightGCN is set to 1, 2, and 3 to conduct experiments respectively, and the representation dimension of the model is set to 64. For comparison, all models were initialized using Xavier's method. Additionally, Adam optimizer was used to optimize the models with learning rate 0.001, batch size 1024, temperature parameter 0.2, λ_1 and ϵ in the range {0.005, 0.01, 0.05, 0.1, 0.5, 1.0, respectively. 2.0} and set the Projection Head layer number to conduct comparative experiments in the range of {0, 1, 2, 3}.

4.3 Baselines

In this paper, we choose four collaborative filtering methods as baselines:

NGCF [21]: This is a graph structure based collaborative filtering algorithm that largely follows the criteria of GCN and performs message passing on the graph structure to achieve recommendation.

LightGCN [22]: This method proposes a lightweight graph convolution for recommendation, which improves the recommendation efficiency and generation ability.

BPR-MF [23]: This method obtains latent vector representations of users and items by factorizing the interaction matrix between users and items. Then, the inner product of these vectors can be calculated to predict the user's rating or preference for the uninteracted item.

Mult-VAE [24]: This is a collaborative filtering method based on Variational Autoencoder (VAE). It is optimized by an additional reconstruction objective, which is also a kind of self-supervised learning task.

4.4 Comparative Experiments

Table 2 shows the comparison of experimental results between the CLMR model proposed in this paper and the baseline model under the Last.FM.

Table 2. Experimental results of the baseline system and the CLMR model

Datasets	Model	Recall@20	NDCG@20
Last.FM	NGCF	0.0616	0.0943
	LightGCN	0.0713	0.1142
	BPR-MF	0.0554	0.0885
	Mult-VAE	0.0728	0.1153
	CLMR	**0.0791**	**0.1212**

In Table 2, it can be seen that the recommendation effect of the proposed CLMR model is better than the baseline model on Last.FM. This is mainly caused by the sparsity of the data set, and the interaction supervision signal between the user and the song item in Last.FM is too sparse to guide the representation learning of the model, which also indicates that the CLMR model is better in the scene with sparse data.

The experimental results also show that compared with the recommendation algorithm BPR-MF based on matrix factorization alone, the way of modeling the interaction relationship by using graph neural network has obvious advantages. This shows that using graph-structured data modeling can better improve the recommendation effect.

In addition, the CLMR model performs better than the LightGCN supervised recommendation model to some extent. This indicates that the use of contrastive learning self-supervised task to supplement the main supervised task can better mine user interests. At the same time, the experimental results also show that enhancing the contrastive learning method by adding Gaussian noise and Projection Head layer can significantly improve the recommendation effect.

4.5 Ablation Experiments

To Explore the Influence of Introducing Different Contrast Learning Methods on the Experimental Results Under Different Number of Graph Convolution Layers.
Table 3 shows the comparison results of the three data-enhanced contrast learning modes (ND, ED, RW) introduced into the LGMR model and the LGMR model without contrast learning under the LightGCN network with different layers, as well as the CLMR model introduced in this paper.

Table 3. Performance comparison of different contrast learning methods on different layers of graph convolutional networks

Datasets		Last.FM	
Layer	Model	Recall@20	NDCG@20
	LGMR	0.0664	0.1043
	ND	0.0717	0.1102
1-layer	ED	0.0728	0.1137
	RW	0.0723	0.1131
	CLMR	0.0732	0.1142
	LGMR	0.0690	0.1074
	ND	0.0731	0.1126
2-layer	ED	0.0768	0.1183
	RW	0.0756	0.1178
	CLMR	0.0780	0.1201
	LGMR	0.0701	0.1103
	ND	0.0746	0.1135
3-layer	ED	0.0779	0.1194
	RW	0.0761	0.1181
	CLMR	0.0791	0.1212

Table 3 demonstrates the superiority of contrastive learning methods with various data augmentation techniques over the LGMR model without contrastive learning.

This highlights the effectiveness of leveraging self-supervised contrastive learning tasks to complement the supervised recommendation task.

Furthermore, the experimental results in Table 3 reveal that increasing the depth of the convolutional layers in the CLMR model from 1 to 3 improves its performance. This is attributed to the aggregation of more neighbor information, enriching node representation and facilitating the identification of highly similar nodes, thus enhancing the recommendation effectiveness.

Among the data augmentation methods, random edge dropping outperforms random node dropping and random walk-in terms of graph structure enhancement. This is because random edge dropping captures inherent patterns in the graph structure and retains more comprehensive interaction information. Although random node dropping yields slightly inferior results compared to random walk, it still maintains connections of low-degree nodes, which is crucial in sparser datasets.

Notably, the CLMR model achieves the best performance across the dataset, confirming the effectiveness of data augmentation using random noise in supporting contrastive learning tasks.

To Explore the Effect of Noise Size on the Experimental Results. The Fig. 5 show the comparison results of the impact of imposing different sizes of Gaussian noise ϵ on the model performance in the contras learning task under the Last.FM.

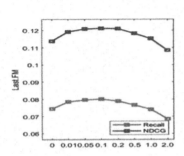

Fig. 5. Results of noise size under dataset Last.FM

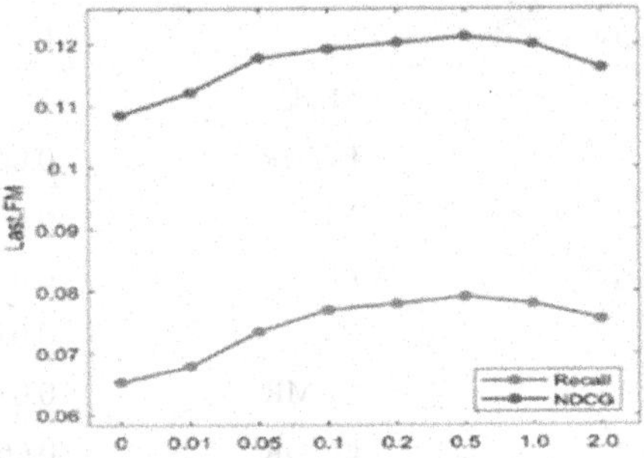

Fig. 6. Comparison results for the size of hyperparameter λ_1 under the Last.FM dataset

From the above results, we can see that adding Gaussian noise ϵ in the dataset can improve the effect of the CLMR model. But, when the noise ϵ is around 0.1, the model has the best performance. When the noise is too large, the model performance decreases instead. This is because moderate noise helps enhance the model's ability to find positive and negative sample pairs; however, excessive noise will make the generated positive and negative samples too difficult, thus interfering with the model learning and making the model unable to correctly distinguish nodes similar to the input node, resulting in the recommendation task being hindered.

To Explore the Influence of the Value of Hyperparameter λ_1 in the Joint Loss Function on the Experimental Results. The Fig. 6 show the comparative experimental results of CLMR model under the Last.FM, when the noise ϵ is set to 0.1, and the research hyperparameter λ_1 takes different values.

The experimental results in the above figure show that the performance of the CLMR model gradually improves as the value of λ_1 increases, and reaches the highest point when λ_1 is set to 0.5 on the Last.FM dataset, and then starts to decline. At the same time, it is able to adjust the values of ϵ and λ_1 in the same range, which indirectly indicates that by adjusting the value of λ_1, the noise ϵ is able to provide a more fine-grained regularized representation, thus improving the model performance.

5   Conclusion

In this paper, we propose a contrastive learning model for music recommendation that addresses the issue of data sparsity in collaborative filtering-based approaches. To improve the robustness and recommendation performance, we introduce disturbance nodes generated by adding random noise, which serve as data augmentation. Additionally, we incorporate a Projection Head layer after the encoder to retain more interactive information for downstream recommendation tasks. This enables a more accurate expression of user preferences and a comprehensive understanding of user preferences and song features. Experimental results confirm the feasibility and effectiveness of the CLMR model. Compared to other graph-based collaborative filtering recommendation algorithms, the CLMR model demonstrates superior accuracy in recommending songs of interest to users, ultimately enhancing the overall user experience.

Acknowledgments. This work was supported by the National Natural Science Foundation of China (No. 61966025, No. 62366038), and Natural Science Foundation of Inner Mongolia (No. 2023MS06010).

References

1. Menghani, G.: Efficient deep learning: a survey on making deep learning models smaller, faster, and better. ACM Comput. Surv. **55**(12), 1–37 (2023)
2. Naumov, M., Mudigere, D., Shi, H.J.M., et al.: Deep learning recommendation model for personalization and recommendation systems. arXiv preprint arXiv:1906.00091 (2019)
3. Fang, H., Zhang, D., Shu, Y., et al.: Deep learning for sequential recommendation: Algorithms, influential factors, and evaluations. ACM Trans. Inf. Syst. (TOIS) **39**(1), 1–42 (2020)
4. Hao, B., Yin, H., Zhang, J., et al.: A multi-strategy-based pre-training method for cold-start recommendation. ACM Trans. Inf. Syst. **41**(2), 1–24 (2023)
5. Ko, H., Lee, S., Park, Y., et al.: A survey of recommendation systems: recommendation models, techniques, and application fields. Electronics **11**(1), 1–41 (2022)
6. Chen, C., Zhang, M., Zhang, Y., et al.: Efficient neural matrix factorization without sampling for recommendation. ACM Trans. Inf. Syst. (TOIS) **38**(2), 1–28 (2020)
7. Huang, J., Zhao, W.X., Dou, H., et al.: Improving sequential recommendation with knowledge-enhanced memory networks. In: The 41st International ACM SIGIR Conference on Research & Development in Information Retrieval, pp. 505–514 (2018)
8. Wang, T.H., Hu, X., Jin, H., et al.: AutoRec: an automated recommender system. In: Proceedings of the 14th ACM Conference on Recommender Systems, pp. 582–584 (2020)
9. Angarita-Zapata, J.S., Masegosa, A.D., Triguero, I.: AutoEn: an AutoML method based on ensembles of predefined machine learning pipelines for supervised traffic forecasting. arXiv preprint arXiv:2303.10732 (2023)
10. Melchiorre, A.B., Zangerle, E., Schedl, M.: Personality bias of music recommendation algorithms. In: Proceedings of the 14th ACM Conference on Recommender Systems, pp. 533–538 (2020)
11. Wang, X., Wang, Y.: Improving content-based and hybrid music recommendation using deep learning. In: Proceedings of the 22nd ACM International Conference on Multimedia, pp. 627–636 (2014)

12. Lee, S., Jeong, H., Ko, H.: Classical music specific mood automatic recognition model proposal. Electronics **10**(20), 2489–2508 (2021)
13. Magron, P., Févotte, C.: Leveraging the structure of musical preference in content-aware music recommendation. In: ICASSP 2021–2021 IEEE International Conference on Acoustics, Speech and Signal Processing (ICASSP), pp. 581–585. IEEE (2021)
14. Yang, Y., Huang, C., Xia, L., et al.: Debiased contrastive learning for sequential recommendation. In: Proceedings of the ACM Web Conference, pp.1063–1073 (2023)
15. Tai, W., Lan, T., Wu, Z., Wang, P., Wang, Y., Zhou, F.: Improving session-based recommendation with contrastive learning. User Model. User-Adap. Interact. **33**(1), 1–42 (2023)
16. Yao, T., Yi, X., Cheng, D.Z., et al.: Self-supervised learning for large-scale item recommendations. In: Proceedings of the 30th ACM International Conference on Information & Knowledge Management, pp. 4321–4330 (2021)
17. Xia, X., Yin, H., Yu, J., et al.: Self-supervised hypergraph convolutional networks for session-based recommendation. In: Proceedings of the AAAI Conference on Artificial Intelligence, vol. 35, no. 5, pp. 4503–4511 (2021)
18. Qiu, R., Huang, Z., Yin, H., et al.: Contrastive learning for representation degeneration problem in sequential recommendation. In: Proceedings of the Fifteenth ACM International Conference on Web Search and Data Mining, pp. 813–823 (2022)
19. Zhang, J., Gao, M., Yu, J., et al.: Double-scale self-supervised hypergraph learning for group recommendation. In: Proceedings of the 30th ACM International Conference on Information & Knowledge Management, pp. 2557–2567 (2021)
20. Gao, C., Wang, X., He, X., et al.: Graph neural networks for recommender system. In: Proceedings of the Fifteenth ACM International Conference on Web Search and Data Mining, pp. 1623–1625 (2022)
21. Wang, X., He, X., Wang, M., et al.: Neural graph collaborative filtering. In: Proceedings of the 42nd International ACM SIGIR Conference on Research and Development in Information Retrieval, pp. 165–174 (2019)
22. He, X., Deng, K., Wang, X., et al.: LightGCN: simplifying and powering graph convolution network for recommendation. In: Proceedings of the 43rd International ACM SIGIR Conference on Research and Development in Information Retrieval, pp. 639–648 (2020)
23. He, X., He, Z, Du, X, et al.: Adversarial personalized ranking for recommendation. In: The 41st International ACM SIGIR Conference on Research & Development in Information Retrieval, pp. 355–364 (2018)
24. Liang, D., Krishnan, R.G., Hoffman, M.D, et al.: Variational autoencoders for collaborative filtering. In: Proceedings of the 2018 World Wide Web Conference, pp. 689–698 (2018)

A Memory Optimization Method for Distributed Training

Tiantian Lv, Lu Wu$^{(\boxtimes)}$, Zhigang Zhao, Chunxiao Wang, and Chuantao Li

Qilu University of Technology (Shandong Academy of Sciences), Shandong Computer Science Center (National Supercomputer Center in Jinan), Jinan, China
13808935325@163.com

Abstract. In recent years, with the continuous development of artificial intelligence technology, the complexity of deep learning algorithms and the scale of model training is also increasing. A series of efficient pipelined parallel training methods emerged to improve the training speed and accuracy. Distributed training becomes an effective way to train large-scale models. To solve this problem, we propose an efficient pipeline-parallel training optimization method. Our approach processes small batches of data in parallel through multiple compute nodes in a pipelined manner. We propose a prefix sum partition algorithm to realize a balanced partition and save the memory of computing resources. At the same time, we also design a clock optimization strategy to limit the number of weight version generations to ensure the model's accuracy. Compared with the current famous pipeline parallel frameworks, our method can achieve about 2 times training acceleration, save about 30% of memory consumption, and improve the model accuracy by about 10% compared with PipeDream.

Keywords: Deep learning · Parallel training · Balanced partition · Clock optimization

1 Introduction

Deep learning has seen significant growth in recent years under optimization methods that make it possible to train DNN networks in a scalable manner. With the increasing number of network layers and parameters of DNN, traditional single-card training can not meet people's computing needs. For example, GPT-2 [1]is a language model based on transformer [2], with 1.5 billion parameters. However, training such a large model is very expensive. Later, many methods were proposed to reduce the size of the model without losing its performance of the model, including conducting an architecture search [3], designing more efficient architecture [4], modifying the model [5,6], etc. The approach to training neural networks is sequential. Some researchers commonly use data parallelism [7] to train large-scale models. In this way, it is equivalent to several workers completing work together. This mode effectively reduces the size of data, thus

© The Author(s), under exclusive license to Springer Nature Singapore Pte Ltd. 2024
B. Luo et al. (Eds.): ICONIP 2023, CCIS 1961, pp. 383–395, 2024.
https://doi.org/10.1007/978-981-99-8126-7_30

speeding up the overall training process [8,9]. One disadvantage of parallel data training [10–12] is that the device that completes the work must wait for the subsequent devices to complete the work and then send all parameters to the parameter server for unified parameter update. So in the process of waiting, computing resources are wasted. When many parameters must be synchronized, it also increases the communication burden. Model parallelism [13–15] is also a popular way to train large models by treating successive layers of the model as a partition, each on a different device. Each device computes only a tiny portion of the model and updates only the parameters for that portion. However, the timeline shows that only one device works at any given time while the others are waiting. So model parallelism can not avoid the problem of computing resource waste. To alleviate this problem, pipelined model parallel (PMP) [16] has been proposed recently; that is, data is divided into small batches and submitted to GPU one by one for pipelined processing [15,17]. But it comes with several challenges. For example, how to ensure load balance in each phase to avoid memory insufficiency of computing resources? How to ensure the training accuracy of the model will not be reduced, and so on.

Given the above problems, combined with the resource environment of the supercomputer platform, we designed a more efficient pipelined parallel training method. The proposed method inherits the pipeline structure of GPipe [4] and PipeDream [18], but it provides a more novel prefix sum partitioning algorithm and clock optimization strategy to partition the model better and ensure the training efficiency and precision of the model. We used an ImageNet ILSVRC2012 and a custom data set to evaluate the improved pipeline framework through two currently complex DNN models, ResNet101 [19] and AmoebaNet [20]. Compared with several pipeline frameworks commonly used, our optimized framework effectively speeds up the convergence of loss function, reaching up to about twice the training acceleration, saving about 30% of the memory consumption, and improving the experimental accuracy by about 10% compared with PipeDream.

The rest of this article is organized as follows. Section 2 discusses the optimization and improvement of pipeline models carried out by previous researchers. Section 3 will discuss the innovative work done in the optimized model. First, we design a prefix sum partition algorithm that can better partition. Secondly, we create a clock optimization strategy to ensure the training accuracy of the model. We then demonstrate that the tuning components suggested in this article are critical to performance and evaluate the performance of all innovative efforts in Sect. 4. Section 5 is a summary of our work.

2   Related Work

In recent years, pipeline-parallel training techniques have been extensively studied and widely applied to improve the speed of network training [21–23]. Combining the advantages of model parallelism and pipelining, pipelining model parallelism (PMP) is proposed to effectively train large-scale models in a model parallel way [24,25]. According to the updating method of weights, the existing pipeline model parallel methods can be roughly divided into two categories: asynchronous pipeline model parallel (ASP) [26] and synchronous pipeline parallel (BSP) [27]. However, the convergence in training may be seriously affected due to inconsistent weight updating of the model. PipeDream [18] keeps a copy of the weights for small batches of each activity in the pipeline. However, the weight store will waste GPU memory, especially if the DNN model has many model parameters. The GPU memory footprint will be extensive. PipeDream, on the other hand, has an absolutization problem because it uses different weights throughout the forward and backward loops. The old issues slow the convergence rate and reduce the model's accuracy. PipeDream-2BW [16] reduces the number of stored versions to 2 but still has the problem of outdated parameters. Chen et al. [28] proposed SpecTrain to simultaneously alleviate the inconsistency and obsolescence in the parallelism of the asynchronous pipeline model. However, SpecTrain still does not fully resolve inconsistencies and obsolescence and often results in poor accuracy. PipeMare [29]tried to solve the convergence problem of PipeDream by introducing new approximation techniques to correct outdated updates. But PipeMare uses 33% more memory than GPipe, a concern for large-scale models. To solve the problem of low GPU utilization of the most original model parallel strategy and overcome the memory limitation of extended DNN, the researchers proposed GPipe. GPipe trains each set of micro-batches in a pipelined fashion, which to some extent, allows multiple GPUs to train simultaneously. Compared with the simple model parallel strategy, GPU utilization is significantly improved. However, since data from the same small batch needs to flow sequentially through all GPUs, GPipe can't always keep all GPUs busy with parallel training models, so it still has a resource wait problem.

We designed a more efficient pipelined parallel training method to solve the above problems. As long as the method we developed achieves the following three goals: First, to achieve balanced partitioning, make the training of the pipeline more smooth, and ensure that each worker does not have to wait for each other, we design a prefix sum partitioning algorithm to ensure that the training duration of model blocks allocated to each computing device is roughly equal, to avoid the problem of computing resource waiting. Secondly, to save memory resources and ensure the consistency of weight versions as much as possible, we designed a clock optimization strategy to limit the buffer size and provide the model's training accuracy. We will elaborate on our work in the next section.

3   Main Work

3.1   Prefix Sum Partition Algorithm

Prefix sum is a commonly used algorithm. Due to its high efficiency and practicability, the prefix sum algorithm is widely used in various scenarios [30,31] [15][16], especially in problems requiring efficient processing of large-scale data and frequent query of interval sum [4,24,26,32]. In this section, we design a more efficient prefix sum algorithm. Compared with the traditional single-card method of calculating prefix sum, we fully use the existing equipment to calculate prefix sum, speeding up the calculation process. We first designed an estimation phase, using existing models to train a custom data set, typically 20,000 images. Through this training process, the training time and occupied memory of each layer of the network in the model are predicted, and two prediction sequences are obtained, namely, the expected duration sequence L={L1, L2... .. Ln}, the amount of memory occupied M={M1, M2... .Mn}. In prefix sum algorithm, we mainly partition the estimated length sequence L. Let's say we have K GPUs, which means we have K partitions. First, the sequence L is evenly divided among K GPUs, and each GPU calculates the prefix sum of the subsequence. The last element of the first prefix sum array are passed to the second GPU so that each component in the second prefix sum array is added to the value sent by the first GPU. Then repeat the process. Finally, all the arrays are merged into one array, the prefix sum array of L, denoted L_prefix_sum. An example of this is shown in Fig. 1.

This is the process of using multiple GPUs to find prefix sum arrays. The next step is to divide the array into successive subarrays and make sure that the sum of each subarray is roughly equal.

Then we need to find the critical point for each partition in L. Starting at the prefix sum the first position of the array, walk backward while maintaining two pointers, one to the start position and one to the end position of the current group. Calculate the sum of the elements in the current group for each pointer. If it equals the sum of the elements in each group, then the current group is legal. Save it, and set the starting pointer to the position next to the ending pointer.

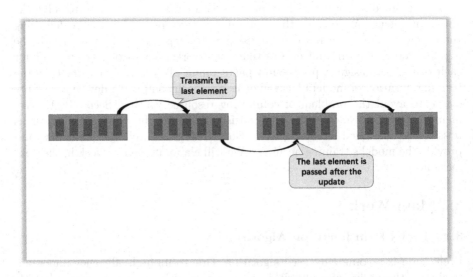

Fig. 1. The process of solving prefix sum array for K GPUs.

If the sum of the currently grouped elements is less than the sum of each set of elements, the end pointer is moved back one bit. If the sum of the elements in the current group is greater than the sum of the elements in each group, then the current group is invalid, and you need to start a new group with the start pointer set one position before the end pointer. Repeat steps 4 to 6 until the entire prefix sum array have been traversed. In the end, each group's resulting sum of elements is equal, and the consecutive elements are grouped together. The specific algorithm steps are shown as follows:

Algorithm 1: Finding partitioning critical point algorithm

```
1: def prefix_sum_grouping(nums, k):
2:    L_prefix_sum = [0] * (len(nums) + 1)
3:    for i in range(1, len(L_prefix_sum)):
4:       L_prefix_sum [i] = L_prefix_sum [i - 1] + nums[i - 1]
5:    total_sum = L_prefix_sum [-1]
6:    target_sum = total_sum // k
7:    result = []
8:    start, end = 0, 1
9:    while end < len(L_prefix_sum):
10:      cur_sum = L_prefix_sum [end] - L_prefix_sum [start]
11:      if cur_sum == target_sum:
12:         result.append(nums[start:end])
13:         start = end
14:         end += 1
15:      elif cur_sum < target_sum:
16:         end += 1
17:      else:
18:         start = end - 1
19:    result.append(nums[start:])
20: return result
```

Nums represent the original long sequence, and k means the number of groups to be divided into. The function returns the result after grouping; each grouping is a list. The time complexity of this algorithm is $O(n)$, and the algorithm is stable. The training duration of submodels assigned to each computing device is similar, ensuring that each pipeline worker can avoid the waiting state.

3.2 Clock Optimization Strategy

To solve the problem of inconsistent weight versions, PipeDream designed a weight storage scheme, which stores the weights that are not used for the time being and then retrieves them from the buffer when needed later. Although this method can ensure the consistency of weight versions when the model scale is large, many weight versions must be stored, significantly saving memory space. To solve this problem, we design a clock optimization strategy. In general, the clock optimization strategy includes two parts, the first part is limited buffer allocation, and the second part is clock task allocation. We will elaborate on these two parts next. To ensure the consistency of weight versions, the version and quantity of weights stored on each computing device should be consistent. According to the preliminary estimation stage, we obtained the time and occupied memory required by each network layer during training. After we partition the model using prefix sum algorithm, the memory occupied by each partition during training can also be estimated. When we weigh buffers for each partition, the size of the buffer is determined by how much memory it occupies. Suppose there are n partitions, and the set of memory occupied by each partition during training is M={M1, M2... .. Mn}. We choose the maximum value in set M, denoted as Max, as the standard for setting the buffer size, which is approximately 1/2Max and cannot exceed the device's physical memory. The buffer can only receive a limited number of micro-batch data weights. We assume it can accept n micro-batch weights, denoted as W1, W2... .Wn. When Wn+1 is about to enter the buffer, W1 can be cleared to make room for Wn+1 storage, when Wn+2 is about to enter, W2 empties, and so on. This way, when the following weights enter in sequence, the initial consequences are cleared in sequence, and there are always n weights in the buffer. An example application is shown in Fig. 2.

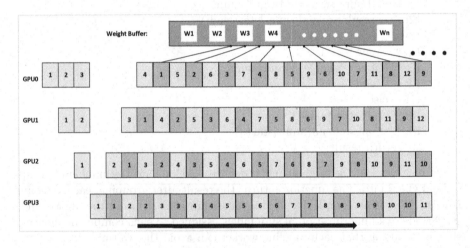

Fig. 2. Sample weight buffer diagram.

The steps of the algorithm to limit buffer allocation are as follows:

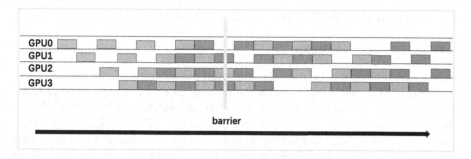

Fig. 3. Clock task optimization strategy.

Algorithm 2: Limit the buffer allocation algorithm

1: function buffer_algorithm(M, n, size):
2:     Step 1:Find the maximum value in M as the buffer size setting criteria
3:     Max = max(M)
4:     Step 2:The buffer size is 1/2Max, but cannot exceed the physical memory of the device
5:     buffer_size = min(Max / 2, device_memory)
6:     Step 3:The buffer can receive n micro-batch data weights
7:     buffer = []
8:     for i = 1 to n:
9:     buffer.append(empty_batch(size))
10:       Step 4:Starting with the n+1 micro-batch data, it is sequentially buffered
11:       i = n + 1
12:       while True:
13:       Wi = get_batch(i, size)
14:       if Wi is None:
15:     break
16:       buffer.pop(0)
17:     buffer.append(Wi)
18:     i = i + 1

To avoid excessive accumulation of weights outside the buffer, we designed a clock task scheduling strategy when sending micro-batch data into the first partition successively for training. A concrete example is shown in Fig. 3.

The practice was to ensure that when the training reached a steady state, the cross-training state could be achieved. When the backward of one micro-batch data was complete, the forward propagation of another micro-batch data began. However, our designed buffer can store limited weights, so we set a time interval. When n micro-batches are sent to the pipeline for training, the next batch of data should be sent to the pipeline for training after L1 (L1 is the estimated training time on the first partition) time interval (the barrier in Fig. 3). This is done to prevent excessive weight accumulation outside the buffer. We should ensure that storing weights in the buffer is sequential and smooth.

4   Experimental Results and Analysis

4.1   Experimental Setup

In the experiment, we compare the improved framework with the parallel data framework (DP) and several pipelining frameworks that are popular at present. To exclude the influence of other factors, we separately test the effect of each improved algorithm on the overall training process. Two representative DNN models AmoebaNet and ResNet101, were used in the experiment. Combined with the resource environment of the supercomputer platform, eight NVIDIA A100 GPUs were configured in the investigation, each with 40GB of GPU device memory. To find out how the prefix sum algorithm and clock optimization strategy affect the model, let's do some simple experiments.

Let's first look at how prefix sum algorithm affect the system. When the variable factors are only prefix sum algorithm, the experimental results are shown in Fig. 4. We used a custom dataset containing 20,000 images, which size are 3 × 224 × 224. When 8 GPUs were used to train AmoebaNet, the training times were 20 rounds, and the addition of prefix sum algorithm improved the training time compared with the non-addition. The experimental results show that the prefix sum algorithm can effectively improve the partition efficiency and make the model have a relatively balanced partition.

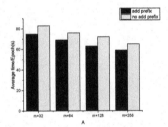   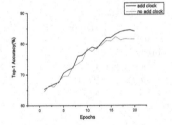

Fig. 4. Training time per round with 8 GPUs.   **Fig. 5.** Accuracy per round with 8 GPUs.

When the variable factor is only the clock optimization strategy, the experimental results are shown in Fig. 5. After adding the clock optimization strategy, the model training accuracy is improved by about 8%. This shows that the weight buffering strategy can correct the convergence direction of the model loss function to a certain extent, and improve the model accuracy.

In the following section, we first made a statistical comparison between training time and device throughput to test the contribution of prefix sum partitioning algorithm to training efficiency. Then the model accuracy and the loss function were counted to push the clock optimization strategy's assistance to save the memory footprint. Next, we will elaborate on the details of the experiment.

4.2 Analysis of Training Time and Equipment Throughput

Table 1. The throughput and memory usage with different number of GPU.

AmoebaNet-256	Throughput	Memory usage
K=1	134sample/s	39.1 GB
K=2	182 sample/s	34.6 GB
K=4	211 sample/s	30.7 GB
K=8	242 sample/s	26.3GB

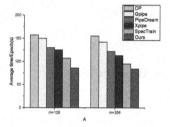

Fig. 6. Training time per round with 4 GPUs.

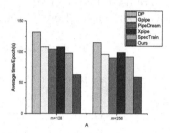

Fig. 7. Training time per round with 8 GPUs.

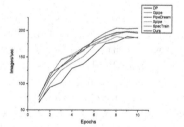

Fig. 8. Device throughput with 4 GPUs.

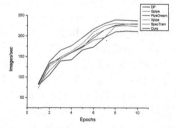

Fig. 9. Device throughput with 8 GPUs.

Figure 6 and 7 show the time required for AmoebaNet to complete a round of training when the number of GPUs was 4 and 8. As can be seen from the figure, the improved framework has improved the training speed. Figure 8 and 9 show the throughput of the device when AmoebaNet conducted a round of training when the number of GPUs was 4 and 8. In the initial stage of training, its throughput is roughly the same as that of other pipeline frameworks, but its growth rate is higher. As the epoch increases, its throughput increases, which is a significant increase compared to other frameworks. The experimental results show that the partition obtained by prefix sum algorithm is more stable, and each worker on the pipeline can work smoothly, reducing the waiting for computing resources. Partitioning takes less time and speeds up training overall (Table 1).

4.3   Analysis of Training Accuracy and Loss Function

This section mainly compares our improved pipelined parallel training framework and several more commonly used pipelined frameworks in terms of the loss function and model accuracy. In all experiments, we used the same hyperparameter settings to compare their performance to reduce interference from other factors. We set the batch size to 256 and initialize the learning rate to 1e-2. Table 2 calculates how each framework compares loss function and model accuracy when eight GPUs are used, and the batch size is set to 256. As you can see from Table 2, our improved framework improves the model accuracy by about 10% compared with PipeDream. Figure 10 depicts the training accuracy curve on the epoch when 8 GPUs are used. Figure 11 shows the convergence curve of the loss function when eight GPUs are used. Here we set the micro-batch size to 32.

Table 2. Loss function and Training accuracy

Approach	Min.Val.Loss	Max.Val.Top-1 Accuracy
DP	0.450	76.82%
Gpipe	0.433	77.23%
PipeDream	0.425	78.65%
SpecTrain	0.396	84.29%
Ours	0.368	88.72%

The results in Table 2 show that the learning curve of our improved framework is about 10% higher in experimental accuracy than that of PipeDream. As can be seen in Figs. 10 and 11, the approach we investigated always achieves smaller loss function values and faster convergence than PipeDream. Using asynchronous weights will cause the loss function to fluctuate at the beginning of training, but this does not affect its confluence. This indicates that the clock optimization strategy is a good solution, which limits the number of weight versions generated, effectively guarantees the model's accuracy, and speeds up the convergence of the loss function. At the same time, the memory usage is saved, so each partition can be load balanced.

Fig. 10. Training accuracy of ResNet101 with 8 GPUs.

Fig. 11. Loss function curve of ResNet101 with 8 GPUs.

5 Conclusions

In recent years, with the continuous development of artificial intelligence technology, the complexity of deep learning algorithms and the scale of model training is also increasing. Distributed training becomes an effective way to train large-scale models. To solve this problem, we propose an efficient pipeline-parallel training optimization method. Our approach processes small batches of data in parallel through multiple compute nodes in a pipelined manner. We offer a prefix sum partition algorithm to realize a balanced partition and save the memory of computing resources. At the same time, we also design a clock optimization strategy to limit the number of weight version generations to ensure the model's accuracy. Compared with the current famous pipeline parallel frameworks, our method can achieve about 2 times training acceleration, save about 30% of memory consumption, and improve the model accuracy by about 10% compared with PipeDream.

Acknowledgment. This work is supported by the R&D and application of key technologies of independent and controllable computing power network (grant No. 2022JBZ01-01) and the Joint Fund of Shandong Natural Science Foundation (No. ZR2022LZH010).

References

1. Radford, A., Wu, J., Child, R., Luan, D., Amodei, D., Sutskever, I., et al.: Language models are unsupervised multitask learners. OpenAI Blog **1**(8), 9 (2019)
2. Vaswani, A., et al.: Attention is all you need. In: Advances in Neural Information Processing Systems, vol. 30 (2017)
3. Cai, H., Zhu, L., Han, S.: Proxylessnas: direct neural architecture search on target task and hardware. arXiv preprint arXiv:1812.00332 (2018)
4. Huang, Y., et al.: GPipe: efficient training of giant neural networks using pipeline parallelism. In: Advances in Neural Information Processing Systems, vol. 32 (2019)
5. Alvarez, J.M., Salzmann, M.: Learning the number of neurons in deep networks. In: Advances in Neural Information Processing Systems, vol. 29 (2016)
6. Han, S., Pool, J., Tran, J., Dally, W.: Learning both weights and connections for efficient neural network. In: Advances in Neural Information Processing Systems, vol. 28 (2015)
7. Krizhevsky, A., Sutskever, I., Hinton, G.E.: Imagenet classification with deep convolutional neural networks. Commun. ACM **60**(6), 84–90 (2017)
8. Goyal, P., et al.: Accurate, large minibatch SGD: training imagenet in 1 hour. arXiv preprint arXiv:1706.02677 (2017)
9. Shallue, C.J., Lee, J., Antognini, J., Sohl-Dickstein, J., Frostig R., Dahl, G.E.: Measuring the effects of data parallelism on neural network training. arXiv preprint arXiv:1811.03600 (2018)
10. Sabet, M.J., Dufter, P., Yvon, F., Schütze, H.: Simalign: high quality word alignments without parallel training data using static and contextualized embeddings. arXiv preprint arXiv:2004.08728 (2020)
11. Fan, S., et al.: Dapple: a pipelined data parallel approach for training large models. In: Proceedings of the 26th ACM SIGPLAN Symposium on Principles and Practice of Parallel Programming, pp. 431–445 (2021)

12. Zhang, M., Zhou, Y., Zhao, L., Li, H.: Transfer learning from speech synthesis to voice conversion with non-parallel training data. IEEE/ACM Trans. Audio Speech Lang. Process. **29**, 1290–1302 (2021)
13. Dean, J., et al.: Large scale distributed deep networks. In: Advances in Neural Information Processing Systems, vol. 25 (2012)
14. Shoeybi, M., Patwary, M., Puri, R., LeGresley, P., Casper, J., Catanzaro, B.: Megatron-LM: training multi-billion parameter language models using model parallelism. arXiv preprint arXiv:1909.08053 (2019)
15. Jia, Z., Zaharia, M., Aiken, A.: Beyond data and model parallelism for deep neural networks. Proc. Mach. Learn. Syst. **1**, 1–13 (2019)
16. Narayanan, D., Phanishayee, A., Shi, K., Chen, X., Zaharia, M.: Memory-efficient pipeline-parallel DNN training. In: International Conference on Machine Learning, pp. 7937–7947. PMLR (2021)
17. Moritz, P., et al.: Ray: a distributed framework for emerging {AI} applications. In: 13th {USENIX} Symposium on Operating Systems Design and Implementation ({OSDI} 18), pp. 561–577 (2018)
18. Narayanan, D., et al.: Pipedream: generalized pipeline parallelism for DNN training. In: Proceedings of the 27th ACM Symposium on Operating Systems Principles, pp. 1–15 (2019)
19. Zhang, Q.: A novel resnet101 model based on dense dilated convolution for image classification. SN Appl. Sci. **4**, 1–13 (2022)
20. Real, E., Aggarwal, A., Huang, Y., Le, Q.V.: Regularized evolution for image classifier architecture search. In: Proceedings of the AAAI Conference on Artificial Intelligence, vol. 33, no. 01, pp. 4780–4789 (2019)
21. Jiang, J., Cui, B., Zhang, C., Yu, L.: Heterogeneity-aware distributed parameter servers. In: Proceedings of the 2017 ACM International Conference on Management of Data, pp. 463–478 (2017)
22. Jia, Z., Lin, S., Qi, C.R., Aiken, A.: Exploring hidden dimensions in accelerating convolutional neural networks. In: International Conference on Machine Learning, pp. 2274–2283. PMLR (2018)
23. Jiang, W., et al.: A novel stochastic gradient descent algorithm based on grouping over heterogeneous cluster systems for distributed deep learning. In: 2019 19th IEEE/ACM International Symposium on Cluster, Cloud and Grid Computing (CCGRID), pp. 391–398. IEEE (2019)
24. Kim, J.K., et al.: STRADS: a distributed framework for scheduled model parallel machine learning. In: Proceedings of the Eleventh European Conference on Computer Systems, pp. 1–16 (2016)
25. Darriba, D., Taboada, G.L., Doallo, R., Posada, D.: jmodeltest 2: more models, new heuristics and parallel computing. Nat. Methods **9**(8), 772–772 (2012)
26. Prosky, L., Asp, N.-G., Schweizer, T.F., Devries, J.W., Furda, I.: Determination of insoluble, soluble, and total dietary fiber in foods and food products: interlaboratory study. J. Assoc. Off. Anal. Chem. **71**(5), 1017–1023 (1988)
27. L. Shen, Y. Mao, Z. Wang, H. Nie, and J. Huang, "Dnn training optimization with pipelined parallel based on feature maps encoding. In: 2022 Tenth International Conference on Advanced Cloud and Big Data (CBD), pp. 36–41. IEEE (2022)
28. Chen, C.-C., Yang, C.-L., Cheng,H.-Y.: Efficient and robust parallel DNN training through model parallelism on multi-GPU platform. arXiv preprint arXiv:1809.02839 (2018)
29. Yang, B., Zhang, J., Li, J., Ré, C., Aberger, C., De Sa, C.: PipeMare: asynchronous pipeline parallel DNN training. In: Proceedings of Machine Learning and Systems, vol. 3, pp. 269–296 (2021)

30. Harris, M., Sengupta, S., Owens, J.D.: Parallel prefix sum (scan) with cuda. GPU Gems **3**(39), 851–876 (2007)
31. Sengupta, S., Lefohn, A., Owens, J.D.: " work-efficient step-efficient prefix sum algorithm (2006)
32. Safari, M., Oortwijn, W., Joosten, S., Huisman, M.: Formal verification of parallel prefix sum. In: Lee, R., Jha, S., Mavridou, A., Giannakopoulou, D. (eds.) NFM 2020. LNCS, vol. 12229, pp. 170–186. Springer, Cham (2020). https://doi.org/10.1007/978-3-030-55754-6_10

Unsupervised Monocular Depth Estimation with Semantic Reconstruction Using Dual-Discriminator Generative Adversarial

Jiwen Li, Shiwen Xie$^{(\boxtimes)}$, Yongfang Xie, Xiaofang Chen, and Xi Chen

School of Automation, Central South University, Changsha, Hunan, China
{jiwenli,sw.xie,yfxie,xiaofangchen,cek_chenxi}@csu.edu.cn

Abstract. Monocular depth estimation is a key issue in the field of computer vision. The unsupervised learning framework has the advantage of not requiring data labels, and has become a hot research topic. Currently, most methods use view synthesis as a supervisory signal, resulting in unclear edges and semantic distortion of predicted results in some situations. We proposed a new framework that introduce a semantic reconstruction loss to provide additional constraints for the network and improve the ability of the depth network to understand scenarios. In addition, we proposed a dual-discriminator adversarial training strategy to further strengthen semantic supervision and improve the accuracy of depth estimation. The test results show that our proposed method has achieved competitive performance on the KITTI dataset.

Keywords: Monocular Depth Estimation · Unsupervised Learning · Semantic Reconstruction · Dual-Discriminator GANs

1 Introduction

Monocular depth estimation refers to recovering the depth information of a scene from an RGB image, which can help improve the computer's perception of the scene and thus restore the 3D structure of the scene. Depth information is usually presented as a depth map in which the pixel value of a point represents the distance from the point in the corresponding real scene to the camera imaging plane. Depth information is critical for many tasks in the field of computer vision, such as target recognition, three-dimensional reconstruction, obstacle detection, auto-driving, and so on. In addition, UAVs, smart robots, and many devices in military operations require sophisticated depth information to help perform complex tasks. At present, there are hardware devices that can directly measure depth information, such as lidar, depth cameras, and so on. However, traditional hardware devices have many drawbacks, for example: lidar has a small spatial measurement range and cannot obtain texture information of objects, depth camera has poor anti-jamming ability and is not suitable for open outdoor scenes. Depth sensors cannot be widely used due to these shortcomings and high cost. Therefore, it is meaningful and necessary to consider restoring the depth information of the scene only through inexpensive RGB cameras.

© The Author(s), under exclusive license to Springer Nature Singapore Pte Ltd. 2024
B. Luo et al. (Eds.): ICONIP 2023, CCIS 1961, pp. 396–406, 2024.
https://doi.org/10.1007/978-981-99-8126-7_31

Traditional monocular depth estimation methods are mainly based on visual cues or machine learning. Common ones are Markov random field (MRF) [1] and structure from motion (SFM) [2]. These traditional algorithms depend on the parameters of the camera, or need other auxiliary devices, and most of them are not accurate enough to be susceptible to environmental interference. Therefore, it is difficult to promote the use. With the development of deep learning technology and the increasingly powerful performance of hardware devices, monocular depth estimation methods combined with deep learning have gradually become the mainstream.

In deep learning based monocular depth estimation methods, unsupervised methods break through the limitations of difficulty in obtaining true depth labels, requiring only cheap RGB image sequences for training, and have become a trend in current research. Unsupervised methods can be divided into two types based on the training dataset: one is stereo image pairs, and the other is monocular videos. When using stereo images to train models, strict calibration of binocular cameras is required, and the accuracy of camera parameters directly affects the performance of the network. The method based on monocular video adopts a new approach to address this limitation. The network not only needs to predict the depth map of the target image, but also needs to train a network to estimate the camera pose transformation between the target frame and adjacent frames. Zhou et al. [3] first proposed a method, which can restore scene depth and camera pose from monocular videos alone. Later, scholars continuously improved this method from different perspectives. [4] used semantic information to guide depth synthesis, which improves the rationality of depth prediction. Godard et al. [5] proposed a new image reconstruction loss that significantly improves the accuracy of network prediction. Aleotti et al. [6] combined unsupervised frameworks with GAN, treating the network as a generator and discriminator, and utilizing the max-min game between the two to improve the performance of the network.

2 Related Works

2.1 Supervised Learning Method

The supervised learning method needs the real depth map as a label to train the mapping relationship between the network learning RGB map and the depth map. Eigen et al. [7] first proposed a network architecture that utilizes global and local scale information through a CNN network, and trains the network to obtain pixel by pixel depth estimation. Kumar et al. [8] utilized the contextual ability of LSTM networks to enable depth networks to extract spatiotemporal related features of image sequences, so that depth networks can focus on more helpful information. Although supervised learning methods can enable networks to efficiently learn and understand the 3D structure of scenes from real depth maps, these methods inevitably require a large number of labels, and the cost of obtaining labels, i.e. real depth maps, is very high, requiring precise instruments and equipment.

2.2 Unsupervised Learning Method

The unsupervised deep learning method can effectively solve the dependence of supervised learning method on depth labels, which is usually based on the assumption of color invariance between consecutive frames. In 2016, Grage et al. [9] first proposed an unsupervised monocular depth estimation method using stereo image pairs to train the networks, which utilizes photometric error as the networks' supervision. However, the image synthesis process proposed by this method is non differentiable, which brings difficulty to the optimization of network parameters. Godard et al. [10] proposed consistency loss between left and right disparity map using epipolar geometric constraints between binocular images. Additionally, bilinear sampling method was used to generate images, which facilitates solving network gradients and improves the accuracy and robustness of monocular depth estimation.

Zhou et al. [3] proposed a method to restore scene depth and camera pose from monocular video sequences only, using the generated depth map and camera motion to rebuild the target view. The loss function is constructed by the photometric difference between the reconstructed image and the original image to train an unsupervised framework. Aleottiet et al. [6] proposed an unsupervised monocular depth estimation framework based on antagonistic training. The model is divided into two parts: a generator and a discriminator, which generates a depth map to distinguish the synthetic image from the real image. Generators tend to produce more realistic depth maps when constrained against loss. On this basis, Almalioglu et al. [11] designed an unsupervised learning framework for antagonism and looping. A two-layer LSTM network was added to the position and posture estimation network. Depth maps, 6-degree-of-freedom camera postures were generated from the source images, and the view reconstruction module was used to synthesize the target images, and discriminators were used to distinguish them. In addition, studies have shown that the introduction of semantic information into the network can also help improve depth prediction. Chen [12] proposed scenenet, introduced left and right semantic consistency constraints of scenes in training, and achieved regional-level depth estimation. [4] use automatic convolution of pixels and redefine the sampling process of depth network using semantic features to improve the clarity of object boundaries in the prediction depth map. We focus on the training process of the network and design semantic reconstruction loss and semantic adversarial loss to constrain training, utilizing semantic information in multiple ways.

Inspired by the above, we propose a new depth estimation framework with GANs, using semantic information to constrain the network to obtain a higher quality depth map.

3   Network Architecture

This section mainly introduces the unsupervised monocular depth estimation network architecture based on the semantic reconstruction and the dual-discriminator GANs proposed in this paper, as well as unsupervised training framework and loss function.

3.1   Architecture Overview

This section mainly introduces the unsupervised monocular depth estimation network architecture based on the semantic reconstruction and the dual-discriminator GANs proposed in this paper, as well as unsupervised training framework and loss function.

The network architecture proposed in this article is shown in Fig. 1. Our network consists of a generator and two different discriminators. Firstly, the generator takes a sequence of three consecutive adjacent images as input signal, which contains a target image I_t and two sources images I_s. The target image is encoded and decoded through a depth network to obtain the corresponding depth map, while the frame snippets are sent to the pose network to output a 6-DoF pose. The depth and pose estimation generated by the generator can warp the source image to match the target view and then obtain a reconstructed image. The reconstructed image and the target image are sent to pretrained semantic network for semantic dense feature extraction, and two semantic feature maps are obtained respectively. Finally the discriminator performs authenticity verification on the synthesized view, while the feature discriminator do the same on the semantic feature map of the synthesized view.

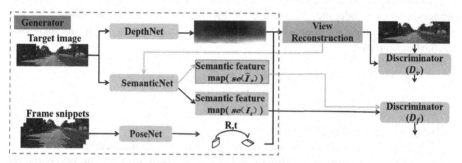

Fig. 1. Illustration of our proposed unsupervised dual-discriminator adversarial framework. DepthNet and PoseNet can be used independently in testing phase.

In previous methods, it was usually only necessary to design a discriminator to verify the authenticity of the target frame and the reconstructed frame. In order to enhance the deep network's understanding of the semantic structure in the scene and avoid semantic distortion, we design another feature discriminator to distinguish the semantic feature map of target frame or reconstruct target frame. We first take a review of the method using GANs for unsupervised monocular training, and then describe the idea of ours for depth predictions.

3.2 Traditional Unsupervised Monocular Training

1) Image Reconstruction Loss
The pose transformation and depth information predicted by the generator can be used to reconstruct the target view, which requires sampling the pixels of the source view. The key to this operation is to obtain the pixel coordinate correspondence from the target view to the source view. The view reconstruction algorithm [3], which has been widely used in unsupervised monocular depth estimation frameworks in the past, can be used to obtain the coordinate index of sampling points, which is described by the following formula:

$$P_s \sim K\widehat{T}_{t\rightarrow s}\widehat{D}_t(P_t)K^{-1}P_t. \tag{1}$$

P_s and P_t are the coordinate indices of the corresponding points in the reconstructed view and source view, respectively, while K is the camera's internal parameter matrix. D is the depth estimation output of the depth network, and T is a 6-DoF camera pose transformation generated by the pose network. After obtaining the corresponding relationship of pixel coordinates, the appearance of target image can be reconstructed from the viewpoint of the source view using method of bilinear sampling, and the view reconstruction error from I_s to I_t can be obtained based on the following formula:

$$E_{rec}(I_t, I_s) = \frac{\theta}{2}(1 - SSIM(I_t, I_s)) + (1 - \theta)|I_t - I_s|. \tag{2}$$

where SSIM [13] is a structural similarity index measure between I_t and I_s, with a balance factor $\theta = 0.85$.

To reduce the adverse effects of non-Lambertian surfaces and occlusion on training images, as in monodepth2 [5], we use the method of calculating the minimum value between the synthesized source views of different viewpoints to obtain the final view reconstruction loss:

$$\mathcal{L}_{serec} = \sum_P \min_{I_s} E_{rec}(I_t, \hat{I}_s). \tag{3}$$

2) Adversarial Training Loss

The original depth estimation model was composed of several different network structures, and through adversarial training, the model can be re divided into two parts: the generator and the discriminator. The function of the generator is to generate images with the same distribution as real images. The discriminator distinguishes between real images and synthesized images by learning data distribution. The max-min game between discriminator and generator has a positive effect on generating more accurate depth predictions in the network. The adversarial training loss is described by the following formula:

$$\mathcal{L}_{GAN} = \mathbb{E}_{I_t, P(I_t)}[\log D(I_t)] + \mathbb{E}_{\hat{I}_s, P(\hat{I}_s)}[\log(1 - D(\hat{I}_s))]. \tag{4}$$

where $P(I_t)$ and $P(\hat{I}_s)$ stand for the data distribution of target images and synthesized images, respectively.

3) Smoothness Loss

In monocular training, we observed that if additional constraints are not added, the predicted depth map often exhibits discontinuous object edges. Referring to [10, 14], we use edge perception smoothing to penalize discontinuous depth predictions:

$$\mathcal{L}_s = \sum_{w,h}^{W,H} |\partial_x D_t^{w,h}|e^{|-\partial_x I_t^{w,h}|} + |\partial_y D_t^{w,h}|e^{|-\partial_y I_t^{w,h}|}. \tag{5}$$

where $D_t^{w,h}$ is the mean-normalized disparity of point $P^{w,h}$ to preventing predictions from falling out of a reasonable range.

3.3 Monocular Training with Semantic Constrains

1) Semantic Reconstruction Loss
In unsupervised monocular depth estimation methods, geometric constraints are the only source of supervision. Although deep networks can learn the mapping relationship from RGB images to depth maps, problems such as semantic distortion and edge blurring often exist and affect the prediction ability of the model due to the lack of additional semantic supervision. In order to solve this problem, we deign a new reconstruction loss to extract the semantic dense features from target image and reconstructed image through the pretrained semantic network to calculate the semantic reconstruction error, and add new constraints to the depth synthesis. The semantic reconstruction error can be described by the following formula:

$$E_{serec}(I_t, \hat{I}_s) = |se(I_t) - se(\hat{I}_s)|. \tag{6}$$

where $se(\cdot)$ describes a mapping relationship from RGB space to dense semantic feature space. The specific process is to extract semantic features of RGB images from semantic network. At each pixel in the feature space, similar to view reconstruction loss, we use the minimum value method to calculate the loss for all source images (After experiments, we find that the model performance is improved more than directly calculating the average value.) and finally calculate the average value for all pixels. The semantic reconstruction loss is formulated by:

$$\mathcal{L}_{serec} = \sum_P \min_{se(I_s)} E_{serec}(I_t, \hat{I}_s). \tag{7}$$

2) Semantic Adversarial Loss
Inspired by the traditional GAN based monocular depth estimation framework, we have designed a new semantic feature discriminator D_f (shown in Fig. 2) for semantic adversarial training of the network, and it shares the same generator as the original GAN. This time, instead of RGB images, we send the feature maps $se(I_t)$ and $se(I_s)$ into the discriminator, enabling it to learn the distribution of different semantically dense features. Therefore, in the game with the discriminator, the generator tends to synthesize reconstructed images that have semantic structure consistency with the target image, and the depth information of which discourage the depth map, as an intermediate variable, to exhibit semantic distortion and blurry object contours. The semantic GAN loss is described by the following formula:

$$\mathcal{L}_{seGAN} = \mathbb{E}_{se(I_t)\ P(se(I_t))}[\log D(se(I_t))] + \mathbb{E}_{se(\hat{I}_s)\ P(se(\hat{I}_s))}[\log(1 - D(se(\hat{I}_s)))]. \tag{8}$$

3) Final Training Loss
Our final training loss is described as by the following formula, α, β, γ, μ and λ are hyper-parameters.

$$\mathcal{L}_{final} = \alpha \mathcal{L}_{rec} + \beta \mathcal{L}_{serec} + \gamma \mathcal{L}_s + \mu \mathcal{L}_{GAN} + \lambda \mathcal{L}_{seGAN}. \tag{9}$$

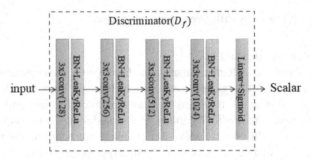

Fig. 2. Structure of our feature discriminator (D_f).

4 Experiment

In this section, we first introduce our unsupervised framework and training details, and then use a series of experimental results to demonstrate the effectiveness of our method.

4.1 Implementation Details

1) Network Structure
Our network can be divided into three parts: generator, RGB discriminator, and feature discriminator. The generator consists of three sub networks: DepthNet, PoseNet, and SemanticNet. For the DepthNet, our architecture is based on U-net [15], where the skip connection between the encoder and decoder enables the network to focus on multiple local details of different resolutions. We take a Resnet-18 architecture [16] as a deep decoder to ensure that the model has good performance and fewer parameter quantities. Instead of random weights, we initialize the encoder using weights pretrained on ImageNet [17]. For the depth decoder, we adopt the framework proposed in [10]. Our PoseNet takes 3 consecutive frames of RGB images as input and outputs a 6-DoF relative pose between the source view and target view. Similar to the encoder, we take Renet-18 as our PoseNet. And for SemanticNet, we directly use a semantic segmentation model FPN [18], which is pretrained on Cityscapes dataset [19]. FPN takes a single RGB image as input, and we add the feature maps of the four prediction layers in FPN by the channel dimension to obtain semantic dense feature. We chose the structure of the RGB discriminator based on work in [11]. As shown in the Fig. 2, we have designed our feature discriminator structure, which outputs a layer based on the channel and size of semantically extracted features. If the channel dimensions of output from SemantitcNet change, it is necessary to adjust the input channel dimension of the semantic discriminator accordingly.

2) Training Details
Our unsupervised framework is implemented in Pytorch [20]. We train our model using Eigen split [21]. After excluding static frames in the scene according to paper [3]'s method, 39810 sequences were left for training. Our model is trained in 20 epochs with

Adam optimizer on a single A40 GPU. The batch size is set to 12, and the learning rate of the first 15 epochs is 0.0001, which is then drop to 0.00001 for rest 5 epochs. For the loss term, we set $\alpha = 1$, $\beta = 0.05$, $\gamma = 0.0001$, $\mu = 0.0001$, and $\lambda = 0.0001$. We crop all training images to a resolution of 640×192. Giving that discriminators has fewer parameters and a faster convergence speed than generator, we introduce a hyper-parameter $k = 5$, which decides the update frequency ratio between generator and both discriminators, in order to ensure that the adversarial game between the two does not lose balance.

4.2 Evaluation on KITTI Dataset

1) Ablation Experiment
In this section, we will report a series of ablation experiments to demonstrate the effectiveness of our proposed modules, as shown in the Table 1. We will rebuild the reconstruction loss $\mathcal{L}_{rec}$, adversarial training loss $\mathcal{L}_{GAN}$ and Smoothness loss $\mathcal{L}_s$ as baseline, expressed by $\mathcal{L}_{basic}$. Then, $\mathcal{L}_{se}$ and $\mathcal{L}_{GAN}$ are introduced into training sequentially to prove the validity of each module. The results show that the performance of the baseline model is the worst. After the introduction of $\mathcal{L}_{serec}$, the depth prediction accuracy of the network has significantly improved, although two indicators have not been obviously improved, indicating that the introduction of semantic reconstruction loss is helpful for depth map generating in some aspects. We analyzed the reasons why some metrics data could not be improved and concluded that our proposed $\mathcal{L}_{serec}$ did not completely describe the differences of semantic features between the target view and synthesized view. This is why, after the final introduction of $\mathcal{L}_{GAN}$, each indicator of our complete model can be improved.

2) Evaluation of Depth Performance
We evaluated our model on split of [21] shown in Table 2. Due to the lack of scale information in monocular depth prediction, we scaled the predicted depth to the same scale as ground truth. All results are obtained by comparing the ground truth with reality. We compare the various indicators of the model with other unsupervised monocular depth estimation methods, and the results show that our model achieves the best results in all indicators. Qualitative comparisons are shown in Fig. 3. It can be seen that our model can produce higher quality depth maps than the compared models. For example, in the predicted depth map of the second RGB image in the first column, our results showed a clear contour of the pole, while other methods don't. And for cars and trees in the scene, our results have clearer edges. Experiments show that our framework can produce more visually appealing depth maps.

Table 1. Quantitative results on the KITTI [22] using Eigen split [21].The best results for each evaluation criterion are presented in bold.

Method	Lower is better				Higher is better		
	Abs Rel	Sq Rel	RMSE	RMSE log	$\delta < 1.25$	$\delta < 1.25^2$	$\delta < 1.25^3$
$\mathcal{L}_{basic}$	0.119	0.898	4.954	0.199	0.864	0.952	0.980
$\mathcal{L}_{basic} + \mathcal{L}_{serec}$	0.119	0.962	4.997	0.198	0.869	0.955	0.980
$\mathcal{L}_{basic} + \mathcal{L}_{serec}$ (Ours) $+\mathcal{L}_{seGAN}$	**0.115**	**0.884**	**4.822**	**0.193**	**0.873**	**0.959**	**0.981**

Table 2. Ablation results on the KITTI [22] using Eigen split [21] with depth capped at 80 m.

Method	Lower is better				Higher is better		
	Abs Rel	Sq Rel	RMSE	RMSE log	$\delta < 1.25$	$\delta < 1.25^2$	$\delta < 1.25^3$
Sfmlearner [3]	0.208	1.768	6.856	0.283	0.678	0.885	0.957
DDVO [14]	0.151	1.257	5.583	0.228	0.810	0.936	0.974
MonoGAN [6]	0.118	0.908	4.978	0.210	0.855	0.948	0.976
Ours	**0.115**	**0.884**	**4.822**	**0.193**	**0.873**	**0.959**	**0.981**

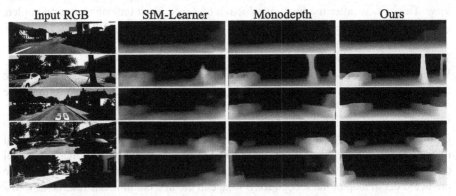

Input RGB   SfM-Learner   Monodepth   Ours

Fig. 3. Comparison of other unsupervised monocular depth estimation methods.

5 Conclusion

In this paper, we propose a new framework of unsupervised monocular depth estimation combining Semantic information. Considering that view reconstruction may fail when encountering occlusions or non-Lambertian surfaces in the scene, and the predicted depth map may not have a clear visual object structure, we design the semantic reconstruction loss to constraint training of the network to overcome this limitation. At the same time, utilizing the strong capabilities of GANS in image processing, we introduce semantic adversarial training loss to strengthen the supervision of semantic consistency on the network. Our improvements have effectively improved the accuracy of depth prediction and can help depth networks produce depth images of higher quality.

Acknowledgements. This work was supported in part by the National Natural Science Foundation of China (62003370), and the Distinguished Youth Foundation of Hunan Nature Science Foundation (2023JJ10079).

References

1. Rajagopalan, A.N., Chaudhuri, S., Mudenagudi, U.: Depth estimation and image restoration using defocused stereo pairs. IEEE Trans. Pattern Anal. Mach. Intell. **26**(11), 1521–1525 (2004)
2. Song, S., Chandraker, M.: Robust scale estimation in real-time monocular SFM for autonomous driving. In: Proceedings of the IEEE Conference on Computer Vision and Pattern Recognition, pp. 1566–1573 (2014)
3. Zhou, T., Brown, M., Snavely, N., Lowe, D.G.: Unsupervised learning of depth and ego-motion from video. In: Proceedings of the IEEE Conference on Computer Vision and Pattern Recognition, pp. 1851–1858 (2017)
4. Guizilini, V., Hou, R., Li, J., Gaidon, A.: Semantically-guided representation learning for self-supervised monocular depth, arXiv preprint arXiv: 2002.12319 (2020)
5. Godard, C., Mac Aodha, O., Firman, M., Brostow, G.J.: Digging into self-supervised monocular depth estimation. In: Proceedings of the IEEE/CVF International Conference on Computer Vision, pp. 3828–3838 (2019)
6. Aleotti, F., Tosi, F., Poggi, M., Mattoccia, S.: Generative adversarial networks for unsupervised monocular depth prediction. In: Leal-Taixé, L., Roth, S. (eds) ECCV 2018. LNCS, vol. 11129, pp. 337–354. Springer, Cham (2019). https://doi.org/10.1007/978-3-030-11009-3_20
7. Eigen, D., Puhrsch, C., Fergus, R.: Depth map prediction from a single image using a multi-scale deep network. In: Advances in Neural Information Processing Systems, pp. 1–9 (2014)
8. Kumar, A., Bhandarkar, S.M., Prasad, M.: Depthnet: a recurrent neural network architecture for monocular depth prediction. In: Proceedings of the IEEE Conference on Computer Vision and Pattern Recognition, pp. 283–291 (2018)
9. Garg, R., Vijay Kumar, B.G., Carneiro, G., Reid, I.: Unsupervised CNN for single view depth estimation: geometry to the rescue. In: Leibe, B., Matas, J., Sebe, N., Welling, M. (eds.) ECCV 2016. LNCS, vol. 9912, pp. 740–756. Springer, Cham (2016). https://doi.org/10.1007/978-3-319-46484-8_45
10. Godard, C., Mac Aodha, O., Brostow, G.J.: Unsupervised monocular depth estimation with left-right consistency. In: Proceedings of the IEEE Conference on Computer Vision and Pattern Recognition, pp. 270–279 (2017)
11. Almalioglu, Y., Saputra, M.R.U., De Gusmao, P.P.B., Markham, A., Trigoni, N.: GANVO: unsupervised deep monocular visual odometry and depth estimation with generative adversarial networks. In: International Conference on Robotics and Automation (ICRA), pp. 5474–5480 (2019)
12. Chen, P., Liu, A., Liu, Y., Wang, Y.: Towards scene understanding: unsupervised monocular depth estimation with semantic-aware representation. In: Proceedings of the IEEE/CVF Conference on Computer Vision and Pattern Recognition, pp. 2624–2632 (2019)

13. Wang, Z., Bovik, A.C., Sheikh, H.R., Simoncelli, E.P.: Image quality assessment: from error visibility to structural similarity. IEEE Trans. Image Process. **13**(4), 600–612 (2004)
14. Wang, C., Buenaposada, J.M., Zhu, R., Lucey, S.: Learning depth from monocular videos using direct methods. In: Proceedings of the IEEE Conference on Computer Vision and Pattern Recognition, pp. 2022–2030 (2018)
15. Ronneberger, O., Fischer, P., Brox, T.: U-Net: convolutional networks for biomedical image segmentation. In: Medical Image Computing and Computer-Assisted Intervention, pp. 234–241 (2015)
16. He, K., Zhang, X., Ren, S., Sun, J.: Deep residual learning for image recognition. In: Proceedings of the IEEE Conference on Computer Vision and Pattern Recognition, pp. 770–778 (2016)
17. Russakovsky, O., et al.: ImageNet largescale visual recognition challenge. Int. J. Comput. Vis. **115**, 211–252 (2015)
18. Kirillov, A., Girshick, R., He, K., Dollár, P.: Panoptic feature pyramid networks. In: Proceedings of the IEEE/CVF Conference on Computer Vision and Pattern Recognition, pp. 6399–6408 (2019)
19. Cordts, M., et al.: The cityscapes dataset for semantic urban scene understanding. In: Proceedings of the IEEE Conference on Computer Vision and Pattern Recognition, pp. 3213–3223 (2016)
20. Paszke, A., et al.: Automatic differentiation in PyTorch. In: Conference and Workshop on Neural Information Processing Systems, pp. 1–7 (2017)
21. Eigen, D., Fergus, R.: Surface normals and semantic labels with a common multi-scale convolutional architecture. In: Proceedings of the IEEE International Conference on Computer Vision, pp. 2650–2658 (2015)
22. Geiger, A., Lenz, P., Urtasun, R.: Are we ready for autonomous driving? The KITTI vision benchmark suite. In: IEEE Conference on Computer Vision & Pattern Recognition, pp. 3354–3361 (2012)

Generating Spatiotemporal Trajectories with GANs and Conditional GANs

Kefan Zhao and Nana Wang[✉]

Jiangsu Normal University, Xuzhou 221116, China
wangnana_5@aliyun.com

Abstract. Modeling the movements of individual and populations, and generating synthetic spatiotemporal trajectory data play an important role in lots of (privacy-aware) analysis and applications, such as urban planning and route navigation. A key challenge in trajectory generation is to best capture the basic characteristics of the long sequences of location points. This is non-trivial considering the inherent sequentiality and high-dimensionality of trajectory data. This paper presents TS-TrajGAN, a two-stage model to generate spatiotemporal trajectory data by combining a Generative Adversarial Network (GAN) and a conditional GAN. We train the GAN of stage I to simulate the distribution of the initial trajectory segments such that the basic characteristics of the length-limited initial trajectory segments can be well depicted. In stage II, the conditional GAN is used to predict the next location point for the current generated trajectory and preserve the variability in individuals' mobility. In addition, a predictor network is added to the GAN of stage I for trajectory length prediction. Experiments on a real-world taxi dataset demonstrate that TS-TrajGAN is not only able to generate trajectories that have similar characteristics with the real ones, but also outperforms the state-of-the-art methods in terms of data utility. Our code is available at https://github.com/kfZhao726/TS-TrajGAN.

Keywords: Trajectory Generation · Two-Stage · Generative Adversarial Network · Conditional GAN

1 Introduction

With the pervasive use of GPS equipped mobile devices, an increasing number of trajectories which describe mobility behaviors have been generated. These data are very useful for a variety of applications ranging from urban planning, traffic flow analysis, mobile advertising, to epidemic modelling [1, 2]. However, due to the concerns of privacy leakage, individuals might be reluctant to reveal their trajectories to any third-parties. The shortage of public mobility datasets hinders most trajectory-driven studies and extensive experimental analyses, and many applications cannot be best served [1, 3]. This issue emphasizes the need for synthetic trajectory generators, which simulate the behavior of moving objects and enable comprehensive performance evaluations.

Typically, the methods on trajectory generation construct some mobility models (e.g., statistical distributions [4], Markov process [5], generative models [6]) to summarize the mobile entities' complete movement behaviors, and then sample synthetic

© The Author(s), under exclusive license to Springer Nature Singapore Pte Ltd. 2024
B. Luo et al. (Eds.): ICONIP 2023, CCIS 1961, pp. 407–421, 2024.
https://doi.org/10.1007/978-981-99-8126-7_32

trajectories from the model. Early works in this field design the models manually. For example, [5] uses a Markov process to predict the next point for each trajectory. However, because of the inherent sequentiality and high-dimensionality of trajectory data, the manually designed rigid models may not be sufficient to depict the individuals' complicated movement patterns [3].

Deep learning technology [7–10] recently has been found to be a highly effective in mobility data generation. They usually leverage generative models like generative adversarial networks (GANs) [11–13] and conditional generative adversarial networks (conditional GANs) [13] or sequential models like recurrent neural networks (RNNs) [1, 2] to simulate the distribution of the real trajectories before sampling synthetic trajectories from that distribution. Although these data-driven, deep learning-based methods might perform better, accurately capturing both the spatial and temporal characteristics of the high dimensional real trajectories is still a challenge. To illustrate our claim, in Fig. 1 we compare our proposed model (TS-TrajGAN) with five most relevant works: RNN [1], SeqGAN [14], TTS-GAN [15], TimeGAN [16], and TS-TrajGAN-v (i.e., a variant of TS-TrajGAN which we present in Sect. 4.1.2). The overall temporal visit frequency distribution in the generated trajectories for each method has been compared to the true dataset (Taxi-1 [17]). The frequencies of visits for each 2 h have been calculated. The red curve and the histogram show the overall temporal visit frequency distribution of the true dataset and the synthetic dataset, respectively. Clearly, TS-TrajGAN and TS-TrajGAN-v have better utility visually. In Sect. 4, we demonstrate their superiority also quantitatively.

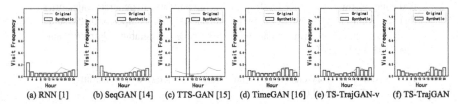

Fig. 1. Overall temporal visit frequency distribution of different methods over Taxi-1: (a) RNN [1], (b) SeqGAN [14], (c) TTS-GAN [15], (d) TimeGAN [16], (e) TS-TrajGAN-v (a variant of TS-TrajGAN which we present in Sect. 4.1.2), and (f) TS-TrajGAN (our model).

To generate trajectories with good data utility, we present TS-TrajGAN, a two-stage generative model for spatiotemporal trajectories based on GAN and conditional GAN in this paper (Fig. 2). Rather than generate one trajectory as a whole in one step, we assume a trajectory is mainly determined by the initial trajectory segment and the next point, and generate a trajectory in two stages. In stage I, we leverage a GAN to simulate the distribution of the initial trajectory segments, and a conditional GAN is used to capture the characteristics of the next point and preserve the variability in individuals' mobility in stage II. During trajectory generation, the initial segment of a trajectory is firstly generated, and then the conditional GAN iteratively predicts the next point for the current generated trajectory by feeding several previously generated location points as a condition label. To decide when the next location point prediction should stop, a trajectory length prediction module can be added to the GAN of stage I.

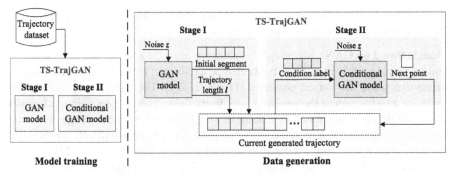

Fig. 2. Model training and data generation of TS-TrajGAN.

In TS-TrajGAN, the features of the initial trajectory segments can be well 'learned' using the GAN, because the length of an initial trajectory segment is very limited. Moreover, using a conditional GAN to predict the next point can produce different predictions in contrast to the single average behavior predicted by a sequential model, such as RNN or LSTM models. This helps generate new trajectory datasets that are disentangled from the real trajectories, and thus reduce the risk of privacy leakage. We conduct extensive experiments on a real-world dataset and demonstrate that TS-TrajGAN outperforms the state-of-the-art works with multiple qualitive metrics.

2 Related Work

Current deep learning-based trajectory generation studies focus on three research streams. In this section, we discuss relevant works under each of them.

Models for Trajectory Reconstruction. The works of this type aim at reconstructing plausible routes using the given points of a trajectory [1, 18, 20]. For instance, Wu et al. [18] modify the RNN structure [7] to model the road network-constrained trajectories, and generate the complete trajectory by taking an initial point as the input. Based on variational autoencoder [19, 20] presents a sequential variational autoencoder, which reconstructs the input trajectories by capturing the salient characteristics of the training data. These models [18, 20] can generate plausible trajectories, but the "average" behavior may be produced in cases where can be diverse outputs. This is undesirable if similar trajectories are generated given the same input, e.g., the same origin. To preserve the variability of trajectories, [1] builds a RNN to generate trajectories between the home and work locations for a corresponding synthetic population, and introduce calibrated randomness to allow variations in the model's output.

Models Based on Next Point Prediction. Methods of this research stream generate a trajectory by sequentially predicting the next location of a trajectory using the previously visited locations [2, 21]. For example, Kulkarni and Garbinato [21] builds a next location prediction model by leveraging a LSTM-based RNN. Starting at some arbitrary location, this method generates the complete trajectory by iteratively feeding the current output trajectory sequence to the trained model. [2] develops a similar model using the BiLSTM neural network. Since these methods [2, 21] may predict the single average behavior, the key features of the original data may not be well captured.

Models Based on GAN. Models of this category learn the original trajectory distribution by playing a minimax game between a generator and a discriminator, and sample trajectories using the trained generator [11–13, 22–24]. These methods primarily differ in the construction of the generator and discriminator, and the encoding techniques applied to the training dataset. For example, [11] leverages convolution layers to build the GAN model, and discretizes continuous geographical space into cells to represent the stays of each trajectory. To simulate the road information within cells, [13] develops a two-stage GAN where the first stage uses the idea of [11] to generate trajectories in cell form and the second stage employs two parallel LSTMs to construct a conditional GAN for the detail trajectory generation in each cell. Rao et al. [23] convert the trajectories employing the one-hot encoder, and feed the encoded trajectories into the GAN model which is constructed based on LSTMs. Cao and Li [12] combine a DCGAN and a Seq2Seq model to capture the distribution of spatial data, and use a non-linear ANN model to sample the timestamp for the first location of each trajectory. In addition, [25, 26] use generative adversarial imitation learning and the adversarial autoencoder (AAE) structure which are similar to GAN to generate trajectories.

Trajectories have similar sequential, temporal correlational characteristics as time-series data. Recently, several GAN-based methods for time-series generation have been proposed. They may construct the architecture by leveraging RNNs [27], CNNs [28] or transformers [15]. TimeGAN [16] is a state-of-the-art method which combines autoencoding components and the adversarial components together to preserve the temporal correlations of time-series data. In our implementation of TS-TrajGAN, TimeGAN is adopted as the backbone of the GAN and the conditional GAN.

3 Our Method

3.1 Overview

The system design is illustrated in Fig. 2. TS-TrajGAN consists of two stages. The GAN of stage I learns the distribution of initial trajectory segments and predicts the trajectory lengths, while the conditional GAN of stage II simulates the distribution of next point by taking an extra condition label as an input.

In the implementation of TS-TrajGAN, we use TimeGAN [16] as the backbone. Some necessary modifications are made to convert TimeGAN to the GAN and the conditional GAN of our model (Fig. 3). Specifically, we add a predictor network to TimeGAN in stage I for initial trajectory segments generation and trajectory length prediction (Fig. 3 (b)). In stage II, to predict the next point for the current generated trajectory and preserve the variability in individuals' mobility, we add a concatenation block (Fig. 4) to each network component of TimeGAN for the embedding of the condition label.

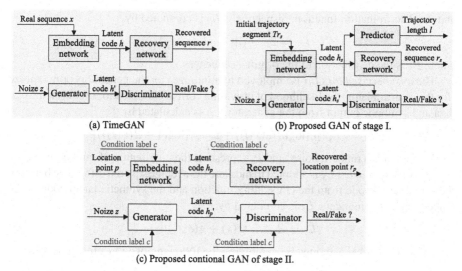

(a) TimeGAN   (b) Proposed GAN of stage I.

(c) Proposed contional GAN of stage II.

Fig. 3. Overview of the architectures of different models: (a) TimeGAN [16], (b) proposed GAN of stage I, and (c) proposed conditional GAN of stage II.

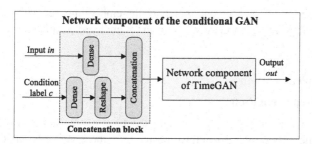

Fig. 4. Block diagram of a network component of the proposed conditional GAN.

3.2 TimeGAN [16]

TimeGAN [16] is a variant of GAN for generating time-series data. It is made up of four network components: an embedding network, recovery network, generator, and discriminator (Fig. 3(a)). It simultaneously learns to encode features, generate representations, and iterate across time by jointly training the autoencoding components (first two) and the adversarial components (latter two). The adversarial network operates within the latent space provided by the embedding network, and a supervised loss synchronizes the latent dynamics of both real and synthetic data.

The four components are trained on three losses: *reconstruction loss* L_R, *unsupervised loss* L_U and *supervised loss* L_S. The *reconstruction loss* L_R is used to measure the average squared difference between the original sequences and the output sequences recovered from their latent representations. Let x be a real sequence, $e(\cdot)$, $r(\cdot)$, $g(\cdot)$ and $d(\cdot)$ represent the embedding function, the recovery function, the generating function, and the discrimination function, respectively. L_R is computed by.

$$L_R = \mathbb{E}_{x \sim P_x}(||x - r(e(x))||_2),$$ (1)

where P_x is the distribution of the original sequences.

The *unsupervised loss* L_U is employed to minimize (for the generator) or maximize (for the discriminator) the likelihood of providing correct classifications for both the real data and synthetic output from the generator. L_U is calculated by

$$L_U = \mathbb{E}_{x \sim P_x}[\log(d(e(x)))] + \mathbb{E}_{z \sim P_z}[\log(1 - d(g(z)))].$$ (2)

where z denotes a random noise, and P_z represents the distribution of random noise.

The *supervised loss* L_S is adopted to measure the average squared difference between the actual latent code from the embedding function and the synthetic latent code of the embeddings of actual data. L_S is computed by

$$L_S = \mathbb{E}_{||x \sim P_x}(e(x) - g(e(x))||_2).$$ (3)

Then, TimeGAN is trained by combining the following two objective functions:

$$\begin{cases} \min\limits_{\theta_e, \theta_r}(\varphi L_S + L_R) \\ \min\limits_{\theta_g}(\beta L_S + \max\limits_{\theta_d} L_U) \end{cases},$$ (4)

where θ_e, θ_r, θ_g, and θ_d are the parameters of the embedding, recovery, generator, and discriminator networks, respectively, and φ and β are hyperparameters that balance the losses.

3.3 Our Implementation

3.3.1 Data Representation

A spatiotemporal trajectory is often a finite sequence of timestamped location points. Maintaining the ascending order of the timestamps in the generated outputs is important for providing good data utility. Here, we represent the temporal data based on time increment.

Let $D = \{Tr^i | i = 0, 1, ..., |D|-1\}$ be a trajectory dataset that has $|D|$ trajectories, and $Tr^i = \{p_j^i(loc_j^i, t_j^i)| j = 0, 1, ..., |Tr^i| - 1\}$ denote the i-th trajectory of D. Tr^i is represented by a sequence of $|Tr^i|$ location points, with each location point p_j^i consisting of a location loc_j^i and a time stamp t_j^i ($t_j^i \leq t_{j+1}^i$). The location loc_j^i is further denoted by ($loc_j^i.lon$, $loc_j^i.lat$), which refers to the longitude and latitude of loc_j^i.

For the j-th location point $p_j^i(loc_j^i, t_j^i)$ of Tr^i, we transform its timestamp t_j^i into its increment form Δ_j^i,

$$\begin{cases} \Delta_j^i = t_j^i - t_{j-1}^i & \text{if } j > 0 \\ \Delta_j^i = t_j^i & \text{if } j = 0 \end{cases}.$$ (5)

After the temporal data transformation, the original dataset D becomes $D_c = \{Tr_c^i | i = 0, 1, ..., |D_c| - 1\}$ ($|D_c|=|D|$), where $Tr_c^i = \{p_{c,j}^i(loc_j^i, \Delta_j^i)| j = 0, 1, ..., |Tr_c^i| - 1\}$ is the i-th trajectory of D_c, and $p_{c,j}^i$ denotes the j-th location point of Tr_c^i.

3.3.2 Stage I: Initial Trajectory Segments Distribution Learning

To learn the initial trajectory segments distribution and predict the trajectory length simultaneously, we design a predictor network, and add it to TimeGAN [16] to construct the GAN of stage I (Fig. 3 (b)). Let Tr_s be the initial segment of the trajectory Tr_p. As shown in Fig. 5, the predictor takes the latent code h_s of Tr_s to the predicted trajectory length l. It consists of a RNN, an average pooling layer and a dense layer. The loss function of the predictor is calculated by

$$L_{pred} = \mathbb{E}_{Tr_s \sim P_{Tr_s}}(\||l_p - pred(e(Tr_s))\||_2), \qquad (6)$$

where P_{Tr_s} denotes the distribution of the real initial trajectory segments, l_p represents the length of Tr_p, and $pred(\cdot)$ is the predicting function.

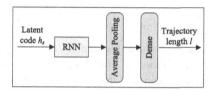

Fig. 5. Block diagram of the predictor network.

During training, the five components of the GAN in stage I are trained to encode features, generate latent representations, predict trajectory length, and iterate simultaneously. The objective function of the predictor network can be expressed as:

$$\min_{\theta_{pred}} (L_{pred}), \qquad (7)$$

where θ_{pred} is the parameter of the predictor network.

For this stage, we derive the training dataset by clipping a length-m initial segment from each trajectory of D_c. Let $Tr_s^i = \left\{ p_{c,j}^i | j = 0, 1, \ldots, m \right\}$ be Tr_c^i's initial segment, and $D_s = \left\{ Tr_s^i | i = 0, 1, \ldots, |D_s| - 1 \right\} (|D_s| = |D_c|)$ that contains all the initial segments of D_c be the training dataset of stage I.

3.3.3 Stage II: Next Point Distribution Learning

In stage II, we aim to model the probability distribution of the next point p given the current generated trajectory $Tr_g = \{p_j | j = 0, 1, \ldots, |Tr_g| - 1\}$, i.e., $Pr(p|Tr_g)$. To achieve this goal, we build a conditional GAN (Fig. 3 (c)) by feeding an extra condition label c, which is composed of the last n ($n < m$) location points of Tr_g, to the four components of TimeGAN. The conditional GAN is trained in a similar fashion to TimeGAN, but learns to generate the next location points from distributions which are conditioned on the condition label.

To embed the conditional label, the four network components of the conditional GAN are built by adding a concatenation block to each of the four components of TimeGAN. The structures of the added concatenation blocks are the same. As shown in Fig. 4, within the concatenation block, the input *in* passes through a dense layer to construct its representation, while the condition label c passes through a dense layer and then is reshaped before obtaining its representation. After that, the representations of the input *in* and the condition label c are concatenated.

During training, the four components of the conditional GAN are trained to encode features and condition labels, generate latent representations, and iterate simultaneously. We use the following three losses to update the parameters: *reconstruction loss* $L_{R'}$, *unsupervised loss* $L_{U'}$ and *supervised loss* $L_{S'}$. $L_{R'}$ is used to measure the average squared difference between the real location point and the output location point recovered from the latent representation under the condition of c. Let P_p be the distribution of the original location points, and $e'(\cdot)$, $r'(\cdot)$, $g'(\cdot)$, and $d'(\cdot)$ denote the embedding function, the recovery function, the generating function, and the discrimination function, respectively. $L_{R'}$ is obtained by

$$L'_R = \mathbb{E}_{p \sim P_p}(||p - r'((e'(p|c))|c)||_2). \tag{8}$$

The *unsupervised loss* $L_{U'}$ is adopted to minimize (for the generator) or maximize (for the discriminator) the likelihood of providing correct classifications for both the latent codes of the real location point and the synthetic output from the generator under the condition of c. We calculate $L_{U'}$ by

$$L'_U = \mathbb{E}_{p \sim P_p}[\log(d'((e'(p|c))|c))] + \mathbb{E}_{z \sim P_z}[\log(1 - d'((g'(z|c))|c))]. \tag{9}$$

The *supervised loss* $L_{S'}$ is leveraged to measure the average squared difference between the actual latent code from the embedding function and the synthetic latent code of the embeddings of actual data under the condition of c. $L_{S'}$ is expressed as:

$$L'_S = \mathbb{E}_{p \sim P_p}(||e'(p|c) - g'e'(p|c)||_2). \tag{10}$$

The objective functions to optimize the four component networks can be written as:

$$\begin{cases} \min_{\theta_{e'}, \theta_{r'}} (\varphi L'_S + L'_R) \\ \min_{\theta_{g'}} (\beta L'_S + \max_{\theta_{d'}} L'_U) \end{cases}, \tag{11}$$

where $\theta_{e'}$, $\theta_{r'}$, $\theta_{g'}$, and $\theta_{d'}$ are the parameters of the embedding, recovery, generator, and discriminator networks of the conditional GAN, respectively.

In our model, each condition label c contains n location points. Therefore, we derive the training dataset of stage II based on $(n + 1)$-gram extraction. We regard $n + 1$ adjacent location points of each trajectory as an $(n + 1)$-gram. For a trajectory $Tr^i_c = \left\{ p^i_{c,j}(loc^i_j, \Delta^i_j)|j = 0, 1, \ldots, |Tr^i_c| - 1 \right\}$ of D_c, we can get $(|Tr^i_c|$-n-$1)$ $(n + 1)$-grams by scanning its location points from $p^i_{c,1}$. Then, each $(n + 1)$-gram is transformed into a sample of the training dataset, with its first n location points being used to form the condition label of its last location point. That is, for an $(n + 1)$-gram $s = \{p_0, p_1, \ldots, p_n\}$ extracted from D_c, the n-gram $s_c = \{p_0, p_1, \ldots, p_{n-1}\}$ is the condition label of p_n.

3.3.4 Trajectory Generation

To generate complete trajectories, we concatenate the GAN and conditional GAN trained in the two stages. Specifically, we use the generator, recovery and predictor networks of the GAN trained in stage I to generate the initial trajectory segments and their corresponding trajectory lengths. For each initial trajectory segment whose length has not reached its predicted length, we leverage the generator and the recovery networks of the conditional GAN trained in stage II to predict the next location point by regarding its last n-gram as the condition label, and append the generated location point to it. This location point prediction and appending process is repeated until the predicted trajectory length is reached.

Let $D_{c}{'} = \{Tr_{c}^{i}{'}|i = 0, 1, \ldots, |Tr_{c}{'}| - 1\}$ be the trajectory dataset generated by TS-TrajGAN, $Tr_{c}^{i}{'} = \{p_{c,j}^{i}{'}(loc_{j}^{i}{'}, \Delta_{j}^{i}{'})|j = 0, 1, \ldots, |Tr_{c}^{i}{'}| - 1\}$ denote the i-th trajectory of D_{c}', and $p_{c,j}^{i}{'}(loc_{j}^{i}{'}, \Delta_{j}^{i}{'})$ represent the j-th location point of $Tr_{c}^{i}{'}$. We get the final generated trajectory dataset D' by transforming each trajectory $Tr_{c}^{i}{'}$ of D_{c}' into the output form $Tr^{i}{'} = \{(p_{j}^{i}{'}(loc_{j}^{i}{'}, t_{j}^{i}{'}) \mid j = 0, 1, \ldots, |Tr^{i}{'}| - 1\}$ ($|Tr^{i}{'}| = |Tr_{c}^{i}{'}|$) by

$$
\begin{aligned}
t_{j}^{i}{'} &= t_{j-1}^{i}{'} + \Delta_{j}^{i}{'} \quad \text{if } j > 0 \\
t_{j}^{i}{'} &= \Delta_{j}^{i}{'} \quad\quad\quad\ \text{if } j = 0
\end{aligned}
\tag{12}
$$

4 Experimental Results and Analysis

4.1 Experiment Setup

4.1.1 Datasets

We use the dataset of the Taxi Service Prediction Challenge at ECML-PKDD2015 [17] to evaluate the performance of TS-TrajGAN. This dataset consists of a complete year (from 01/07/2013 to 30/06/2014) data of 1.72 million traces for 442 taxis running in Porto. We extract two sub-datasets from it (namely Taxi-1 and Taxi-2) using two bounding boxes: ((41.104N, 8.665W), (41.250N, 8.528W)) and ((41.064N, 8.662W), (41.210N, 8.525W)). Both Taxi-1 and Taxi-2 is 11.49 km × 16.23 km in area. For each of them, the first 20000 trajectories are used to construct the original dataset D in the experiments.

4.1.2 Baselines

We compare TS-TrajGAN with four state-of-the-art baselines [1, 14–16] and a variant of it (TS-TrajGAN-v). The implementations of [1, 14–16] were obtained from the respective authors.

- **RNN** [1]. This work trains a RNN to generate realistic spatiotemporal trajectories by taking home and work locations as input.
- **SeqGAN** [14]. It combines reinforcement learning with GAN for generating sequences of discrete tokens. We directly apply this method to generate trajectories.
- **TTS-GAN** [15]. It constructs the generator and discriminator networks of the GAN model by using a pure transformer encoder architecture for time-series generation.

- **TimeGAN** [16]. It is the backbone of the GAN and conditional GAN in the implementation of TS-TrajGAN. It preserves the temporal correlations of time-series data by combining autoencoding components and the adversarial components together.
- **TS-TrajGAN-v.** It is a variant of TS-TrajGAN. Compared with TS-TrajGAN, TS-TrajGAN-v does not have the predictor network. The trajectory generation stops when the maximum trajectory length is reached or an unreasonable location point is generated, e.g., the time stamp of the location point is earlier than its previous location point. We design this model to validate the efficacy of the predictor network we designed for stage I.

To provide home and work locations for [1] and ensure fair comparison, we preprocess the datasets for all the methods. We impose a 20×20 uniform grid over the spatial domain of the datasets, map all the location points into the cells of the grid, and select representative location points for discrete place representation. For the neighboring location points of a trajectory which are located in the same cell, the first of them is regarded as a representative location point. For [1], we regard the start and end cells of each trajectory as the home and work location, respectively.

4.1.3 Evaluation Metrics

We evaluate TS-TrajGAN using qualitative visualizations and quantitative metrics, and compare it with the baselines.

- **Visualization**. To illustrate how closely the distribution of generated trajectories resembles that of the real in 2-dimensional space, we apply t-SNE analysis on both the real data and synthetic data.
- **Discriminative score** [16]. We train a classifier (by optimizing a 2-layer LSTM) to distinguish between the real and synthetic trajectories as a standard supervised task. The discriminative score is calculated on the held-out test set. A score close to 0 indicating the generated trajectories are hard to distinguish from the real ones, i.e., the lower the better.
- **Predictive score** [16]. We train a trajectory-prediction model (by optimizing a 2-layer LSTM) using the synthetic trajectory dataset to predict the next point over each input trajectory segment. The predictive score is obtained using the mean absolute error on the real trajectory dataset. A score close to 0 is better.
- **Frequent pattern average relative error (FP AvRE)** [29]. To evaluate the frequent pattern preservation ability, we impose a uniform grid G over the spatial domain of D to compute FP AvRE. Let $C = \{c_i | i = 0, 1, \ldots, \Upsilon\}$ be the cell set of G, and Υ be the number of C's elements. Assume a pattern P is represented by an ordered list of the cells of C. Let $o(P, D)$ denote the number of occurrences of P in D, and $FP_G^k(D) = \{P_i | i = 0, 1, \ldots, k - 1\}$ be the top-k patterns in D. The FP AvRE is obtained by

$$FPAvRE = \frac{\sum_{P_i \in FP_G^k(D)} \frac{|o(P_i,D)-o(P_i,D')|}{o(P_i,D)}}{k}. \tag{13}$$

In our experiments, the scale of G is 20×20, the parameter $k = 200$, and the length of each pattern is from 2 to 8.

- **Trip error** [29]. We use this metric to measure how well the correlations between the start and end regions of the original trajectories are preserved. Let $dt(G, D)$ denote the distribution of all possible pairs of start and end cells of D. The trip error is calculated as the Jensen-Shannon divergence between $dt(G, D)$ and $dt(G, D')$: $JSD(dt(G, D), dt(G, D'))$.
- **Length error** [29]. Let $Len(D)$ be the empirical distribution of the trip lengths on D, where the trip lengths are quantized into 20 equal width buckets. Then, the length error is calculated as:$JSD(Len(D), Len(D'))$.
- **Overall temporal visit frequency distribution** [23]. This metric can be used to evaluate the temporal similarity of different models. In our experiments, we count the frequencies of visits for each time interval (e.g. 2 h) in D and D' and convert them into probability distribution for the overall temporal visit frequency distribution evaluation.

In the experiments, we repeat each experiment five times and report the average results for the FP $AvRE$, trip error, and length error.

4.2   Comparison with Baselines

We compare TS-TrajGAN with five state-of-the-art baselines on Taxi-1 and Taxi-2 using the evaluation metrics in Sect. 4.1.3. The comparison results are presented in Fig. 1, Fig. 6, Table 1, Table 2, and Fig. 7. We have two general observations. First, TS-TrajGAN provides better data utility than the baselines in general. Second, the performance of TS-TrajGAN is superior to TS-TrajGAN-v, which indicates that the predictor network we designed for stage I could help to improve the data utility.

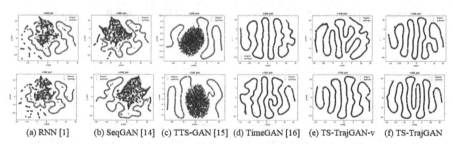

(a) RNN [1]   (b) SeqGAN [14]   (c) TTS-GAN [15]   (d) TimeGAN [16]   (e) TS-TrajGAN-v   (f) TS-TrajGAN

Fig. 6. t-SNE visualization on Taxi-1 (1st row) and Taxi-2 (2nd row). Each column provides the visualization for each of the 6 methods. Red dots denote original data, and blue dots denote synthetic data.

- **Visualization with t-SNE.** In Fig. 6, each dot represents a real or synthetic trajectory value after dimensionality reduction. We observe that the dots from the synthetic datasets generated by TS-TrajGAN overlap well with the dots from the real trajectories, which indicates that the synthetic trajectories generated by TS-TrajGAN are similar to the original ones. We also notice that the performance of TS-TrajGAN and TS-TrajGAN-v are comparable to TimeGAN [16], and remarkably better than RNN [1], SeqGAN [14] and TTS-GAN [15].

Table 1. Comparing TS-TrajGAN with the baselines on Taxi-1. Best result in each category is shown in bold. For all the metrics, lower values are better.

Metrics	Discriminative score	Predictive score	FP *AvRE*	Trip error	Length error
RNN [1]	0.181 ± 0.043	0.086 ± 0.000	0.675	**0.355**	0.035
SeqGAN [14]	0.188 ± 0.053	0.089 ± 0.001	0.837	0.358	0.024
TTS-GAN [15]	0.446 ± 0.046	0.130 ± 0.006	1.852	0.665	0.090
TimeGAN [16]	0.272 ± 0.064	0.251 ± 0.012	0.836	0.687	0.425
TS-TrajGAN-v	0.124 ± 0.025	**0.085 ± 0.000**	**0.629**	0.391	0.022
TS-TrajGAN	**0.077 ± 0.024**	**0.085 ± 0.000**	**0.629**	0.377	**0.012**

Table 2. Comparing TS-TrajGAN with the baselines on Taxi-2. Best result in each category is shown in bold. For all the metrics, lower values are better.

Metrics	Discriminative score	Predictive score	FP *AvRE*	Trip error	Length error
RNN [1]	0.191 ± 0.046	**0.082 ± 0.000**	0.648	0.369	0.041
SeqGAN [14]	0.218 ± 0.046	0.083 ± 0.001	0.811	**0.360**	0.046
TTS-GAN [15]	0.433 ± 0.033	0.381 ± 0.017	3.325	0.682	0.380
TimeGAN [16]	0.250 ± 0.069	0.147 ± 0.003	1.100	0.601	0.399
TS-TrajGAN-v	0.118 ± 0.028	0.084 ± 0.001	0.498	0.368	0.032
TS-TrajGAN	**0.083 ± 0.013**	0.084 ± 0.000	**0.497**	**0.360**	**0.014**

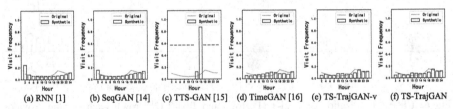

(a) RNN [1]   (b) SeqGAN [14]   (c) TTS-GAN [15]   (d) TimeGAN [16]   (e) TS-TrajGAN-v   (f) TS-TrajGAN

Fig. 7. Overall temporal visit frequency distribution of different methods over Taxi-2: (a) RNN [1], (b) SeqGAN [14], (c) TTS-GAN [15], (d) TimeGAN [16], (e) TS-TrajGAN-v, and (f) TS-TrajGAN (our model).

- **Discriminative and predictive scores.** From Table 1 and Table 2, we can see TS-TrajGAN produces lower discriminative score than the baseline models on both Taxi-1 and Taxi-2. The predictive scores of TS-TrajGAN and TS-TrajGAN-v are better than the other four baselines on Taxi-1, and are very close to the best provided by RNN [1] (the difference between them is less than 0.003) on Taxi-2. That is, the synthetic trajectories generated by TS-TrajGAN are more similar to the original ones than the baselines in general. In addition, we can find TS-TrajGAN outperforms TS-TrajGAN-v in terms of discriminative score. This demonstrates the efficacy of the predictor network we proposed for stage I.

- **FP $AvRE$, trip error, and length error.** It can be observed from Table 1 and Table 2 that TS-TrajGAN outperforms the baselines in terms of FP $AvRE$ and length error on both Taxi-1 and Taxi-2. The trip error value of TS-TrajGAN on Taxi-1 is a little higher than the best case (no more than 0.025). That is, TS-TrajGAN outperforms the baselines in the preservation of the spatial characteristics in general. Moreover, we can find TS-TrajGAN-v is a little inferior to TS-TrajGAN in terms of length error and trip error, which also demonstrates the efficacy the predictor network of stage I.
- **Overall temporal visit frequency distribution.** Figure 1 and Fig. 7 illustrate the overall temporal visit frequency distribution of different methods over Taxi-1 and Taxi-2, respectively. The i-th ($i = 1, 2, ..., 12$) bin indicates the visit frequency of the time interval from the $(2i - 1)$-th hour to the $2i$-th hour of the day. We can see the overall temporal visit frequency distributions from both TS-TrajGAN and TS-TrajGAN-v can better fit the original data than the baselines, which indicates TS-TrajGAN and TS-TrajGAN-v preserve more temporal characteristics of the original data. Moreover, we notice that the performance of TTS-GAN [15] is not good. The reason might be that the pure transformer encoder architecture adopted by TTS-GAN is not quite effective in preserving the correlational characteristics of the location points in Taxi-1 and Taxi-2.

5 Conclusions

We present TS-TrajGAN, a two-stage generative model for spatiotemporal trajectories based on GAN and conditional GAN. The GAN of stage I generates the initial trajectory segments and predicts the trajectory length, while the conditional GAN predicts the next location point for the current generated trajectory and preserves the variability in individuals' mobility behaviors. Experimental results on a real-world taxi dataset show that the trajectories generated by TS-TrajGAN provide similar distribution with the real ones, and preserve more essential spatial and temporal characteristics of the real data than state-of-the-art techniques [1, 14–16].

References

1. Berke, A., Doorley, R., Larson, K., Moro, E.: Generating synthetic mobility data for a realistic population with RNNs to improve utility and privacy. In: Proceedings of the 37th ACM/SIGAPP Symposium on Applied Computing, pp. 964–967 (2022)
2. Blanco-Justicia, A., Jebreel, N.M., Manjón, J.A., Domingo-Ferrer, J.: Generation of synthetic trajectory microdata from language models. In: Domingo-Ferrer, J., Laurent, M. (eds.) Privacy in Statistical Databases. PSD 2022. Lecture Notes in Computer Science, vol. 13463, pp. 172–187. Springer, Cham (2022). https://doi.org/10.1007/978-3-031-13945-1_13
3. Kulkarni, V., Tagasovska, N., Vatter, T., Garbinato, B.: Generative models for simulating mobility trajectories. arXiv preprint arXiv:1811.12801 (2018)
4. Theodoridis, Y., Silva, J.R.O., Nascimento, M.A.: On the generation of spatiotemporal datasets. In: Güting, R.H., Papadias, D., Lochovsky, F. (eds.) SSD 1999. LNCS, vol. 1651, pp. 147–164. Springer, Heidelberg (1999). https://doi.org/10.1007/3-540-48482-5_11

5. Bindschaedler, V., Shokri, R.: Synthesizing plausible privacy-preserving location traces. In: 2016 IEEE Symposium on Security and Privacy (SP), pp. 546–563. IEEE (2016)
6. Shin, S., Jeon, H., Cho, C., Yoon, S., Kim, T.: User mobility synthesis based on generative adversarial networks: a survey. In: 2020 22nd International Conference on Advanced Communication Technology (ICACT), pp. 94–103. IEEE (2020)
7. Elman, J.L.: Finding structure in time. Cogn. Sci. **14**(2), 179–211 (1990)
8. Hochreiter, S., Schmidhuber, J.: Long short-term memory. Neural Comput. **9**(8), 1735–1780 (1997)
9. Goodfellow, I.J., et al.: Generative adversarial nets. Adv. Neural Inf. Process. Syst. 27 (2014)
10. Mirza, M., Osindero, S.: Conditional generative adversarial nets. arXiv preprint arXiv:1411. 1784 (2014)
11. Ouyang, K., Shokri, R., Rosenblum, D.S., Yang, W.: A non-parametric generative model for human trajectories. In: International Joint Conferences on Artificial Intelligence, vol. 18, pp. 3812–3817 (2018)
12. Cao, C., Li, M.: Generating mobility trajectories with retained data utility. In: Proceedings of the 27th ACM SIGKDD Conference on Knowledge Discovery & Data Mining, pp. 2610–2620 (2021)
13. Wang, X., Liu, X., Lu, Z., Yang, H.: Large scale GPS trajectory generation using map based on two stage GAN. J. Data Sci. **19**(1), 126–141 (2021)
14. Yu, L., Zhang, W., Wang, J., Yu, Y.: SeqGAN: sequence generative adversarial nets with policy gradient. In: Proceedings of the AAAI Conference on Artificial Intelligence, vol. 31, no. 1 (2017)
15. Li, X., Metsis, V., Wang, H., Ngu, A.H.H.: TTS-GAN: a transformer-based time-series generative adversarial network. In: 20th International Conference on Artificial Intelligence in Medicine, pp. 133–143 (2022)
16. Yoon, J., Jarrett, D., Schaar, M.V.: Time-series generative adversarial networks. Adv. Neural Inf. Process. Syst. **32** (2019)
17. https://www.kaggle.com/competitions/pkdd-15-predict-taxi-service-trajectory-i/data
18. Wu, H., Chen, Z., Sun, W., Zheng, B., Wang, W.: Modeling trajectories with recurrent neural networks. In: Proceedings of the 26th International Joint Conference on Artificial Intelligence, vol. 25, pp. 3083–3090 (2017)
19. Kingma, D.P., Welling, M.: Auto-encoding variational Bayes. arXiv preprint arXiv:1312. 6114 (2013)
20. Huang, D., et al.: A variational autoencoder based generative model of urban human mobility. In: 2019 IEEE Conference on Multimedia Information Processing and Retrieval, pp. 425–430. IEEE (2019)
21. Kulkarni, V., Garbinato, B.: Generating synthetic mobility traffic using RNNs. In: Proceedings of the 1st Workshop on Artificial Intelligence and Deep Learning for Geographic Knowledge Discovery, pp. 1–4 (2017)
22. Liu, X., Chen, H., Andris, C.: trajGANs: using generative adversarial networks for geo-privacy protection of trajectory data (Vision paper). In: Location Privacy and Security Workshop, pp. 1–7 (2018)
23. Rao, J., Gao, S., Kang, Y., Huang, Q.: LSTM-TrajGAN: A deep learning approach to trajectory privacy protection. arXiv preprint arXiv:2006.10521 (2020)
24. Rossi, L., Paolanti, M., Pierdicca, R., Frontoni, E.: Human trajectory prediction and generation using LSTM models and GANs. Pattern Recogn. **120**, 108136 (2021)
25. Choi, S., Kim, J., Yeo, H.: TrajGAIL: Generating urban vehicle trajectories using generative adversarial imitation learning. Transp. Res. Part C Emerg. Technol. **128**, 103091 (2021)
26. Kim, J.W., Jang, B.: Deep learning-based privacy-preserving framework for synthetic trajectory generation. J. Netw. Comput. Appl. **206**, 103459 (2022)

27. Esteban, C., Hyland, S.L., Rätsch, G.: Real-valued (medical) time series generation with recurrent conditional GANs. arXiv preprint arXiv:1706.02633 (2017)
28. Ramponi, G., Protopapas, P., Brambilla, M., Janssen, R.: T-CGAN: Conditional generative adversarial network for data augmentation in noisy time series with irregular sam-pling. arXiv preprint arXiv:1811.08295 (2018)
29. Gursoy, M.E., Liu, L., Truex, S., Yu, L., Wei, W.: Utility-aware synthesis of differentially private and attack-resilient location traces. In: Proceedings of the 2018 ACM SIGSAC Conference on Computer and Communications Security, pp. 196–211. ACM (2018)

Visual Navigation of Target-Driven Memory-Augmented Reinforcement Learning

Weimin Li[1], Xing Wu[1(✉)], Zhongshi He[1(✉)], and Bo Zhou[2]

[1] College of Computer Science, Chongqing University, Chongqing 401400, China
{wuxing,zshe}@cqu.edu.cn
[2] China National Postal and Telecommunications Appliances Co., Beijing 100032, China

Abstract. Compared to visual navigation methods based on reinforcement learning that rely on auxiliary information such as depth images, semantic segmentation, object detection, and relational graphs, methods that solely utilize RGB images do not require additional equipment and have better flexibility. However, these methods often suffer from underutilization of RGB image information, resulting in poor generalization performance of the model. To address this limitation, we present the Target-Driven Memory-Augmented (TDMA) framework. This framework utilizes an external memory to store fused Target-Scene features obtained from the observed and target images. To capture and leverage long-term dependencies within this stored data, we employ the Transformer model to process historical information. Additionally, we introduce a self-attention sub-layer in the Decoder section of the Transformer to enhance the model's focus on similar regions between the observed and target images. Experimental evaluations conducted on the AI2-THOR dataset demonstrate that our proposed method achieves an 8% improvement in success rate and a 16% improvement in success weighted by path length compared to methods in the same experimental setup.

Keywords: Reinforcement Learning · Visual Navigation · Transformer · AI2-THOR

1 Introduction

Visual navigation entails the task of an agent locating a target object within a given environment. Traditional visual navigation methods involve extracting feature points from images and matching them with predefined feature points on a map, subsequently generating actions based on the matching results [1, 2]. However, this approach necessitates the manual creation of maps, making the process laborious. The Deep Reinforcement Learning (DRL) method employs neural networks to learn the features in images, enabling direct action prediction. This eliminates the need for acquiring maps and matching key points, resulting in a faster and more convenient process.

The primary challenge in visual navigation methods based on DRL lies in effectively utilizing limited information to comprehend complex environments. During continuous navigation, agents must comprehend the relationship between local information observed

© The Author(s), under exclusive license to Springer Nature Singapore Pte Ltd. 2024
B. Luo et al. (Eds.): ICONIP 2023, CCIS 1961, pp. 422–432, 2024.
https://doi.org/10.1007/978-981-99-8126-7_33

in the environment and the target information. Furthermore, they must utilize historical information to overcome disorientation dilemmas that may arise. In order to enhance the understanding of complex environments, it is common practice to mine internal information or combine external auxiliary information such as depth images, semantic segmentation, object detection, and relational graphs. Compared with the method of using auxiliary information [4–13], the method of mining internal information [14–17] does not require additional equipment and is more flexible in application. However, current methods of mining internal information are deficient in fusing information and processing historical information. For instance, when fusing information, a single linear layer struggles to emphasize the similar parts of the observed and target information. Additionally, recurrent neural networks face challenges in capturing long-term dependencies when handling historical data.

To address these challenges, we propose the Target-Driven Memory-Augmented (TDMA) framework. Aiming at the problem of gradual loss of historical memory in recurrent neural networks, this framework fuses observation images and target images to create target scene features, stores them in chronological order, and finally employs an attention mechanism to extract features from the stored data. To address the problem of insufficient attention to similar information when fusing observed and target images, we design a method to enhance feature fusion using an attention mechanism. The specific method is to add a self-attention sublayer of the Target-Scene feature of the current time step to the Decoder of Transformer, and also serve as the query layer of the subsequent attention sublayer.

The contributions of our work can be summarized as follows:

- We introduce the Target-Driven Memory-Augmentation (TDMA) framework. The framework fuses observation and target images as Target-Scene features, stores them in chronological order and adopts attention mechanism for feature extraction, which better stores and utilizes historical information.
- We propose methods to enhance feature fusion using attention mechanisms. Specifically, we incorporate the self-attention sublayer of the Target-Scene features into the Transformer's Decoder, where it acts as a query for subsequent attention sublayers. This approach allows for a better emphasis on the similar parts between the observed image and the target image.
- Through experiments conducted in the AI2-THOR indoor environment [3], we demonstrate that our proposed method improves the success rate and success weighted by path length (SPL) of the model in navigating unknown targets by 8% and 16%, respectively.

2   Related Work

2.1   Methods With Auxiliary Information

In visual navigation methods based on DRL, auxiliary information plays a crucial role, and it primarily includes four types: depth images, semantic segmentation, object detection, and relationship graphs. Depth images and semantic segmentations are commonly employed as inputs to provide additional depth and semantic information [4–7]. However, this approach is difficult to directly obtain the category information of the object from the observed image. To address this, Mousavian et al. [8] developed a navigation strategy based on object detection and semantic segmentation. Du et al. [9] proposed a new Visual Transformer Network (VTNet), which not only encodes the relationship between objects detected by the object detection, but also establishes a strong correlation with the navigation signal.

However, both methods [8, 9] lack correlation information between objects and targets. Fukushima et al. [10] addressed this by calculating GloVe vector similarity between objects in observed images and the target object. Nevertheless, the correlation information obtained in [10] does not account for the positional relationship of objects in the environment. Yang et al. [11] utilized graph neural networks to incorporate prior knowledge of the positional relations of objects into the DRL framework. Wu et al. [12] captured prior knowledge of the scene layout from the training environment. Additionally, Lyu et al. [13] improved training efficiency by introducing attention mechanisms and target skill extension modules on a 3D knowledge graph in the classic DRL framework.

2.2 Methods Without Auxiliary Information

Effectively utilizing limited visual observation information is the key challenge in methods without auxiliary information. Jaderberg et al. [14] addressed this by adding three auxiliary reward functions to enable the model to adapt to navigation tasks more quickly. Zhu et al. [15] introduced a method where both the observation image and the target image are used as inputs, allowing navigation to different targets by changing the target image. However, this method [15] lacks a memory module for historical information, resulting in limitations in long-distance navigation. To overcome this, Chen et al. [16] incorporated LSTM to process historical information based on [15].

However, effectively handling the diversity of complex environments solely through LSTM is challenging. Xiao et al. [17] introduced a memory prediction mechanism that integrates the predicted next state with the current state to better adapt to complex environments. Most similar to this study is Mezghani et al. [18], but their method adopts both LSTM and Transformer structure, wherein LSTM is used to deal with continuous scene memory and Transformer is used to deal with discontinuous and distant observation features. In our method, we employ a self-attention sub-layer to enhance the similarity features of the observed image and the target image, and directly use the Transformer to handle continuous scene memory, effectively capturing the dependencies of continuous information. Finally, the navigation effects of Transformer and LSTM models dealing with historical information on unknown targets in a known environment and unknown targets in an unknown environment are evaluated by experiments.

3 Method

In this section, we first outline the problem setup studied in this paper, then introduce the model structure, and finally present the loss function.

3.1 Problem Setup

Visual navigation problems can be expressed as partially observable Markov decision processes (POMDP) $\left(S, A, O, R(s), T\left(s'|s, a\right), P(o, s)\right)$ where S, A, O are state, action and observation space, $R(s)$ is the reward function for state $s \in S$, $T\left(s'|s, a\right)$ is state transition probability, $P(o, s)$ is observation probability.

The state space S is the set of all reachable positions of the agent in the environment. We adopt a discrete action space defined as $A = \{Move\ Forward, Rotate\ Right, Rotate\ Left, Move\ Backward\}$, where each move corresponds to a distance of 0.5 m, and a rotation of 90 degrees. The Observations in O consist of RGB images representing the observed image and the target image. The reward function $R(s)$ is shown in Eq. (1).

$$R(s) = \begin{cases} 10 & if\ success \\ -0.01 & otherwise \end{cases} \tag{1}$$

The goal of each round is to have the agent reach the observation position that corresponds to the target image, and this condition is considered as the measure of success in the navigation task.

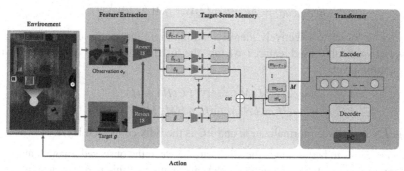

Fig. 1. Target-Driven Memory-Augmented (TDMA) framework. The leftmost is a schematic diagram of the agent in the environment, and the rest are the three stages of the framework, from left to right: Feature Extraction, Target-Scene Memory, and Transformer. The double-headed arrows in the figure indicate the weight sharing of the two networks referred to.

3.2 Model Architecture

In order to navigate to unknown targets without auxiliary information and minimize information loss from historical data, our proposed approach involves fusing observed and target images into Target-Scene features stored in chronological order. The architecture of our model is illustrated in Fig. 1, consisting of three main stages: Feature Extraction, Target-Scene Memory, and Transformer.

Feature Extraction. As shown in the blue area in Fig. 1, the current observation image o_t and target image g are input to a pre-trained ResNet18. This produces feature vectors

$\overline{o}_t, \overline{g} \in R^{512*7*7}$. During the whole training process, we do not update the parameters of ResNet18.

Target-Scene Memory. In the Target-Scene Memory stage, we aim to retain and fuse relevant information from the observed image and the target image. As depicted in the yellow block in Fig. 1, $\overline{o}_t$ and $\overline{g}$ are processed through a convolutional layer and fully connected layer, resulting in feature vectors that are concatenated and passed through another fully connected layer. Finally, the fused feature vector $m_t \in R^{1*128}$ is output and stored in the external memory. In order to preserve the similar information of the observed image and the target image, we employ convolutional and fully connected layers with the same parameters. We perform this operation on T time frames, producing a set of fused features $M = \{m_t, m_{t-1}, \ldots, m_{t-T-1}\}$.

Transformer. This stage aims to capture long-term dependencies and extract meaningful features from the fused Target-Scene Memory. As illustrated in the green area in Fig. 1, the Transformer consists of an Encoder and a Decoder. The Encoder incorporates a multi-head attention sublayer and a feedforward sublayer. The Decoder comprises two multi-head attention sublayers and a feedforward sublayer. We refer to [5, 19] and adopt a multi-head attention layer with 8 heads, which is expressed as:

$$MultiHead(Q, K, V) = Concat(head_1, \ldots, head_8)$$
$$where head_i = Attention\left(QW_i^Q, KW_i^K, VW_i^V\right) \tag{2}$$

Encoder. In order to effectively identify and highlight the spatio-temporal correlation between historically fused features, we take the fused feature set M as the two inputs of the attention layer.

$$Encoder(M, M) = LN(FC(H) + H)$$
$$where H = LN(MultiHead(M, M, M) + M) \tag{3}$$

where LN is the layer normalization and FC is the fully connected layer.

Decoder. So as to better notice the similarity between the observed image features and the target image features and enhance the information of this part, we design the first attention sublayer of the Decoder as a self-attention mechanism for the current fusion feature m_t, which output is C. The second attention sublayer takes C as the query and the output E of the Encoder as the key-value pair. The formula of the entire Decoder is expressed as:

$$Decoder(m_t, E) = LN(FC(A) + A)$$
$$where A = LN(MultiHead(C, E, E) + C) \tag{4}$$
$$where C = LN(MultiHead(m_t, m_t, m_t) + m_t)$$

The useful features extracted by the Transformer are then passed through the fully connected layer for calculation. The final actions are obtained through a classification distribution as follows:

$$\pi(a|m_t, M) = Cat(softmax(Q))$$
$$where Q = FC(Decoder(m_t, Encoder(M))) \tag{5}$$

where Cat is classification distribution.

3.3 Loss Function

To facilitate the rapid acquisition of useful experience, this study incorporates Dijkstra's shortest path algorithm to calculate the optimal action from the current position to the target position. By imitating these optimal actions, the agent can adapt to the environment more efficiently. The loss function for imitation learning (IL) is defined as follows:

$$L_{il} = CE\left(a_t, \hat{a}_t\right) \tag{6}$$

where a_t is the predicted movement probability distribution based on our navigation strategy at time t, $\hat{a}_t$ is the expert's action at time t, CE is cross-entropy loss.

Additionally, the navigation loss in reinforcement learning is represented by $L_{nav} = L_{actor} + L_{critic}$. The overall training loss function for the navigation strategy network is defined as:

$$L = L_{nav} + \alpha L_{il} \tag{7}$$

To prevent the model from over-adapting to the seen environment based on expert strategies, the expert strategies are only used as guidance during the initial 100,000 episodes of training. The parameter α gradually decreases from 0.3 to 0, controlling the contribution of the imitation learning loss in the total loss function.

4 Experiments

4.1 Datasets

In this study, the AI2-THOR platform is utilized as the experimental environment. AI2-THOR provides a diverse range of room categories, including kitchens, living rooms, bedrooms, and bathrooms, each consisting of 30 different rooms. Among the 30 rooms in each category, 7 distinct targets are randomly chosen. The target selection process follows the methodology presented in [18], focusing on selecting targets that can effectively represent the robot's position while avoiding ambiguous targets. Moreover, efforts are made to minimize overlap among the selected targets, as depicted in Fig. 2.

Out of the 7 targets, 5 are randomly designated as training targets (referred to as "seen targets"), while the remaining 2 are used as test targets (referred to as "unseen targets"). For the training phase, 20 rooms from each category are selected as training environments (referred to as "seen scenes"), while 10 different rooms are reserved as test environments (referred to as "unseen scenes").

Fig. 2. Seven Randomly selected targets in the same environment

4.2 Evaluation Metrics

In this research field, the success rate and the weighted path length success rate (success weighted by path length, SPL) [20] are usually used to evaluate the model, the success rate refers to the ratio of the number of times the agent successfully navigates to the target and the total number of times, SPL according to the optimal path length The index after weighting the success rate by the ratio of the actual path length to the actual path length, the calculation formula is as follows:

$$Success = \frac{1}{N} \sum_{n=0}^{N} S_n \qquad (8)$$

$$SPL = \frac{1}{N} \sum_{n=0}^{N} S_n \frac{L_n}{max(L_n, P_n)} \qquad (9)$$

where N is the number of rounds, S_n is the success tag of the n episode, L_n is the step size of the n round, and P_n is the optimal step size of the n round.

4.3 Models

When selecting the comparison models, we specifically focus on visual navigation models without auxiliary information. Therefore, the following models are selected for comparison on the AI2-THOR environment:

Random: This model selects actions randomly from the action space at each step during visual navigation.

Zhu [15]: The scenario-specific layer from the model is removed.

Chen [16]: Similar to Zhu [15], the scenario-specific layer is removed from the model. Additionally, the hidden layer size of the LSTM, which is added before calculating the policy network, is set to 256.

Xiao [17]: Using a model with a memory prediction mechanism. This study demonstrates that its model is superior to a navigation model that uses semantic information to represent targets and combines visual and semantic information.

Ours: The proposed model in this paper utilizes external storage and Transformer to process historical information.

Except for the Random model, the A3C algorithm is employed to accelerate training for the other models. An attenuation parameter of 0.99 is used, and a shared RMSProp optimizer with a learning rate of 0.0007 is employed to train the network.

4.4 Comparative Experiments

To assess the effectiveness of the proposed method, experiments are conducted under three different conditions: seen targets in seen scenes, unseen targets in seen scenes, and unseen targets in unseen scenes. The experiment with seen targets in seen scenes is designed to evaluate the convergence speed of the model. On the other hand, two experiment with unseen targets are conducted to assess the generalization ability of the model.

Following the test method described in [18], each training target is navigated from a random starting position. Each target is tested 100 times, and the average values are computed. The maximum number of navigation steps allowed is set to 500. If the number of steps taken is less than 500, the navigation is considered successful. However, if the number of steps reaches 500, the navigation task is deemed as failed and terminated, and the next round of testing begins.

The experimental results for the seen targets in seen scenes, unseen targets in seen scenes, and unseen targets in unseen scenes are presented in Table 1, Table 2, and Table 3, respectively.

Table 1. Seen targets in Seen scenes (SPL/Success (%))

	Kitchen	Living room	Bedroom	Bathroom	Avg
Random	1.8/2.7	1.5/2.1	2.4/3.5	3.6/4.2	2.3/3.1
Zhu [15]	38.5/51.2	38.3/57.6	56.8/61.5	69.1/81.7	50.7/63.0
Chen [16]	52.9/68.7	47.6/63.4	60.9/76.2	64.8/78.7	56.5/71.8
Xiao [17]	58.6/74.6	51.9/65.2	64.4/79.7	**74.6**/88.4	61.2/77.0
Ours	**60.4/97.4**	**69.1/98.2**	**74.3/99.0**	68.6/**98.7**	**68.1/98.3**

In Table 1, the proposed method achieves higher performance levels in all four types of scenes, except for the SPL in the bathroom environment. The primary reason behind this improvement is the adoption of imitation learning as guidance during the initial stages of training. By leveraging imitation learning, the model can quickly understand the target, thereby reducing early failures. However, since the bathroom environment is relatively small and does not require extensive exploration, the convergence rate improvement facilitated by imitation learning is relatively smaller in this specific environment.

Table 2. Unseen targets in Seen scenes (SPL/Success (%))

	Kitchen	Living room	Bedroom	Bathroom	Avg
Random	0.9/1.3	0.8/1.2	2.7/3.4	1.5/2.3	1.5/2.1
Zhu [15]	1.6/5.8	3.2/4.8	3.2/4.9	27.1/40.5	8.5/14.0
Chen [16]	6.7/9.1	7.7/13.2	9.7/14.2	24.5/37.6	12.2/18.5
Xiao [17]	6.4/10.2	**10.4**/15.9	8.4/13.5	25.2/39.7	12.6/19.8
Ours	**8.3/14.3**	10.2/**16.5**	**11.4/15.3**	**27.8/41.2**	**14.4/21.8**

Table 3. Unseen targets in Unseen scenes (SPL/Success (%))

	Kitchen	Living room	Bedroom	Bathroom	Avg
Random	1.9/2.8	0.6/1.0	2.0/3.8	2.7/3.9	1.8/2.9
Zhu [15]	2.2/7.5	2.5/4.2	1.3/4.4	3.4/9.3	2.4/6.4
Chen [16]	6.5/10.7	2.0/5.0	3.7/5.8	7.9/12.5	5.0/8.5
Xiao [17]	6.8/**13.6**	4.7/8.8	4.8/11.6	7.6/15.5	6.0/12.4
Ours	**8.1**/12.1	**5.2/10.5**	**6.5/13.7**	**8.3/17.8**	**7.0/13.5**

Table 2 presents the results of experiments conducted on unseen targets in seen scenes. The results show that in seen scenes, our method has a better ability to generalize to unseen targets. In the kitchen, living room, and bedroom environments, models with memory networks outperformed models without memory networks. However, in the bathroom environment, there was little difference observed between the two types of models. The reason behind this discrepancy is that in a small environment with low complexity and a limited number of object types, the model without memory networks can learn a direct mapping between the observed image and the target image to determine the appropriate action. Conversely, in a more complex environment, historical information becomes crucial for accurate target localization and action decision-making.

Table 3 presents the results of experiments conducted on unseen targets in unseen scenes, which further evaluate the model's understanding of navigation tasks. The proposed method in this paper demonstrates superior performance in terms of both success rate and SPL. This indicates that the Transformer architecture effectively leverages historical information and utilizes the self-attention sub-layer to fuse the features of the current observation image and target image in a more efficient manner. Furthermore, when comparing the results in Table 2 with those in Table 3, it can be observed that the success rate and SPL of each method are generally lower in the unknown environment. This reduction is attributed to the fact that the feature information of the scenes learned from the original training in the seen scenes becomes less valuable and less applicable in the unknown environment.

4.5 Ablation Experiment

The experiment on unseen targets in unseen scenes evaluates the impact of Transformer processing historical information and the self-attention sub-layer on improving generalization performance. The following experimental settings were considered:

A3C + LSTM: Using LSTM instead of Transformer to process historical information.

No SA: Transformer is used, but the first self-attention sub-layer in this paper's Decoder is removed.

Ours: Our proposed model.

The results of the ablation experiments conducted on unseen targets in unseen scenes are presented in Table 4.

Table 4. Results of ablation experiments (SPL/Success (%))

	Kitchen	Living room	Bedroom	Bathroom	Avg
A3C + LSTM	6.5/10.7	2.0/5.0	3.7/5.8	7.9/12.5	5.0/8.5
No SA	7.2/11.0	4.8/8.6	6.3/11.7	7.5/15.2	6.5/11.6
Ours	**8.1/12.1**	**5.2/10.5**	**6.5/13.7**	**8.3/17.8**	**7.0/13.5**

As shown in Table 4, both the method of using Transformer to process historical information and the self-attention sublayer for the fusion features of current observation image and target image are helpful to the generalization performance of the model.

5 Conclusions

This paper presents an approach that addresses visual navigation challenges without relying on auxiliary information. By encoding and fusing the observed image and target image, and subsequently processing the stored historical data using the Transformer, the proposed method enhances the similarity between the observed and target images using the self-attention sub-layer. This approach improves the generalization ability of the visual navigation model in unknown environments, as demonstrated by the experimental results. However, it is important to note that the navigation performance of our method on unknown targets still falls far short compared to that of humans. Future research should explore ways to incorporate additional information and techniques to achieve higher levels of navigation capability for unknown targets.

Acknowledgments. This work is supported by the Fundamental Research Funds for the Central Universities (No. 2022CDJYGRH-001) and the Chongqing Technology Innovation & Application Development Key Project (cstc2020jscx-dxwtBX0055; cstb2022tiad-kpx0148).

References

1. Oriolo, G., Vendittelli, M., Ulivi, G.: Online map building and navigation for autonomous mobile robots. In: Proceedings of 1995 IEEE International Conference on Robotics and Automation, vol. 3, pp. 2900–2906 (1995)
2. Davison, A.J.: Real-time simultaneous localisation and mapping with a single camera. In: Proceedings Ninth IEEE International Conference on Computer Vision, vol. 2, pp. 1403–1410 (2003)
3. Kolve, E., et al.: AI2-THOR: An Interactive 3D Environment for Visual AI. ArXiv, abs/1712.05474 (2017)
4. Zhang, Y., Yang, Z., Zhu, Z., Feng, W., Zhou, Z., Wang, W.: Visual navigation of mobile robots in complex environments based on distributed deep reinforcement learning. In: 2022 6th Asian Conference on Artificial Intelligence Technology (ACAIT), pp. 1–5 (2022)
5. Fang, K., Toshev, A., Fei-Fei, L., Savarese, S.: Scene memory transformer for embodied agents in long-horizon tasks. In: 2019 IEEE/CVF Conference on Computer Vision and Pattern Recognition (CVPR), pp. 538−547 (2019)

6. Li, W., Hong, R., Shen, J., Yuan, L., Lu, Y.: Transformer memory for interactive visual navigation in cluttered environments. IEEE Robot. Autom. Lett. **8**, 1731–1738 (2023)
7. Kulhánek, J., Derner, E., Babuška, R.: Visual navigation in real-world indoor environments using end-to-end deep reinforcement learning. IEEE Robot. Autom. Lett. **6**, 4345–4352 (2020)
8. Mousavian, A., Toshev, A., Fiser, M., Kosecka, J., Davidson, J.: Visual representations for semantic target driven navigation. In: 2019 International Conference on Robotics and Automation (ICRA), pp. 8846−8852 (2018)
9. Du, H., Yu, X., Zheng, L.: VTNet: Visual Transformer Network for Object Goal Navigation. ArXiv, abs/2105.09447 (2021)
10. Fukushima, R., Ota, K., Kanezaki, A., Sasaki, Y., Yoshiyasu, Y.: Object memory transformer for object goal navigation. In: 2022 International Conference on Robotics and Automation (ICRA), pp. 11288−11294 (2022)
11. Yang, W., Wang, X., Farhadi, A., Gupta, A.K., Mottaghi, R.: Visual Semantic Navigation using Scene Priors. ArXiv, abs/1810.06543 (2018)
12. Wu, Y., Wu, Y., Tamar, A., Russell, S.J., Gkioxari, G., Tian, Y.: Bayesian relational memory for semantic visual navigation. In: 2019 IEEE/CVF International Conference on Computer Vision (ICCV), pp. 2769−2779 (2019)
13. Lyu, Y., Shi, Y., Zhang, X.: Improving target-driven visual navigation with attention on 3D spatial relationships. Neural. Process. Lett. **54**, 3979–3998 (2020)
14. Jaderberg, M., et al.: Reinforcement Learning with Unsupervised Auxiliary Tasks. ArXiv, abs/1611.05397 (2016)
15. Zhu, Y., et al.: Target-driven visual navigation in indoor scenes using deep reinforcement learning. In: 2017 IEEE International Conference on Robotics and Automation (ICRA), pp. 3357−3364 (2016)
16. Chen, L., Moorthy, M.P., Sharma, P., Kawthekar, P.: Imitating shortest paths for visual navigation with trajectory-aware deep reinforcement learning. Comput. Sci. (2017)
17. Xiao, Q., Yi, P., Liu, R., Dong, J., Zhou, D., Zhang, Q.: Deep reinforcement learning visual navigation model integrating memory-prediction mechanism. In: 2021 IEEE 24th International Conference on Computer Supported Cooperative Work in Design (CSCWD), pp. 109–114 (2021)
18. Mezghani, L., et al.: Memory-augmented reinforcement learning for image-goal navigation. In: 2022 IEEE/RSJ International Conference on Intelligent Robots and Systems (IROS), pp. 3316−3323 (2021)
19. Vaswani, A., et al.: Attention is All you Need. NIPS (2017)
20. Anderson, P., et al.: On Evaluation of Embodied Navigation Agents. ArXiv, abs/1807.06757 (2018)

Recursive Constrained Maximum Versoria Criterion Algorithm for Adaptive Filtering

Lvyu Li[1], Ji Zhao[1(✉)] (iD), Qiang Li[1(✉)] (iD), Lingli Tang[2], and Hongbin Zhang[3] (iD)

[1] School of Information Engineering, Southwest University of Science and Technology, Mianyang 621010, China
zhaoji@swust.edu.cn, liqiangsir@swust.edu.cn
[2] Department of Social Sciences, Southwest University of Science and Technology, Mianyang 621010, China
[3] School of Information and Communication Engineering, University of Electronic Science and Technology of China, Chengdu 611731, China

Abstract. This paper proposes a recursive constrained maximum Versoria criterion (RCMVC) algorithm. In comparison with recursive competing methods, our proposed RCMVC can achieve smaller steady-state misalignment in non-Gaussian noisy environments. Specifically, we use the maximum Versoria criterion (MVC) to derive a new robust recursive constrained adaptive filtering within the least-squares framework for solving linearly constrained problems. For RCMVC, we analyze the mean-square stability and characterize the theoretical transient mean square deviation (MSD) performance. Furthermore, we conduct some simulations to validate the consistency between the analytical and simulation results and show the effectiveness of RCMVC in non-Gaussian noisy environments.

Keywords: Maximum Versoria criterion · Constrained Adaptive filtering · System identification · Robustness

1 Introduction

Recently, constrained adaptive filtering (CAF) algorithms have been extensively studied in the domain of signal processing in spectrum analysis, spatial-temporal processing, antenna arrays, and interference suppression in multiple access communications and others [1]. The greatest merit of CAF is the error correction feature preventing the accumulation of errors, e.g., the quantization errors in a digital implementation, thus realizing good filtering performance [2,3].

This work is supported in part by the National Natural Science Foundation of China (Grant no. 62201478 and 61971100), in part by the Southwest University of Science and Technology Doctor Fund (Grant no. 20zx7119), in part by the Sichuan Science and Technology Program (Grant no. 2022YFG0148), and in part by the Heilongjiang Provincial Science and Technology Program (No. 2022ZX01A16).

© The Author(s), under exclusive license to Springer Nature Singapore Pte Ltd. 2024
B. Luo et al. (Eds.): ICONIP 2023, CCIS 1961, pp. 433–445, 2024.
https://doi.org/10.1007/978-981-99-8126-7_34

In the assumption of Gaussian noise environment, some classic ℓ_2-norm based CAF algorithms were investigated, such as the constrained least mean square [4], the constrained recursive least squares algorithm [3], and the constrained affine projection algorithm [5]. However, in many engineering applications, various non-Gaussian noisy interference can dramatically damage the filtering performance of these ℓ_2-norm based CAFs. Therefore, it's necessary to develop robust CAF against interference, especially impulsive interference.

In recent years, various non-ℓ_2-norm cost functions have been proposed. Examples include the minimum error entropy (MEE), the generalized maximum correntropy (GMC), the maximum q-Renyi kernel, the M-estimate, and least lncosh [6]. Therefore, several robust CAF algorithms are developed by injecting these criteria into the gradient and least-squares frameworks for solving linearly constrained problems. Examples include the robust MEE based constrained adaptive filtering algorithm [7], the recursive constrained maximum q-řenyi kernel (RCMqR) algorithm [8], the recursive constrained least lncosh (RCLL) algorithm [9], the recursive constrained generalized maximum correntropy (RCGMC) algorithm [10], and the constrained recursive least M-estimate (RCRLM) algorithm [11].

In this work, we propose a new recursive CAF algorithm called the recursive constrained maximum Versoria criterion (RCMVC) algorithm because the gradient method is affected by gradient noise and color input signals, resulting in a slow convergence rate. RCMVC is robust against some large errors because MVC combats outliers. For RCMVC, we provide analytical performance analysis on mean square convergence and transient mean square deviation (MSD). Simulation results also demonstrate that our proposed algorithm provides better filtering performance than competing CAF algorithms for system identification under non-Gaussian noises.

Outline of the paper: In Sect. 2, after briefly introducing the system model and optimal problem, we derive RCMVC. Section 3 conducts the mean square convergence analysis, and derives the theoretical transient MSD. Simulation results are given in Sect. 4. Finally, Sect. 5 gives the conclusion.

2   The Proposed RCMVC Algorithm

2.1   System Model

Consider a linear system identification problem, in which the input signal is $\boldsymbol{x}(n)$ and the desired signal satisfies $d(n) = \boldsymbol{x}(n)^T \boldsymbol{w}_0 + v(n)$, where $\boldsymbol{w}_0 = [w_0, w_1, ..., w_{L-1}]^T$ is the unknown system weight vector needed to be estimated, $\boldsymbol{x}(n) = [x(n), x(n-1), ..., x(n-L+1)]^T$ is the input vector, and $v(n)$ represents system background noise. Denote $e(n)$ as system estimation error, i.e., $e(n) = d(n) - \boldsymbol{w}(n)^T \boldsymbol{x}(n)$, where $\boldsymbol{w}(n)$ is the adaptive weight vector. To avoid error accumulation, based on some prior knowledge, the linear constraints can be applied to $\boldsymbol{w}(n)$, i.e., $\boldsymbol{C}^T \boldsymbol{w}(n) = \boldsymbol{f}$, where $\boldsymbol{C}^T$ is the $N \times L$ constrained matrix, and $\boldsymbol{f}$ is a vector of size $N \times 1$ containing the N constrained values. In this work, our main aim is to develop a new robust adaptive filtering algorithm within the constrained framework under non-Gaussian noisy environments.

2.2   Optimal Problem

Motivated by the generalized Gaussian probability density function, the generalized Versoria function $f(e(n)) = 2a(1 + \lambda|e(n)|^\alpha)^{-1}$ as an optimal criterion is received much attention [12,13]. Introducing MVC into the constrained framework [14], a new robust CAF is derived by solving the following optimal problem

$$\begin{cases} \min\limits_{w} & \gamma\left\{1 - E\left[\dfrac{1}{(1 + \lambda|e(n)|^\alpha)}\right]\right\}, \\ s.t. & C^T w = f, \end{cases} \tag{1}$$

where γ is a normalization constant, $E[\cdot]$ is an expectation operation, $\alpha > 0$ is the shape parameter, and λ controls the width of the Versoria kernel [12,15]. In a special case, it includes the original Versoria function when $\alpha = 2$.

2.3   Constrained Maximum Versoria Criterion Algorithm

For (1), based on the Lagrange method, it becomes

$$J(n) = \gamma\left\{1 - E\left[\frac{1}{(1 + \lambda|e(n)|^\alpha)}\right]\right\} + \boldsymbol{\xi}^T\left(C^T w - f\right), \tag{2}$$

where $\boldsymbol{\xi}$ is the Lagrange multiplier. For (2), using the stochastic gradient method, yields $w(n) = w(n-1) + \mu g(e(n))e(n)x(n) + \mu C\boldsymbol{\xi}$. Then after some simple calculations, we can obtain the weight update formula of the constrained Maximum Versoria criterion (CMVC) algorithm as

$$\begin{cases} w(n+1) = \boldsymbol{\Theta}\left[w(n) + \mu g(e(n))e(n)x(n)\right] + C(C^T C)^{-1}f, \\ g(e(n)) = \dfrac{|e(n)|^{\alpha-2}}{(1 + \lambda|e(n)|^\alpha)^2}, \quad \boldsymbol{\Theta} = I - C(C^T C)^{-1}C^T, \end{cases} \tag{3}$$

where $\mu > 0$ is a step size balancing filtering accuracy and convergence speed [16]. More details about CMVC are shown in [14].

2.4   Recursive Constrained Maximum Versoria Criterion Algorithm

In addition, for (2), setting $\dfrac{\partial J(n)}{\partial w} = 0$, we obtain an optimal weight vector as

$$E[g(e(n))e(n)x(n)] + \boldsymbol{\xi}C = 0$$
$$\Rightarrow E[g(e(n))d(n)x(n)] + \boldsymbol{\xi}C = E[g(e(n))x(n)x(n)^T]w_{opt} \tag{4}$$
$$\Rightarrow w_{opt} = R_g^{-1}p_g + R_g^{-1}C\boldsymbol{\xi},$$

where $R_g = E[g(e(n))x(n)x(n)^T]$ is a weighted autocorrelation matrix and $p_g = E[g(e(n))d(n)x(n)]$ is a weighted cross-correlation vector. Consider $C^T w_{opt} = f$, the Lagrange multiplier is obtained as

$$C^T w_{opt} = f \Rightarrow C^T R_g^{-1} p_g + C^T R_g^{-1} C \xi = f$$

$$\Rightarrow \xi = (C^T R_g^{-1} C)^{-1}(f - C^T R_g^{-1} p_g). \tag{5}$$

Injecting (5) into (4), yields

$$w_{opt} = R_g^{-1} p_g + R_g^{-1} C (C^T R_g^{-1} C)^{-1}(f - C^T R_g^{-1} p_g). \tag{6}$$

In practice, R_g and p_g can be estimated as

$$\begin{cases} \hat{R}_g(n) = \sum_{i=1}^{n} g(e(i))x(i)x(i)^T = \hat{R}_g(n-1) + g(e(n))x(n)x(n)^T, \\ \hat{p}_g(n) = \sum_{i=1}^{n} g(e(i))d(i)x(i) = \hat{p}_g(n-1) + g(e(n))d(n)x(n). \end{cases} \tag{7}$$

Equation (7) means that w_{opt} can be iteratively estimated as

$$w(n) = \hat{R}_g^{-1}(n)\hat{p}_g(n) + \hat{R}_g^{-1}(n)C(C^T \hat{R}_g^{-1}(n)C)^{-1}(f - C^T \hat{R}_g^{-1}(n)\hat{p}_g(n)). \tag{8}$$

For (8), let $Q_g(n) = \hat{R}_g^{-1}(n)$, based on the matrix inversion lemma, we obtain

$$\begin{cases} Q_g(n) = Q_g(n-1) - m(n)x(n)^T Q_g(n-1), \\ m(n) = \dfrac{\hat{R}_g^{-1}(n-1)x(n)}{g(e(n))^{-1} + x(n)^T \hat{R}_g^{-1}(n-1)x(n)} = g(e(n))Q_g(n)x(n). \end{cases} \tag{9}$$

In addition, let $\bar{w}(n)$ be a recursive solution to (2) without constraint at instance n. After some calculations, the update formula of $\bar{w}(n)$ is

$$\begin{aligned} \bar{w}(n) &= \hat{R}_{\bar{g}}^{-1}(n)\hat{p}_{\bar{g}}(n) = Q_{\bar{g}}(n)\hat{p}_{\bar{g}}(n) \\ &= \bar{w}(n-1) + g(\bar{e}(n))d(n)x(n)Q_{\bar{g}}(n) - \bar{m}(n)x(n)^T Q_{\bar{g}}(n-1)\hat{p}_{\bar{g}}(n-1) \quad (10) \\ &= \bar{w}(n-1) + \bar{m}(n)\bar{e}(n), \end{aligned}$$

in which

$$\begin{cases} \bar{e}(n) = d(n) - x(n)^T \bar{w}(n-1), \quad g(\bar{e}(n)) = \dfrac{|\bar{e}(n)|^{\alpha-2}}{(1 + \lambda|\bar{e}(n)|)^2}, \\ \hat{R}_{\bar{g}}(n) = \hat{R}_{\bar{g}}(n-1) + g(\bar{e}(n))x(n)x(n)^T, \\ \hat{p}_{\bar{g}}(n) = \hat{p}_{\bar{g}}(n-1) + g(\bar{e}(n))d(n)x(n), \\ Q_{\bar{g}}(n) = Q_{\bar{g}}(n-1) - \bar{m}(n)x(n)^T Q_{\bar{g}}(n-1), \\ \bar{m}(n) = \dfrac{g(\bar{e}(n))Q_{\bar{g}}(n-1)x(n)}{1 + g(\bar{e}(n))x(n)^T Q_{\bar{g}}(n-1)x(n)} = g(\bar{e}(n))Q_{\bar{g}}(n)x(n). \end{cases} \tag{11}$$

Algorithm 1: The RCMVC algorithm

---

parameters: $\alpha > 0$, $\lambda > 0$, $\mu > 0$, C, f

initiation: $\bar{w}(0) = C(C^T C)^{-1} f$

while $\{x(n), d(n)\}, (n \geq 1)$ do

  1) $\bar{e}(n) = d(n) - \bar{w}^T(n-1)x(n)$,   2) $g(\bar{e}(n)) = \frac{|\bar{e}(n)|^{\alpha-2}}{(1+\lambda|\bar{e}(n)|)^2}$

  3) $\bar{m}(n) = \dfrac{g(\bar{e}(n))Q_{\bar{g}}(n-1)x(n)}{1 + g(\bar{e}(n))x(n)^T Q_{\bar{g}}(n-1)x(n)}$,   4) $\bar{w}(n) = \bar{w}(n-1) + \bar{m}(n)\bar{e}(n)$

  5) $Q_{\bar{g}}(n) = Q_{\bar{g}}(n-1) - \bar{m}(n)x(n)^T Q_{\bar{g}}(n-1)$,

  6) $h(n) = C(C^T Q_{\bar{g}}(n)C)^{-1}(f - C^T\bar{w}(n))$,   7) $w(n) = \bar{w}(n) + Q_{\bar{g}}(n)h(n)$

end

Substituting (9) and (10) into (8), yields the adaptive version of (8) as

$$w(n) = \bar{w}(n) + Q_{\bar{g}}(n)h(n), \tag{12}$$

where $h(n) = C(C^T Q_{\bar{g}}(n)C)^{-1}(f - C^T\bar{w}(n))$.

Remark: According to (12), this work derives the *recursive constrained maximum Versoria criterion* (RCMVC) algorithm, which is listed in **Algorithm 1**. Observing (12), one can get that the solution w_{opt} in (6) consists of two terms, namely, the solution $\bar{w} = R_g^{-1}p_g$ without constraints, and an auxiliary term $R_g^{-1}C(C^T R_g^{-1}C)^{-1}(f - C^T R_g^{-1}p_g)$ that helps $\bar{w}$ to fulfill the linearly constrain $C^T w = f$. Hence, in (12), the estimation $w(n)$ is updated by the learning of $\bar{w}(n)$. In addition, compared with CMVC, the proposed RCMVC achieves superior filtering performance in terms of convergence rate and filtering accuracy, irrespective of the variance of impulsive noise shown in Fig. 2. Furthermore, from the perspective of cost functions, the computational complexity of MVC is lower than GMC and MEE. Moreover, as shown in Fig. 3, we can find the advantages of RCMVC, which has faster convergence speed and better filtering accuracy than other competing algorithms.

3 Performance Analysis for RCMVC

Before conducting the mean square convergence analysis of RCMVC, the following assumptions are made [4, 8, 17]

A1: The input $x(n)$ is a zero-mean multivariate Gaussian with statistically independent elements, and its positive definite covariance matrix $R = E\{x(n)x(n)^T\}$.

A2: The noise signal $v(n)$ is zero-mean and independent identically distributed, and it is independent of any other signals in the system.

A3: The priori error $\bar{e}_a(n) = \Delta w_{0o}^T x(n)$ is zero-mean, and $\Delta w_{0o} = w_0 - w_{opt}$ represents the intrinsic weight parameter between w_0 and w_{opt}.

A4: The nonlinear error $g(e(n))$ is uncorrelated with $\|x(n)\|^2$.

3.1　Convergence Analysis

Let $\bar{e}(n) = d(n) - x(n)^T \bar{w}(n-1)$ and $e(n) = d(n) - x(n)^T w(n-1)$ be the unconstrained prediction error and the constrained prediction error, respectively. From (12), we obtain

$$\bar{e}(n) - e(n) = x(n)^T Q_{\bar{g}}(n-1)Ck(n), \tag{13}$$

where $k(n) = (C^T Q_{\bar{g}}(n)C)^{-1}(f - C^T \bar{w}(n))$. Based on (12), (13) and $k(n)$, yields

$$w(n) - w(n-1) = \bar{e}(n)\bar{m}(n) + Q_{\bar{g}}(n)Ck(n) - Q_{\bar{g}}(n-1)Ck(n-1) \tag{14}$$
$$= e(n)\bar{m}(n) + Q_{\bar{g}}(n)C(C^T Q_{\bar{g}}(n)C)^{-1}(f - C^T(w(n-1) + \bar{m}(n)e(n))).$$

According to **A4**, we obtain

$$\hat{R}_g^{-1}(n) = \left(n\left(\frac{1}{n}\sum_{i=1}^{n} g(\bar{e}(i))x(i)x(i)^T \right) \right)^{-1}$$
$$\approx \frac{1}{n}\left(E\left\{ g(\bar{e}(n))x(n)x(n)^T \right\} \right)^{-1} = \frac{\beta}{n}R^{-1}, \tag{15}$$

where $\beta = (E\{g(\bar{e}(n))\})^{-1}$. Therefore, injecting (15) into (14), yields

$$w(n) \approx w(n-1) + \frac{\beta}{n}g(\bar{e}(n))e(n)R^{-1}x(n) + R^{-1}C(C^T R^{-1}C)^{-1}$$
$$\times (f - C^T(w(n-1) + \frac{\beta}{n}g(\bar{e}(n))e(n)R^{-1}x(n))) \tag{16}$$
$$= P\left(w(n-1) + \frac{\beta}{n}g(\bar{e}(n))e(n)R^{-1}x(n) \right) + q,$$

where $P = I - R^{-1}C(C^T R^{-1}C)^{-1}C^T$, and $q = R^{-1}C(C^T R^{-1}C)^{-1}f$. Denote $\Delta w_{opt}(n) = w(n) - w_{opt}(n)$, from (16), we have

$$\Delta w_{opt}(n) = P\left(w(n-1) + \frac{\beta}{n}g(\bar{e}(n))e(n)R^{-1}x(n) \right) + q - w_{opt}. \tag{17}$$

Since $Pw_{opt} - w_{opt} + q = 0$, (17) is represented as

$$\Delta w_{opt}(n) = P\left[I - \frac{\beta}{n}g(\bar{e}(n))R^{-1}x(n)x(n)^T \right] \Delta w_{opt}(n-1)$$
$$+ \frac{\beta}{n}g(\bar{e}(n))P\left(R^{-1}x(n)v(n) + R^{-1}x(n)x(n)^T \Delta w_{0o} \right). \tag{18}$$

Taking the expectation of l_2-norm on the both sides of (18), yields

$$E\left\{\|\ \Delta w_{opt}(n)\ \|^2\right\} = E\left\{\|\ \Delta w_{opt}(n-1)\ \|^2_{W(n)}\right\} + \frac{\beta^2 E\left\{g^2(\bar{e}(n))\right\} E\left\{v^2(n)\right\}}{n^2}t$$
$$+ \frac{\beta^2 E\{g^2\bar{e}(n))\}}{n^2}\Delta w_{0o}^T E\left\{tR\right\}\Delta w_{0o}, \tag{19}$$

and

$$W(n) = I - \frac{\beta g(\bar{e}(n))}{n}(P^T + P) + \frac{\beta^2 g^2(\bar{e}(n))}{n^2}(2P^T P + tR), \tag{20}$$

where $t = \mathrm{Tr}(PR^{-1}P^T)$. Let r_i and p_i be the eigenvalues of R and P, respectively. Then, after some manipulations, the stability condition is

$$n > \max_{i\in[1,L]} \frac{\beta E\{g^2(\bar{e}(n))(2p_i^2) + tr_i\}}{2p_i E\{g(\bar{e}(n))\}}. \tag{21}$$

Equation (21) means that, for RCMVC, after a limited number of iterations, it converges. Simulation results show that the n value can be reasonably small.

3.2   Mean Square Deviation

For a symmetric positive definite matrix T, (19) can be rewritten as

$$E\left\{\|\Delta w_{opt}(n)\|^2_T\right\} = E\left\{\|\Delta w_{opt}(n-1)\|^2_{\Pi(n)}\right\}$$
$$+ \frac{\beta^2 E\{g^2(\bar{e}(n))\}}{n^2}(E\{v^2(n)\}\phi + \|\Delta w_{0o}\|^2_\Omega), \tag{22}$$

where

$$\begin{cases} \Pi(n) = \left(I - \dfrac{\beta E\{g(\bar{e}(n))\}}{n}P\right)^T T\left(I - \dfrac{\beta E\{g(\bar{e}(n))\}}{n}P\right) + \\ \quad \dfrac{\beta^2 E\{g^2(\bar{e}(n))\}}{n^2}\Omega - \dfrac{\beta^2 E^2\{g(\bar{e}(n))\}}{n^2}P^T T P \\ \Omega = 2P^T T P + \phi R, \quad \phi = Tr(TPR^{-1}P^T). \end{cases} \tag{23}$$

Based on the fact that, for compatible matrices, $vec(XYZ) = (Z^T \otimes X)vec(Y)$, $\mathrm{Tr}(X^T Y) = vec(Y)^T vec(X)$, where $vec(\cdot)$ is the vectorization operator and $\otimes$ is the Kronecker product. Thus we have

$$F(n) = \left(I - \frac{\beta E\{g(\bar{e}(n))\}}{n}P\right)^T \otimes \left(I - \frac{\beta E\{g(\bar{e}(n))\}}{n}P\right) + \frac{\beta^2 E\{g^2(\bar{e}(n))\}}{n^2}$$
$$\times \left(2(P^T \otimes P) + vec(R)vec(PR^{-1}P^T)^T\right) - \frac{\beta^2 E^2\{g(\bar{e}(n))\}}{n^2}(P^T \otimes P). \tag{24}$$

Therefore, (22) can be transformed as

$$E\left\{\|\Delta w_{opt}(n)\|_{\rho}^2\right\} = E\left\{\|\Delta w_{opt}(n-1)\|_{F(n)\rho}^2\right\}$$
$$+ \frac{\beta^2 E\{g^2(\bar{e}(n))\}}{n^2}\left(E\{v^2(n)\} + \|\Delta w_{0o}\|_R^2\right)vec(PR^{-1}P^T)^T \rho, \qquad (25)$$

where $\rho = vec(T)$, and $E\{\|\Delta w_{opt}(n)\|_{\rho}^2\} = E\{\|\Delta w_{opt}(n)\|_T^2\}$. From [17], we get

$$\frac{E\{\|\Delta w_{opt}(n)\|^2\}}{E\{\|\Delta w_{opt}(n-1)\|^2\}} = \frac{n-1}{n}, \qquad (26)$$

then (25) is changed as

$$E\left\{\|\Delta w_{opt}(n)\|_{(I-\frac{n}{n-1}F(n)\rho)}^2\right\}$$
$$= \frac{\beta^2 E\{g^2(\bar{e}(n))\}}{n^2}\left(E\{v^2(n)\} + \|\Delta w_{0o}\|_R^2\right)vec(PR^{-1}P^T)^T\rho, \qquad (27)$$

which means that, when $\rho = (I - \frac{n}{n-1}F(n))^{-1}vec(I)$, we can obtain

$$E\{\|\Delta w_{opt}(n)\|^2\} = \frac{\beta^2 E\{g^2(\bar{e}(n))\}}{n^2}(E\{v^2(n)\} + \|\Delta w_{0o}\|_R^2)$$
$$\times\ vec(PR^{-1}P^T)^T(I - \frac{n}{n-1}F(n))^{-1}vec(I). \qquad (28)$$

Furthermore, for $E\{g^2(\bar{e}(n))\}$ in (28), taking the second-order Taylor expansion of $g(\bar{e}(n))$ with respect to $\bar{e}_a(n)$ at $v(n)$ as

$$g(\bar{e}(n)) = g(v(n)) + g'(v(n))\bar{e}_a(n) + \frac{1}{2}g''(v(n))(\bar{e}_a(n))^2 + o(\bar{e}_a^2(n)), \qquad (29)$$

where

$$\begin{cases} g(v(n)) = \dfrac{1}{(1+\lambda|v(n)|^\alpha)^2}|v(n)|^{\alpha-2}, \\[2mm] g'(v(n)) = \dfrac{g(v(n))}{|v(n)|^2}(\alpha - 2 - 2\alpha\varkappa(1+\varkappa)^{-1})v(n), \\[2mm] g''(v(n)) = \dfrac{g(v(n))}{|v(n)|^2}((\alpha-2)(\alpha-3) + \alpha\varkappa(10-6a)(1+\varkappa)^{-1}, \\[2mm] + 6\alpha^2\varkappa^2(1+\varkappa)^{-2}), \quad \varkappa = \lambda|v(n)|^\alpha. \end{cases} \qquad (30)$$

Since the higher order term $o(\bar{e}_a^2(n))$ is too small, it can be neglected [18], and we have

$$\begin{cases} E\{g(\bar{e}(n))\} \approx E\{g(v(n))\} + \dfrac{1}{2}E\{g''(v(n))\}E\{\bar{e}_a^2(n)\}, \\[2mm] E\{g^2(\bar{e}(n))\} \approx E\{g^2(v(n))\} + E\{g(v(n))g''(v(n)) + g'(v(n))^2\}E\{\bar{e}_a^2(n)\}. \end{cases} \qquad (31)$$

Substituting (31) into (28), we get the theoretical mean square deviation (MSD) of RCMVC as

$$E\{\|\Delta w_{opt}(n)\|^2\} = \frac{\beta^2}{n^2}\left(E\{v^2(n)\} + \|\Delta w_{0o}\|_R^2\right)vec(PR^{-1}P^T)^T\left(I - \frac{n}{n-1}F(n)\right)^{-1}$$
$$vec(I)E\{g^2(v(n))\} + E\{g(v(n))g''(v(n)) + g'(v(n))^2\}E\{\bar{e}_a^2(n)\}. \tag{32}$$

So far, we have finished the theoretical analyses in terms of convergence and mean square deviation, and the main results are (21) and (32). And Sect. 4 shows the consistency between the analytical and simulation results.

4 Simulation Results

In this section, some simulations are conducted to validate the effectiveness of theoretical analysis and the better filtering behavior of the proposed RCMVC. In addition, all simulation results are averaged over 100 independent runs.

4.1 Validation of Theoretical Transient MSD

In this part, the validity of the theoretical transient MSD expression in (32) is texted for a linearly constrained system identification problem with the following constraint vector f and the matrix C [4,10,14], i.e.,

$$\begin{cases} w_0 = [0.3328, -0.0329, -0.0944, 0.7174, -0.652, -0.0721, 0.58]^T \\ \\ f = [1.0767, -0.5783, -0.5993]^T, \ C = \begin{bmatrix} -0.1058 & -0.2778 & -0.115 \\ 1.5423 & 1.6399 & 0.3357 \\ 0.2614 & -0.7101 & -0.0217 \\ 1.855 & 0.49 & 0.6389 \\ -0.718 & 0.4914 & -0.5332 \\ -0.3816 & -0.00427 & -1.0077 \end{bmatrix}. \end{cases} \tag{33}$$

For comparison, the simulated numerical MSD is calculated as MSD = $20\log(\|w_{opt} - w(n)\|_2)$, and the input signal is generated from a multivariate Gaussian distribution with zero-mean and the same covariance as the one used in [4]. And four types of noise are considered to model noisy interference, namely: 1) Gaussian noise (GN): $v(n) \in \mathcal{N}(v_\mu, \delta_v^2)$ and the parameter vector $p_G = [v_\mu, \delta_v^2]^T$; 2) Binary noise (BN): $v(n) \in [v_a, v_b]$ and the parameter vector $p_B = [v_a, v_b, p_r]^T$, where $p(v(n) = v_a) = p_r$ and $p(v(n) = v_b) = 1 - p_r$; 3) Uniform noise (UN): $v(n) \in \{v_a, v_b\}$ with parameter vector $p_U = [v_a, v_b]^T$; 4) Mixture Gaussian noise (MGN): $v(n) = (1 - \varphi(n))v_1(n) + \psi(n)v_2(n)$, where $v_1(n) \in \mathcal{N}(v_{1\mu}, \delta_{1v}), v_2(n) \in \mathcal{N}(v_{2\mu}, \delta_{2v})$, and $\varphi(n) \in$ BN with parameter vector $p_M = [v_{1\mu}, \delta_{1v}^2, v_{2\mu}, \delta_{2v}^2, v_a, v_b, pr]^T$. Figure 1 shows the theoretical and simulated MSD results under different noise environments. From this figure, we can get that there is a good match between the simulation and theoretical results.

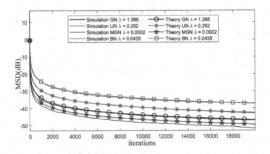

Fig. 1. Theoretical and simulated transient MSD of RCMVC with $\alpha = 2$ and different noisy environments: $\boldsymbol{p}_G = [0, 0.1]^T$, $\boldsymbol{p}_B = [-1, 1, 0.5]^T$, $\boldsymbol{p}_U = [-0.9487, 0.9487]^T$, $\boldsymbol{p}_M = [0, 0.01, 0, 1, 1, 0, 0.05]^T$.

4.2 Linear System Identification

In this experiment, to test the filtering performance of RCMVC, we consider a linear system identification problem, where the length of weight vector $\boldsymbol{w}_0$ is $L = 31$, $C = \mathbb{R}^{31 \times 1}$, and $f = \mathbb{R}^{1 \times 1}$, as were used in [4]. The colored input data $x(n)$ is generated by filtering GN with $\boldsymbol{p}_G = [0, 1]^T$ through a first-order system as $H(z) = \left(1 - 0.7z^{-1}\right)^{-1}$. In addition, the normalized MSD is used to make filtering performance comparison, i.e., NMSD $= 20\log\left(\|\boldsymbol{w}_0 - \boldsymbol{w}(n)\|_2 / \|\boldsymbol{w}_0\|_2\right)$.

Firstly, we compare the filtering performance between CMVC and RCMVC with $\alpha \in \{2, 2.5, 3, 3.5, 4\}$, where CMVC is superior to other gradient-based CAF algorithms [14]. In this trial, we use the standard symmetric α-stable (SSαS) distribution to model impulsive noise and set $\kappa \in \{0.5, 1.5\}$ for the characteristic exponent factor and $\epsilon = 0.3$ for the dispersion. The NMSD results are plotted in Fig. 2, which shows that: 1) As expected, RCMVC is superior to CMVC in terms of convergence speed under colored input and impulsive noisy situations; 2) Armed with different values of α and λ, CMVC and RCMVC can achieve their corresponding fastest convergence speed. That is to say, MVC can provide more flexibility between filtering accuracy and convergence rate, by adjusting the values of α and λ. Then, we compare RCMVC with RCGMC, RCLL, RCMqR, and RCRLM under different non-Gaussian noisy situations. The parameters, as shown in Fig. 3, are set to realize a similar initial convergence rate. And a time-varying system is also considered by changing $\boldsymbol{w}_0$ to $-\boldsymbol{w}_0$ at middle iteration. Figure 3 plots the corresponding NMSD curves and reveals that: 1) With SSαS noise, RCMVC is superior to RCMqR and RCLL; With uniform noise, RCLL is the worst, and the filtering performance of RCMVC is slightly better than that of RCMqR, RCGMC, and RCRML; With binary noise, RCMVC outperforms others; With Laplacian noise, RCMVC is slightly better than that of RCGMC, RCMqR, RCLL, and RCRLM; 2) Totally speaking, no matter what kind of non-Gaussian noisy environments, RCMVC has a faster convergence rate and a better filtering accuracy.

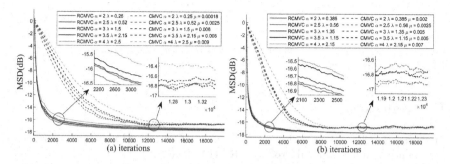

Fig. 2. The NMSD curves of RCMVC and CMVC algorithms with different parameters for linear system identification under SSαS environments. (a) $\kappa = 0.5$; (b) $\kappa = 1.5$.

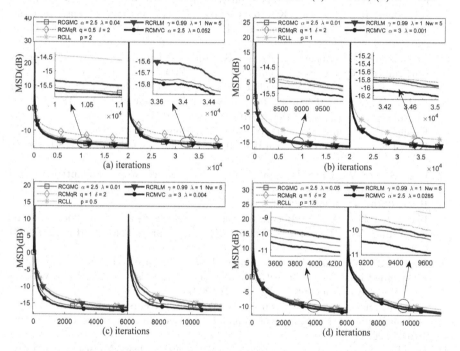

Fig. 3. The NMSD curves of various algorithms for linear system identification under different non-Gaussian noisy environments. (a) SSαS with $\kappa = 1.5$ and $\epsilon = 0.5$; (b) UN noise with $p_U = [-\sqrt{3}, \sqrt{3}]^T$; (c) BN noise with $p_B = [-1, 1, 0.5]^T$; (d) Laplacian noise with zero-mean and 8 unit variance.

5 Conclusion

This paper has proposed a recursive constrained maximum Versoria criterion (RCMVC) algorithm by maximizing the MVC cost function. The mean square convergence and the transient MSD result have been theoretically analyzed and numerically studied to demonstrate the analysis. The presented simulation results validate the effectiveness of our theoretical results and show that RCMVC achieves a smaller MSD and faster convergence rate compared to competitors under various noises. Furthermore, we will develop a low-complexity version of RCMVC in future work.

References

1. Peng, S., Chen, B., Sun, L., Ser, W., Lin, Z.: Constrained maximum correntropy adaptive filtering. Signal Process. **140**, 116–126 (2017)
2. de Campos, M.L.R., Werner, S., Apolinário, J.A.: Constrained adaptive filters. In: Chandran, S. (eds.) Adaptive Antenna Arrays. Signals and Communication Technology. Springer, Heidelberg (2004). https://doi.org/10.1007/978-3-662-05592-2_3
3. Arablouei, R., Dogancay, K.: Reduced-complexity constrained recursive least-squares adaptive filtering algorithm. IEEE Trans. Signal Process. **60**(12), 6687–6692 (2012)
4. Arablouei, R., Dogancay, K., Werner, S.: On the mean-square performance of the constrained LMS algorithm. Signal Process. **117**(Dec.), 192–197 (2015)
5. Lee, K., Baek, Y., Park, Y.: Nonlinear acoustic echo cancellation using a nonlinear postprocessor with a linearly constrained affine projection algorithm. IEEE Trans. Circuits Syst. II Express Briefs **62**(9), 881–885 (2015)
6. Kumar, K., Pandey, R., Karthik, M.L.N.S., Bhattacharjee, S.S., George, N.V.: Robust and sparsity-aware adaptive filters: a review. Signal Process. **189**, 108276 (2021)
7. Peng, S., Ser, W., Chen, B., Sun, L., Lin, Z.: Robust constrained adaptive filtering under minimum error entropy criterion. IEEE Trans. Circuits Syst. II Express Briefs **65**(8), 1119–1123 (2018)
8. Liang, T., Li, Y., Xia, Y.: Recursive constrained adaptive algorithm under q-Renyi Kernel function. IEEE Trans. Circuits Syst. II Express Briefs **68**(6), 2227–2231 (2021)
9. Liang, T., Li, Y., Xue, W., Li, Y., Jiang, T.: Performance and analysis of recursive constrained least lncosh algorithm under impulsive noises. IEEE Trans. Circuits Syst. II Express Briefs **68**(6), 2217–2221 (2021)
10. Zhao, J., Zhang, J.A., Li, Q., Zhang, H., Wang, X.: Recursive constrained generalized maximum correntropy algorithms for adaptive filtering. Signal Process. **199**, 108611 (2022)
11. Wenjing, X., Zhao, H.: Robust constrained recursive least M-estimate adaptive filtering algorithm. Signal Process. **194**, 108433 (2022)
12. Huang, F., Zhang, J., Zhang, S.: Maximum Versoria criterion-based robust adaptive filtering algorithm. IEEE Trans. Circuits Syst. II Express Briefs **64**(10), 1252–1256 (2017)
13. Chen, B., Xing, L., Zhao, H., Zheng, N., Príncipe, J.C.: Generalized correntropy for robust adaptive filtering. IEEE Trans. Signal Process. **64**(13), 3376–3387 (2016)
14. Bhattacharjee, S.S., Shaikh, M.A., Kumar, K., George, N.V.: Robust constrained generalized correntropy and maximum Versoria criterion adaptive filters. IEEE Trans. Circuits Syst. II Express Briefs **68**(8), 3002–3006 (2021)
15. Akhtar, M.T., Albu, F., Nishihara, A.: Maximum Versoria-criterion (MVC)-based adaptive filtering method for mitigating acoustic feedback in hearing-aid devices. Appl. Acoust. **181**, 108156 (2021)

16. Ren, C., Wang, Z., Zhao, Z.: A new variable step-size affine projection sign algorithm based on a posteriori estimation error analysis. Circuits Syst. Signal Process. **36**(5), 1989–2011 (2017)
17. Qian, G., Ning, X., Wang, S.: Recursive constrained maximum correntropy criterion algorithm for adaptive filtering. IEEE Trans. Circuits Syst. II Express Briefs **67**(10), 2229–2233 (2020)
18. Radhika, S., Albu, F., Chandrasekar, A.: Steady state mean square analysis of standard maximum Versoria criterion based adaptive algorithm. IEEE Trans. Circuits Syst. II Express Briefs **68**(4), 1547–1551 (2021)

Graph Pointer Network and Reinforcement Learning for Thinnest Path Problem

Jingjing Li[1,2], Yang Wang[1,2]([✉])[ID], and Chuang Zhang[1,2]

[1] Software Engineering Institute, East China Normal University, Shanghai, China
ywang@sei.ecnu.edu.cn
[2] Shanghai Key Lab of Trustworthy Computing, Shanghai, China

Abstract. The complexity and NP-hard nature make finding optimal solutions challenging for combinatorial optimization problems (COPs) using traditional methods, especially for the large-scale problem. Recently, deep learning-based approaches have shown promise in solving COPs. Pointer Network (PN) has become a popular choice due to its ability to handle variable-length sequences and generate variable-sized outputs. Graph Pointer Network (GPN), which incorporates graph embedding layers in PN, can be well-suited for problems with graph structures. Additionally, Reinforcement Learning (RL) has great potential in enhancing scalability for solving large-scale instances. In this paper, we focus on Thinnest Path Problem (TPP). We propose an approach using RL to train GPN with constraints (GPN-c) to solve TPP. Our approach outperforms traditional solutions by providing faster and more efficient solving strategies. Specifically, we achieved significant improvements in solution quality, runtime, and scalability, and successfully extended our approach to instances with up to 500 nodes. Furthermore, RL and GPN can provide more flexible and adaptive solving strategies, making them highly applicable to real-world scenarios.

Keywords: Thinnest Path Problem (TPP) · Graph Pointer Network (GPN) · Reinforcement Learning (RL)

1 Introduction

Eavesdropping on communication networks has become a critical issue in today's world, and extensive research has been conducted to address this problem. In this paper, we focus on Thinnest Path Problem (TPP) [1], which was proposed by Gao et al. in 2012 as an NP-hard problem to address the eavesdropping problem. TPP aims to find a path from a source node to a destination node with the least number of nodes that can overhear the message. Figure 1 illustrates an example, node 1 represents the source node, and node 20 represents the destination node. Here, the transmission range is dynamically adjusted according to the distance between nodes. A node becomes a potential eavesdropping node if its distance to

© The Author(s), under exclusive license to Springer Nature Singapore Pte Ltd. 2024
B. Luo et al. (Eds.): ICONIP 2023, CCIS 1961, pp. 446–457, 2024.
https://doi.org/10.1007/978-981-99-8126-7_35

the currently selected node is closer than the distance from the currently selected node to the next selected node. The influenced nodes include both the selected nodes and the potential eavesdropping nodes. In Fig. 1 (b) and (c), two different routes from the source node to the destination node are shown, and the circles show the transmission range from the currently selected node to the next selected node. Although Fig. 1 (c) has the least selected nodes, it has more influenced nodes compared to Fig. 1 (b). Therefore, the route in Fig. 1 (b) is considered the best route since it minimizes the possibility of eavesdropping. TPP cannot be considered as the shortest path problem with the fixed weight because each node can only be influenced once. Alternatively, TPP can be treated as the shortest path problem with dynamic weights. Detailed problem definitions are provided in Sect. 3.1.

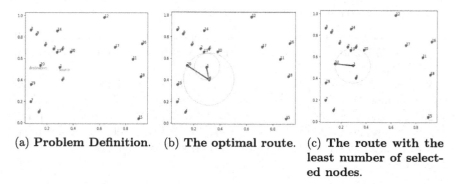

(a) **Problem Definition.**   (b) **The optimal route.**   (c) **The route with the least number of select-ed nodes.**

Fig. 1. The TPP instance. Node 1 is the source node and node 20 is the destination node. Blue nodes are uninfluenced nodes, red nodes with the line are selected nodes, and other red nodes are not selected but within the transmission range. The green circle in (b) is the transmission range from node 1 to node 8, and the orange circle is the transmission range from node 8 to node 20. The green circle in (c) is the transmission range from node 1 to node 20. (Color figure online)

Previous studies on TPP primarily relied on traditional algorithms [1–4]. NP-hard problems are known challenging to solve using traditional methods. Generally, there are three traditional methods: exact algorithms, approximation algorithms, and heuristics. Exact algorithms can obtain the optimal solution, but they are often computationally prohibitive for large-scale problems. Approximation algorithms can solve the problem in polynomial time, but they may not guarantee optimality or fail to approximate certain problems. Heuristics can quickly obtain near-optimal solutions, but they are still time-consuming on large-scale problems, and the solutions obtained may be unstable.

Recently, deep learning-based methods have shown promising results in solving combinatorial optimization problems (COPs), as they have a good extension in addressing large-scale problems. Among these methods, Pointer network (PN) [5] that was proposed by Vinyals et al. in 2015 becomes a popular choice. Therefore, Bello et al. [6], Nazari et al. [7] and Dai et al. [8] used Reinforcement

Learning (RL) to train PN to solve COPs and achieved better results than the original method. Furthermore, Ma et al. [9] proposed Graph Pointer Network (GPN) and applied a hierarchical RL approach to solving TSP with constraints.

Despite the success of these deep learning-based methods in other work, they have shown poor performance on TPP, particularly in large-scale instances. These methods often tend to select nearby nodes without considering their distance to the destination, when there are many nodes close to the source node but opposite to the destination nodes, it can result in a high number of nodes being selected. In this paper, we reframe the model architecture for TPP and propose a dynamic constraint tailored to the characteristics to improve the scalability of the model. This constraint can also be used in traditional methods. As a result, our model can obtain sub-optimal solutions for large-scale problems with up to 500 nodes with good time efficiency. The specific results are shown in Sect. 4.

The main contributions of our work are as follows:

- We propose the first deep learning-based approach that uses RL to train GPN to solve TPP and achieve significant time reduction compared with traditional methods.
- We add a constraint that improves the performance of the original COP-solving models and demonstrates good generalizability. We train the model in 20-node instances and it can extend to 500-node instances which the optimal solver Gurobi cannot solve. Alternatively, We first solve TPP using a heuristic algorithm ACO (Ant Colony Optimization) with the same constraints as the baseline.

2   Related Work

Thinnest Path Problem (TPP) was first proposed by Gao et al. [1] in 2012, and subsequent studies have mainly focused on traditional algorithmic approaches. Gao et al. [1,3] utilized a hypergraph structure to describe this problem, assuming fixed maximum transmission ranges for all vertices, which differs from our problem setting. Moberly et al. [2] addressed a similar problem where nodes can choose their transmission ranges randomly, aligning with our approach. Wu et al. [4] presented the problem using a grid graph representation.

In the domain of solving combinatorial optimization problems with deep learning, Pointer network (PN) [5] that was proposed by Vinyals et al. in 2015 has served as a fundamental method due to its ability to handle variable-length sequences and output variable-sized outputs. Furthermore, Guo et al. [10] proposed a multi-pointer network. Building on the PN, Bello et al. [6], Nazari et al. [7], and Kool et al. [11] combined Reinforcement Learning (RL) methods to solve COPs and achieved better results than before. However, this gradient-based approach suffered from low sample efficiency. To address the issue, Dai et al. [8] proposed S2V-DQN that does not use a separate encoder and decoder, but introduced a graph embedding network to achieve better results on several problems such as Traveling Salesman Problem (TSP). Barrett et al. [12] proposed ECO-DQN that combines RL and deep graph networks to solve graph-based combinatorial problems and got better results than S2V-DQN. Expanding on these advancements, Ma et al. [9] introduced Graph Pointer Network (GPN) by incorporating a graph embedding layer into PN and applying a hierarchical RL approach to solve the TSP with constraints. The graph embedding layer transforms the graph data into continuous vector representations, enhancing the representation of relationships between nodes.

In the PN-independent methods, Joshi et al. [13] used Graph Convolutional Network (GCN) and beam search to solve TSP, achieving good sample efficiency and solution quality, but was limited to instances with up to 100 nodes. Li et al. [14] used GCN and guided tree search to solve questions similar to real social network graphs. Bengio et al. [15] explored how RL can be well combined to solve COPs. Schuetz et al. [16] used Graph Neural Networks (GNNs) to propose a versatile and scalable general-purpose solver, it can solve NP-hard problems in the form of quadratic unconstrained binary optimization problems very well. Ma et al. [17] proposed a deep bidirectional competitive learning method to address the problem of the time-consuming of GNN training, the effectiveness of the method is verified on the TSP but only up to 100 nodes.

3   Problem Definitions

3.1   Thinnest Path Problem (TPP)

In this paper, we focus on solving the Thinnest Path Problem (TPP) in a 2-D Euclidean space. Different from [1,3], we do not use the hypergraph structure. Given a complete and undirected graph $G(V, E)$, where V is the vertex set and E is the edge set, and a list of N node coordinates $v_1, v_2, ..., v_N \subset \mathbb{R}^2$, our goal is to find the path from the source node s to the destination node d with the least influenced nodes, denoted as $\min \sum_{i=1}^{N} x_i$. Assuming the current node is s and the selected node is d, if the distance from v_i to s is shorter than the distance from s to d, v_i will also be influenced. This can be represented by considering the transmission range as a circle with s as the center and the distance from s to d as the radius. To achieve this, we need to satisfy the following constraints:

1. Each edge y_{ij} can only be influenced once. $y_{ij} \leqslant 1$ for $i, j = 1, 2, ..., N$
2. Each node x_i can only be influenced once, ensuring that the selection of nodes is unique. $x_i \leqslant 1$ for $i = 1, 2, ...N$
3. If an edge is selected, then its corresponding node must be selected. $x_j \geqslant \sum y_{jk}$ for $k = 1, 2, ..., N$
4. If $e_{ik} > e_{il}$, then $y_{il} \cdot x_k \geqslant y_{il} \cdot x_l$ for $l = 1, 2, ..., N$

Table 1 provides an overview of the notations used in our formulation.

By formulating and solving the TPP with these constraints, we aim to find an optimal solution that minimizes the number of influenced nodes, providing an effective approach for addressing eavesdropping concerns in communication networks.

Table 1. Notations.

Symbol	Description
s	the source node
d	the destination node
V	the vertex set
E	the edge set
v_i	the i_{th} node
e_{ij}	the distance of the edge linked with v_i and v_j
x_i	equal to 1 if v_i is influenced, else equal to 0
y_{ij}	equal to 1 if e_{ij} is selected, else equal to 0

3.2   RL for TPP

In the context of Reinforcement Learning (RL), we define the states, actions, and rewards. Let S be the state space, each $s_t \in S$ represents the state of all nodes at time t; let A be the action space, each $a_t \in A$ represents the next selected route; let R be the reward function, $R = -\sum_{i=1}^{N} x_i$.

Denote a policy π with parameter θ (i.e. GPN in this paper), and the probability distribution $p(\pi_\theta)$, we need to find a route σ that minimizes the influenced nodes:

$$L(\sigma, V) = \sum g(v_{\sigma(i)}, v_{\sigma(i+1)}) \tag{1}$$

where L is the loss function, $\sigma(i)$ represents the i_{th} node in the route, and $g(\cdot, \cdot)$ is the function that calculates the increased number of influenced nodes from the currently selected node to the next selected node. That is, the sum of the current time t influenced nodes minus the sum of the previous time $t-1$ influenced nodes.

$$g(v_{\sigma(i)}, v_{\sigma(i+1)}) = \sum_{i=1}^{N} (x_{i(t)} - x_{i(t-1)}) \tag{2}$$

We optimize it with gradient descent, using a baseline b. In this paper, we use REINFORCE algorithm [18]. TPP cannot achieve good efficiency in large batches, so we simply use the greedy policy as the baseline to satisfy the single batch, i.e. the action is sampled greedily.

$$\nabla L(\theta|s) = \sum_{i=1}^{N} \mathbb{E}_{p(\pi_\theta|s_i)} [L(\sigma, V) - b_{greedy}(s_i) \nabla \log p(\pi_\theta|s_i)] \tag{3}$$

The Algorithm 1 gives the detailed pseudo-code. The training set V includes all nodes information, h is the hidden state of the neural network, $\tilde{a}$ and $\tilde{h}$ are the action and hidden state provided by the greedy policy π^{greedy}. Here we use an array $mask$ to record whether the node is selected. $mask$ initializes to 0, if a node v is selected, $mask_v$ set to $-\infty$ indicates the node cannot be selected again.

Algorithm 1: REINFORCE for TPP

Input: training set V, destination node d, current node v, learning rate λ

Initialize network parameters θ

for i *in epoch* **do**

    for $v\ != d$ **do**

        $a_t, h_t \sim \pi_\theta(\cdot | s_t, h_{t-1})$

        $\tilde{a}_t, \tilde{h}_t \sim \pi_\theta^{greedy}(\cdot | \tilde{s}_t, \tilde{h}_{t-1})$

        v sampled from $a_t, \tilde{a}_t$

        $mask_v \leftarrow -\infty$

    Compute $L(\theta)$, $\nabla L(\theta)$

    $\theta \leftarrow \theta + \lambda \nabla L(\theta)$

return π_θ

In this paper, to minimize the number of eavesdropping nodes, we propose a constraint named *unsafe* to keep the algorithm from selecting the node that is nearer to the current node but far from the destination. As Fig. 2 shows, node 1 is the source node, node 20 is the destination node, the unconstrained model chose a node much closer to itself, but lead to more nodes be influenced. The green arrow is the optimal route and the green circle is the transmission range of the route, in this route, node 9 will not be influenced because node 1 is closer to node 20 than node 4. Therefore, we add the constraint to avoid this situation. The following equations need to be satisfied for the currently selected node v, and *unsafe* will participate in the calculation of GPN, affecting the probability of the next selected node.

$$if\ e_{vd} - e_{id} < 0,\ unsafe(i)+ = e_{vd} - e_{id} \tag{4}$$

$$if\ e_{dv} - e_{iv} < 0,\ unsafe(i)+ = e_{dv} - e_{iv} \tag{5}$$

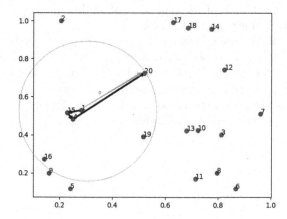

Fig. 2. Counter-example. 1 is the source node, and 20 is the destination node. The green circle is the transmission range from node 1 to node 20. (Color figure online)

3.3   Graph Pointer Network (GPN)

Graph pointer network (GPN) was first proposed by Ma et al. [9] for solving Traveling Salesman Problem (TSP). It consists of an encoder and decoder component.

Encoder. The encoder consists of two components: the pointer encoder and the graph encoder. The pointer encoder uses Long Short-Term Memory (LSTM) [19], encoding the hidden state h of all nodes V, where V includes node coordinates. The graph encoder encodes coordinate relations of all nodes through the graph embedding layer. The graph embedding layer is crucial for selecting a good route when the source node and the destination node are far apart.

In TPP, we need to find a node that is relatively close to the currently selected node k and also has a tendency to approach the destination node d. Previous work [9] used a Graph Neural Network (GNN) in GPN to learn the context information and better represent the relationships between nodes. Each layer of the GNN can be expressed as:

$$V^l = \gamma V^{l-1}\Theta + (1 - \gamma)\Phi_\theta(V^{l-1}/N(i)) \tag{6}$$

where V^l is the lth layer variable, γ is a trainable parameter, Θ is a trainable weight matrix, Φ_θ is the aggregation function, and $N(i)$ is the adjacency set of node i.

Decoder. The decoder employs the Attention mechanism to output a distribution of the next selected node, denoted as u. The calculation of u is given by the equation:

$$u = w \cdot tanh(W_r r_j + W_q q) + unsafe, \ j \notin \sigma \tag{7}$$

where w is the trainable parameter, W_r and W_q are trainable matrices, q is a query vector from the hidden variable of the LSTM, and r_i is a reference vector containing the information of the context of all nodes. $unsafe$ is the constraint to guarantee the performance of the model.

The model architecture is shown in Fig. 3. Inputting the node coordinates V and the previously selected node coordinate v_{i-1}, the encoder encodes the coordinates using LSTM and the graph embedding layer which uses GNN to obtain the query vector q and the context information $context$ of nodes. The decoder generates the distribution u_i based on q and $context$. Finally, the next selected node v_i can be sampled from the probability distribution $p_i = softmax(u_i)$.

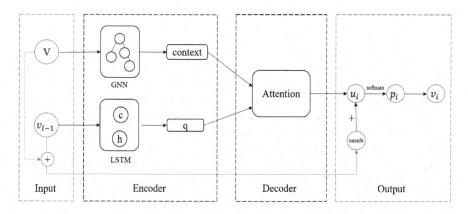

Fig. 3. The model architecture of GPN-c.

Algorithm 2 gives the implementation details of the model architecture.

Algorithm 2: GPN

Input: all node coordinate V, the previously selected node coordinate
v_{i-1}
encoder: LSTM, GNN
decoder: Attention
for i *in* σ **do**
$\quad$ *context* $\sim GNN(V)$
$\quad$ $q \sim LSTM(v_{i-1}, h)$
$\quad$ $u_i \sim decoder(q, context)$
$\quad$ $u_i \leftarrow tanh(u) + unsafe$
$\quad$ $p_i \sim softmax(u_i)$
return p_i

---

4   Experiments

We use Gurobi [20], Ant Colony Optimization (ACO) [21], ACO with the *unsafe* constraint (ACO-c), Graph Pointer Network (GPN) [9], Pointer Network (PN) [5] with the constraint (PN-c) and as a comparison for GPN with the constraint (GPN-c). Compared to GPN-c, GPN without constraint, PN-c without graph embedding layer and Reinforcement Learning (RL).

Hyperparameters. The training data is randomly generated from a uniform distribution $[0,1]^2$. All models were trained in the 20-node situation for 100 epochs, with each epoch consisting of 5000 training steps. For PN-c, GPN GPN-c, each epoch takes around 10:50 min, 4:20 min and 3:50 min respectively (NVIDIA GeForce RTX 3070). The learning rate for GPN and GPN-c was set to 10^{-4} with a decay factor of 0.98 per epoch. For PN-c, a learning rate of 2×10^{-4} with a decay factor of 0.96 was used.

Table 2 shows the average number of influenced nodes and average run-time for 20-node, 50-node, and 100-node situations. Each situation includes 1000 experiment data. As the table shows, Gurobi can achieve the optimal result as a control. However, as the size of the problem increases, the time-consuming also increases significantly. ACO can get near-optimal solutions, but it still requires significant computation time. ACO-c just slightly improves results. And they are less secure and time-consuming than Gurobi in both 20-node and 50-node situations. The deep learning-based model generates solutions that are not as good as Gurobi and ACO in terms of quality, but it is capable of obtaining sub-optimal results in less than one second. GPN-c is slightly better than other deep learning methods in small-scale instances, as the size increases, the advantages are becoming more and more obvious. It is also clear to see the importance of the constraint and graph embedding layer in this model.

Table 2. Mean value of data from 1000 experiments

Method	20		50		100	
	Node	Time (s)	Node	Time (s)	Node	Time (s)
gurobi	6.846	0.110	10.422	5.067	14.416	110.524
ACO	6.861	4.293	10.626	18.058	15.106	64.825
ACO-c	6.856	4.208	10.558	18.031	15.022	59.403
PN-c	7.72	0.011	13.62	0.021	23.423	0.044
GPN	7.629	0.010	12.642	0.025	20.592	0.061
GPN-c	7.478	0.011	11.978	0.025	18.278	0.060

Due to time constraints, for Gurobi, we only compare the results for the first 15 instances in the 200-node situation and it cannot run in the 500-node situation. For ACO, due to the same reason, we only compare the results for the first 20 instances in the 500-node situation. The results are shown in Fig. 4 and Fig. 5.

As figures show, Gurobi can obtain optimal results, but the time consumption and resource consumption is significant in large-scale instances. And it cannot run in over 200-node situation. In Fig. 4, the 200-node situation, ACO can get near-optimal results and have much less time-consuming than Gurobi, but is still more than deep learning-based methods. ACO and ACO-c affect about the same number of nodes, GPN-c is the best in the deep learning-based methods and nearly half as small as GPN. In Fig. 5, the 500-node situation, Gurobi cannot be running. ACO and ACO-c also affect about the same number of nodes, but ACO is almost twice more time-consuming as ACO-c. The advantage of GPN-c is becoming more obvious. GPN is almost twice more time-consuming and influenced nodes as GPN-c. Therefore, although GPN-c generates solutions that are not as good as Gurobi and ACO, it has a great advantage in obtaining results in less than one second even for instances with up to 500 nodes. In practice, the appropriate method can be selected for different types of problems.

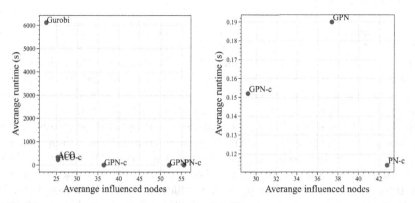

(a) The comparison of six methods from 15 experiments. (b) The comparison of three methods from 1000 experiments.

Fig. 4. Experiment results in 200-node. Figure (a) shows the comparison of six methods (Gurobi, ACO, ACO-c, PN-c, GPN, GPN-c) in terms of average influenced nodes and average run-time from 15 experiments. Figure (b) shows the comparison of three methods (PN-c, GPN, GPN-c) in terms of average influenced nodes and average run-time from 1000 experiments. The closer to the bottom left of the chart, the better.

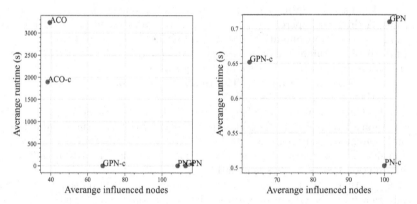

(a) The comparison of five methods from 20 experiments. (b) The comparison of three methods from 1000 experiments.

Fig. 5. Experiment results in 500-node. Guribi cannot run in this situation. Figure (a) shows the comparison of five methods (ACO, ACO-c, PN-c, GPN, GPN-c) in terms of average influenced nodes and average run-time from 20 experiments. Figure (b) shows the comparison of three methods (PN-c, GPN, GPN-c) in terms of average influenced nodes and average run-time from 1000 experiments. The closer to the bottom left of the chart, the better.

5   Conclusions

In this paper, we propose the first deep learning-based approach to solve Thinnest Path Problem (TPP) using Graph Pointer Network with constraint (GPN-c) and Reinforcement Learning (RL). By leveraging the unique characteristics of TPP, our method addresses the challenge of finding thin paths in large-scale instances. Through extensive experiments, we have demonstrated the effectiveness of our approach. The results show that our method can successfully solve TPP instances within seconds, even for larger-scale instances where optimal solvers like Gurobi fail. While the obtained results are sub-optimal compared to the baseline models like Gurobi and Ant Colony Optimization (ACO), they still offer valuable solutions for decision-making and optimization purposes. For future work, we plan to explore more efficient methods for model training, aiming to further enhance the training stability of the proposed approach. Additionally, we intend to find safer methods on this basis to improve the practical applicability of deep learning-based methods.

Acknowledgment. This work was supported in part by the Shanghai Artificial Intelligence Innovation and Development Fund grant 2020-RGZN-02026 and in part by the Shanghai Key Lab of Trustworthy Computing Chairman Fund 2022.

References

1. Gao, J., Zhao, Q., Swami, A.: The thinnest path problem for secure communications. A directed hypergraph approach. In: Allerton Conference on Communication, Control, and Computing, pp. 847–852 (2012)
2. Moberly, R.: A thinner thinnest path using directional transmissions in a network. In: Military Communications Conference, pp. 1026–1031 (2013)
3. Gao, J., Zhao, Q., Swami, A.: The thinnest path problem. IEEE-ACM Trans. Networking **23**(4), 1176–1189 (2015)
4. Wu, S., Chen, Z., Wang, Y., Gao, X., Wu, F., Chen, G.: Efficient approximations for thinnest path problem in grid graph. In: Ubiquitous Information Management and Communication, pp. 1–8 (2018)
5. Vinyals, O., Fortunato, M., Jaitly, N.: Pointer networks. In: Advances in Neural Information Processing Systems (2015)
6. Bello, I., Pham, H., Le, Q.V., Norouzi, M., Bengio, S.: Neural combinatorial optimization with reinforcement learning. In: International Conference on Learning Representations, pp. 1–8 (2017)
7. Nazari, M., Oroojlooy, A., Snyder, L., Takác, M.: Reinforcement learning for solving the vehicle routing problem. In: Advances in Neural Information Processing Systems (2018)
8. Dai, H., Khalil, E., Zhang, Y., Dilkina, B., Song, L.: Learning combinatorial optimization algorithms over graphs. In: Advances in Neural Information Processing Systems, pp. 6348–6358 (2017)
9. Ma, Q., Ge, S., He, D., Thaker, D., Drori, I.: Combinatorial optimization by graph pointer networks and hierarchical reinforcement learning. arXiv preprint arXiv:1911.04936 (2019)

10. Guo, T., Han, C., Tang, S., Ding, M.: Solving combinatorial problems with machine learning methods. In: Du, D.-Z., Pardalos, P.M., Zhang, Z. (eds.) Nonlinear Combinatorial Optimization. SOIA, vol. 147, pp. 207–229. Springer, Cham (2019). https://doi.org/10.1007/978-3-030-16194-1_9

11. Kool, W., Hoof, H., Welling, M.: Attention, learn to solve routing problems!. In: International Conference on Learning Representations (2019)

12. Barrett, T., Clements, W., Foerster, J., Lvovsky, A.: Exploratory combinatorial optimization with reinforcement learning. In: AAAI Conference on Artificial Intelligence, vol. 34, pp. 3243–3250 (2020)

13. Joshi, C., Laurent, T., Bresson, X.: An efficient graph convolutional network technique for the travelling salesman problem. arXiv preprint arXiv:1906.01227 (2019)

14. Li, Z., Chen, Q., Koltun, V.: Combinatorial optimization with graph convolutional networks and guided tree search. In: Advances in Neural Information Processing Systems, pp. 539–548 (2018)

15. Bengio, Y., Lodi, A., Prouvost, A.: Machine learning for combinatorial optimization: a methodological tour d'horizon. Eur. J. Oper. Res. **290**(2), 405–421 (2021)

16. Schuetz, M., Brubaker, J., Katzgraber, H.: Combinatorial optimization with physics-inspired graph neural networks. Nat. Mach. Intell. **4**(4), 367–377 (2022)

17. Ma, H., Tu, S., Xu, L.: IA-CL: a deep bidirectional competitive learning method for traveling salesman problem. In: Tanveer, M., Agarwal, S., Ozawa, S., Ekbal, A., Jatowt, A. (eds.) Neural Information Processing, ICONIP 2022. LNCS, vol. 13623, pp. 525–536. Springer, Cham (2022). https://doi.org/10.1007/978-3-031-30105-6_44

18. Williams, R.J.: Simple statistical gradient-following algorithms for connectionist reinforcement learning. Mach. Learn. **8**(3–4), 229–256 (1992)

19. Hochreiter, S., Schmidhuber, J.: Long short-term memory. Neural Comput. **9**(8), 1735–1780 (1997)

20. Gurobi Optimization, LLC: Gurobi optimizer reference manual (2022). Retrieved from http://www.gurobi.com

21. Dorigo, M.: Optimization, learning and natural algorithms. In: Thesis Politecnico Di Milano Italy (1992)

Multi-neuron Information Fusion for Direct Training Spiking Neural Networks

Jinze Wang[1]([✉]), Jiaqiang Jiang[1], Shuang Lian[2], and Rui Yan[1]([✉])

[1] College of Computer Science and Technology,
Zhejiang University of Technology, Hangzhou, China
{2112112017,ryan}@zjut.edu.cn
[2] College of Computer Science and Technology,
Zhejiang University, Hangzhou, China

Abstract. Spiking neural networks (SNNs) are currently receiving increasing research attention. Most existing SNNs utilize a single class of neuron models. These approaches fail to consider features such as diversity and connectivity of biological neurons, thus limiting their adaptability to different image datasets. Inspired by the gap junctions in neuroscience, we propose a multi-neuron information fusion (MIF) model. This model incorporates multiple neuron models, forming neuron groups that can reflect biological plausibility while aiming improving experimental performance. We evaluate the proposed model on the MNIST, Fashion-MNIST, CIFAR10, and N-MNIST datasets, and the experimental results show that it can achieve competitive results with fewer time steps.

Keywords: Multi-neuron Information Fusion (MIF) · Spiking neurons · Information fusion · Spiking neural networks

1 Introduction

Spiking neural network (SNN), as the third generation of neural network [1], has been widely applied and studied for its advantages of low power consumption and high robustness [2]. SNNs use binary spike signals to transmit information and has the ability to process information in both the spatial domain (SD) and the temporal domain (TD). SNNs are promising for energy-efficient deployment on neuromorphic hardwares, but they still face great challenges due to the neuronal dynamics and non-differentiable spiking activation function involved [3].

The learning processes now used to train high-performance SNNs are mainly divided into two main approaches. The first category is the conversion from ANN to SNN [4–6], mainly by converting a pre-trained ANN model into a similarly structured SNN model. Previous researches have demonstrated that this approach can make deep SNN training with competitive results, but the large number of time steps also bring the problem of computational surge. The second

© The Author(s), under exclusive license to Springer Nature Singapore Pte Ltd. 2024
B. Luo et al. (Eds.): ICONIP 2023, CCIS 1961, pp. 458–469, 2024.
https://doi.org/10.1007/978-981-99-8126-7_36

category is direct supervised learning [7–9], where a widely used learning algorithm is Spatio-temporal backpropagation (STBP) [10]. STBP enables effective backpropagation (BP) learning by processing non-differentiable firing functions. This approach does not require the use of ANN, and reduces the use of time steps by making full use of spatio-temporal information for training. Researches on this method now focus on the improvement of surrogate gradients [3], the deepen of network layers [6], and the modification of neuron models [11], etc. It has been experimentally demonstrated that these ways allow SNN to have results that can compete with ANN. However, these methods mainly consider the effect of a single class of neuron model, and do not consider the diversity and connectivity of neurons in conjunction with basic biological principles.

Recently, there have been a number of studies addressing inter-neuronal interactions, multiple neurons acting together, and other aspects that reflect the biological principles to some extent. Because the SNN model mainly uses inter-layer connections without intra-layer connections, inspired by lateral interactions, Cheng et al. [12] suggested that local lateral interaction connections can effectively process spike training information, and thus propose a new SNN-based lateral interaction model, named LISNN, which takes full advantage of such connections and improves noise robustness. For deeper SNN training, existing direct supervised learning still has problems to be solved, such as the gradient vanishing problem. For this reason, Feng et al. [13] proposed the MLF method, which aims to alleviate the problem by allocating the coverage of approximate derivatives at each level to extend the non-zero region of the rectangular approximate derivatives, in which the membrane potential of neurons has a higher probability of falling into the region where the gradient is not zero. There are many other attempts to make use of multiple neurons, and this paper is also investigated them by such ideas.

Biologists have discovered a special type of intercellular connection among animal neurons called the gap junction [14]. It can directly connect two neurons, acting as a regulatory gate for the direct transmission of various molecules, ions, and electrical spikes between neurons. Gap junctions allow multiple classes of neurons to connect, enabling the formation of a considerable number of intercellular channels with different physiological properties. Moreover intercellular channels of gap junctions are dynamically regulated, and the communication that takes place between neurons is regulated at multiple levels. Inspired by the gap junctions, we believe that this information interaction and regulation rules can be used to represent the effect of intercombination between different classes of neurons. Because most previous models used single class of neurons, there is only one representation of the information. By implementing information fusion between different classes of neurons, more information can be expressed by way of neuron groups. Therefore, we propose the multi-neuron information fusion model named MIF, which consists of multiple information channels, all receiving the same input. This can better reflect the diversity and combination of biological neurons. Also, it is experimentally demonstrated that simple combinations bring the opposite effect, in order to improve the ability to represent information accurately, we introduce selection rules to automatically adjust the combinations. It is experimentally proved that we are able to obtain better accuracy on MNIST, N-MNIST, Fashion-MNIST and CIFAR10 datasets while improving the spiking firing rate.

2   Methods

In this section, we introduce the proposed MIF model. First, we introduce the iterative Leaky Integrate-and-Fire (LIF) neuron and Integrate-and-Fire (IF) neuron as well as their advantages and disadvantages. Then we describe the MIF models in detail, including the selection rule, fusion method, etc.

2.1   Neuron Model

The IF neuron model, as a common neuron model, was widely used initially and applied to SNNs because of its computational simplicity and easy deployment. It can also analyze and deal with many problems at the same time. The formulation of the IF neuron can be described as follows:

$$V_{i,IF}^{n+1}(t) = V_{i,IF}^{n+1}(t-1)\left(1 - o_{i,IF}^{n+1}(t-1)\right) + I_{i,IF}^{n+1}(t) \tag{1}$$

It can be seen that the IF neuron model does not have a decay property and therefore acts only as a simple accumulator, lacking the biological interpretability required by spiking neural networks. Also its spiking sparsity is low.

The ensuing LIF neuron model [15], is currently the most widely used model in computational neuroscience. It has a certain bionic nature, while taking into consideration the high computational efficiency of ordinary artificial neurons. Thus it can better describe neuronal activity. Based on the traditional LIF model, Wu et al. [16] proposed an iterative LIF model that accelerates the direct training of SNNs, whose forward process can be described as:

$$I_{i,LIF}^{n+1}(t) = \sum_{j=1}^{l(n)} w_{i,j}^{n+1} o_{j,LIF}^{n}(t) \tag{2}$$

$$V_{i,LIF}^{n+1}(t) = \alpha V_{i,LIF}^{n+1}(t-1)\left(1 - o_{i,LIF}^{n+1}(t-1)\right) + I_{i,LIF}^{n+1}(t) \tag{3}$$

$$o_{i,LIF}^{n+1}(t) = f\left(V_{i,LIF}^{n+1}(t)\right) \tag{4}$$

$$f(V) = \begin{cases} 1 & V \geq V_{th1} \\ 0 & \text{otherwise} \end{cases} \tag{5}$$

where $I_i^{n+1}(t)$ denotes the external input current received by the i-th neuron in the $(n+1)$-th layer at t-th time step. $l(n)$ represents the number of neurons in the n-th layer. $w_{i,j}^{n+1}$ denotes the synapse weight of the j-th neuron in the n-th layer to the i-th neuron in the $(n+1)$-th layer. α is the decay factor, which is different from IF neurons. $V_i^{n+1}(t)$ and $o_i^{n+1}(t)$ represent the membrane potential and binary spiking output of the i-th neuron in the $(n + 1)$-th layer at t-th time step, respectively. $f(\cdot)$ is the activation function used to determine whether to fire a spike or not, and $f(V) = 1$ when $V \geq V_{th1}$, otherwise $f(V) = 0$.

Because of the decay property of LIF neuron model, it inevitably causes a certain degree of information loss. And its computational process is more complicated than the IF neuron model.

Most of the current SNNs use a single class of neuron models, which lack information representation between different classes of neurons. Combined with the idea of gap connection existing in biological connectivity, in order to make full use of the strengths of both neuron models, we proposed a MIF model to form a neuronal group by integrating the advantages of both to achieve better experimental results.

2.2   The Multi-neuron Information Fusion Model

Here we firstly detail the relevant rules used in the MIF model which as a neuron group consists of primary and auxiliary neurons, and we use LIF neurons as primary neurons and IF neurons as auxiliary neurons, enabling the model to achieve better results through information fusion while fully reflecting biological rationality.

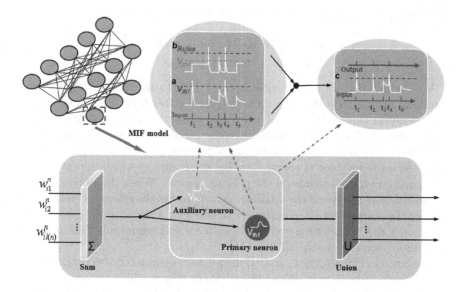

Fig. 1. Illustration of MIF model. Under certain conditions, auxiliary neurons can make neuron groups more active and fire more spikes. Here the primary neuron denotes the LIF neuron, whose corresponding membrane potential curve is (a), and the auxiliary neuron denotes the IF neuron, whose corresponding membrane potential curve is (b). (c) indicates the membrane potential curve and output results of the spiking neuron groups.

To begin with, the IF neuron, as an auxiliary neuron, cannot unconditionally deliver spikes to the model. So there are certain limitations and constraints imposed on this neuron. First of all, neuron can only fire spikes when the membrane potential exceeds the predefined firing threshold, which is the basis. But it is not desirable to act on the primary neuron as soon as the spike is generated, as we have found experimentally that excessive spikes will have the opposite effect. Therefore, we employ a correlation-based approach to determine whether the IF neuron should act or not, based on the membrane potential of the primary neuron. Specifically, we consider the average membrane potential of primary neurons within the same layer at a given time. When one of these points is not satisfied, the IF neuron enters a waiting queue. By accumulating membrane potentials over time, it functions when the condition is satisfied.

As shown in Fig. 1(b), the model receives inputs at t_1, t_2, t_3, t_4 and t_5. When determining the input membrane potential according to the threshold, the auxiliary neuron will fire spikes at t_1, t_2, t_3 and t_4. However, under the constraint of the relevant rules, the membrane potential at t_1, t_3 cannot reach this requirement, and therefore, the spikes cannot act on the neuron group. By applying the selection rule, spikes are successfully applied to the model at t_2, t_4. Experimental results demonstrate that with a well-defined selection rule, the actual number of fired spikes is reduced due to constraints. As a result, the auxiliary neurons can deliver precise and accurate information and better serve the neuron group. The corresponding equation is as follows:

$$x = \begin{cases} 1 & V_{i,IF}^{n+1} \geq \max\left(V_{\text{th2}}, \text{avg}\left(V_{i,LIF}^{n+1}\right)\right) \\ 0 & \text{otherwise} \end{cases} \tag{6}$$

We present the proposed fusion method. Due to the binary output nature of spiking neural networks, where the output is either 0 or 1, a simple summation operation cannot be applied when two classes of neurons generate spikes simultaneously. The relevant method seems necessary to fuse the information of different neurons in a reasonable way. When the primary neuron generates a spike, the auxiliary neuron enters a resting state, during which the spikes firing of the neuron group depends on the primary neuron. When the primary neuron fails to generate a spike, if the auxiliary neuron meets the relevant conditions for firing at the corresponding moment, it can transmit a signal to the primary neuron and act as a trigger to directly wake up it for spiking output, thus improving the neuron's expressive capacity.

As shown in Fig. 1(a), at t_1 and t_4, the primary neuron receives the inputs and produces the spike output. While the auxiliary neuron also performs the signal transmission at t_2 and t_4 as shown in Fig. 1(b) in the above analysis. Through the corresponding fusion method, as shown in Fig. 1(c), at t_2, the auxiliary neuron is successfully activated, generating a spike that directly influences the primary neuron. It is evident that the auxiliary neuron does not dominate the neuron group but serves as a compensatory element, particularly for capturing non-sharp features with small values. This allows the neuron group to express a greater amount of information, thereby improving its overall capacity. The corresponding equation is as follow:

$$o_i^{n+1}(t) = f\left(V_{i,LIF}^{n+1}(t)\right) \| x \tag{7}$$

In summary, as shown in Fig. 1, we use the proposed MIF model instead of the conventional LIF neuron model. The MIF model includes two different classes of neurons that have different level thresholds. Once the input is received, these neurons will update the membrane potentials. The final output of the MIF unit is composed of the two types of spikes mentioned above.

3 Experiment

We test our proposed MIF model on four image datasets. And compare the accuracy of MIF with several excellent models with the same or similar network structure, and the experiments prove that the accuracy obtained by our model is competitive among many models.

3.1 Datasets Processing

For different types of datasets we have different ways of processing the data to make the images into spike trains. For static datasets, we use Poisson coding, which requires first to normalize the data and generate matching random numbers with the same dimension as the input data. Then make a value comparison and if the input data is larger than the random number, the input signal is 1 and the opposite is 0. For neuromorphic datasets, since this class of datasets has temporal properties, we adopt the approach of accumulating time stream information over time steps. The event stream data needs to be sliced throughout the time window and spikes are accumulated at each time step, during which the operation is stopped if a spike is generated. The output event stream information is 0 or 1.

3.2 Network Structure

The network structures of SNNs for different datasets are shown in Table 1. For all Conv2d layers, kernel size is set to 3 and stride and padding are both set to 1. For all pooling layers, kernel size and stride are set to 2. We set the initial threshold for the primary neuron to 0.5 and the threshold for the auxiliary neuron to 1.6.

Table 1. Network structures of different datasets.

Dataset	Hidden Layers
MNIST	128C3-AP2-128C3-AP2-128FC-10FC-10
N-MNIST	128C3-AP2-128C3-AP2-128FC-10FC-10
Fashion-MNIST	128C3-AP2-128C3-AP2-128FC-10FC-10
CIFAR10	128C3-BN-256C3-BN-AP2-512C3-BN-AP2- 1024C3-BN-512C3-BN-1024FC-512FC-10FC-10

3.3 Static Dataset

Here we use MNIST, Fashion-MNIST and CIFAR10 datasets. The training set of the first two datasets contains 60,000 images and the testing set contains 10,000 images. They are both 28 * 28 pixels. The former is 10 classes of grayscale handwritten digital images from 0 to 9, and the latter is more complex compared to the former and consists of 10 classes of clothing images. CIFAR10 is a very challenging dataset currently. It has a training set of 50,000 images and a test set of 10,000 images with 34 * 34 pixels. It consists of 10 classes of RGB images, including truck, car, plane, boat, frog, bird, dog, cat, deer, horse.

We compared the accuracy of our model with the best models of recent years on these datasets. The specific details are shown in Table 2. It is important to note that when test on the CIFAR10 dataset, the MIF model and the original STBP algorithm use the same time step and network structure, obtaining more than 1% accuracy which leads to significant improvement.

Here, to justify our proposed model, we add the model to the TET algorithm [22] to compare the difference in accuracy before and after. We use the VGG11 network structure. As shown in the Table 2, compared to the initial use single LIF neuron model, the accuracy obtained is superior when using MIF model to form neuron groups acting together.

The accuracy obtained by our model achieves competitive results in all methods except for the LTMD [21]. However, compared to LTMD, our model uses a simpler network structure.

3.4 Neuromorphic Dataset

Here we use N-MNIST. This dataset is an extension of MNIST and is captured by a dynamic vision sensor and has the same number and image pixels. This dataset is more challenging compared to MNIST because it may have more noise.

We compared the accuracy of our and other models on the N-MNIST dataset, and the results are shown in Table 2. It can be found that compared with most models, our model can obtain better N-MNIST recognition accuracy with fewer time steps.

Table 2. Performance comparison between the proposed method and the state-of-the-art methods on different datasets. Number of timesteps are highlighted in italics (in parentheses).

Method	MNIST	Fashion-MNIST	CIFAR10	N-MNIST
STBP [16]	–	–	90.53%(*12*)	99.53%(*12*)
TSSL-BP [17]	99.53%(*5*)	92.83%(*5*)	91.41%(*5*)	99.40%(*100*)
LISNN [12]	99.50%(*20*)	92.07%(*20*)	–	99.45%(*20*)
ASF-BP [18]	99.60%	–	91.35%(*400*)	–
ST-RSBP [19]	99.62%(*400*)	90.13%	–	–
Spike-based BP [20]	99.59%(*50*)	–	90.95%(*100*)	99.09%(*100*)
ANN-to-SNN [4]	–	–	91.55%	–
LTMD [21]	99.60%(*4*)	–	94.19%(*4*)	99.65%(*15*)
TET [22]	–	–	92.89%(*4*)	–
MIF+TET	–	–	93.23%(*4*)	–
MIF+STBP	99.55%(*5*)	92.38%(*20*)	91.62%(*12*)	99.58%(*10*)

3.5   Ablation Study

We conduct an extensive ablation study in the four datasets mentioned above to evaluate the role of selection rules and fusion method in the proposed model. Here we named the neuron group without selection rules as MC. In our experiments, we trained all models under the same parameters, network structure, and compared the accuracy of the experimental results, respectively. As shown in Table 3, it can be seen that the accuracy obtained by the neuron group with reasonable selection rules is always better than that of MC. Also compared to the STBP algorithm that uses only LIF neurons or IF neurons, our model achieves better results. Figure 2 shows the results obtained during the training period for the four cases mentioned above. It can be seen that our MIF model achieves better results throughout the training process for most of the datasets. The above results further demonstrate the rationality of the selection rule, showing that more spikes firing may not be better, and when too much useless information is received, the effect will be counter-productive. It also demonstrates that information fusion between neurons can transmit more accurate information, thus improving the accuracy of recognition.

Table 3. Accuracy of using different methods.

Method	MNIST	Fashion-MNIST	CIFAR10	N-MNIST
STBP+MIF	**99.54%**	**91.84%**	**91.52%**	**99.49%**
STBP+MC	99.51%	91.60%	91.31%	99.40%
STBP+LIF	99.47%	91.62%	91.17%	99.40%
STBP+IF	99.44%	91.41%	91.04%	99.10%

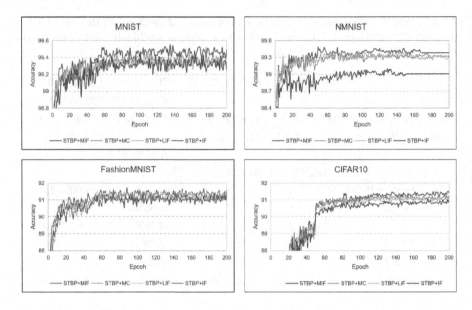

Fig. 2. Test accuracy of different methods on different datasets during training.

3.6 Analysis of Firing Rates

As mentioned above, our proposed MIF model can help improve the accuracy of recognition, while we know the importance of spikes in spiking neural networks, and the spiking firing rate is a key evaluation metric. To justify the model, we take the Fashion-MNIST dataset as an example.

As shown in Fig. 3, the spiking firing rate is calculated by accumulating the firing spikes in all channels for each of the 8 pre-defined time steps. It can be seen that the firing activity of the network is sparse, and a very large proportion of neurons do not participate in firing the spikes, but remain silent at all times.

As we know, with deeper convolutional layers, the information represented decreases and becomes more susceptible to interference, so it is meaningful to analyze the feature maps formed by the firing spikes. As shown in Fig. 4, we accumulate the firing spikes of the channels in the second convolutional layer. It can be seen that the feature maps using the MC model and the STBP algorithm with IF neuron model generate a relatively large number of spikes, but these spikes do not act inside the outline of the recognized images, and are scattered around, acting as interferences. Thus, it can be demonstrated that with reasonable selection rules, the MIF model can better utilize the spikes generated by IF neurons, while not generating too much disturbing spike information as the MC model does. Compared with the feature maps obtained by the STBP algorithm using the LIF neuron model, our model is more like a further refinement tool that can complement the spike information inside the images and improve the image recognizability.

In summary, the MIF model can make better use of a certain number of neurons capable of firing spikes to achieve better results.

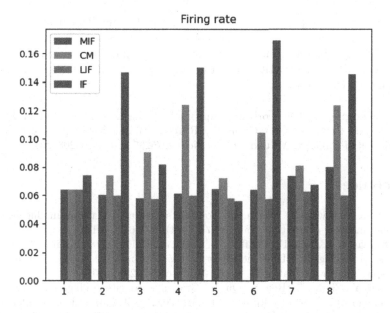

Fig. 3. Firing rate on Fashion-MNIST.

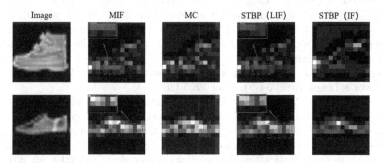

Fig. 4. Feature maps on Fashion-MNIST. The brighter the color, the larger the gray value means the better the recognition effect. The image in the red box is the part of the feature map with significant differences between the MIF model and the single LIF model. The MIF model can utilize the additional spike information well. (Color figure online)

4 Conclusion

Spiking neural networks usually consist of a single neuron model, which may be insufficient for information representation. In this paper, we propose a multi-neuron information fusion model with the help of the idea of gap connection to train SNNs by combining information from different neurons to improve the representation ability. Firstly, we apply the model to the algorithm and successfully prove its usefulness. Secondly, we propose a selection rule and justify it by comparing the training results with and without the rule. By applying the auxiliary neuron to the primary neuron, we find that the auxiliary neuron helps the primary neuron to distinguish some non-sharp features with small values, thus improving the spiking firing rate. Experiments demonstrate that our proposed model can achieve competitive accuracy rates when applied to both static and neuromorphic datasets.

Acknowledgment. This work was supported by the Leading Innovation Team of the Zhejiang Province under Grant 2021R01002 and the National Natural Science Foundation of China NSAF under Grant No. 62276235 and No. U2030204.

References

1. Roy, K., Jaiswal, A., Panda, P.: Towards spike-based machine intelligence with neuromorphic computing. Nature **575**(7784), 607–617 (2019)
2. Stromatias, E., Neil, D., Pfeiffer, M., Galluppi, F., Furber, S.B., Liu, S.C.: Robustness of spiking deep belief networks to noise and reduced bit precision of neuro-inspired hardware platforms. Front. Neurosci. **9**, 222 (2015)
3. Chen, Y., Zhang, S., Ren, S., Qu, H.: Gradual surrogate gradient learning in deep spiking neural networks. In: ICASSP 2022–2022 IEEE International Conference on Acoustics, Speech and Signal Processing (ICASSP), pp. 8927–8931. IEEE (2022)
4. Sengupta, A., Ye, Y., Wang, R., Liu, C., Roy, K.: Going deeper in spiking neural networks: VGG and residual architectures. Front. Neurosci. **13**, 95 (2019)
5. Yan, Z., Zhou, J., Wong, W.F.: Near lossless transfer learning for spiking neural networks. In: Proceedings of the AAAI Conference on Artificial Intelligence, vol. 35, pp. 10577–10584 (2021)
6. Hu, Y., Tang, H., Pan, G.: Spiking deep residual networks. IEEE Trans. Neural Networks Learn. Syst. **34**(8), 5200–5205 (2021)
7. Gu, P., Xiao, R., Pan, G., Tang, H.: STCA: spatio-temporal credit assignment with delayed feedback in deep spiking neural networks. In: IJCAI, pp. 1366–1372 (2019)
8. Liu, Q., Xing, D., Feng, L., Tang, H., Pan, G.: Event-based multimodal spiking neural network with attention mechanism. In: ICASSP 2022–2022 IEEE International Conference on Acoustics, Speech and Signal Processing (ICASSP), pp. 8922–8926. IEEE (2022)
9. Zheng, H., Wu, Y., Deng, L., Hu, Y., Li, G.: Going deeper with directly-trained larger spiking neural networks. In: Proceedings of the AAAI Conference on Artificial Intelligence, vol. 35, pp. 11062–11070 (2021)
10. Wu, Y., Deng, L., Li, G., Zhu, J., Shi, L.: Spatio-temporal backpropagation for training high-performance spiking neural networks. Front. Neurosci. **12**, 331 (2018)
11. Yao, X., Li, F., Mo, Z., Cheng, J.: GLIF: a unified gated leaky integrate-and-fire neuron for spiking neural networks. arXiv preprint arXiv:2210.13768 (2022)
12. Cheng, X., Hao, Y., Xu, J., Xu, B.: LISNN: improving spiking neural networks with lateral interactions for robust object recognition. In: IJCAI, pp. 1519–1525 (2020)
13. Feng, L., Liu, Q., Tang, H., Ma, D., Pan, G.: Multi-level firing with spiking DS-ResNet: enabling better and deeper directly-trained spiking neural networks. arXiv preprint arXiv:2210.06386 (2022)
14. Goodenough, D.A., Paul, D.L.: Gap junctions. Cold Spring Harb. Perspect. Biol. **1**(1), a002576 (2009)
15. Orhan, E.: The leaky integrate-and-fire neuron model, vol. 3, pp. 1–6 (2012)

16. Wu, Y., Deng, L., Li, G., Zhu, J., Xie, Y., Shi, L.: Direct training for spiking neural networks: faster, larger, better. In: Proceedings of the AAAI Conference on Artificial Intelligence, vol. 33, pp. 1311–1318 (2019)
17. Zhang, W., Li, P.: Temporal spike sequence learning via backpropagation for deep spiking neural networks. Adv. Neural. Inf. Process. Syst. **33**, 12022–12033 (2020)
18. Wu, H., et al.: Training spiking neural networks with accumulated spiking flow. In: Proceedings of the AAAI Conference on Artificial Intelligence, vol. 35, pp. 10320–10328 (2021)
19. Zhang, W., Li, P.: Spike-train level backpropagation for training deep recurrent spiking neural networks. In: Advances in Neural Information Processing Systems, vol. 32 (2019)
20. Lee, C., Sarwar, S.S., Panda, P., Srinivasan, G., Roy, K.: Enabling spike-based backpropagation for training deep neural network architectures. Front. Neurosci. 119 (2020)
21. Wang, S., Cheng, T.H., Lim, M.H.: LTMD: learning improvement of spiking neural networks with learnable thresholding neurons and moderate dropout. Adv. Neural. Inf. Process. Syst. **35**, 28350–28362 (2022)
22. Deng, S., Li, Y., Zhang, S., Gu, S.: Temporal efficient training of spiking neural network via gradient re-weighting. arXiv preprint arXiv:2202.11946 (2022)

Event-Based Object Recognition Using Feature Fusion and Spiking Neural Networks

Menghao Su[1(✉)], Panpan Yang[1], Runhao Jiang[2], and Rui Yan[1(✉)]

[1] College of Computer Science and Technology, Zhejiang University of Technology, Hangzhou, China
{2112112289,ryan}@zjut.edu.cn
[2] College of Computer Science and Technology, Zhejiang University, Hangzhou, China

Abstract. Event-based cameras have garnered growing interest in computer vision due to the advantages of sparsed spatio-temporal representation. Spiking neural networks (SNNs), as representative brain-inspired computing models, are inherently suitable for event-driven processing. However, event-based SNNs still have shortcomings in using multiple feature extraction methods, such as the loss of feature information. In this work, we propose an event-based hierarchical model using feature fusion and SNNs for object recognition. In the proposed model, input event stream is adaptively sliced into segment stream for the subsequent feature extraction and SNNs with Tempotron rule. And the model utilizes feature mapping to realize the fusion of the orientation features extracted by Gabor filter and spatio-temporal correlation features extracted by the clustering algorithm considering the surrounding past events within the time window. The experiments conducted on several event-based datasets (i.e., N-MNIST, MNIST-DVS, DVS128Gesture and DailyAction-DVS) show superior performance of the proposed model and the ablation study demonstrates the effectiveness of feature fusion for object recognition.

Keywords: Neuromorphic computing · Address event representation (AER) · Event driven · Spiking neural networks (SNNs) · Object recognition

1 Introduction

Event-based camera is a novel type of imaging sensor that captures visual information in the form of asynchronous and sparse events. Unlike the traditional framed-based camera, each pixel in the event-based camera can individually monitor the relative change of logarithmic light intensity and generates an event when the change exceeds a certain threshold [1]. The event-based camera encodes

© The Author(s), under exclusive license to Springer Nature Singapore Pte Ltd. 2024
B. Luo et al. (Eds.): ICONIP 2023, CCIS 1961, pp. 470–482, 2024.
https://doi.org/10.1007/978-981-99-8126-7_37

the visual information as a continuous spatio-temporal event stream, and its output only depends on the dynamic changes of the scene, thus it has less information redundancy and lower computational power. With microsecond-level asynchronous event output, the event-based camera offers faster perception speed for visual information processing, making it of great potential in real-time object recognition and tracking. However, due to the difference in output data between the two visual sensors, the traditional frame-based object recognition methods cannot be directly used to handle the event data.

To solve this problem, some event-based feature extraction and learning methods for AER recognition have been proposed. Inspired by HMAX [2], a visual cortex-like object recognition model, Zhao et al. proposed an event-driven feedforward classification model [3] and Orchard et al. also proposed an event-driven hierarchical SNN architecture (HFirst) [4]. Both utilize Gabor filters and temporal max-pooling operation, but differently Zhao's method applies a single-spike Tempotron algorithm for classification, while HFirst employs multi-spike encoding strategy and a statistics-based classification. However, both the single-spike Tempotron and statistical method lose time information. To solve the issue, [5] proposed a muti-spike encoding and learning model using event-driven Tempotron networks. Based on [5,6] proposed an event-based classification model using an activated connected domain (ACD) location method, and [7] proposed an effective SNN learning algorithm, Segmented Probability-Maximization (SPA). All of the above models use Gabor filters for feature extraction. Lagorce et al. proposed a hierarchical model of event-based time surfaces (HOTS) [8], which encodes the last event information of the pixels in the receptive field and utilizes clustering algorithm to extract spatio-temporal correlation features. Sironi et al. presented histograms of averaged time surfaces (HATS) [9], which uses local memory units to efficiently leverage past temporal information and build a robust event-based representation. Inspired by the above two types of feature extraction methods, [10] proposed an event-based hierarchical model using Gabor filters and time surfaces, but it loses the orientation features since the two methods are only executed sequentially without feature fusion.

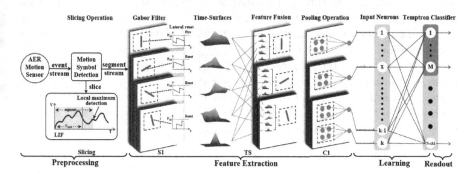

Fig. 1. A schematic of the event-driven hierarchical system for object recognition. It can be divided into four parts: preprocessing, feature extraction, learning and readout.

To address the issue of feature loss in combining multiple feature extraction methods, we propose an event-based hierarchical SNN model for object recognition. The proposed model comprises four components: preprocessing layer, feature extraction layer, learning layer and readout layer. The first layer mainly slices the input event stream using Leaky Integrate-and-Fire (LIF) neurons and local maximum detection. There are three sub-layers in the feature extraction layer, namely S1 layer, TS layer and C1 layer. S1 layer performs orientation-selective edge detection spatio-temporally to extract orientation features by Gabor filters and LIF neurons. In the TS layer, all past events within the time window in the receptive field are used to generate time surfaces, and time-surface prototypes are learned through the clustering algorithm. Furthermore, feature fusion is applied to combine spatio-temporal correlation features output by prototype match in the TS layer and orientation features output in the S1 layer. C1 layer primarily performs pooling operation to reduce computational complexity. In the last two layers, the SNNs utilize the event-driven Tempotron rule for learning and classification tasks (Fig. 1).

2   Methodology

2.1   Preprocessing

For the AER event stream output by the neuromorphic vision sensor, it can be expressed as

$$E = \{e_i \mid e_i = (x_i, y_i, t_i, p_i), i \in \mathbb{N}\} \tag{1}$$

where e_i is the ith event in the stream. x_i and y_i are two spatial coordinates of pixel generating the event, and t_i is the timestamp at which the event is generated. p_i is the polarity of the event, which indicates an increase (ON) or decrease (OFF) of brightness, where ON/OFF can be represented via +1/-1 values.

For the input event stream, the proposed model utilizes motion symbol detection (MSD) [3] for slicing adaptively, which consists of LIF model and local maximum detection. Figure 2 shows how to perform slicing adaptively and the three color regions represent three consecutive segments. Firstly, the minimum time window T_{min} needs to be determined, which represents the minimum duration for all but the last segment. Secondly, the first input event is selected as the starting point I_{start}, and the first point at which the membrane voltage reaches the local maximum after T_{min} time is found as the end position I_{end} of the segment. The output segment consists of the events within the interval $[I_{start}, I_{end}]$. Then, the starting position of the next segment is set to $I_{end} + 1$, and the above operation will be repeated until all events of input stream are processed. Finally, if the duration of the last segment is less than half of T_{min}, it will be merged into the previous segment.

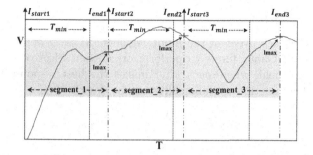

Fig. 2. The membrane voltage curve without threshold of a sample, where three color regions represent three consecutive segments and $lmax$ deonotes the local maximum.

After the preprocessing of slicing adaptively for the event stream, the output segment stream S can be expressed as

$$S = \{s_i \mid s_i = E_i, i \in \mathbb{N}\} \tag{2}$$

where s_i is the ith segment in the segment stream. E_i denotes the event stream in the ith segment. Figure 3 shows the three-dimensional visualization of segment stream, with the X-axis representing the timestamp and the others are two spatial coordinates. ON and OFF events are represented by blue and red dots, respectively. The output of slicing operation are seven segments, and the classification result of input event stream is determined by the seven segments together. This method, which combines fixed time window and local maximum detection, can not only ensure that the duration of each segment is not too short, but also effectively capture the dynamic characteristics of event stream.

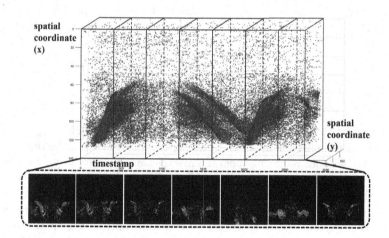

Fig. 3. The output of a sample in the DVS128Gesture dataset after slicing operation. The top is the three-dimensional visualization of segment stream and the bottom is the reconstructed images of each segment.

2.2 Feature Extraction

S1 Layer. For each event of the input segment, S1 layer applies convolution using a group of Gabor filters to perform orientation-selective edge detection in the receptive field, where each filter makes the best response to a certain orientation. The function of Gabor filter can be described as:

$$G(x, y) = \exp(-\frac{(X^2 + \gamma^2 Y^2)}{2\sigma^2})cos(\frac{2\pi}{\lambda}X) \tag{3}$$

where $X = xcos\theta + ysin\theta, Y = -xsin\theta + ycos\theta$. θ is the orientation. γ is the aspect ratio. λ is the wavelength of the cosine factor. σ is the effective width. Each filter corresponds to a specific feature map with the LIF neurons. When the event comes in, each filter kernel will be covered to the corresponding position(i.e., the receptive field centered on the address of the input event) of its feature map, which updates the membrane voltage at the center point. And the membrane potential will decay over time. Once the membrane potential surpasses the threshold, it will output a spike, which is a feature event $S1feat_i = (x_i, y_i, t_i, p_i)$ with $p_i \in \{1, ..., O\}$ representing the orientation feature, and trigger the lateral inhibition and refractory period of the pixel. O denotes the number of orientation.

TS Layer. HOTS [8] introduces the concept of time surface to describe local spatio-temporal patterns around an event, but it only considers last event with the same polarity of each pixel in the receptive field, which makes it too sensitive to noise or small variations in the event stream. To address the issue, HATS [9] utilizes local memory units to generate time surfaces, but it does not output the AER feature events. Inspired by both HOTS and HATS, we make a modification to the online clustering algorithm by incorporateing local memory time surfaces, which consider all past events of every pixel in the receptive field within the time window Δt. Firstly, we define $\mathcal{T}_{e_i}$ as the local memory time surface of event e_i, and each pixel value $\mathcal{T}_{e_i}(\mathbf{x}, \mathbf{y})$ of $\mathcal{T}_{e_i}$ is expressed as

$$\mathcal{T}_{e_i}(\mathbf{x}, \mathbf{y}) = \sum_{e_j \in \mathcal{N}_{(\mathbf{x},\mathbf{y})}(e_i)} e^{-\frac{t_i - t_j}{\tau}} \tag{4}$$

$$\mathcal{N}_{(\mathbf{x},\mathbf{y})}(e_i) = \{e_j : x_j = x_i + \mathbf{x}, y_j = y_i + \mathbf{y}, t_j \in [t_i - \Delta t, t_i)\} \tag{5}$$

where $\mathbf{x}, \mathbf{y} \in [-R, R]$ and R is the radius of the receptive field. $\mathcal{N}_{(\mathbf{x},\mathbf{y})}(e_i)$ is a set including the past events of the pixel $(x_i + \mathbf{x}, y_i + \mathbf{y})$ within the time window $[t_i - \Delta t, t_i)$, without requiring them to have the same polarity as p_i. τ is the time constant of the exponential decay kernel for reducing the influence of past events. The local memory time surface $\mathcal{T}_{e_i}$ is obtained by calculating the value $\mathcal{T}_{e_i}(\mathbf{x}, \mathbf{y})$ of each pixel in the receptive field around (x_i, y_i). Local memory time surface encodes the dynamic spatio-temporal context and captures the spatio-temporal information about the past events within the time window in the neighborhood, so we can utilize its spatio-temporal information to extract features for AER events. The input feature events from S1 layer are divided into different types by comparing their time surfaces with all the time-surface prototypes, which can be learned through unsupervised online clustering algorithm, which is proposed by HOTS and modified by local memory method.

Algorithm 1. Online clustering of local memory time surfaces

Require: K cluster centers $\mathbf{C}_{arr}$, Memory $\mathcal{M}$, τ, R, Δt, $arr = [1, 2, .., K]$

1: Use the first K events' $\mathcal{T}_e$ as initial values for $\mathbf{C}_{arr}$, initialize $p_{arr} \leftarrow 1$
2: **for** every input event $e_i \in$ segment **do**
3:     $\mathcal{M}_{e_i} \leftarrow$ updateAndGetMemoryFiled $(e_i, R, \Delta t)$
4:     $\mathcal{T}_{e_i} \leftarrow$ computeTimeSurface $(e_i, \mathcal{M}_{e_i}, \tau)$
5:     Find closest cluster center $\mathbf{C}_k$
6:     $\alpha \leftarrow 0.01 / (1 + p_k / 20000)$
7:     $\beta \leftarrow \mathbf{C}_k \cdot \mathcal{T}_{e_i} / (\|\mathbf{C}_k\| \cdot \|\mathcal{T}_{e_i}\|)$
8:     $\mathbf{C}_k \leftarrow \mathbf{C}_k + \alpha (\mathcal{T}_{e_i} - \beta \mathbf{C}_k)$
9:     $p_k \leftarrow p_k + 1$
10: **end for**

The complete modified clustering process is showed in Algorithm 1, which is based on K-means algorithm. K is the number of cluster centers. $\mathbf{C}$ is cluster center, which takes the same form as the time surface. $\mathcal{M}$ records the timestamp of past events within the time window for all pixels in the DVS resolution. $\mathcal{M}_{e_i}$ records the timestamp of past events for the pixels in the receptive field around e_i. p_k records the count of time surfaces which have already been assigned to $\mathbf{C}_k$. Firstly, $\mathcal{T}_e$ of the first K input events are selected as initial values for $\mathbf{C}_{arr}$ and the counter p_{arr} is set to 1. For each subsequent input event e_i, we update only the memory $\mathcal{M}_{e_i}$ instead of the memories of all pixels $\mathcal{M}$, and compute the local memory time surface $\mathcal{T}_{e_i}$. Then $\mathcal{T}_{e_i}$ is compared with all time-surface prototypes $\mathbf{C}_{arr}$ to find the closest cluster center $\mathbf{C}_k$, which is updated using these formulas in Algorithm 1. In the learning stage of the time-surface prototypes, the time-surface prototypes are learned using training set, and in the next stage, they are applied to the time-surface match of training and test sets.

Feature Fusion. In the time-surface match stage, for each input feature event $S1feat_i$, the closest time-surface prototype $\mathbf{C}_{k_i}$ is found by comparing time surface $\mathcal{T}_{S1feat_i}$ of event $S1feat_i$ with all time-surface prototypes $\mathbf{C}_{arr}$, and then the input feature event $S1feat_i$ is converted to the output feature event $TSfeat_i$ by the following feature fusion operation.

$$p = K(p_i - 1) + k_i \tag{6}$$

where p_i is the feature polarity of input feature event $S1feat_i$. $p_i \in \{1, .., O\}$, which represents the edge orientation features extracted in the S1 layer. K is the total number of prototypes. k_i is the index of the closest time-surface prototype $\mathbf{C}_{k_i}$ for $\mathcal{T}_{S1feat_i}$. Figure 4 shows the process of feature fusion in the time-surface match stage. O is the number of orientation in the S1 layer. There are O orientation feature mapping buckets and each bucket contains K time-surface prototypes. As shown in the Fig. 4, the K time-surface prototypes in the gray region belong to the p_ith feature mapping bucket and have the same orientation feature. The index k_i of the closest prototype is found through prototype match and Eq. (6) performs the feature mapping to put the time-surface match result k_i in the p_ith orientation feature mapping bucket. Through the feature fusion, we can preserve the orientation features extracted in the S1 layer and combine them with the spatio-temporal correlation features extracted in the TS layer, and output the feature event $TSfeat_i = (x_i, y_i, t_i, p_i)$ with $p_i \in \{1, .., K \times O\}$.

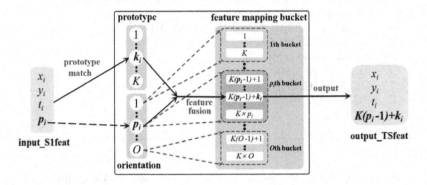

Fig. 4. A schematic of the feature fusion in the time-surface match stage.

C1 Layer. C1 layer functions to perform pooling and regulate output size. The neurons in the C1 layer divide the output of the TS layer into non-overlapping 4×4 regions, where neurons from the same region in the TS layer will project their output spikes to the corresponding one neurons in the C1 layer. When a neuron in the TS layer produces a output spike, the corresponding neuron in the C1 layer will generate an output spike and enter a refractory period, waiting for the arrival of the next spike after the refractory period.

2.3 Learning Layer

Tempotron [13] has two expected output states: firing or not firing. Tempotron learning rule aims to train the weights so that the output neuron can correctly fire or not according to its class label. If it is consistent with the expectation, the weights do not need to be changed. Otherwise, the weights need to be modified according to the following rules:

$$\Delta w_i = \begin{cases} \lambda \sum_{t_i < t_{max}} K(t_{max} - t_i), & \text{if fails to fire;} \\ -\lambda \sum_{t_i < t_{max}} K(t_{max} - t_i), & \text{if fires wrongly;} \\ 0, & \text{otherwise.} \end{cases} \qquad (7)$$

where t_{max} denotes the timestamp at which the neuron reaches its maximum potential in the time domain. λ is the learning rate. In the learning rule, if the neuron is supposed to fire but it actually fails to do so, the corresponding synaptic weights will be increased. And if the neuron fires wrongly, the associated synaptic weights will be decreased. In this work, the model utilizes the event-driven Tempotron networks with Integrate-and-Fire (IF) model.

2.4   Readout Layer

The purpose of the readout layer is to perform classification operations based on the response of the spiking neural networks. In the Tempotron rule, the ideal state is that if the event stream belongs to the specific class, the corresponding neurons fire, and the others don't fire, but it is difficult to achieve the ideal state every time. To improve the classification performance, we use the majority of voting scheme of group coding for each class.

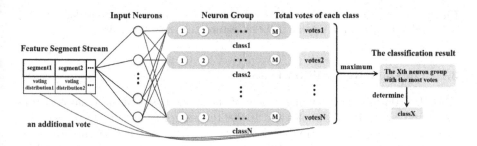

Fig. 5. A schematic of the majority of voting scheme of group coding for each class.

As is shown in Fig. 5, for a N-class categorization task, we allocate M decision neurons for each category, which vote by firing, and each neuron group record own total votes. Moreover, since the classification of each input event stream is determined by all its segments, we add an additional form of voting, which is that each segment independently records its own voting distribution and the neuron group with the most votes in each segment will be cast an additional vote. And the classification result is the class represented by neuron group with the most total votes.

3   Experimental Setup

To analyse the performance of the proposed method, the chosen datasets for experiments are two event-based digit recognition datasets (N-MNIST dataset [14] and MNIST-DVS dataset [15]) and two event-based action recognition datasets (DVS128Gesture dataset [16] and DailyAction-DVS dataset [12]) (Fig. 6).

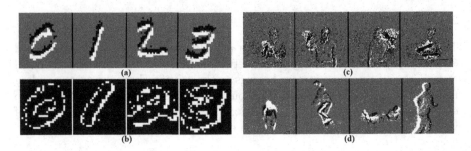

Fig. 6. Reconstructed images of four categories of samples in the four datasets. (a) N-MNIST; (b) MNIST-DVS; (c) DVS128Gesture; (d) DailyAction-DVS.

3.1   On Digit Recognition Datasets

N-MNIST dataset, the spiking version of the frame-based MNIST dataset, consists of 60,000 training samples and 10,000 test samples, and the pixel resolution of each sample is 34×34. MNIST-DVS dataset converts 10,000 samples from the original MNIST dataset into AER data by scanning each image with specified repeated movement in front of a DVS camera at three different scales (scale-4, scale-8 and scale-16), and the pixel resolution of each sample at scale-4 is 32×32.

Table 1. Main parameter settings for different layers of different datasets.

Dataset	slicing		S1			TS				Learning	
	$minT$	τ_m	O	V_{th}	τ_{decay}	R	τ	K	Δt	V_{th}	λ
N-MNIST	80	20	12	100	10	3	20	3	50	20	0.01
MNIST-DVS	500	20	12	100	10	3	30	4	120	20	0.01
DVS128Gesture	400	20	12	100	10	3	40	5	150	20	0.01
DailyAction-DVS	600	20	12	100	10	3	50	5	100	20	0.01
unit	ms	ms		mV	mV/ms		ms		ms	mV	

In our work, we choose the MNIST-DVS dataset at scale-4, and 90% of the MNIST-DVS dataset is randomly allocated for training and the others are used for testing. As shown in the in Table 1, the main different parameters among the datasets are the minimum time window $minT$ in the slicing layer, the radius of the receptive field R and time window Δt in the TS layer. In addition, for all datasets, the refractory period t_{refer} is set to 5ms in the S1 layer and C1 layer, and the number of Tempotron neurons for each category is set to 10. Also, we adopt the $7 \times 7 \times 1$ filter size in the S1 layer. Table 2 shows that our model achieves better performance on two datasets than [7] using the HFirst-based features and SPA learning algorithm. For the N-MNIST dataset, our model

Table 2. Performance comparison in N-MNIST dataset and MNIST-DVS dataset between the proposed method and others with the single training layer.

Method	Structure	N-MNIST	MNIST-DVS
[3]	Gabor-Tempotron SNNs	85.84%	88.14%
HFirst [4]	Gabor-Statistics SNNs	79.22%	82.40%
HOTS [8]	HOTS-based SNNs	80.84%	80.32%
[11]	Gabor-STDP SNNs	89.70%	89.96%
[5]	Gabor-Tempotron SNNs	94.02%	91.51%
[6]	Gabor-Tempotron SNNs	93.96%	95.87%
SPA [7]	Gabor-SPA SNNs	96.30%	96.70%
Our	Feature Fusion-Tempotron SNNs	**96.76%**	**98.20%**

achieves the recognition accuracy of 96.76%, which is significantly higher than HFirst [4] and HOTS [8]. For the MNIST-DVS dataset, the recognition accuracy of our model is 6.69% and 2.33% higher than [5,6], respectively, both of which use HFirst-based features and Tempotron learning algorithm.

3.2   On Action Recognition Datasets

DVS128Gesture dataset contains 11 gesture actions of 29 subjects recorded by DVS128 event camera under three different lighting conditions, and each gesture action contains 122 samples. And DailyAction-DVS dataset contains 12 actions of 15 subjects recorded by DVS128 event camera under two shooting positions and two lighting conditions, and each action contains 120 samples.

Table 3. Performance comparison in DVS128Gesture dataset and DailyAction-DVS dataset between the proposed method and others. The first four methods are deep SNNs and the others are SNNs with single training layer.

Method	Structure	DVS128Gesture	DailyAction-DVS
[16]	Deep SNNs (16 layers)	91.77%	–
SCRNN [17]	Conv-RNN SNNs (5 layers)	92.01%	–
SLAYER [18]	Deep SNNs (8 layers)	93.64%	–
STBP [19]	Deep SNNs (8 layers)	93.40%	–
[5]	Gabor-Tempotron SNNs	61.50%	68.30%
SPA [7]	Gabor-SPA SNNs	70.10%	76.90%
[12]	Motion Feature-SPA SNNs	92.70%	90.30%
Our	Feature Fusion-Tempotron SNNs	**94.65%**	**92.65%**

In our experiments, the training and test sets are randomly divided in a ratio of 6:1 for the DVS128Gesture dataset and DailyAction-DVS dataset. The main parameters about the two datasets are summarized in Table 1. For the DVS128Gesture dataset, Table 3 shows our model achieves the recognition accuracy of 94.65%, which substantially outperforms [5] and SPA [7] by 33.15% and 24.55%, respectively, both of which use Gabor filters and Tempotron learning algorithm. And our model with only one training layer also achieves better performance than some deep SNNs, for examples, the classification accuracy of our model is 2.88% and 2.64% higher than [16] with 16 training layers and SCRNN [17] with 5 training layers, respectively. For the DailyAction-DVS dataset, although the diverse recording conditions of this dataset and more categories of actions increase the difficulty of recognition, the proposed model still achieves the recognition accuracy of 92.65%, which substantially outperforms [5] and SPA [7] with the single training layer by 24.35% and 15.75%, respectively. There are some opposite classes in both datasets, such as hand waving clockwise versus hand waving counterclockwise in the DVS128Gesture dataset, and standing versus sitting and lying versus getting up in the DailyAction-DVS dataset, which has a high requirement for the spatio-temporal association among the events in the action recognition method. And [12] using motion feature information achieves great performance on event-based action datasets. HFirst makes use of time to encode signal strength of spatial orientation features rather than capturing scene dynamics, and time-surface uses both spatial and temporal information to build the motion association information of events in the receptive field within the time window. The proposed method utilizes feature mapping to achieve the fusion of spatial orientation features and spatio-temporal event correlation features, and our model achieves better performance in the both datasets than [12] using the motion information and SPA learning algorithm.

3.3   Ablation Study of the Feature Extraction

In this section, we conduct the ablation study of feature extraction in our proposed model on DVS128Gesture dataset and DailyAction-DVS dataset. Based on the full model, we utilize only S1 layer, S1 layer & TS layer and S1 layer & TS layer & feature fusion to verify their effects on the original architecture.

Table 4. Recognition accuracy and feature number for different feature extraction methods on DVS128Gesture dataset and DailyAction-DVS dataset based on the full model. It keeps C1 layer for pooling after feature extraction.

Feature Method	DVS128Gesture	DailyAction-DVS	FeatureNumber
only S1	84.49%	84.31%	O
S1 & TS	88.24%	85.78%	K
S1 & TS & Feature Fusion	94.65%	92.65%	$O \times K$

In the proposed hierarchical event-based model, the extracted features are recorded by changing the polarity of input event. The polarity of feature event represents its feature and the maximum value of the polarity represents the number of features that can be extracted by feature extraction. Table 4 shows the results of ablation study of feature extraction in proposed model on action recognition datasets. K is the total number of time-surface prototypes and O is the number of orientation features. For our proposed model using the only S1 layer, the recognition accuracy on two datasets drops by 10.16% and 8.34%, respectively. For the proposed model bypassing the feature fusion, less computation is required since it doesn't preserve the features extracted in the S1 layer. However, the recognition accuracy drops significantly, which in turn indicates the effectiveness of feature fusion for action recognition. S1 layer makes use of Gabor filters to capture the spatial orientation features and time surface encodes the dynamics information of past events in the spatial neighborhood. By the feature fusion, the model can extract the spatial orientation features and spatio-temporal correlation features, which significantly improves the accuracy of action recognition.

4 Conclusion

In this paper, we propose an event-based hierarchical recognition method using feature fusion and SNNs. The method utilizes MSD to adaptively slice the input event stream into the segment stream. All past events within the time window in the receptive field are used to generate time surface and the time-surface prototypes are learned through the clustering algorithm. The model extracts the spatio-temporal correlation features through time-surface prototype match and makes use of feature fusion to combine them with the edge orientation features extracted by Gabor filters, resulting in more detailed spatio-temporal features. The SNNs using event-driven Tempotron rule are applied for learning and classification with the majority of voting scheme. Finally, the proposed model demonstrates superior performance in experiments conducted on several AER datasets, including N-MNIST, MNIST-DVS, DVS128Gesture and DailyAction-DVS, and ablation study indicates the effectiveness of feature fusion in the model.

Acknowledgment. This work was supported by the National Natural Science Foundation of China sNSAF under Grant No. U2030204 and No. 62276235, and by the Leading Innovation Team of the Zhejiang Province under Grant 2021R01002.

References

1. Delbrück, T., Linares-Barranco, B., Culurciello, E., Posch, C.: Activity-driven, event-based vision sensors. In: Proceedings of 2010 IEEE International Symposium on Circuits and Systems, pp. 2426-2429. IEEE (2010)
2. Riesenhuber, M., Poggio, T.: Hierarchical models of object recognition in cortex. Nat. Neurosci. **2**(11), 1019–1025 (1999)
3. Zhao, B., Ding, R., Chen, S., Linares-Barranco, B., Tang, H.: Feedforward categorization on AER motion events using cortex-like features in a spiking neural network. IEEE Trans. Neural Netw. Learn. Syst. **26**(9), 1963–1978 (2014)

4. Orchard, G., Meyer, C., Etienne-Cummings, R., Posch, C., Thakor, N., Benosman, R.: Hfirst: a temporal approach to object recognition. IEEE Trans. Pattern Anal. Mach. Intell. **37**(10), 2028–2040 (2015)
5. Xiao, R., Tang, H., Ma, Y., Yan, R., Orchard, G.: An event-driven categorization model for AER image sensors using multispike encoding and learning. IEEE Trans. Neural Netw. Learn. Syst. **31**(9), 3649–3657 (2019)
6. Tang, T., Jiang, R., Yan, R., Tang, H.: An event-driven object recognition model using activated connected domain detection. In: 2020 IEEE Symposium Series on Computational Intelligence (SSCI), pp. 3049-3056. IEEE (2020)
7. Liu, Q., Ruan, H., Xing, D., Tang, H., Pan, G.: Effective AER object classification using segmented probability-maximization learning in spiking neural networks. Proc. AAAI Conf. Artif. Intell. **34**, 1308–1315 (2020)
8. Lagorce, X., Orchard, G., Galluppi, F., Shi, B.E., Benosman, R.B.: Hots: a hierarchy of event-based time-surfaces for pattern recognition. IEEE Trans. Pattern Anal. Mach. Intell. **39**(7), 1346–1359 (2016)
9. Sironi, A., Brambilla, M., Bourdis, N., Lagorce, X., Benosman, R.: Hats: histograms of averaged time surfaces for robust event-based object classification. In: Proceedings of the IEEE Conference on Computer Vision and Pattern Recognition, pp. 1731-1740 (2018)
10. Nan, Y., Xiao, R., Gao, S., Yan, R.: An event-based hierarchy model for object recognition. In: 2019 IEEE Symposium Series on Computational Intelligence (SSCI), pp. 2342-2347. IEEE (2019)
11. Liu, Q., Pan, G., Ruan, H., Xing, D., Xu, Q., Tang, H.: Unsupervised AER object recognition based on multiscale spatio-temporal features and spiking neurons. IEEE Trans. Neural Netw. Learn. Syst. **31**(12), 5300–5311 (2020)
12. Liu, Q., Xing, D., Tang, H., Ma, D., Pan, G.: Event-based action recognition using motion information and spiking neural networks. In: IJCAI, pp. 1743–1749 (2021)
13. Gütig, R., Sompolinsky, H.: The tempotron: a neuron that learns spike timing-based decisions. Nat. Neurosci. **9**(3), 420–428 (2006)
14. Orchard, G., Jayawant, A., Cohen, G.K., Thakor, N.: Converting static image datasets to spiking neuromorphic datasets using saccades. Front. Neurosci. **9**, 437 (2015)
15. Serrano-Gotarredona, T., Linares-Barranco, B.: Poker-DVS and mnist-DVS their history, how they were made, and other details. Front. Neurosci. **9**, 481 (2015)
16. Amir, A., et al.: A low power, fully event-based gesture recognition system. In: Proceedings of the IEEE Conference on Computer Vision and Pattern Recognition, pp. 7243–7252 (2017)
17. Xing, Y., Di Caterina, G., Soraghan, J.: A new spiking convolutional recurrent neural network (scrnn) with applications to event-based hand gesture recognition. Front. Neurosci. **14**, 590164 (2020)
18. Shrestha, S.B., Orchard, G.: Slayer: spike layer error reassignment in time. Adv. Neural Inf. Process. Syst. **31** (2018)
19. He, W., et al.: Comparing SNNs and RNNs on neuromorphic vision datasets: similarities and differences. Neural Netw. **132**, 108–120 (2020)

Circular FC: Fast Fourier Transform Meets Fully Connected Layer for Convolutional Neural Network

Dengjie Yang[1], Junjie Cao[2], YingZhi Ma[3], Jiawei Yu[4], Shikun Jiang[1(✉)], and Liang Zhou[5]

[1] Government and Enterprise Customer Business Group, China United Network Communication Group Co., Ltd., Beijing, China
{yangdj35,jiangsk3}@chinaunicom.cn

[2] National University of Singapore, Singapore, Singapore

[3] New York University, New York, USA

[4] University of Electronic Science and Technology of China, Chengdu, China
202222010619@std.uestc.edu.cn

[5] Gansu Provincial Branch Government Enterprise Customer Business Group, China United Network Communication Group Co., Ltd., Beijing, China
zhoul48@chinaunicom.cn

Abstract. The fully connected (FC) layer is generally located behind the global pooling layer in the convolutional neural network (CNN). Its essence is the weighted summation of the features extracted from the previous convolutional layers, that is, feature remapping. However, the FC layer with close internal correlation inevitably brings parameter redundancy. In order to alleviate this problem, in this paper, we propose a novel lightweight FC-like module, dubbed as Circular FC, by constructing weight parameters in a circular manner. Inspired by digital signal processing theories, we implement Circular FC by fast Fourier transform (FFT) based on the circular convolution theorem of discrete signals. Circular FC is designed to be a plug-and-play classification head and can be easily embedded into existing CNNs such as VGG, Xception, DenseNet, and ResNets. Extensive experiments on CIFAR-10, CIFAR-100, and ImageNet datasets illustrate that the above networks equipped with Circular FC reduce the number of parameters while maintaining comparable image classification performance.

Keywords: CNN · Circular FC · FFT · Image Classification

1 Introduction

Image classification is a fundamental visual recognition task, taking an important part in computer vision. AlexNet [1] has made tremendous progress in image

D. Yang, J. Cao, and Y.Z. Ma—Equal contribution.

© The Author(s), under exclusive license to Springer Nature Singapore Pte Ltd. 2024
B. Luo et al. (Eds.): ICONIP 2023, CCIS 1961, pp. 483–494, 2024.
https://doi.org/10.1007/978-981-99-8126-7_38

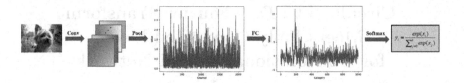

Fig. 1. A general CNN architecture for the image classification task. **Conv** stands for a series of convolutional layers used to extract features. **Pool** refers to global average pooling or global max pooling.

classification accuracy by replacing the traditional hand-crafted features with features learned from CNN. After that, researchers have further explored the depth, width, and scale of the network, bringing about numerous CNNs with better performance, e.g., VGG [4], Xception [10], DenseNet [3], and ResNets [2,6,8,12].

As shown in Fig. 1, in general, CNN is a hierarchical model whose primary components are the convolutional layer, pooling layer and FC Layer. Specifically, the convolution operation extracts high-level semantic features from the original input image by stacking a series of convolutional layers. Then, the pooling layer and FC layer aggregate and lower the dimension of the features extracted by the convolution operation. Eventually, the softmax layer is widely used to classify the informative features. Empirically, CNNs typically suffer from parameter redundancy. Redundancy among feature maps has become a common phenomenon of the convolution operation. Therefore, many optimized convolution operators [5,9,13,15,17] are proposed to alleviate the redundancy problem and reduce the complexity of the model. As a classification head, the FC layer is usually located at the last layer of CNNs to match features to labels. Similarly, the FC layer has parameter redundancy due to the intrinsic interaction among all nodes. Exhaustive analyses in [16] indicate that shallow CNNs need more nodes in the FC layer, while deeper CNNs need fewer neurons in the FC layer regardless of the type of dataset. Therefore, we intuitively think that an ideal classification head should have three characteristics: 1) lightweight with few parameters and FLOPs; 2) spatial mapping to match the number of categories; 3) better separability of aggregated features. In view of these, we assume that the weight parameters of vanilla FC can be constrained in a circular fashion (illustrated in Fig. 2), namely Circular FC, to mitigate the parameter burden and redundancy with remaining the classification performance of the model. However, to realize the process of parameter loop, it is unavoidable to increase flops. Fortunately, inspired by the theory of digital signal processing, we spur FFT to meet FC, which gracefully implements Circular FC.

Overall, our contributions are as follows:

1) We propose a novel variant of FC, namely Circular FC, with fewer parameters compared to vanilla FC.

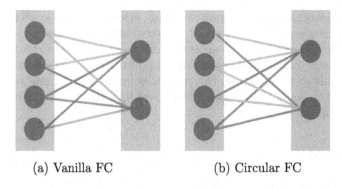

(a) Vanilla FC   (b) Circular FC

Fig. 2. An illustration of Vanilla FC and Circular FC. The same slash color means the same weight parameter value.

2) We realize Circular FC by FFT based on the circular convolution theorem, aiming to provide a more effective and efficient alternative for the classification head of CNN.

3) We verify the effectiveness of Circular FC on three large-scale benchmark datasets in terms of image classicification.

2   Related Work

For image classification, the FC layer is one of the most important basic units of CNN. According to the publicly available literature, CNNs with different architectures have one or multiple FC layers. Both AlexNet [1] and VGG [4] have three FC layers. Furthermore, ResNet [2] changes multiple FC layers to one FC layer, which has become a classification head paradigm used by deep CNNs for image classification. This paper is the further optimization of the FC layer.

With the increase of network layers, a large number of trainable parameters can readily lead to the overfitting problem. Therefore, Dropout [1] and Batch-Normalization [14] are proposed as regularization methods for the FC layer and convolutional layer, respectively, to prevent deep CNNs from overfitting. Additionally, there exist many other variants of Dropout regularization [18,19,21–23]. With regard to the FC layer, [18] presents a block named DropConnect, where the connections to the FC layer are sparse. Unlike Dropout [1] and DropConnect [18], our proposed Circular FC explicitly performs circular constraints on weight parameters with less manual modification, rather than randomly zeroing some weight parameters with a fixed probability. From this perspective, as the classification head of deep CNNs, Circular FC itself has a certain regularization effect.

Then, how to establish Circular FC concisely has become our concern. Recently, many works have been devoted to applying some theories in digital signal processing, especially frequency domain transform, to the task of deep learning models for visual recognition. Some methods [25, 26] convert the input image into frequency domain through discrete cosine transform, and then utilize the frequency domain information as network input to improve the performance of specific tasks. In addition, others [24, 28] exploit FFT to accelerate neural networks. In this paper, based on the circular convolution theorem, we efficiently implement Circular FC through FFT without increasing the number of parameters and computational cost.

3 Method

3.1 DFT and FFT

We first introduce DFT which is a crucial mathematical tool in the field of digital signal processing. Given a 1D discrete sequence $x[n]$ of length N, its 1D DFT can be expressed as follows:

$$X[k] = \sum_{n=0}^{N-1} x[n]e^{-j(2\pi/N)kn} = \sum_{n=0}^{N-1} x[n]W_N^{kn} \tag{1}$$

where j is the imaginary part and $W_N^{kn} = e^{-j(2\pi/N)kn}$. On the whole, each output node value of 1D DFT is associated with all values of the input sequence. Hence, DFT is the weighted summation operation of the input. For DFT $X[k]$, $k = 0, 1, ..., N-1$, the original input can be restored via the inverse DFT (IDFT):

$$x[n] = \frac{1}{N} \sum_{n=0}^{N-1} X[k]e^{j(2\pi/N)kn} \tag{2}$$

Due to the periodicity and symmetry of the rotation factor W_N^{kn}, DFT and IDFT have fast calculation algorithms known as FFT and IFFT, respectively. Both FFT and IFFT are extensively used to reduce computational complexity from $\mathcal{O}(N^2)$ to $\mathcal{O}(N \log N)$.

3.2 FC and Circular FC

The FC layer performs a weighted summation on the features generated by the previous layers, mapping the feature space to the sample label space through a linear transformation. For a vanilla FC layer, given the aggregated features $\mathbf{x} = [x_1, x_2, \cdots, x_C]^T \in \mathbb{R}^{C \times 1}$ of the global pooling layer, the output of the FC layer can be defined as:

$$\mathbf{y}_1 = \boldsymbol{W}_1\mathbf{x} = \begin{bmatrix} w_{1,1} & w_{1,2} & \cdots & w_{1,C} \\ w_{2,1} & w_{2,2} & \cdots & w_{2,C} \\ \vdots & \vdots & \ddots & \vdots \\ w_{C,1} & w_{C,2} & \cdots & w_{C,C} \end{bmatrix} \begin{bmatrix} x_1 \\ x_2 \\ \vdots \\ x_C \end{bmatrix} \tag{3}$$

where $\boldsymbol{W}_1$ is a $C \times C$ weight matrix and $\mathbf{y}_1$ is the same dimension as $\mathbf{x}$. Here, for simplicity, we omit the bias.

Circular FC is a regular constraint or binding of parameters, regarded as a particular case of parameter sharing. As the name implies, Circular FC is represenetd as follows:

$$\mathbf{y}_2 = \boldsymbol{W}_2 \mathbf{x} = \begin{bmatrix} w_1 & w_2 & \cdots & w_C \\ w_C & w_1 & \cdots & w_{C-1} \\ \vdots & \vdots & \ddots & \vdots \\ w_2 & w_3 & \cdots & w_1 \end{bmatrix} \begin{bmatrix} x_1 \\ x_2 \\ \vdots \\ x_C \end{bmatrix} \tag{4}$$

where $\boldsymbol{W}_2 \in \mathbb{R}^{C \times C}$ involves C trainable parameters (only $1/C$ of $\boldsymbol{W}_1$). Clearly, $\boldsymbol{W}_2$ is constructed by a row vector $\mathbf{w} = [w_1, w_2, \cdots, w_C] \in \mathbb{R}^{1 \times C}$ in the form of circular shift to the right. Significantly, a matrix like $\boldsymbol{W}_2$ is termed as a circular matrix. According to the property of the circular matrix, $\boldsymbol{W}_2^T$ is also a circular matrix.

We randomly generate a matrix (1000×1000) according to Kaiming initialization method [29], and build a circular matrix based on the first row of this matrix. We calculate and normalize the singular values of two matrices, and then visualize their distribution. As shown in Fig. 3, the singular value distribution of the circular matrix is more uniform than that of the ordinary matrix, so the circular matrix, as a weight matrix, has stronger anti-interference ability in the CNN training process, and can achieve the ideal classification effect with fewer learnable parameters.

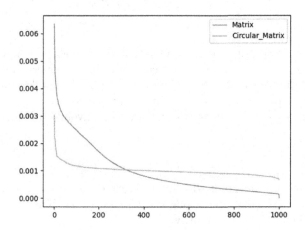

Fig. 3. Singular value distribution of the matrix and circular matrix.

3.3   From FFT to Circular FC

DFT has many essential properties among which the circular convolution theorem has great engineering application value. The circular convolution theorem indicates that the DFT of the circular convolution of two discrete sequences is equal to the product of the two sequences in the frequency domain. We first describe the circular convolution. Given two discrete sequences $z[n]$ and $w[n]$, $0 \leq n \leq N - 1$, we have N-point circular convolution:

$$y[n] = z[n] \otimes w[n] = [\sum_{m=0}^{N-1} z[m]w[((n - m))_N]]R_N(n) \tag{5}$$

where $\otimes$ denotes the circular convolution and $((n))_N$ represents n modulo N. Given the above equation, the process of the circular convolution can be described by matrix operation:

$$\mathbf{y} = \mathbf{W}\mathbf{z} = \begin{bmatrix} w[0] & w[N-1] & \cdots & w[1] \\ w[1] & w[0] & \cdots & w[2] \\ \vdots & \vdots & \ddots & \vdots \\ w[N-1] & w[N-2] & \cdots & w[0] \end{bmatrix} \begin{bmatrix} z[0] \\ z[1] \\ \vdots \\ z[N-1] \end{bmatrix} \tag{6}$$

Note that the structure of $\mathbf{W}$ is similar to that of $\mathbf{W}_2^T$, which indicates that Circular FC can be characterized by the circular convolution. Furthermore, considering the DFT of $y[n]$, the circular convolution theorem is formalized as follows:

$$\begin{aligned} Y[k] &= DFT(y[n]) \\ &= DFT(x[n]) \cdot DFT(w[n]) \\ &= FFT(x[n]) \cdot FFT(w[n]) \\ &= X[k] \cdot W[k] \end{aligned} \tag{7}$$

where $\cdot$ denotes the product of $x[n]$ and $w[n]$ in the frequency domain. Normally, in order to match the number of image categories in the dataset, FC layer needs to reduce the dimension of the input features. In our method, we implement a dimensionality reduced Circular FC. Given an input $\mathbf{x} \in \mathbb{R}^{1 \times C_1}$ and a learnable weight matrix $\mathbf{w} \in \mathbb{R}^{1 \times C_1}$, we utilize the circular convolution theorem to obtain $\mathbf{y} \in \mathbb{R}^{1 \times C_1}$:

$$\mathbf{y} = IFFT(FFT(\mathbf{x}) \odot FFT(\mathbf{w})) \tag{8}$$

where $\odot$ means the point-wise multiplication operation. Specifically, both FFT and IFFT are computed along the C_1 dimension. Since the real output dimension is lower than the input, we have

$$\hat{\mathbf{y}} = \mathbf{y}[:, 0 : C_2] \tag{9}$$

where C_2 ($C_2 < C_1$) is label categories. Taking the bias $\mathbf{b} \in \mathbb{R}^{1 \times C_2}$ into consideration, the final output of reduced dimensional Circular FC is expressed as follows:

$$\hat{\mathbf{y}} = \hat{\mathbf{y}} + \mathbf{b} \tag{10}$$

As shown in Algorithm 1, Circular FC is easily performed in deep learning frameworks (e.g., PyTorch [27]). It is worth noting that we refer to the algorithm description style in [20]. Compared with vanilla FC, the reduced dimension Circular FC can significantly reduce parameters and calculations in most cases. Since both of them have the bias, we temporarily ignore it during analysis. On the one hand, Circular FC has C_1 learnable parameters that is only $1/C_2$ of vanilla FC. On the other hand, Circular FC is mainly composed of two FFTs $(\mathcal{O}(C_1 \log C_1))$, a point-wise multiplication $(\mathcal{O}(C_1))$ and an IFFT $(\mathcal{O}(C_1 \log C_1))$, which states that the total computational complexity is $\mathcal{O}(C_1 \log C_1)$. However, the complexity of FC is $\mathcal{O}(C_1 C_2)$. When C_2 is large enough (such as 1000 for ImageNet), the complexity of FC is much higher than that of Circular FC.

Algorithm 1. Pseudocode of Circular FC.

```
# x: input features with shape [1,C_1]
# w: weight parameters with shape [1,C_1]
# b: bias with shape [1,C_2] (C_2 is label categories)
# rfft/irfft: 1D FFT/IFFT for real signal

x_fft = rfft(x, dim=-1)
w_fft = rfft(w, dim=-1)
y = x_fft * w_fft
y = irfft(y)
y = y[:,0:C_2] + b
```

---

4 Experiments

4.1 Evaluation Details

We apply our proposed Circular FC to CNNs with different architectures, and evaluate its effectiveness for image classification on CIFAR-10 [11], CIFAR-100 [11], and ImageNet [7] datasets. The CIFAR-10 and CIFAR-100 datasets correspond to 10 and 100 categories, respectively. For both datasets, they contain 50000 training images and 10000 test images, with 32×32 pixels. The ImageNet dataset contains 1.28M training images and 50K evaluation images with 1000 categories.

In our experiments, we replace the FC of CNNs with the Circular FC. In particular, we only retain the first FC at the end of the traditional VGG-16, referring to the mainstream classifier structure. Similarly, all networks are trained using stochastic gradient descent (SGD), with an initial learning rate of 0.1, a weight decay of 1e-4, and a momentum of 0.9. All images are flipped horizontally at random in the training stage. For CIFAR datasets, we zero-pad each image with 4 pixels and then crop it to 32×32. We train networks 200 epochs with a batch size of 128 on a single GPU. The initial learning rate is divided by 10 at 100 and 150 epochs. Evaluation is carried out on original images. For the ImageNet dataset, input images are randomly cropped to 224×224. Networks are trained 100 epochs with a batch size of 256 on four 2080ti GPUs. The initial learning rate is decreased by a factor of 10 after every 30 epochs. During validation, we first resize each input image to 256×256 and then adopt a center crop of 224×224.

4.2   Results on CIFAR-10

We present the image classification results on the CIFAR-10 dataset. It should be noted that, due to the small number of label categories (only 10 categories), the difference between Circular FC and FC in terms of parameters and FLOPs is very small and can be ignored. In Table 1, we can find that with Circular FC as the classification head, VGG-16 and Xception improve classification accuracy by 0.51% and 0.50%, respectively. Furthermore, DensNet-121 and ResNet-50 obtain higher accuracy of 1.20% and 1.07% than the original ones.

4.3   Comparisons with Dropout Regularization

We compare the regularization effect of Circular FC with the classical Dropout method in Table 2. Zeroing the input of FC with the probability of 0.3 (p = 0.3), VGG-16 with Dropout achieves an accuracy of 93.46% on the CIFAR-10 dataset, which is slightly higher than that of VGG-16 equipped with vanilla FC. Nonetheless, Circular FC achieves a higher accuracy of 93.77%. In addition, ResNet-50 with Circular FC has the consistent effect as VGG-16. Notably, Dropout method drops the performance of VGG-16 on the CIFAR-100 dataset. On the contrary, Circular FC improves the classification accuracy to a certain extent, which shows that it has a better generalization and regularization effect than Dropout.

4.4   Resluts on ImageNet

As shown in Table 3, we further verify the validity of Circular FC with several commonly used deep CNNs on the ImageNet dataset. Since the network is relatively large, computational savings from Circular FC are negligible when counted as GFLOPs. Nevertheless, benefiting from Circular FC, SCNet, ResNet, ResNeXt, and Res2Net save about 2M parameters and achieve slightly higher Top-1 and Top-5 accuracy. For instance, with a depth of 50 layers, the accuracy of ResNeXt for Top-1 and Top-5 has increased by 0.33% and 0.28%, respectively. Remarkably, when going to the deeper 101 layers, equipped with Circular FC, both ResNet and ResNeXt have improved the Top-1 accuracy by 0.5%.

Table 1. Comparisons with vanilla FC on CIFAR-10

Models	Head	Params	FLOPs	Top-1
DenseNet-121 [3]	FC	6.96M	58.47M	88.40
	Circular FC			**89.60**
VGG-16 [4]	FC	14.73M	314.03M	93.26
	Circular FC			**93.77**
Xception [10]	FC	20.83M	1.13G	94.20
	Circular FC			**94.70**
ResNet-50 [2]	FC	23.52M	1.31G	94.74
	Circular FC			**95.81**

Table 2. Comparisons with Dropout regularization on CIFAR

Dataset	VGG-16 [4]/ResNet-50 [2]				
	FC	Dropout [1] + FC			Circular FC
		p = 0.3	p = 0.5	p = 0.7	
CIFAR-10	93.26/94.74	93.46/95.05	93.29/94.28	93.03/94.08	**93.77/95.81**
CIFAR-100	72.78/75.84	71.02/75.92	72.32/75.99	71.91/75.72	**73.35/77.22**

Table 3. Comparisons with vanilla FC on ImageNet

Models	Head	Params	GFLOPs	Top-1	Top-5
50-layer					
SCNet [12]	FC	25.56M	3.95	77.52	93.70
	Circular FC	**23.51M**		**77.54**	**93.80**
ResNet [2]	FC	25.56M	4.11	76.42	93.04
	Circular FC	**23.50M**		**76.57**	**93.29**
ResNeXt [6]	FC	25.02M	4.25	77.45	93.63
	Circular FC	**22.98M**		**77.78**	**93.91**
Res2Net [8]	FC	25.69M	4.28	77.67	93.74
	Circular FC	**23.65M**		**77.96**	**93.89**
101-layer					
SCNet [12]	FC	44.56M	7.20	78.37	94.00
	Circular FC	**42.51M**		**78.83**	**94.26**
ResNet [2]	FC	44.54M	7.83	77.96	93.99
	Circular FC	**42.50M**		**78.46**	**94.15**
ResNeXt [6]	FC	44.17M	8.01	78.35	94.21
	Circular FC	**42.12M**		**78.85**	**94.39**
Res2Net [8]	FC	45.21M	8.10	78.41	94.18
	Circular FC	**43.15M**		**78.80**	**94.45**

Furthermore, we visualize the output response values of the trained ResNet-101 and ResNeXt-101. In Fig. 4, it is consistent that the response values are large at the correct category and its small neighborhood. When FC is adopted as the classification head, the response values at most of the wrong categories fluctuate around the zero value, and the oscillation is prominent. However, Circular FC greatly reduces the oscillation amplitude of the response values at the error categories. It is observable that Circular FC effectively suppresses the feature expression ability of the wrong categories, which can lower the possibility of network misclassification. As the classification head of CNNs, especially deep CNNs, Circular FC outperforms FC.

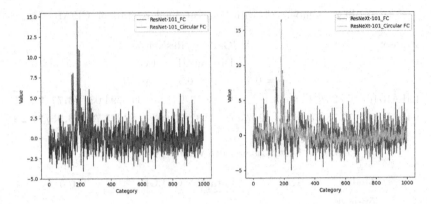

Fig. 4. Visualization of the output response values of different classification heads (FC and Circular FC).

5   Conclusion

In this paper, we propose a novel plug-and-play FC-like module termed as Circular FC. Circular FC aggregates semantic information in the manner of parameter loop and can be regarded as a regularization method. Based on the circular convolution theorem, we realize Circular FC with fewer parameters. Extensive experimental results on multiple image benchmark datasets demonstrate the effectiveness of Circular FC for classification tasks. Circular FC aims to serve as a replacement for vanilla FC for CNNs classification head. In the future, we will further explore and extend Circular FC to other visual recognition architectures and tasks.

References

1. Krizhevsky, A., Sutskever, I., Hinton, G.E.: Imagenet classification with deep convolutional neural networks. In: Neural Information Processing Systems (NeurIPS), pp. 1097–1105 (2012)
2. He, K., Zhang, X., Ren, S., Sun, J.: Deep residual learning for image recognition. In: IEEE Conference on Computer Vision and Pattern Recognition (CVPR), pp. 770–778 (2016)
3. Huang, G., Liu, Z., Maaten, L., Weinberger, K.Q.: Densely connected convolutional networks. In: IEEE Conference on Computer Vision and Pattern Recognition (CVPR), pp. 4700–4708 (2017)
4. Simonyan, K., Zisserman, A.: Very deep convolutional networks for large-scale image recognition. In: International Conference on Learning Representations (ICLR), pp. 1–14 (2015)
5. Han, K., et al.: GhostNets on heterogeneous devices via cheap operations. In: International Journal of Computer Vision (IJCV), pp. 1050–1069 (2022)
6. Xie, S., Girshick, R., Dollar, P., Tu, Z., He, K.: Aggregated residual transformations for deep neural networks. In: IEEE Conference on Computer Vision and Pattern Recognition (CVPR), pp. 1492–1500 (2017)

7. Deng, J., Dong, W., Socher, R., Li, L.-J., Li, K., Fei-Fei, L.: Imagenet: a large-scale hierarchical image database. In: IEEE Conference on Computer Vision and Pattern Recognition (CVPR), pp. 248–255 (2009)
8. Gao, S., Cheng, M., Zhao, K., Zhang, X., Yang, M., Torr, P.: Res2Net: a new multi-scale backbone architecture: In: IEEE Transactions on Pattern Analysis and Machine Intelligence (TPAMI), pp. 652–662 (2021)
9. Zhang, C., Xu, Y., Shen, Y.: CompConv: a compact convolution module for efficient feature learning. In: IEEE Conference on Computer Vision and Pattern Recognition Workshops (CVPR Workshops), pp. 3006–3015 (2021)
10. Chollet, F.: Xception: deep learning with depthwise separable convolutions. In: IEEE Conference on Computer Vision and Pattern Recognition (CVPR), pp. 1800–1807 (2017)
11. Krizhevsky, A., Hinton, G.: Learning multiple layers of features from tiny images. Citeseer, Tech. Rep. TR-2009, University of Toronto (2009)
12. Liu, J.-J., et al.: Improving convolutional networks with self-calibrated convolutions. In: IEEE Conference on Computer Vision and Pattern Recognition (CVPR), pp. 10093–10102 (2020)
13. Yang, D., et al.: BFConv: improving convolutional neural networks with butterfly convolution. In: International Conference on Neural Information Processing (ICONIP), pp. 40–50 (2021)
14. Ioffe, S., Szegedy, C.: Batch normalization: accelerating deep network training by reducing internal covariate shift. In: International Conference on Machine Learning (ICML), pp. 448–456 (2015)
15. Han, K., Wang, Y., Tian, Q., Guo, J., Xu, C., Xu, C.: GhostNet: more features from cheap operations. In: IEEE Conference on Computer Vision and Pattern Recognition (CVPR), pp. 1577–1586 (2020)
16. Basha, S.H.S., Dubey, S.R., Pulabaigari, V., Mukherjee, S.: Impact of fully connected layers on performance of convolutional neural networks for image classification. In: Neurocomputing, pp. 112–119 (2020)
17. Zhang, Q., et al: Split to be slim: an overlooked redundancy in vanilla convolution. In: International Joint Conference on Artificial Intelligence (IJCAI), pp. 3195–3201 (2020)
18. Wan, L., et al.: Regularization of neural networks using DropConnect. In: International Conference on Machine Learning (ICML), pp. 1058–1066 (2013)
19. Ghiasi, G., Lin, T., Le, Q.V.: DropBlock: a regularization method for convolutional networks. In: Neural Information Processing Systems (NeurIPS), pp. 10750–10760 (2018)
20. Rao, Y., Zhao, W., Zhu, Z., Lu, J., Zhou, J.: Global filter networks for image classification. In: Neural Information Processing Systems (NeurIPS), pp. 980–993 (2021)
21. Quyang, Z., Feng, Y., He, Z., Hao, T., Dai, T., Xia, S.: Attentiondrop for convolutional neural networks. In: International Conference on Multimedia and Expo (ICME), pp. 1342–1347 (2019)
22. Goodfellow, I.J., Warde-Farley, D., Mirza, M., et al.: Maxout networks. In: International Conference on Machine Learning (ICML), pp. 1319–1327 (2013)
23. Zhang, H., Qu, D., Shao, K., Yang, X.: DropDim: a regularization method for transformer networks. In: IEEE Signal Processing Letters (SPL), pp. 474–478 (2022)
24. Li, S., et al.: Falcon: a fourier transform based approach for fast and secure convolutional neural network predictions. In: IEEE Conference on Computer Vision and Pattern Recognition (CVPR), pp. 8705–8714 (2020)

25. Gueguen, L., Sergeev, A., Kadlec, B., Liu, R., Yosinski, J.: Faster neural networks straight from JPEG. In: Neural Information Processing Systems (NeurIPS), pp. 3937–3948 (2018)
26. Xu, K., Qin, M., Sun, F., Wang, Y., Chen, Y., Ren, F.: Learning in the frequency domain. In: IEEE Conference on Computer Vision and Pattern Recognition (CVPR), pp. 1737–1746 (2020)
27. Paszke, A., et al.: Pytorch: an imperative style, high-performance deep learning library. In: Neural Information Processing Systems (NeurIPS), pp. 8024–8035 (2019)
28. Vahid, K.A., Prabhu, A., Farhadi, A., Rastegari, M.: Butterfly transform: an efficient FFT based neural architecture design. In: IEEE Conference on Computer Vision and Pattern Recognition (CVPR), pp. 12021–12030 (2020)
29. He, K., Zhang, X., Ren, S., Sun, J.: Delving deep into rectifiers: surpassing human-level performance on ImageNet classification. In: IEEE International Conference on Computer Vision (ICCV), pp. 1026–1034 (2015)

Accurate Latency Prediction of Deep Learning Model Inference Under Dynamic Runtime Resource

Haihong She[1], Yigui Luo[1(✉)], Zhaohong Xiang[1], Weiming Liang[2], and Yin Xie[1]

[1] School of Software Engineering, Tongji University, Shanghai, China
{haihong_she,05097,2231528,yiyexy}@tongji.edu.cn
[2] School of Automotive, Tsinghua University, Beijng, China
lwm20@mails.tsinghua.edu.cn

Abstract. Accurate prediction of inference time can effectively accelerate model design and deployment in neural architecture search (NAS) algorithms, and provide hints for process scheduling in intelligent systems. Most of current latency prediction methods assume that all hardware resources are dominated by a single inference task. Due to the neglect of inter-process resource contention, these methods will fail in multi-tasking parallel scenarios. As an attempt to fill this gap, a hardware-aware prediction method based on double feature embedding (HADE) for deep learning model inference time is proposed. In HADE, operator-wise switchable Multilayer Perceptron and GNN-based graph embedding are used to encode node features, which are generated by unified nodes vectorization. Then, relying on the Resource Aware Latency (RAL) formula based on processor modeling techniques, the model inference time under dynamic runtime resource can be predicted. The experiments show that the accuracy of standard inference time prediction can reach 96.4%, and of resource dependent inference time can reach 79.7%.

Keywords: inference latency prediction · graph neural network · feature embedding · performance modeling · deep neural network

1 Introduction

Due to the excellent performance, deep learning (DL) models have entered the era of large-scale deployment and application in tasks such as pose evaluation [1], image generation [2], and human-computer interaction dialogues [3]. In practical products, DL models are usually constrained by efficiency constraints (e.g., latency) while emphasizing high accuracy performance. Therefore, algorithm design nowadays often utilizes model compression [4,5] and NAS [6,7], which use inference latency as a hard constraint, to improve model efficiency while maintaining the accuracy. However, involving complex pipelines, actually measuring and collecting the latency of a model is expensive. Especially for NAS

© The Author(s), under exclusive license to Springer Nature Singapore Pte Ltd. 2024
B. Luo et al. (Eds.): ICONIP 2023, CCIS 1961, pp. 495–510, 2024.
https://doi.org/10.1007/978-981-99-8126-7_39

tasks, there are a large number of candidate models to be measured for each round of search. In order to obtain model inference latency more efficiently, predicting the latency with a meticulously designed predictor is a feasible approach.

In general, there are two challenges for inference latency prediction: 1) Remarkable differences in network structure and computational properties exist between different DL models. This means that we need a good strategy to describe the structural and computational information of one model. Simple metrics (e.g., FLOPs) are not enough. 2) Many hidden factors have a significant impact on the final inference latency, such as the garbage collection mechanism, the optimizations of low-level computational operations and the hardware performance. It is difficult to accurately predict the latency by simply learning from historical latency data even for a single device.

To characterize the extensive configuration space of DL models, existing hierarchical prediction methods divide the whole model inference into multiple kernels, based on the hypothesis that total latency can be obtained by summing up the latency of kernels. However, due to the fusion of computational graph, this assumption may be invalid. Some other studies propose the direct prediction of model latency by utilizing a simplified model representation specifically tailored to the particular NAS dataset. Nevertheless, the feasibility of extending these approaches to a broader space of model structures remains limited. Liu et al. [9] introduces a unified node feature design for model embedding, yet it lacks the precise encoding of inherent differences within operators. As for capturing hidden factors in hardware, currently researches [10–13] on hardware-aware inference latency prediction mainly emphasize the generalization for unknown devices, they premise that the hardware resources for inferring are sufficient, but when multiple tasks run simultaneously, resource contention may lead to an unsaturated state of critical computational resources required for neural network inference, significantly increasing the inference latency. Sectum [14] proposes a method for predicting neural network latency when memory over-commitment is triggered. DGL [10] predicts inference latency from network computation parameters and hardware configuration parameters. This indicates the feasibility of constructing a mapping of hardware features to latency. Multi-task parallelism is just the load state of the system, the saturation state of the key computational resources is the actual factor that affects the neural network inference latency. But the relationship between them has not been fully investigated, which leads to a gap in the current research.

To address the aforementioned problems, we propose a hardware-aware neural network inference time prediction method based on double feature embedding. To implement double feature embedding, HADE first use an operator-level Multilayer Perceptron (MLP), which is switchable depending on the type of operator, to embed the node features obtained by unified nodes vectorization. It adds additional feature mapping to better capture the inherent distinctions among operators. Subsequently, the model's graph embedding representation is obtained through the encoder based on GraphSAGE [15], it supports implicitly aggregating the topological information after the optimization of the operator.

This double feature embedding strategy better captures the overall structure and characteristics of the model. Considering the cost of predicting itself, a lightweight prediction head is used to process embedded vectors and output prediction latency. By combining processor modeling techniques and correlations between inference latency and critical computational resources, HADE establishes Resource-Aware Latency (RAL) formulas, which enable the prediction of latency under dynamic runtime resource at a low computational cost.

Our main contributions can be summarized as follows:

- To address the two challenges in model inference latency prediction, we propose a novel hardware-aware prediction method called HADE, it fills the research gap in resource-aware latency prediction.
- To enhance the characterization of neural networks, a double feature embedding strategy is proposed. This strategy utilizes switchable multilayer perceptron and graph neural network to capture both local and global information. Experimental results demonstrate that comprehensive model information leads to improved accuracy in predicting latency.
- We propose RAL formula to predict the inference time under dynamic runtime resource. RAL formula possesses the advantages of low computational cost and flexibility, making it suitable for integration into various existing delay prediction methods as a standalone component.

2 Related Work

The emergence of NAS algorithms has engendered a surge in scholarly investigations dedicated to predicting inference latency of deep learning models. Depending on whether hardware modeling is used to achieve cross-device latency prediction, inference latency prediction can be divided into hardware-aware latency prediction and standard latency prediction

2.1 Hardware-Aware Latency Prediction.

Wu et al. [16] reveals the resource sensitivity of operators, shows that inference latency is constrained by hardware conditions. The fundamental concept behind hardware-aware latency prediction entails acquiring hardware characteristics through modeling and subsequently establishing a correlation between inference latency and hardware features. This correlation is leveraged to minimize the computational expense of cross-device prediction. MAPLE [12] and MAPLE-Edge [13] describe the characteristics of the underlying microprocessor by measuring relevant hardware performance metrics. This approach predominantly rely on CPU multi-level cache metrics as descriptive indicators of the hardware and program state, aiming to accomplish hardware-aware functionality. Sectum [14] discusses the impact of memory over-commitment (MOC) on deep learning models inference in TEE servers, and uses a graph neural network based model to detect whether a given model will trigger MOC. Habitat [17] records information about DNN training iterations on a given GPU, and predicts the execution time of model training on different GPUs through wave scaling. These two methods explore the factors influencing inference latency from a higher hardware perspective. Differently, taking the perspective of processes, we investigate the impact patterns of their critical resources on the inference latency of neural networks.

2.2 Standard Latency Prediction

Early NAS and model compression [18–20] methods use Floating-Point Operations (FLOPs), parameter quantities or Memory Access Costs (MAC) as substitutes for inference latency to measure the effectiveness of a method. Paleo [21] directly constructs a performance model with FLOPs. However, the proxy metric exhibits significant errors and demonstrates a low correlation with latency [22]. Building upon the assumption of sequential execution of model operators, several studies have proposed hierarchical prediction models. Samuel et al. [23,24] and Zhang et al. [8] divide the neural network model into multiple kernels and sum up kernels latency as the total inference time. While the hierarchical prediction model facilitates the decomposition of the model, it encounters challenges in accurately predicting latency in scenarios characterized by operator parallelism and intricate graph optimization. In recent years, end-to-end prediction models get attention, Lukasz et al. [25] proposes the first end-to-end inference time prediction method based on Graph Convolutional Network (GCN), but it is only applicable to the static dataset NAS-Bench-201; Liu et al. [9] proposes NNLQP system, which automatically deploys models and measures their inference times to construct an evolving dataset. Through a unified node feature design for graph embedding, the prediction ability is gradually improved; Gao et al. [26] uses the proposed self-attention based node-edge encoder to extract model features.

Given that few studies focus on the inference latency under dynamic hardware resource, and single feature embedding in end-to-end prediction method may ignore the difference between operators, we proposes a hardware-aware neural network inference time prediction method based on double feature embedding.

3 Methodology

The remainder of this section is organized as follows. In Sect. 3.1, we introduce the workflow of HADE method. In Sect. 3.2, we define a unified nodes vectorization to parse the computation graph of input model. In Sect. 3.3, we propose a double feature embedding to aggregate model inference information. The theory of resource aware latency prediction is shown in Sect. 3.4.

3.1 Workflow

The workflow of HADE method proposed in this article is shown in Fig. 1, which comprises three procedures: unified nodes vectorization, double feature embedding and resource aware latency prediction.

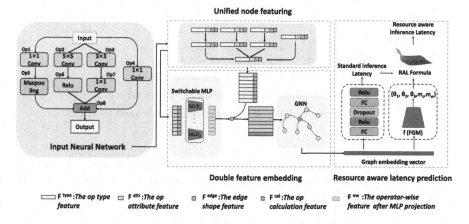

Fig. 1. The workflow of HADE latency prediction model.

HADE accepts a neural network model file in ONNX format, parses and transforms the computation graph to unified feature vector proposed in Sect. 3.2. In the double feature embedding strategy, the unified node feature vector is first updated using a switchable operator-wise MLP. Subsequently, a GNN-based encoder is employed to map the updated node feature vectors into a one-dimensional graph embedding representation. The graph embedding vector bifurcates into two pathways: the first pathway directs towards a lightweight predictor, enabling the estimation of standard inference latency. Meanwhile, the second pathway traverses an additional RAL formula, enabling the estimation of inference latency under the dynamic variations of runtime resources.

3.2 Unified Nodes Vectorization

Essentially, a neural network model can be represented as a directed acyclic computation graph:

$$M = \{\mu_i, e_{ij} = (\mu_i, \mu_j) | i, j \subseteq [1, n]\} \tag{1}$$

Each node μ_i represents an operator in the model (such as Conv2D), each edge e_{ij} represents the data delivery between two operators, and n is the number of operators. Through a meticulous examination of the data structure, the operators information of the model in ONNX format is meticulously extracted and stored within a directed graph structure defined by NetworkX.

Subsequently, the operator information should be transformed into a vectorized representation to form a unified input form for inference prediction. This paper fully considers the factors that may affect the inference latency and proposes following node features:

Table 1. Node Features.

Symbol	Description
F^{type}	$F^{type} \subseteq N+^{40}$. Indicating operator type
F^{attr}	$F^{attr} \subseteq R^9$. Indicating operator attributes
F^{edge}	$F^{edge} \subseteq R^4$. Indicating the shape of output tensor
F^{cal}	$F^{cal} \subseteq R^3$. FLOPs, MAC and parameter amount of operator

We traverse the directed graph obtained by model parsing and vectorize the node and edge information according to Table 1. Finally, the unified node feature vector generates by concatenating all above features:

$$F_{opi} = F_i^{type} \oplus F_i^{attr} \oplus F_i^{edge} \oplus F_i^{cal}, i \subseteq [1, n] \qquad (2)$$

The set of node vectors for the entire model, which is the input value of HADE, is defined as follow:

$$F_{op} = \{F_{opi}, i \subseteq [1, n]\}, F_{op} \subseteq R^{n \times 56} \qquad (3)$$

3.3   Double Feature Embedding

Hierarchical prediction models have difficulties in capturing the topological relationship between operators and the potential graph optimization. In end-to-end prediction models, GNN encoder utilize a shared projection matrix to handle all operators, neglecting the intrinsic differences between them.

In this paper, a novel double feature embedding strategy is introduced. Specifically, two distinct features, F^{attr} and F^{cal}, closely associated with operator calculations, are extracted from the node feature vector F_{op}. Following that, switchable MLP are selected according to operator type to obtain operator-level embedding vectors F_{opi}^{ow}, which are subsequently utilized to update the node feature vector F_{opi}. It's notable that the type of op_i equals δ:

$$F_{opi}^{ow} = MLP_\delta(F^{attr} \oplus F^{cal}), i \subseteq [1, n] \qquad (4)$$

$$F_{opi} = F_i^{type} \oplus F_i^{edge} \oplus F_{opi}^{ow}, i \subseteq [1, n] \qquad (5)$$

In the second stage of feature embedding, the updated node feature vectors and adjacency matrix A are passed to a GraphSAGE-based encoder to accomplish graph embedding. This process results in the generation of the model-level embedding vector F_{GM}, which encapsulates a comprehensive representation of the entire graph structure and node features:

$$F_{GM} = GNN(F_{op}, A), F_{GM} \subseteq R^{512} \qquad (6)$$

The two-stage embedding design is expected to enrich the global information through refined operator-level feature embedding.

3.4 Resource Aware Latency Prediction

Based on the interval analysis theory, DGL [10] believes that the total latency of program is composed of the time without missing events and the penalty delay caused by missing events (Eq. 7):

$$L = T_{base} + T_{L1miss} + T_{L2miss} + T_{brmiss} \tag{7}$$

where T_{base} is the ideal model inference time when there is no cache miss, T_{L1miss} and T_{L2miss} are the time penalty caused by the first-level and second-level cache misses, and T_{brmiss} represents the time penalty when the branch prediction is wrong. Unlike interval analysis theory, which mainly considers the working mechanism of CPU multi-level cache, this paper aims to expound the negative impact on inference latency caused by the unsaturated state of critical computing resources due to inter-process resource contention. Various hardware configurations affect inference latency, such as bus bandwidth and cache size. However, in terms of process, the key resources corresponding to computation, data transfer, and storage are CPU utilization, disk IO rate, and physical memory, respectively, other factors may be of little concern. The findings regarding the correlation between inference latency and the three crucial resources are depicted in Fig. 2.

As illustrated in Fig. 2(c) and Fig. 2(b), when the value of another resource is fixed, inference latency exhibits approximate a logarithmic correlation with the CPU utilization. Regarding the disk IO rate, as depicted in Fig. 2(c), when the physical memory is not constrained, the inference time is not relevant with the disk IO rate due to the absence of data exchange with disk. However, in Fig. 2(b), when the available physical memory falls below the critical threshold, the inference latency displays a negative correlation with the IO rate.

The behavior of inference latency in response to the variation of physical memory exhibits two distinct patterns in Fig. 2(a): platform pattern and monotonic pattern. Keeping CPU utilization unchanged, in the measured range of memory, models exhibiting the monotonic pattern (Densenet121 and efficient-net_b0) consistently demonstrate an approximately linear increase in inference time as the memory capacity decreases. Conversely, models displaying the platform pattern (Alexnet and efficientnet_v2_m) initially experience an increase in inference time with declining memory, but as the memory amount further decreases, the correlation between the two disappears.

According to the discovery of exploration, inference latency includes computation latency and IO latency. IO latency involves the data exchange between disk and memory, the time penalty it brings is much greater than the T_{L1miss}, T_{L2miss} and T_{brmiss} in interval theory. Therefore, Eq. 7 is rewritten as follow:

$$T_{RD} = T_{cal} + T_{io} \tag{8}$$

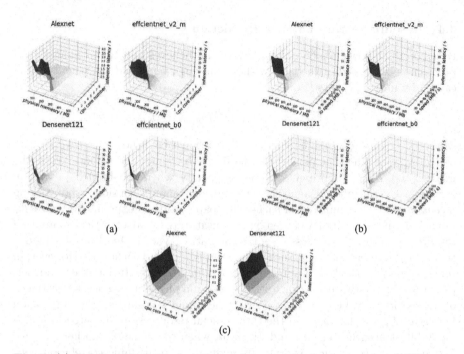

Fig. 2. (a) Relationship between inference latency, physical memory and CPU core number. (b) Relationship between inference latency, physical memory and IO rate. (c) Relationship between inference latency, the number of CPU cores and IO rate.

Where T_{RD} means the model inference time under specific resources state, T_{cal} represents the computation time during the inference process, and T_{io} signifies the duration of data exchange between the disk and memory. To verify observed patterns between latency and CPU utilization, a power function (Eq. 9) is employed to fit the relationship between T_{RD} and CPU utilization U_C. a and b are fitting coefficients, the coefficient of determination R^2 can reach 0.99:

$$T_{RD} = a + U_C^b \tag{9}$$

Since CPU undertakes all computational tasks of model inference in this paper, The effect of CPU utilization variations is captured by T_{cal}:

$$T_{cal} = T_{std} \times (1 + \theta_1 \times (U_c)^{\theta_2}) \tag{10}$$

T_{std} is the standard prediction inference time without resource constraints. By this way, we establish a connection between the two latency prediction tasks within a single equation. Then, the second term T_{io} of Eq. 8 will be examined. T_{io} equals the time penalty caused by data transfer between disk and memory:

$$T_{io} = \frac{D}{V_{io}} \tag{11}$$

Where D is defined as the quantity of data exchange necessitated to execute inference for a particular model, under a specific available memory capacity. It is an observation value that is challenging to obtain through theoretical calculations. This is due to its dependence on factors such as the model's operator execution, framework support, and system garbage recycling mechanisms, among others. Why we introduce D is to explain the two correlation patterns observed between inference time and the physical memory. In the case of the platform pattern, once the memory delay phenomenon emerges (namely the available physical memory falls below the critical memory threshold m_c), the initially observed negative linear correlation between inference time and physical memory ceases to exist when the memory capacity Continues declining to the "window threshold". We attribute this phenomenon to the hypothesis that the amount of data exchange, that is D, remains constant when the physical memory varies below the "window threshold". Constant D leads to a constant T_{io}, resulting in the disappearance of the linear correlation. This behavior is typically observed in models that lack residual connections, as they do not rely on earlier data and therefore have a lower probability of encountering memory page faults. Consequently, a certain memory window is sufficient to facilitate data swapping in and out. In the case of the monotonic mode, the "window threshold" can be observed. Summarizing the two patterns, we can formulate the data exchange amount D as follows:

$$D = \theta_3 \times min(max(0, m_c - m_r), m_w) \tag{12}$$

Where θ_3 represents linearity coefficient, m_c represents the critical memory at which data exchange between disk and memory begins to occur, m_r is the actual physical memory allocated to the model, and m_w represents the window memory. At this juncture, we can define the "window threshold" using a symbolic representation $m_c - m_w$. Equation 12 reveals that, when the difference between m_c and m_r is greater than the window memory m_w, the data exchange amount D is controlled by m_w. Combined with the previous analysis, the resource-aware latency can be formulate as bellow:

$$T_{RD} = T_{std} \times (1 + \theta_1 \times (U_c)^{\theta_2}) + \frac{\theta_3 \times min(max(0, m_c - m_r), m_w)}{V_{io}} \tag{13}$$

In addition, when there is not enough available memory, the CPU cannot be fully used for computation. We assign this correlation between physical memory and CPU utilization to T_{io}. The final resource-aware latency prediction formula is as follow:

$$T_{RD} = T_{std} \times (1 + \theta_1 \times (U_c)^{\theta_2}) + (U_c)^{\theta_2} \times \frac{\theta_3 \times min(max(0, m_c - m_r), m_w)}{V_{io}} \tag{14}$$

There are five unknown parameters $\theta_1, \theta_2, \theta_3, m_c, m_w$ contained in Eq. 14. We represent the set of unknown parameters as follow:

$$P = \{\theta_1, \theta_2, \theta_3, m_c, m_w | M\} \tag{15}$$

To enhance the generalization of RAL formula, we utilize the network parameter predictor $f(F_{GM})$, trained on known samples, to directly predict the parameter set for unknown neural networks. A two-stage training method is employed for the network parameter predictor. In the first stage, the inference latency of sampled neural network models is collected within a specific resource space. This latency information is then used to obtain the corresponding parameter set P_i through UGO algorithm fitting. In the second stage, the model graph embedding vector F_{GMi} and parameter set P_i are served as the data and labels to train predictor $f(F_{GM})$.

4   Experiment

4.1   Dataset Description

We create two datasets to conduct experiments: standard latency dataset and resource dependent latency dataset.

Standard Latency Dataset. We adopt similar construction method in nn-meter [8]. We choose 6 classic CNN models: AlexNet, EfficientNet, MobileNetV3, ResNet50, MnasNet and SqueezeNet. 3000 variants for each classic model were generated by re-sampling the channels number and kernel size of the convolution operator. Additionally, we generated the 15625 models defined by NAS-bench-201. A total of 33,625 models with seven different topologies were measured following the rule of warming up the models five times and inferring them 50 times. The average inference time was considered as the actual inference latency.

Resource Dependent Latency Dataset. We adopt same construction method as standard latency dataset and sampled 500 variants for each base model. The resources space is defined as Table 2. The inference latency of a total of 3000 variants is measured under 1664 resource configurations separately. The measurement follows the rule of warming up the model 2 times and inferring 20 times, and the average time is taken as the actual inference latency. By fitting the inference time of each model under all resource configurations, we generate a set of unknown parameters. In total, we generate 3000 samples for train and validate network parameter predictors.

Table 2. Resource allocation sampling space.

Type of resource	Value space
CPU shares	$[100, 800, 100]$ %
physical memory	$[80, 600, 20]$ MB
IO rate	$[20, 300, 40]$ MB/s

4.2   Evaluation Metric

The evaluation metrics in this paper are Mean Absolute Percentage Error (MAPE) and Error-Bound Accuracy. MAPE represents the average value of the relative error between the predicted latency of all samples and the actual measured latency:

$$MAPE = \frac{\sum_{i=1}^{N} \left| \frac{\hat{y}_i - y_i}{y_i} \right|}{N} \times 100\% \tag{16}$$

Error-Bound Accuracy, which is defined as the percentage of samples that falls within a given error-bound. This paper adopts $\pm 10\%$ Error-Bound Accuracy. When evaluating the method effectiveness, a smaller mean absolute percentage error and a larger the Error-bound accuracy value are preferred.

4.3   Baseline

In order to verify the effectiveness of HADE, the following baseline model are used to compare the standard latency prediction accuracy:

nn-Meter [8]. nn-Meter is considered the state-of-the-art hierarchical prediction model. It partitions the entire model inference process into kernels and provides latency predictions at the kernel level. It incorporates an automatic detection mechanism for kernel fusion rules during model inference, utilizing a set of carefully designed test cases.

NNLQP [9]. NNLQP uses a unified GNN-based graph embedding vector to represent neural networks, followed by latency prediction. Through the continuous acquisition of latency data, the progressive learning of prediction model is realized. As a representative end-to-end predictor, it has achieved state-of-the-art performance in standard latency prediction.

FLOPs+MAC. As a proxy metric, FLOPs+MAC is a widely used predictor that reflects the computational and memory access requirements during the inference process. The predictor is trained using a simple linear regression.

4.4   Comparative Experiments

We initially conducted a comparison of latency prediction accuracy and robustness across four methods. The standard latency dataset was divided into training and testing sets using an 4:1 ratio for each model family. Table 3 shows the MAPE and 10% Error-Bound Accuracy results of all four approaches for predicting standard inference latency on testing set.

On average, HADE achieved MAPE of 3.95% and $\pm 10\%$ Error-Bound Accuracy of 96.37% on standard latency testing set, which exceed the best results of the baseline model by 0.52% and 2.11%, respectively. While the performance of HADE on the Alexnet and SqueezeNet model family is lower than that of the nn-meter, HADE demonstrates better prediction performance overall, particularly for models containing depth-wise separable convolution operators. This can be attributed to the hierarchical nature of nn-meter, where the overall prediction accuracy heavily relies on the accuracy of operator-level latency predictions.

Table 3. Results of comparison methods on standard latency testing dataset.

Model Family	Metric	HADE	FLOPs+MAC	nn-meter	NNLQP
AlexNet	MAPE (%)	4.12	9.22	4.74	**3.92**
	±10% ACC (%)	96.33	73.33	95.83	**97.50**
EffcientNet	MAPE (%)	**4.72**	13.14	6.49	5.84
	±10% ACC (%)	**92.17**	55.67	86.83	89.83
MnasNet	MAPE (%)	3.85	7.22	**3.29**	4.57
	±10% ACC (%)	98.50	84.50	**98.67**	94.33
MObileNetV3	MAPE (%)	**5.40**	11.28	8.12	6.47
	±10% ACC (%)	**92.83**	61.83	82.67	86.00
ResNet50	MAPE (%)	**2.55**	6.47	3.64	3.11
	±10% ACC (%)	**99.16**	85.83	97.16	98.33
SqueezeNet	MAPE (%)	3.62	8.06	4.27	**3.39**
	±10% ACC (%)	96.67	81.67	95.33	**97.33**
NAS-Bench-201	MAPE (%)	**3.41**	7.25	4.61	4.01
	±10% ACC (%)	**98.67**	84.67	95.17	96.50
Average	MAPE (%)	**3.95**	8.95	5.02	4.47
	±10% ACC (%)	**96.37**	75.36	93.11	94.26

In our experiment, nn-meter exhibits a high MAPE of 10.24% on latency prediction of depth-separable convolution. Compared to NNLQP, which only conducts model-level graph embedding, HADE incorporates operator-level feature embedding. By considering global and local time characteristics, HADE achieves superior results. On the other hand, FLOPs+MAC performs poorly in general, as the linear models overlook important details of model structure and operator calculation differences.

Furthermore, this paper assesses the generalization of prediction capability for these four methods on unseen models. The testing set is created by sampling variants from VGG, GoogleNet, and ShuffleNet, while the training set does not include any variant of them. The comparison results show in Table 4.

Table 4. Generalization results of comparison methods on unseen models.

Model Family	Metric	HADE	FLOPs+MAC	NNLQP	nn-meter
VGG	MAPE (%)	7.54	15.47	**6.97**	8.53
	±10% ACC (%)	79.67	45.83	**82.17**	72.67
GoogleNet	MAPE (%)	**7.31**	22.86	9.23	8.37
	±10% ACC (%)	**77.50**	21.83	69.33	74.50
ShuffleNet	MAPE (%)	**11.50**	33.19	15.13	13.67
	±10% ACC (%)	**59.33**	12.17	46.33	53.83
Average	MAPE (%)	**8.78**	23.84	10.44	10.19
	±10% ACC (%)	**72.16**	30.5	65.94	67.00

The overall generalization ability of HADE surpass other baseline models. The MAPE and ±10% Error-Bound Accuracy exceed the best results of the baseline model by 1.41% and 5.16%, respectively. nn-meter exhibits a generalization error when faced with new models due to changes in operator configurations. NNLQP summarize the overall information of the model by embedding, when a model with new topology appears, the prediction accuracy of the model will decrease. FLOPs+MAC has a poor performance on generalization, because there is a significant difference in computing and memory access density bwtween models. However, HADE summarizes the overall and local information of the model through double feature embedding, achieves better generalization performance.

4.5   Validation of RAL Formula

We fit and validate the RAL formula on resource dependent latency dataset. Table 5 shows the average fitting and prediction MAPE.

Table 5. Train and evaluation results of RAL formula.

Model Family	Avg.fitting MAPE (%)	Avg.predction MAPE (%)
AlexNet	13.91	18.52
EffcientNet	17.63	23.63
MobileNet	20.41	27.71
ResNet50	14.76	16.44
Mnasnet	15.25	18.58
SqueezeNet	16.17	16.96
Average	16.36	20.31

For 500 variants of each model family in resource dependent latency dataset, the training set and testing set are divided by 4:1. Avg.fitting HAPE represents the average MAPE between the real latency of the training samples and the RAL formula prediction time that using the parameter set obtained by fitting training data. Avg.prediction HAPE represents the average MAPE between the the real latency of testing sample and the RAL formula prediction time that using unknown parameter set obtained by network parameter predictor. It can be seen from Table 5 that the RAL formula is capable of effectively capturing the resource dependence of inference latency and producing reliable prediction results. When applying the network parameter predictor on the testing set, there is a decrease in prediction accuracy by an average of 3.95%. But this trade-off is justified by the significant reduction in the cost of describing resource dependencies. Due to the strong correlation between hardware performance and device type, current RAL formula lacks cross-device generalizability. Inspired by [17], we will extend the cross-device performance of RAL in future work.

4.6   Ablation Experiment

In this section, we conduct additional experiments to study the impact of HADE components, including: (1) Whether operator-level feature embedding is used. (2) Whether the RAL formula is employed.

Rejecting Operator-Level Feature Embedding. The operator-level feature embedding is removed, keeping other modules and parameters unchanged. The node feature vectors obtained by the pre-processing module is directly fed into the graph neural network. The experimental results are shown in Table 6.

Table 6. Ablation experiment result of removing operator-level feature embedding.

Model Family	Evaluation Metric	Use	Reject
ResNet50	MAPE (%)	2.55	3.17
	±10% ACC (%)	99.16	98.33
EffcientNet	MAPE (%)	4.72	5.91
	±10% ACC (%)	93.17	89.83
MnasNet	MAPE (%)	3.85	4.51
	±10% ACC (%)	98.50	94.13
MobileNetV3	MAPE (%)	5.40	6.59
	±10% ACC (%)	92.83	86.67
GoogleNet (unseen)	MAPE (%)	7.31	8.26
	±10% ACC (%)	78.50	73.33
ShuffleNet (unseen)	MAPE (%)	11.58	13.89
	±10% ACC (%)	59.33	51.50

The operator-level feature embedding brings an average 1.15% MAPE reduction and 4.61% ±10% Error-bound accuracy improvement on four models family with known topology. Furthermore, it results in an average 1.73% MAPE reduction and 6.5% ±10% Error-bound accuracy improvement on two unknown models family. From the experimental results, it can be seen that the operator-level feature embedding effectively extracts the local features of the neural network model, provides richer information for the prediction module, and improves the prediction robustness to unknown models.

Rejecting RAL Formula. Considering the powerful feature extraction and function fitting capabilities of the neural network, we take the resource parameters (CPU utilization, physical memory, and IO rate) as model static features. The Resource dependent latency dataset is divided into training set and testing set by 4:1. Although each model family only contains 500 models, the resource sample space is abundant and can generate a large amount of training data. The MAPE results on testing set are shown in Table 7.

During the training process of all model families, we observed a significant overfitting phenomenon. To mitigate this, we employed an early exit training

Table 7. MAPE (%) results of featurizing resource parameters on testing dataset.

Model Family	AlexNet	EffcientNet	MobileNet	ResNet50	MnasNet	SqueezeNet	Average
w/o RAL	60.27	52.34	64.15	48.16	41.93	38.17	50.83
w/ RAL	18.52	23.63	27.71	16.44	18.58	16.96	20.31

strategy to record the best prediction accuracy and obtain the MAPE results presented in Table 7. Comparing the MAPE results when featurizing resource parameters with those when using the RAL formula, we found that the former yielded significantly higher values. This may be attributed to the excessively long length of the model graph embedding vector or the late introduction of the resource vector timing. However, within the framework of the end-to-end method, it is essential to maintain a certain length for the model graph embedding vector to extract sufficient information. Consequently, directly introducing resource vectors based on the proposed feature embedding method failed to yield satisfactory prediction results.

5 Conclusion

This paper presents HADE, a hardware-aware and double feature embedding prediction method for neural network inference time. By utilizing dedicated operators and model encoders, HADE provides a more comprehensive vectorized representation of neural networks, which improves the accuracy of prediction. HADE employs the RAL formula to accurately describe the impact of three critical computational resource insaturation states on inference latency, simplifying multitasking parallel scenarios from a resource perspective, enabling predictions of standard inference time and resource-dependent inference time for any given resource configuration. Through extensive experiments, the proposed method is thoroughly investigated, and its effectiveness in prediction is validated.

References

1. Xiao, Y., et al.: Adaptivepose: human parts as adaptive points. In: Proceedings of the AAAI Conference on Artificial Intelligence, vol. 36, no. 3 (2022)
2. Ho, J., Jain, A., Abbeel, P.: Denoising diffusion probabilistic models. Adv. Neural. Inf. Process. Syst. **33**, 6840–6851 (2020)
3. Dai, D., et al.: Why can GPT learn in-context? Language models secretly perform gradient descent as meta optimizers. arXiv preprint arXiv:2212.10559 (2022)
4. Fang, G., et al.: Depgraph: towards any structural pruning. In: Proceedings of the IEEE/CVF Conference on Computer Vision and Pattern Recognition (2023)
5. Whitaker, T., Whitley, D.: Prune and tune ensembles: low-cost ensemble learning with sparse independent subnetworks. In: Proceedings of the AAAI Conference on Artificial Intelligence, vol. 36, no. 8 (2022)
6. Cai, H., Zhu, L., Han, S.: Proxylessnas: direct neural architecture search on target task and hardware. arXiv preprint arXiv:1812.00332 (2018)
7. Cai, H., et al.: Once-for-all: train one network and specialize it for efficient deployment. arXiv preprint arXiv:1908.09791 (2019)

8. Zhang, L.L., et al.: Nn-meter: towards accurate latency prediction of deep-learning model inference on diverse edge devices. In: Proceedings of the 19th Annual International Conference on Mobile Systems, Applications, and Services (2021)
9. Liu, L., et al.: NNLQP: a multi-platform neural network latency query and prediction system with an evolving database. In: Proceedings of the 51st International Conference on Parallel Processing (2022)
10. Wang, Q., Zhang, S.: DGL: device generic latency model for neural architecture search on mobile devices. IEEE Trans. Mob. Comput. (2023)
11. Lee, H., et al.: Help: hardware-adaptive efficient latency prediction for NAS via meta-learning. arXiv preprint arXiv:2106.08630 (2021)
12. Abbasi, S., Wong, A., Shafiee, M.J.: Maple: microprocessor a priori for latency estimation. In: Proceedings of the IEEE/CVF Conference on Computer Vision and Pattern Recognition (2022)
13. Nair, S., et al.: Maple-edge: a runtime latency predictor for edge devices. In: Proceedings of the IEEE/CVF Conference on Computer Vision and Pattern Recognition (2022)
14. Li, Y., et al.: Sectum: accurate latency prediction for TEE-hosted deep learning inference. In: 2022 IEEE 42nd International Conference on Distributed Computing Systems (ICDCS). IEEE (2022)
15. Hamilton, W., Ying, Z., Leskovec, J.: Inductive representation learning on large graphs. Adv. Neural Inf. Process. Syst. **30** (2017)
16. Wu, Y., et al.: Serving unseen deep learning models with near-optimal configurations: a fast adaptive search approach. In: Proceedings of the 13th Symposium on Cloud Computing (2022)
17. Geoffrey, X.Y., et al.: Habitat: a runtime-based computational performance predictor for deep neural network training. USENIX Annual Technical Conference (2021)
18. He, Y., et al.: AMC: automl for model compression and acceleration on mobile devices. In: Proceedings of the European Conference on Computer Vision (ECCV) (2018)
19. Tan, M., et al.: Mnasnet: platform-aware neural architecture search for mobile. In: Proceedings of the IEEE/CVF Conference on Computer Vision and Pattern Recognition (2019)
20. Li, H., et al.: Pruning filters for efficient convnets. arXiv preprint arXiv:1608.08710 (2016)
21. Qi, H., Sparks, E.R., Talwalkar, A.: Paleo: a performance model for deep neural networks. In: International Conference on Learning Representations (2017)
22. Ponomarev, E., Matveev, S., Oseledets, I.: LETI: latency estimation tool and investigation of neural networks inference on mobile GPU. arXiv preprint arXiv:2010.02871 (2020)
23. Kaufman, S., Phothilimthana, P.M., Burrows, M.: Learned TPU cost model for XLA tensor programs. In: Proceedings of the Workshop ML Systems NeurIPS (2019)
24. Kaufman, S., et al.: A learned performance model for tensor processing units. Proc. Mach. Learn. Syst. **3**, 387–400 (2021)
25. Dudziak, L., et al.: BRP-NAS: prediction-based NAS using GCNs. Adv. Neural Inf. Process. Syst. **33**, 10480–10490 (2020)
26. Gao, Y., et al.: Runtime performance prediction for deep learning models with graph neural network. Microsoft, Tech. Rep. MSR-TR-2021-3 (2021)

Robust LS-QSVM Implementation via Efficient Matrix Factorization and Eigenvalue Estimation

Junchen Shen[1][(✉)] and Jiawei Ying[2]

[1] East China Normal University, Shanghai 200062, China
kazuyaecnu@gmail.com
[2] Nanjing University of Posts and Telecommunications, Nanjing 210023, China
1021010601@njupt.edu.cn

Abstract. Traditional SVM and LSSVM face challenges in handling large datasets due to the exponential increase in computational complexity. Quantum Support Vector Machines (QSVM) based on the HHL algorithm offer a potential solution, but practical implementation in the Noisy Intermediate-Scale Quantum(NISQ) era remains limited. Our work Least Square-QSVM with Ridge Regression(LS-QSVM-RR) addresses these limitations by efficiently decomposing the LSSVM coefficient matrix and encoding it into quantum circuits via a matrix factorization algorithm. We also leverage quantum circuit properties to compute optimal ridge regression constraints by efficient eigenvalue estimation, achieving a balance between prediction accuracy and robustness in QSVM. Our contributions demonstrate the potential of Variational Quantum Algorithm(VQA)-based QSVM simulations and provide insights into handling large condition numbers in LSSVM coefficient matrices.

Keywords: SVM · Variational Quantum Algorithm · Ridge Regression

1 Introduction

Recently, the development of quantum computers and their applications has gained significant attention in contemporary physics and computer science [1–3]. One prospective area of research in quantum computing is the use of quantum algorithms to overcome the complexity problem faced by classical computers [4,5]. Grover search, for instance, can solve the target retrieval problem in $O(\sqrt{N})$ time complexity, which is much faster than the $O(N)$ complexity of classical algorithms [6]. Due to the exceptional parallelism and computational potential of quantum computing, numerous fields have witnessed the emergence of research based on quantum circuits, such as image and speech recognition [7], recommendation systems [8], and financial forecasting [9], etc.

© The Author(s), under exclusive license to Springer Nature Singapore Pte Ltd. 2024
B. Luo et al. (Eds.): ICONIP 2023, CCIS 1961, pp. 511–523, 2024.
https://doi.org/10.1007/978-981-99-8126-7_40

In classical computer science, support vector machines (SVMs) are a popular algorithm used for classification problems [10,11]. SVMs construct a hyperplane to classify vectors into one of two groups in the original or kernel space. Kernel-based SVMs [12] have been successfully applied in supervised learning for various applications, such as regression problems [13] and training on large data [14]. There have also been many variants of SVMs proposed in recent years [14–16]. Least Squares Support Vector Machines (LSSVM) is one such variant [17]. By transforming inequalities into equality constraints, LSSVM reduces the computational load of SVMs from a quadratic problem to a linear equation problem. It offers certain advantages over SVM, however, it also introduces the common issue of a large condition number in solving linear equations [18].

Quantum Support Vector Machines (QSVM) [19] is an application of quantum computing in machine learning. The QSVM algorithm uses the principles of quantum computing to improve the computational efficiency of SVMs [20]. By encoding classical data into quantum states and applying quantum operations, QSVMs can handle exponentially large feature spaces, which makes it more efficient than classical SVMs for certain applications. The HHL algorithm [5] is a quantum algorithm that can be used to solve systems of linear equations, a fundamental problem in many fields. Since the LSSVM can be viewed as a linear equation problem, using the HHL algorithm to accelerate the calculation of LSSVM is a promising technical path [21–23]. However, the limitations of the Noisy Intermediate-Scale Quantum(NISQ) era [24,25] make implementing HHL challenging [26].

To address these challenges, Variational Quantum Algorithms (VQAs) [27] provide a more realistic technical means to solve problems in the NISQ era. VQAs replace the neural network components in classical deep learning with parameterized quantum circuits, and the parameters on quantum circuits are optimized by classical optimizers. In this paper, we introduce a VQA-based LSSVM named LS-QSVM with Ridge Regression(LS-QSVM-RR), which has been developed to address the challenges posed by the NISQ era while incorporating appropriate regularization constraints that enhance its predictive performance compared to traditional LSSVM methods.

Our contributions in this research are summarized as follows:

Firstly, a quantum circuit-based implementation of LSSVM is achieved under the support of mindspore [28], providing a practical solution to the computational challenges posed by classical SVMs when processing large-scale datasets. Additionally, by incorporating LSSVM with Variational Quantum Algorithms (VQAs), the limitations of NISQ-era constraints are mitigated, thereby facilitating the realization of QSVM in practical machine-learning applications.

Furthermore, a novel matrix decomposition algorithm is introduced, enabling the encoding of the coefficient matrix and ridge regression constraints of LSSVM into quantum circuits. Exploiting the intrinsic properties of the relevant matrices, this algorithm significantly reduces the computational complexity of the decomposition process by a factor of one-half. This advancement paves the way for efficient and feasible implementations of LSSVM on quantum circuits.

Moreover, leveraging the distinctive characteristics of quantum circuits, the power iteration process is accelerated, leading to a more efficient estimation of the maximum eigenvalue of the involved matrices. By accurately determining this critical parameter, the challenge of selecting appropriate regularization coefficients, which has been a persistent issue, is partially addressed. This achievement contributes to the enhancement of LSSVM's performance and provides a valuable solution for regularization parameter determination.

These contributions demonstrate the potential of our work in facilitating efficient QSVM simulation, enabling efficient encoding of LSSVM coefficients, and effectively incorporating ridge regression constraints in quantum computing.

2   Background

2.1   Least Square Support Vector Machine

Least squares support vector machines (LS-SVM) and classical support vector machines (SVM) are both widely used machine learning algorithms for classification. The difference is that LSSVM converts the classification problem into a linear system of equations and seeks to minimize the sum of squared errors between the predicted and actual outputs, which can be stated as minimize

$$\frac{1}{2}||\omega||^2 \tag{1}$$

subject to

$$y_i(\omega^T x_i + b) \geq 1, \quad i = 1, 2, ...m \tag{2}$$

where m is the number of training samples, x_i is the ith sample, y_i is the corresponding target value, ω and b are the parameters of the hyperplane. Unlike standard SVM, LSSVM introduces the errors e of all data points. Thus, the inequality constraints have been converted into the following equality constraint problem.

$$\frac{1}{2}||\omega||^2 + \frac{1}{2}\gamma \sum_{i=1}^{m} e_i^2 \tag{3}$$

subject to

$$\omega^T x_i + b = y_i - y_i e_i \tag{4}$$

where e_i is the deviation of the ith sample from the hyperplane, γ is a regularization parameter that controls the trade-off between the magnitude of the normal vector ω and the sum of squared errors. In fact, the formulation of LSSVM with Lagrange multipliers can be expressed as the following optimization problem:

$$\arg\min(\frac{1}{2}||\omega||^T + \frac{1}{2}\gamma \sum_{i=1}^{m} e_i^2 - \sum_{i=1}^{m} \alpha_i(\omega^T x_i + b + e_i - y_i)) \tag{5}$$

By taking partial derivatives of each variable, a linear equation system can be obtained:

$$\begin{pmatrix} 0 & 1 \\ 1 & K + I\gamma^{-1} \end{pmatrix} \begin{pmatrix} b \\ a \end{pmatrix} = \begin{pmatrix} 0 \\ y \end{pmatrix} \tag{6}$$

while K is a kernel matrix written as

$$
\begin{bmatrix}
k_{11} & \cdots & k_{1m} \\
\vdots & \ddots & \vdots \\
k_{m1} & \cdots & k_{mm}
\end{bmatrix}
\tag{7}
$$

each element in the kernel matrix K is represented as:

$$
k_{ij} = k(x_i, x_j)
\tag{8}
$$

where $k(x_i, x_j)$ is the kernel function, x_i and x_j are 2 inputs samples. The choice of kernel function determines the form of the dot product and influences the performance of the LSSVM model. Commonly used kernel functions include linear, polynomial, radial basis function(RBF), and sigmoid.

Thus, the classification problem based on LS-SVM has been transformed into a linear system of equations in the form of

$$
Mx = b
\tag{9}
$$

to be solved.

2.2   Ridge Regression

One common issue encountered in LSSVM models is the problem of having an excessively large condition number for the coefficient matrix. This makes solving the linear system of equations exceptionally difficult. This problem can be attributed to strong collinearities among the columns of the coefficient matrix [29–31]. If two samples are in close proximity within the original feature space, it can be inferred from Eq.(6,7,8) that the corresponding column vectors of the kernel matrix K will exhibit significant similarity, giving rise to the strong collinearity of M. Furthermore, in the case where two identical samples exist in the original space, the column vectors of the kernel matrix associated with these samples will be completely identical, thereby resulting in a singular coefficient matrix M. This strong correlation leads to the coefficient matrix M approaching singularity and ill-conditioned,and this problem is unaffected by variations in the penalty coefficient γ in SVM.

Standardizing the features of the original data is considered as one promising method to alleviate collinearity. However, since the distribution of the data being measured is unknown, standardizing based on training data can potentially lead to the overfitting of the predictive model to the distribution of the training data. This appears to pose a dilemma. However, ridge regression regularization provides a solution that achieves a balance between mitigating collinearity and promoting model generalization.

Ridge regression constraint [35] is a popular approach for analyzing collinear data. Incorporating a regularized term with an L2 norm into the general linear regression can alleviate the problem of excessive feature dimensions and collinearity among features to a certain extent. Collinearity can lead to an ill-conditioned coefficient matrix in linear equations, making it difficult to obtain an accurate inverse. This problem is commonly encountered when solving linear equations derived from LS-SVM, as they often suffer from a large condition number. To overcome this challenge, we introduce the mechanism of ridge regression, which takes the following form:

$$(M + \eta I)x' = V((\Sigma + \eta I))U^T b \tag{10}$$

where η represents the weight of the ridge regression constraint. Since the dataset used in this study does not contain overlapping samples, there is no pure collinearity among the vectors in the coefficient matrix M. This implies that M is always invertible. Therefore, we can obtain from the Eq. (9, 10) that:

$$
\begin{aligned}
x &= M^{-1}b = (U\Sigma V^T)^{-1}b = V\Sigma^{-1}U^T b \\
x' &= (M + \eta I)^{-1}b = (U\Sigma V^T + \eta I)^{-1}b = V((\Sigma + \eta I)^{-1})U^T b
\end{aligned}
\tag{11}
$$

where U and V are orthogonal matrices composed of the eigenvectors of the two spaces in matrix M, while Σ is a diagonal matrix composed of the singular values of M.

The Eq. (11) can be viewed from two perspectives. Firstly, from the perspective of solving linear equation systems, a large condition number of the coefficient matrix will make it difficult to solve. However, by applying this form of ridge regression constraint, the condition number of the original coefficient matrix can be significantly reduced, making the solution much easier. Moreover, from the perspective of classification problems, the norm of the new solution (i.e. x') after adding the ridge regression constraint will be lower than that of the original linear equation system (i.e. x), which to some extent achieves feature selection and improves the generalization ability of the classification model. In fact, subsequent experiments have also demonstrated that the reduction of the condition number can make the results based on variational algorithms more stable, and the improvement of generalization also enhances the performance of the model (Fig. 1).

3   LS-QSVM with Ridge Regression

3.1   Loss Function Based on Hamiltonian

A commonly used method for solving a system of linear equations is to construct a Hamiltonian whose ground state corresponds to the solution of the problem. Specifically, the Hamiltonian takes the form as follows:

$$H = M^+(I - |b\rangle \langle b|)M \tag{12}$$

for any input status $|\psi\rangle$:

$$
\begin{aligned}
\langle\psi|H|\psi\rangle &= \langle\psi|M^+(I - |b\rangle \langle b|)M|\psi\rangle \\
&= ||M|\psi\rangle||^2 - ||\langle b|M|\psi\rangle||^2 \geq 0
\end{aligned}
\tag{13}
$$

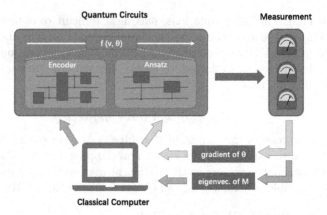

Fig. 1. The Encoder module encodes the coefficient matrix M of LSSVM. The algorithm consists of two main processes: preprocessing (green arrows) and training (orange arrows). During preprocessing, the encoded module outputs the eigenvector corresponding to the largest eigenvalue of M. The classical computer processes the eigenvector to determine the largest eigenvalue and compute the appropriate ridge regression strength. The obtained ridge regression strength is then encoded into the Encoder module. In the training process, the orange arrows represent the calculation of gradients for the parameters in the Ansatz circuit. These gradients are used by the classical computer to update the parameters in the Ansatz circuit iteratively until the training process is completed. (Color figure online)

Obviously, the $|\psi\rangle$ corresponding to the ground state of the Hamiltonian H is the solution to this linear system of equations.

Similar to the gradient descent in classical machine learning, a quantum loss function can be constructed as:

$$L(\theta) = \langle\psi(\theta)|H|\psi\theta\rangle = ||M|\psi\theta\rangle||^2 - ||\langle b|M|\psi\theta\rangle||^2 \qquad (14)$$

where the θ are parameters needed to be optimized. It can be obtained that the optimization direction for θ_i is:

$$\frac{\partial L(\theta)}{\partial \theta_i} = 2\langle\psi(\theta)|M^+M\frac{\partial|\psi(\theta)\rangle}{\partial\theta_i} - 2||\langle b|M|\psi(\theta)\rangle||\langle b|M\frac{\partial|\psi(\theta)\rangle}{\partial\theta_i}$$

3.2   Matrix Factorization

To encode a matrix into a quantum circuit for computation, we employ an encoding method based on matrix decomposition. The matrix's information can be encoded using a series of quantum gates, which can be represented by corresponding Pauli matrices. Therefore, by decomposing the target matrix A into a weighted sum of Pauli matrices, we can successfully encode the matrix's information into the quantum circuit. The decomposition algorithm we utilize is as follows:

$$A = \sum_{k=0}^{2^n-1} \sum_{k_i,\alpha} C_{x,y}^\alpha \bigotimes_{i=1}^{n} \sigma_\alpha^i = \frac{1}{2^n} \sum_{k=0}^{2^n-1} \sum_{k_i,\alpha} \bigotimes_{i=1}^{n} \sigma_\alpha^i (\sum_{x=0}^{2^n-1} \sum_{y=0}^{2^n-1} (-1)^f M_{x,y}^k) \quad (15)$$

where n is the number of qubits, x and y correspond to the indices of the matrix elements, C represents the decomposition coefficients, and the σ corresponds to the four Pauli matrices.: Pauli-X (σ_x), Pauli-Y (σ_y), Pauli-Z (σ_z), and the identity matrix (I). f is the power indicator. More details can be seen in [32].

Factorization on M Since our target matrix M is symmetric, we can compress nearly half of the decomposition time, significantly enhancing the efficiency of matrix decomposition.

Factorization on Ridge Regression. Since the ridge regression constraint is applied to the model in matrix form, we can also encode this constraint using matrix decomposition. It is worth noting that the ridge regression constraint is a diagonal matrix, and we have optimized the decomposition algorithm to the following form:

$$\eta I = \sum_{k_i,\alpha} C^\alpha \bigotimes_{i=1}^{n} \sigma_\alpha^i \quad (16)$$

This algorithm for decomposing the ridge regression constraint achieves a complexity as low as $\frac{1}{N^3}, (N = 2^n)$ of the original factorization, greatly enhancing the efficiency of decomposition and encoding processes.

3.3 Ridge Regression Coefficient Selection Based on Eigenvalue Estimation

As shown in Eq. (10), we alleviate the problem of ill-conditioning caused by a large condition number by applying diagonal matrix regularization constraint. Through singular value decomposition, it is observed that a larger regularization constraint leads to a coefficient matrix condition number closer to 1. This results in matrix M approaching the identity matrix and losing all information, rendering it unsuitable for classification tasks. Conversely, if the regularization coefficient is too small, regularization has little effect. Therefore, the selection of an appropriate regularization coefficient is crucial.

Previous work on ridge regression constraints often employed cross-validation [33,34] and ridge trace analysis [35,36] to select the regularization parameter. However, cross-validation involves an exhaustive process of progressively narrowing down the range of regularization coefficients, which is time-consuming. Using ridge trace analysis, which involves plotting the variation of the dependent variable with respect to the regularization coefficient and selecting the region with lower curve overlap as the range of regularization coefficients, is highly subjective and has not been widely accepted.

Appropriate ridge regression coefficients should satisfy the requirement of preserving the original matrix information (i.e., not exceeding the maximum eigenvalue) while effectively reducing the matrix's condition number. An intuitive observation is that as the condition number of a matrix increases, the strength of the applied ridge regression constraint should also increase. However, obtaining an accurate condition number for selecting suitable constraint coefficients would entail significant computational costs (as it necessitates computing the matrix's maximum and minimum eigenvalues). Hence, we opt to select the constraint coefficients through eigenvalue estimation.

Specifically, we estimate the maximum eigenvalue, denoted as Σ_1, of matrix M. Then, we estimate the maximum eigenvalue, denoted as Σ_2, of $M - \Sigma_1$, which provides us with the matrix's maximum and second-largest eigenvalues. In fact, if these two eigenvalues exhibit a substantial difference, it indicates a high condition number for the matrix, thereby suggesting the need for a larger corresponding ridge regression constraint. Based on this notion, we model the relationship between the two largest eigenvalues of the matrix and the appropriate ridge regression coefficient to determine the most suitable coefficient for ridge regression. The specific formulation is as follows:

$$\eta = \frac{\Sigma_1}{\Sigma_2 * 100} \tag{17}$$

Eigenvalue Estimation via Power Interation. We adopt the power iteration algorithm to estimate the maximum eigenvalue of matrix M, following these steps:

– Step 1: Select a non-zero vector as the initial estimate of the eigenvector. This vector can be obtained through random selection or other heuristic methods.
– Step 2: Normalize the selected vector to become a unit vector.
– Step 3: Start the iteration process. In each iteration, multiply the selected vector with the matrix to obtain a new vector.
– Step 4: Normalize the new vector to become a unit vector.
– Step 5: Repeat steps 3 and 4 until a convergence condition is reached. The convergence condition can be determined by setting a threshold, such as stopping the iteration when the difference between the new vector and the vector from the previous iteration is below a predefined threshold.
– Step 6: At convergence, the estimated eigenvector approximates the eigenvector of the matrix. At this point, compute the result of multiplying the estimated eigenvector with the matrix to obtain a numerical value, which is an approximation of the corresponding eigenvalue.

Due to the properties of quantum superposition states, the vectors in the power iteration process are always normalized, significantly improving the convergence speed. After obtaining the estimate of the maximum eigenvalue, we can determine the range of regularization coefficients (Fig. 2).

4   Simulation

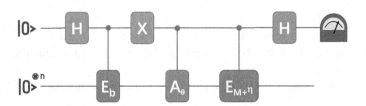

Fig. 2. This figure illustrates a schematic diagram of the LS-QSVM-RR quantum circuit. The gray blocks represent quantum gates. The light blue blocks labeled with the letter E denote the encoding circuits, representing the quantum circuits corresponding to the vector b, matrix M, and ridge regression constraint ηI, respectively. The orange blocks labeled with the letter A depict the Ansatz circuit, which encompasses the parameters to be optimized. (Color figure online)

Simulation on our algorithm follows the following steps:

- Step 1: Decompose and encode the coefficient matrix M corresponding to LSSVM into a quantum circuit.
- Step 2: Use power iteration to find the eigenvector corresponding to the two largest eigenvalues of the coefficient matrix M. Utilizing the characteristics of quantum states, the output vector from the quantum circuit is always normalized, significantly reducing the time overhead of power iteration.
- Step 3: Determine the coefficients for ridge regression constraints by Eq. (17).
- Step 4: Factorize and encode the ridge regression constraints into the quantum circuit.
- Step 5: Input the initial quantum state and measure the gradients of the parameters in the Ansatz circuit.
- Step 6: Pass the gradients to the classical computer, which updates the parameters in the Ansatz circuit. Repeat Steps 5 and 6 until the training is complete.
- Step 7: Perform classification on input data using the trained model.

5   Discussion and Conclusion

In the experiments, apart from controlling the strength parameter for ridge regression constraint η, we set the same set of hyperparameters for SVM, LS-SVM, and QSVM. Five benchmark datasets were selected, and each dataset was randomly divided into training and testing sets in fixed proportions. The experiments were repeated 10 times for each dataset. The averaged accuracy (ACC) and its standard deviation under the linear and RBF kernels were recorded in Tables 1, 2.

Table 1. Averaged ACC using the linear kernel function across the five datasets.

Alg.\Datasets	Iris	Breast Cancer	Wine	Heart Disease	Stroke
SVM	85.80 ± 6.21	83.32 ± 5.24	85.81 ± 4.07	53.50 ± 5.62	57.34 ± 7.07
LSSVM	76.00 ± 13.01	69.01 ± 13.54	84.88 ± 4.19	55.45 ± 2.48	60.43 ± 5.71
LS-QSVM	88.60 ± 8.19	81.58 ± 6.29	84.58 ± 2.99	57.07 ± 3.91	63.13 ± 7.98
LS-QSVM-RR	**94.4 ± 5.39**	**86.12 ± 3.80**	**86.54 ± 3.01**	**57.12 ± 3.20**	**63.27 ± 4.66**

Table 2. Averaged ACC using the RBF kernel function across the five datasets.

Alg.\Datasets	Iris	Breast Cancer	Wine	Heart Disease	Stroke
SVM	86.80 ± 8.71	79.33 ± 3.62	69.68 ± 13.99	57.14 ± 2.40	41.36 ± 9.41
LSSVM	86.80 ± 10.92	81.75 ± 6.88	89.11 ± 4.97	61.78 ± 5.37	54.10 ± 1.61
LS-QSVM	90.80 ± 9.66	84.94 ± 2.08	88.18 ± 5.41	63.78 ± 5.88	**56.75 ± 4.00**
LS-QSVM-RR	**91.20 ± 7.57**	**85.71 ± 0.81**	**90.93 ± 3.85**	**63.98 ± 5.19**	55.79 ± 2.81

5.1   Analysis on Benchmarks

From Tables 1, 2, it is evident that our method consistently achieves the best classification results, with a significant advantage of nearly 6%. The LSSVM-based approach generally outperforms the original SVM, possibly due to its ability to consider a broader range of sample distributions in large datasets. However, classical LSSVM methods exhibit high standard deviations, with the highest standard deviation in predicting ACC across the majority of datasets. This indicates their unstable prediction capabilities, likely attributed to the issue of a large condition number. In such cases, classical computations can significantly impact the prediction results if high computational precision is not provided. In contrast, our QSVM leverages gradient descent optimization for parameter adjustment, enabling finer precision in adjusting the prediction results. Additionally, by imposing appropriate ridge regression constraints, the condition number of the matrix is brought within an ideal range. This empowers LS-QSVM-RR to achieve enhanced robustness in prediction capabilities while ensuring accuracy.

5.2   Analysis on the Ridge Regression Constraint Coefficient

From Fig. 3(a), it can be observed that certain datasets exhibit noticeable variations in predictive performance within a specific range of ridge regression constraints. For example, the Iris and Breast Cancer datasets demonstrate suboptimal performance when the ridge regression coefficients are below a certain threshold. However, when the ridge regression coefficients exceed another threshold, QSVM performance on these datasets approximates random prediction. This empirical evidence supports our hypothesis that ridge regression coefficients within a certain range can enhance model performance.

To evaluate the rationality of the ridge regression coefficients selected in our approach, we compared it with the conventional method of selecting coefficients

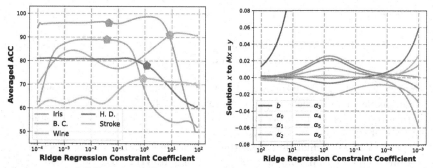

(a) Averaged ACC under different Ridge Regression Coefficients. Pentagon markers correspond to the ACC obtained by LS-QSVM-RR.

(b) The ridge trace of Breast Cancer. b and α correspond to the b and α in Eq. (6).

Fig. 3. The average ACC of various datasets under different ridge regression coefficients, as well as the ridge trace plot for the breast cancer dataset within the same range of ridge regression coefficients. B. C., H. D. represents Breast Cancer and Heart Disease respectively.

based on the ridge trace plot, as Fig. 3(b). We conducted an analysis using the Breast Cancer dataset. It can be observed that if ridge regression coefficients were selected based on the ridge trace plot, they would fall within the range of 0.1 to 10. However, experimental results indicate that the optimal ridge regression coefficient lies between 0.001 and 0.1, which contradicts the information provided by the ridge trace plot. Our approach, which involves selecting ridge regression coefficients based on the maximum eigenvalue, aligns precisely with the appropriate range of ridge regression coefficients, as observed in other datasets.

5.3   Conclusion

In this work, we have successfully implemented an LS-QSVM-RR algorithm based on efficient matrix decomposition and variational quantum algorithms (VQA). This breakthrough overcomes the limitations of the HHL algorithm, which is confined to theoretical scenarios, by enabling simulation and implementation in the NISQ era. Leveraging the remarkable parallelism of quantum circuits, our approach accelerates the training process compared to classical computers. Furthermore, we have successfully encoded the ridge regression constraint in matrix form into the quantum circuit. By utilizing the power iteration on the quantum circuit, we have estimated ridge regression coefficients that outperform those obtained from the traditional ridge trace plot method. This not only reduces the matrix condition number but also demonstrates the accuracy and robustness of our approach, as supported by experimental results. This work expands the horizons of quantum machine learning and offers promising avenues for leveraging quantum computing capabilities to enhance predictive accuracy and computational efficiency.

References

1. Shor, P.W.: Algorithms for quantum computation: discrete logarithms and factoring. In: 35th Annual Symposium on Foundations of Computer Science, pp. 124–134. IEEE, Santa Fe, NM, USA (1994)
2. Arute, F., Arya, K., Babbush, R., et al.: Quantum supremacy using a programmable superconducting processor. Nature **574**(7779), 505–510 (2019)
3. Prakash, A.: Quantum Algorithms for Linear Algebra and Machine Learning. University of California, Berkeley (2014)
4. Shor, P.W.: Polynomial-time algorithms for prime factorization and discrete logarithms on a quantum computer. SIAM J. Comput. **41**(2), 303–332 (1999)
5. Harrow, A.W., Hassidim, A., Lloyd, S.: Quantum algorithm for linear systems of equations. Phys. Rev. Lett. **103**(15), 150502–150506 (2009)
6. Grover L. K.: A fast quantum mechanical algorithm for database search. In: 28th Annual ACM Symposium on Theory of Computing, pp. 212–219. ACM, New York, NY, USA (1996)
7. Henderson, M., Shakya, S., Pradhan, S., et al.: Quanvolutional neural networks: powering image recognition with quantum circuits. Quantum Mach. Intell. **2**(1), 2 (2020)
8. Kerenidis I., Prakash A.: Quantum recommendation systems. In: 8th Innovations in Theoretical Computer Science Conference, pp. 1–21. Schloss Dagstuhl-Leibniz-Zentrum fuer Informatik, Dagstuhl, Germany (2017)
9. Gharipour, A., Jazi, A.Y., Sameti, M.: Forecast combination with optimized SVM based on quantum-inspired hybrid evolutionary method for complex systems prediction. In: 2011 IEEE Symposium on Computational Intelligence for Financial Engineering and Economics, pp. 1–6. IEEE, Paris, France (2011)
10. Boser B.E., Guyon I.M., Vapnik V.N.: A training algorithm for optimal margin classifiers. In: 5th Annual Workshop on Computational Learning Theory, pp. 144–152. ACM, New York, NY, USA. (1992)
11. Cortes, C., Vapnik, V.: Support-vector networks. Mach. Learn. **20**(3), 273–297 (1995)
12. Cristianini, N., Shawe-Taylor, J.: An Introduction to Support Vector Machines and Other Kernel-Based Learning methods. Cambridge University Press, Cambridge (2000)
13. Smola, A.J., Schölkopf, B.: A tutorial on support vector regression. Stat. Comput. **14**, 199–222 (2004)
14. Schölkopf, B., Smola, A.J., Williamson, R.C., et al.: New support vector algorithms. Neural Comput. **12**(5), 1207–1245 (2000)
15. Hsu C. W., Chang C., Lin C.: A practical guide to support vector classification. Bioinformatics **1** (2003)
16. Crammer, K., Singer, Y.: On the algorithmic implementation of multiclass kernel-based vector machines. J. Mach. Learn. Res. **2**(11), 265–292 (2001)
17. Suykens, J.A.K., Vandewalle, J.: Least squares support vector machine classifiers. Neural Process. Lett. **9**, 293–300 (1999)
18. Hamers, B., Suykens., J.A.K., De Moor, B.: Compactly supported RBF kernels for Sparsifying the gram matrix in LS-SVM regression models. In: Dorronsoro, J.R. (ed.) ICANN 2002. LNCS, vol. 2415, pp. 720–726. Springer, Heidelberg (2002). https://doi.org/10.1007/3-540-46084-5_117
19. Rebentrost, P., Mohseni, M., Lloyd, S.: Quantum support vector machine for big data classification. Phys. Rev. Lett. **113**(13), 130503 (2014)

20. Ahmed, S.: Pattern Recognition with Quantum Support Vector Machine (QSVM) on Near Term Quantum Processors. Brac University, Dhaka (2019)
21. Childs, A.M., Kothari, R., Somma, R.D.: Quantum algorithm for systems of linear equations with exponentially improved dependence on precision. SIAM J. Comput. **46**(6), 1920–1950 (2017)
22. Xin, T., Wei, S., Cui, J., et al.: Quantum algorithm for solving linear differential equations: theory and experiment. Phys. Rev. A **101**(3), 032307 (2020)
23. Nghiem, N.A., Wei, T.C.: Quantum Algorithm For Estimating Eigenvalue. arXiv preprint (2022)
24. Preskill, J.: Quantum computing in the NISQ era and beyond. Quantum **2**, 79 (2018)
25. Gill S. S., Kumar A., Singh H., et al.: Quantum computing: a taxonomy, systematic review and future directions. Softw. Pract. Exp. **52**(1), 66–114 (2022)
26. Leonard, W., Zhao, Z., Prakash, A.: Quantum linear system algorithm for dense matrices. Phys. Rev. Lett. **120**(5), 050502–050507 (2018)
27. Cerezo, M., Arrasmith, A., Babbush, R., et al.: Variational quantum algorithms. Nat. Rev. Phys. **3**, 625–644 (2021)
28. Chen, L.: Deep Learning and Practice with MindSpore. CIR, Springer, Singapore (2021). https://doi.org/10.1007/978-981-16-2233-5
29. García, C., García, J., López Martín, M., Salmerón, R.: Collinearity: revisiting the variance inflation factor in ridge regression. J. Appl. Stat. **42**, 648–661 (2015)
30. Wold, S., Ruhe, A., Wold, H., Dunn, III, W. J.: The collinearity problem in linear regression. the partial least squares (PLS) approach to generalized inverses. SIAM J. Sci. Stat. Comput. **3**(5), 735–743 (1984)
31. Wan, A.T.: On generalized ridge regression estimators under collinearity and balanced loss. Appl. Math. Comput. **129**(2–3), 455–467 (2002)
32. Ying, J., et al.: Reparing a fast Pauli decomposition for variational quantum solving linear equations. Annalen der Physik (2023)
33. Bergstra, J., Bengio, Y.: Random search for hyper-parameter optimization. J. Mach. Learn. Res. **13**(2), 281–305 (2012)
34. Soper, D.S.: Greed is good: rapid hyperparameter optimization and model selection using greedy k-fold cross validation. Electronics **10**(16), 1973 (2021)
35. Hoerl, A.E., Kennard, R.W.: Ridge regression: biased estimation for nonorthogonal problems. Technometrics **12**(1), 55–67 (1970)
36. Rokem A, Kay K.: Fractional ridge regression: a fast, interpretable reparameterization of ridge regression. GigaScience **9**(12), giaa133 (2020)

An Adaptive Auxiliary Training Method of Autoencoders and Its Application in Anomaly Detection

Li Niu$^{(\boxtimes)}$ ⓘ, Jiachun Liaoⓘ, Feng Shaⓘ, Zhaokun Chengⓘ, and Yicheng Qiuⓘ

Nanhu Laboratory, Jiaxing, China

{niuli,jliao,fengsha,zhaokunc,qiu-yicheng}@nanhulab.ac.cn

Abstract. In various applications of autoencoders, an auxiliary subnetwork is used to improve the performance of a neural network with an autoencoder as the key component. For the specific task of anomaly detection, we have observed that in certain cases, when the reconstruction performance reaches a high level, the auxiliary subnetwork becomes ineffective in further improving the autoencoder's performance. This phenomenon results in oscillation and degradation of the overall system. To address this issue, we propose an adaptive auxiliary training method (AAT) that ensures continuous improvement in the autoencoder's reconstruction performance throughout the entire training procedure. AAT enhances the monitoring of the autoencoder's training, enabling adaptive adjustment of the training strategy without a validation set. Additionally, an anomaly detection scheme is devised based on the proposed adaptive auxiliary training method. Experimental results on multiple datasets prove that the proposed methods produce autoencoders with better reconstruction and detection performances comparing to the state-of-the-art (SOTA) methods.

Keywords: Adaptive auxiliary training · Anomaly Detection · Autoencoder

1 Introduction

Autoencoders and their variants have garnered significant attention from researchers due to their outstanding performance in various applications such as image processing, pricing models, anomaly detection, and recommender systems [1–14]. One important advancement in autoencoders is the auxiliary training mechanism with an auxiliary subnetwork [15–27]. This development has gained popularity and continues to be refined by researchers, finding wide practical applications in various fields including anomaly detection [15, 16, 20–23, 25], speech-based emotion recognition [17, 26], top-N recommendation [18] and some other fields.

Training autoencoders with an auxiliary subnetwork improves reconstruction and provides control over the latent space [15]. The design has also been found to accelerate the training process [15–21]. However, through our experiments and applications, we have observed that the stability of the training process can be significantly affected by the auxiliary training mechanism. This indicates that the auxiliary training mechanism is

© The Author(s), under exclusive license to Springer Nature Singapore Pte Ltd. 2024
B. Luo et al. (Eds.): ICONIP 2023, CCIS 1961, pp. 524–540, 2024.
https://doi.org/10.1007/978-981-99-8126-7_41

limited to certain cases. This instability results in a decline in the performance of anomaly detection on the MVTec AD [22] dataset. In detail, we have observed that during the training process, when the reconstruction performance reaches a particular good point, it does not converge to the best value. Instead, it begins to oscillate and decrease greatly, and later dropped to a level far below the aforementioned relatively high value when training stopped. We have also eliminated cause factors like network structure design issues or loss function improvements by experiments of replacing different network structures (U-net block, Resnet block, etc.) and the reconstruction loss (L1, L2, Structural Similarity (SSIM) [28], etc.). Based on these experiments, we conclude that the auxiliary training mechanism should be enhanced to effectively address the challenges encountered in datasets such as MVTec AD.

In this paper, we propose an adaptive auxiliary training method (AAT) for autoencoders. The novel contributions of this paper are as follows:

1) An adaptive auxiliary training method for autoencoders is proposed, which effectively enhances the stability of the existing auxiliary training mechanism and improves the performance.
2) The proposed method enhances the monitoring of the autoencoder's training process without the need for a validation set. This method allows for adaptive adjustment of the training strategy and enables termination of training at the appropriate time.
3) The proposed method is applied to improve the performance of anomaly detection, which further proves its effectiveness.

The rest of this paper is organized as follows: we provide a review of recent auxiliary training methods for autoencoders in Sect. 2, and in Sect. 3 we present the details of our proposed method. We then analyze our experimental results in Sect. 4 and give our conclusion in Sect. 5. In the remaining sections of this paper, we will refer to the autoencoder as G and the auxiliary subnetwork as D in some paragraphs. The code used in our experiments is available at https://github.com/yesdtrx/AATAE.

2 Related Work

In related literature, the auxiliary subnetwork is utilized to expedite or enhance the training performance of the autoencoder [15–27]. This new approach is referred to as the autoencoder's auxiliary training (AT) mechanism. In AT, the loss function for training G comprises both the reconstruction loss and the auxiliary loss. G and D are trained in an alternating manner, meaning that during each epoch, D is trained followed by the training of G. This alternating training process allows G and D to learn from each other iteratively. Therefore, this mutual training process enhances the overall performance of the autoencoder.

Based on the described setting [15–21], we implemented the AT approach and conducted experiments on the MVTec AD dataset, and comparing it with the AE approach. Our findings demonstrate that the AT method led to performance improvements and training acceleration on certain subsets, while encountering performance degradation on others. Figure 1 illustrates the performance trends during the training processes on 6 subsets. The average Structural Similarity (SSIM) on the test set was used to evaluate the reconstruction performance of the autoencoder. The x-axis represents the training epoch, while the y-axis represents the SSIM on the test set. The orange line corresponds to the performance of AE, while the blue line corresponds to the performance of AT.

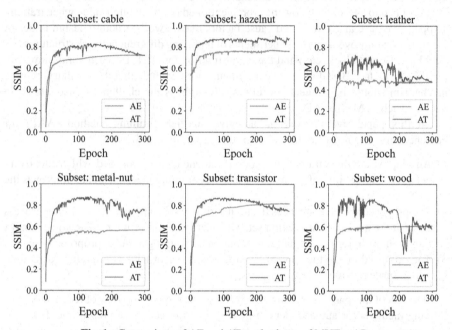

Fig. 1. Comparison of AE and AT on 6 subsets of MVTec AD.

In these experiments, it was observed that when using AE, the SSIM steadily grows but the upper limit is generally low. On the other hand, when utilizing AT, the SSIM improves faster and reaches a higher upper limit. However, for certain subsets such as cable, leather, transistor, and wood, the SSIM exhibits severe oscillation or decline in the middle and late stages of the training process. Comparing the performance of AE and AT, this phenomenon can be summarized as follows: D initially aids in the training of G but eventually interferes with it. Once the reconstruction performance reaches a relatively high level, the auxiliary subnetwork becomes ineffective in further enhancing the performance of the autoencoder. This phenomenon leads to oscillation and degradation of the overall system.

To address this issue, recent literature [10–13, 16] introduces a weight parameter to the auxiliary loss and adjusts it as a hyperparameter. These studies use the hyperparameter to control the regularization strength of the auxiliary loss on the overall loss of the autoencoder. However, a disadvantage of this approach is that the weight hyperparameters need to be manually set and cannot adaptively match the loss changes during training. We provide a performance comparison between our method and existing approaches in the experimental section.

In the aforementioned research work, there are no methods that can effectively supervise network training without relying on a validation set. Typically, a common approach is to set aside a validation set and calculate the performance indicators on the validation set after each epoch. However, this method requires a sufficiently large and representative validation set. In cases where the dataset, such as MVTec AD, does not include a validation set, models are usually trained for a fixed number of epochs. Instead, we innovatively use the average reconstruction performance of all samples in the training set to supervise the training of the autoencoder.

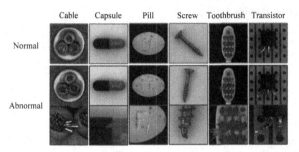

Fig. 2. Example images of the MVTec AD dataset. It is important to note that the training set only consists of normal samples, while the test set contains both normal and abnormal samples.

The MVTec AD dataset is one of the most commonly used datasets in anomaly detection (Fig. 2). The autoencoder and auxiliary subnetwork are trained using only normal samples. During the inference stage, the reconstruction scores and the auxiliary scores of abnormal samples are lower than those of normal samples. This enables the detection of anomalies. The evaluation metric commonly used is the area under the receiver operating characteristic curve (AUROC).

3 Proposed Method

3.1 Phenomenon Analysis

To address the aforementioned issues with the auxiliary training mechanism [15–27], we conducted an in-depth analysis. At a particular moment or stage during the training of the autoencoder network, we observed oscillations or declines in the network's reconstruction accuracy. This led us to infer significant changes in the network's parameters, implying substantial alterations in their gradients.

Within the auxiliary training mechanism, the autoencoder network's loss function comprises both the reconstruction loss and the auxiliary loss. The details of the loss function can be found in Sect. 3.2. We initiated our analysis by scrutinizing the gradient vectors of the reconstruction loss and the auxiliary loss. We maintained consistency with the network architecture and auxiliary training methodology outlined in prior research [16].

This autoencoder network consists of a total of 70 layers, with a trainable parameter count of 7,845,123. Upon the completion of each epoch, we separately computed gradients for the reconstruction loss and the auxiliary loss, subsequently flattening all gradients into a single vector. Therefore, following each epoch, we acquired gradient vectors for both the reconstruction loss and the auxiliary loss, enabling us to calculate the norms of the two vectors as well as the angle between them. Experimental results of the gradient norms are depicted in Fig. 3. And experimental results indicate that, during the training of the autoencoder network, the angle between the two gradient vectors varies in the range of approximately 50 to 150°. This suggests a dynamic interplay between the gradient vectors of the reconstruction loss and the auxiliary loss, involving both cooperation and competition. The gradient vector with the larger norm has a greater impact on the training of the autoencoder.

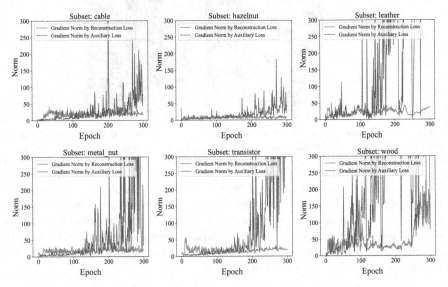

Fig. 3. Gradient norms of the reconstruction loss and auxiliary loss during the training process

From Fig. 3, it can be observed that in the early stage of training, the magnitudes of the gradient norms for both the reconstruction loss and the auxiliary loss are relatively close. As shown in Fig. 1, during this phase, the autoencoder network's reconstruction accuracy exhibits a stable upward trend. However, combining the observations from Fig. 1 and Fig. 3, in the later stages of network training, the gradient norm of the auxiliary loss significantly surpasses that of the reconstruction loss. This disrupts the equilibrium in the cooperative training of the reconstruction and auxiliary losses, leading to adverse effects on the training of the autoencoder network.

3.2 Adaptive Auxiliary Training for Autoencoder

This section proposes a solution called AAT (Adaptive Auxiliary Training) as shown in Fig. 4. The training process of AAT is illustrated in Fig. 5. The original sample to G is denoted as x, and the reconstructed sample is denoted as $\hat{x}$, $\hat{x} = G(x)$. $D(x)$ represents the auxiliary score of the original sample, and $D(\hat{x})$ represents the auxiliary score of the reconstructed sample. $D(x) \in [0, 1]$, $D(\hat{x}) \in [0, 1]$. The loss of G is:

$$L_G = \lambda_{recon} \times L_{recon} + \lambda_{adv} \times L_{adv} \tag{1}$$

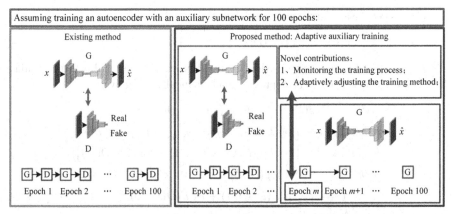

Fig. 4. In existing methods[15–21], the auxiliary training method is maintained throughout 100 epochs. Within each epoch, the following steps are executed: firstly, the parameters of D are fixed, and G is trained once; subsequently, the parameters of G are fixed, and D is trained once. AAT supervises the entire training process and automatically removes the auxiliary subnetwork and adjusts the training method at the *m-th* epoch. Please note that the detailed network structures of the autoencoder and the auxiliary subnetwork are not the primary focus of this study and will not be extensively elaborated upon. For more information, please refer to our open-source code.

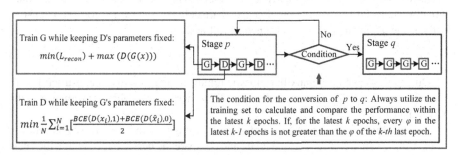

Fig. 5. The training process of AAT.

L_{recon} represents the reconstruction loss, which measures the difference between the reconstructed sample and the original sample. L_{adv} represents the auxiliary loss, which aims to align the reconstructed samples with the original samples based on the evaluation method of the auxiliary subnetwork:

$$L_{adv} = \mathbb{E}_{\hat{x} \sim p_{\hat{x}}}[\log(1 - D(\hat{x}))] \tag{2}$$

And λ_{recon} and λ_{adv} represent weight coefficients, which are used to control the regularization strength of the auxiliary loss on the overall loss of G. And their specific values are discussed in the experimental section. The training objective of G can be expressed mathematically as follows:

$$min(\lambda_{recon} \times L_{recon} + \lambda_{adv} \times L_{adv})$$
$$\Rightarrow min(L_{recon}) + max(\mathbb{E}_{\hat{x} \sim p_{\hat{x}}}[\log(D(\hat{x}))]) \tag{3}$$
$$\Rightarrow min(L_{recon}) + max(D(G(x)))$$

The loss of the auxiliary subnetwork is:

$$L_D = \mathbb{E}_{x \sim p_x}\left[\log(D(x))\right] + \mathbb{E}_{\hat{x} \sim p_{\hat{x}}}[\log(1 - D(\hat{x}))] \tag{4}$$

The training objective for the auxiliary network is to enhance its ability to differentiate between x and $\hat{x}$, which can be expressed mathematically as follows:

$$
\begin{aligned}
&max\left(\mathbb{E}_{x \sim p_x}\left[\log(D(x))\right] + \mathbb{E}_{\hat{x} \sim p_{\hat{x}}}[\log(1 - D(\hat{x}))]\right)\\
&\Rightarrow min\left(\mathbb{E}_{x \sim p_x} - \left[\log(D(x))\right]\right) + min(\mathbb{E}_{\hat{x} \sim p_{\hat{x}}} - [\log(1 - D(\hat{x}))])\\
&\Rightarrow min\left(\mathbb{E}_{x \sim p_x} - \left[1 \times \log(D(x)) + (1 - 1) \times \log(1 - D(x))\right]\right)\\
&+min\left(\mathbb{E}_{\hat{x} \sim p_{\hat{x}}} - [0 \times \log(D(\hat{x})) + (1 - 0) \times \log(1 - D(\hat{x}))]\right)\\
&\Rightarrow min\frac{1}{N} \sum\nolimits_{i=1}^{N} [\frac{BCE(D(x_i), 1) + BCE(D(\hat{x}_i), 0)}{2}]
\end{aligned}
\tag{5}
$$

In this formula, $BCE()$ represents the calculation of binary cross entropy.

In AAT, we use φ to represent the performance of the generator (G) on the training set. The stage where φ shows an upward trend is denoted as p, while the stage where φ oscillates or decreases in AT is denoted as q. Instead of running our optimization algorithm until we reach a local maximum of φ, we run it until the φ has not improved for some amount of time, and then we transition the training from stage p to stage q. During the stage p, the autoencoder is trained with the assistance of the auxiliary subnetwork, and the loss function of G and D are defined as shown in formulas (1)–(5). In the stage q, the autoencoder is trained as follows:

$$L_G = L_{recon} \tag{6}$$

During the stage q, the auxiliary loss is removed, and only the reconstruction loss is used to fine-tune G, aiming to achieve a slight increase or stable convergence in the performance of G.

The condition for the conversion of stage p to stage q can be expressed as follows: Always utilize the training set to calculate and compare the performance within the latest k epochs. If, for the latest k epochs, every φ in the latest k-1 epochs is not greater than the φ of the k-th last epoch. In general, the value of k could be set to 10% of the total training epochs. To implement this mechanism, the proposed approach suggests utilizing the average reconstruction performance on the training set as a supervisory signal during the training process.

It is important to clarify that during the training process φ represents the performance calculated on the training set, while the performance of G presented in all figures and tables in this paper refers to the performance on the test set.

In this section, we propose AAT to address the issues encountered by AT [15–21]. The experimental evaluation in Sect. 4.3 demonstrates that AAT brings significant performance improvements to the autoencoder compared to AT across multiple datasets.

3.3 AAT for Anomaly Detection

Anomaly detection (AD) is a typical application of autoencoders. This section studies how to apply AAT to improve the performance of autoencoders on anomaly detection.

In the field of anomaly detection, the literature [15–18] not only employs the auxiliary subnetwork to assist the training of G but also leverages its capabilities to aid in abnormal judgement. Anomaly detection (AD) score represents the probability that a test sample is classified as a normal sample. Generally, the AD score is calculated as a combination of two components:

$$s = \alpha \times s_{recon} + \beta \times s_d, \tag{7}$$

where s_{recon} represents the reconstruction score, s_d represents the auxiliary score, and α and β are weights assigned to each component. Typically, set $\alpha = \beta = 0.5$, and $s_{recon} \in (0, 1), s_d \in (0, 1), s \in (0, 1)$.

Obviously, in order to adapt to the anomaly detection task, the convergence of the auxiliary subnetwork needs to be guaranteed. We introduce the stage r after stage q of AAT, as shown in Fig. 6. In the stage r, the parameters of G are fixed, and only D is trained. And the loss function is shown in formula (4)−(5).

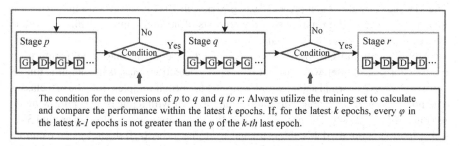

Fig. 6. The training process of AAT for anomaly detection.

The condition of q to r is the same as the condition of p to q. During the training, we use the average auxiliary score of all samples in the training set as the performance of D, denoted by ξ, which is used to supervise the training of D. The early stopping condition for stage r is that ξ exceeds a set threshold D_{th}, which is assigned a value of 0.99.

When the early stopping condition is not met, the network training stops after a fixed number of iterations, for example, training for a total of 100 epochs across stage p, stage q, and stage r. However, in some cases, after stage r, ξ may have a significant difference from the expected threshold:

$$\xi < \gamma \times D_{th} \tag{8}$$

γ represents the ratio, for example, γ can be set to 0.98. In such scenarios, we exclude it from the computation of anomaly detection scores. Consequently, the formula (7) for calculating the AD score simplifies to:

$$s = s_{recon} \tag{9}$$

In our proposed method, in addition to the early stopping strategy in stage r, stage p implements an early stopping strategy by adaptively transitioning to stage q, and stage q implements an early stopping strategy by adaptively transitioning to stage r. Compared to AT, AAT can select models with better performance and significantly reduce training time. We find that our proposed strategies align with existing research on training algorithms for deep learning[29]. Previous studies have suggested using validation set supervision to terminate training when appropriate so as to save training costs. Our approach differs in that we address the scenario where the validation set is unavailable, and we terminate training at a specific stage to transition to the next stage.

In Sect. 4.4, we will present the comparison of average AUROC scores and the average numbers of training epochs between AT and AAT on multiple datasets and their respective subsets. The experimental results provide clear evidence of the effectiveness and advancement of AAT compared to existing methods.

4 Experiments and Discussion

4.1 Datasets

The MVTec AD dataset comprises 15 categories with 3629 images for training and 1725 images for testing. The training set only consists of normal samples, while the test set contains both normal and abnormal samples.

MNIST [30] and CIFAR-10 [31], both of which contain images belonging to 10 classes. For each dataset, we construct 10 anomaly detection datasets by sampling images from each class as normal samples and sampling abnormal samples from the rest classes. The normal samples are split into training and test set with a rate of 2:1. Following the setting used in [32–34]. The proposition of abnormal samples is controlled around 30%. Take the first anomaly dataset constructed from the CIFAR-10 dataset as an example: 6000 pictures of airplanes, of which 4000 are used as training set and 2000 are used as test set. The abnormal samples are randomly selected from the remaining 9 categories, and the abnormal rate is kept at 30%. For this anomalous dataset, the training set has 4000 samples and the test set has 2857 samples. The anomaly detection datasets derived from MNIST and CIFAR-10 are referred to as MNIST AD and CIFAR-10 AD, respectively.

The wildfire dataset, publicly available from the University of Science and Technology of China, consists of images captured by drones in natural settings. The dataset comprises more than 3600 normal images and 1000 abnormal images, including images of wild smoke and wild fire.

4.2 Experimental Details

The Structural Similarity Index (SSIM) was employed to measure the similarity between the reconstructed image and the original image. The average SSIM value across all samples in the test set was used as the reconstruction similarity metric for G. In Formula (1), λ_{recon} and λ_{adv} are the weights of reconstruction loss and auxiliary loss, respectively, and take the value of 10 and 1. Then the loss of the autoencoder in the stage p is as follows:

$$L_G = 10 \times (1 - \text{SSIM}(x, \hat{x}) + \text{BCE}(D(\hat{x}), y) \tag{10}$$

In this formula, $\hat{x} = G(x)$, $\text{SSIM}(x, \hat{x})$ represents the calculation of the SSIM value between the original image and the reconstructed image. $\text{BCE}(D(\hat{x}), y)$ represents the calculation of binary cross entropy, one of the inputs is the auxiliary score of the reconstructed image by the auxiliary subnetwork, and the second input is an array y whose elements are all 1. And the loss of G in the stage q is as follows:

$$L_G = 1 - \text{SSIM}(x, \hat{x}) \qquad (11)$$

The loss of the auxiliary subnetwork is:

$$L_D = \text{BCE}(D(x), y) + \text{BCE}(D(\hat{x}), \hat{y}) \qquad (12)$$

In the first half, $D(x)$ represents the auxiliary score of the auxiliary subnetwork on the original images, y represents the labels of the original images whose elements are all integers 1. In the second half, $D(\hat{x})$ represents the auxiliary score of the reconstructed image by the auxiliary subnetwork, $\hat{y}$ represents the labels of the original images whose elements are all 0.

4.3 Validating the Enhancement of Reconstruction with AAT

Table 1. Summary of average SSIM.

Method	MVTec AD	MNIST AD	Wildfire	CIFAR-10 AD
AT[15–21]	0.7540	0.3748	0.8450	0.891
AAT	**0.8804**	**0.7278**	**0.9036**	0.893

Table 1 presents the SSIM comparison between AT and AAT on multiple datasets. The results in Table 1, Table 2, Table 3, and Table 4 are obtained by averaging the results from 5 experimental runs. This approach provides a more robust and reliable evaluation by considering the variability across multiple trials.

Figure 7 illustrates the performance trends during the training processes on 6 subsets of the MVTec AD dataset. The orange solid line represents the conventional autoencoder training method using only the reconstruction error (AE). The blue solid line corresponds to AT[15–21], while the red solid line represents AAT. The green vertical dotted line represents the moment when AAT adaptively changes the training strategy. At the epoch marked by the green vertical dotted line, AAT discontinues auxiliary training and proceeds with G training solely based on the reconstruction loss. Figure 8 illustrates the performance trends during the training processes on 10 subsets of the MNIST AD dataset. Figure 9 exhibits the SSIM comparison between AT and AAT on the MVTec AD dataset and the MNIST AD dataset.

It is important to note that, as described in the early stopping strategy of AAT in Sect. 3.2, the training processes often terminates before reaching the maximum number of epochs. However, for the purpose of comparison, Fig. 7 and Fig. 8 were intentionally trained for 300 epochs to align with AT.

From Table 1, Fig. 7, Fig. 8, and Fig. 9, it is evident that AAT brings significant improvements to the performance of the autoencoder compared to AT. AAT not only

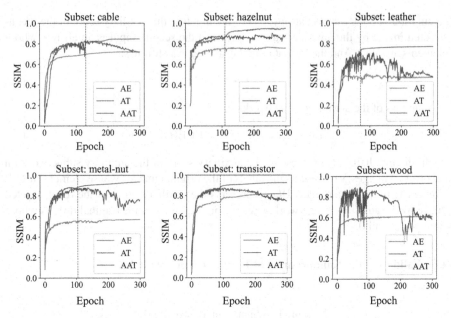

Fig. 7. Comparison of AE, AT and AAT on 6 subsets of MVTec AD.

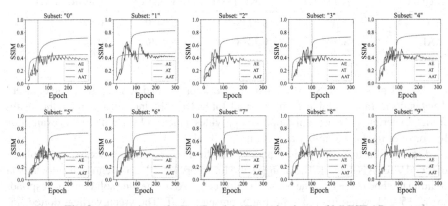

Fig. 8. Comparison of AE, AT and AAT on 10 subsets of MNIST AD.

preserves the advantages of auxiliary training but also addresses its limitations. By adaptive adjustment of the training strategy based on the training process, AAT achieves superior performance.

On the CIFAR-10 AD dataset, the training processes using the AT method all exhibit consistently upward trends, and the condition for transitioning from stage p to stage q in AAT is not triggered. Therefore, in terms of the reconstruction performance of the autoencoder, AAT and AT exhibit consistency.

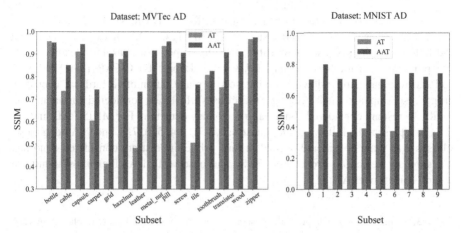

Fig. 9. SSIM comparison of AT and AAT on MVTec AD and MNIST AD.

4.4 Validating the Enhancement of Anomaly Detection with AAT

Table 2 displays the AUROC scores of anomaly detection achieved by AAT and existing methods on multiple datasets. In constructing the anomaly detection datasets from MNIST and CIFAR-10, we followed the guidelines provided by the other methods listed in the table. In addition to MNIST AD and CIFAR-10 AD, we further conducted experimental evaluations using MVTec AD and Wildfire datasets. These additional experiments provided further validation of the advanced and reliable performance of AAT. And the evaluation experiments for AT and AAT differed only in the training strategy, while all other settings remained the same. Moreover, all experiments were repeated 5 times, and the average values were taken as the final experimental results.

Table 2. Comparison of AUROC scores of anomaly detection between AAT and existing methods on multiple datasets.

Method	MNIST AD	CIFAR-10 AD	MVTec AD	Wildfire
OC-SVM [35]	0.9499	0.5619	–	–
KDE [32]	0.8116	0.5756	–	–
VAE [36]	0.9643	0.5725	–	–
PixCNN [37]	0.6141	0.5450	–	–
DSEBM [34]	0.9554	0.5725	–	–
AE [32]	0.9619	0.5706	–	–
MemAE [32]	0.9751	0.6088	–	–
AT [15–21]	0.9653	0.7955	0.6527	0.8176
AAT	**0.9834**	**0.8641**	**0.8256**	**0.9917**

Table 3 presents the comparison of average AUROC between AT [15–21] and AAT on various datasets and their respective subsets. For MVTEC AD, indices 1 to 15 represent the following sub-datasets: bottle, cable, capsule, carpet, grid, hazelnut, leather, metal nut, pill, screw, tile, toothbrush, transistor, wood, and zipper. For CIFAR-10 AD, indices 1 to 10 correspond to the following sub-datasets: airplane, automobile, bird, cat, deer, dog, frog, horse, ship, and truck. Among these, classes 1 to 10 are considered normal in turn, while the rest are considered abnormal for anomaly detection. For MNIST AD, indices 1 to 10 represent the following sub-datasets: 0, 1, 2, 3, 4, 5, 6, 7, 8, and 9. Among these, classes 0 to 9 are considered normal in turn, while the rest are considered abnormal for anomaly detection. From Table 3, we can observe that AAT exhibits significant improvements over AT.

Table 3. Comparison of average AUROC scores between AT and AAT on multiple datasets and their respective subsets.

Dataset	MVTEC AD		CIFAR-10 AD		MNIST AD	
Subset	AT[15–21]	AAT	AT[15–21]	AAT	AT[15–21]	AAT
1	0.9530	**0.9654**	0.8428	**0.9565**	0.9942	**0.9991**
2	0.6557	**0.9003**	0.7616	**0.8978**	0.9880	**0.9988**
3	**0.5962**	0.5773	**0.8579**	0.7421	0.8842	**0.9619**
4	**0.4518**	0.4092	0.7505	**0.7789**	0.9560	**0.9867**
5	**0.9612**	0.9063	0.6833	**0.8330**	**0.9752**	0.9638
6	0.4551	**0.8010**	0.7953	**0.8481**	0.9820	**0.9875**
7	0.2855	**0.9887**	0.7918	**0.9093**	0.9977	**0.9978**
8	0.5447	**0.8691**	0.6678	**0.8208**	**0.9838**	0.9835
9	0.8351	**0.8969**	0.8863	**0.9505**	0.8948	**0.9725**
10	0.3742	**0.6247**	**0.9180**	0.9037	**0.9973**	0.9827
11	0.7024	**0.9815**	–	–	–	–
12	0.8211	**0.8550**	–	–	–	–
13	0.5218	**0.8308**	–	–	–	–
14	0.6844	**0.8339**	–	–	–	–
15	**0.9477**	0.9445	–	–	–	–
Average	0.6527	**0.8256**	0.7955	**0.8641**	0.9653	**0.9834**

It is noteworthy that for the CIFAR-10 AD dataset, as discussed in Sect. 4.3, the SSIM values obtained by AT and AAT are similar. However, there are significant differences in their anomaly detection performance, as highlighted in Sect. 4.4. The reason behind this disparity is that in AAT, we also supervise the training of the auxiliary subnetwork. Therefore, if the auxiliary subnetwork is insufficiently trained, we exclude it from the computation of anomaly detection scores, as demonstrated in formulas (8) and (9).

Table 4 presents the average numbers of epochs required for AT and AAT on multiple datasets and their respective subsets. From Table 4, we can see that on the MVTec AD and MNIST AD datasets, AAT achieves an average training epochs reduction of 44.41% and 62.5% respectively compared to AT. For the Wildfire dataset, which has only one subset, it is not listed in Table 4. On the Wildfire dataset, the average numbers of training epochs for AT and AAT are 500 and 241 respectively. On the CIFAR-10 AD dataset, the numbers of training epochs for AAT remains the same as that of AT.

Table 4. The average numbers of epochs required for AT and AAT on multiple datasets and their respective subsets.

Dataset	MVTEC AD		MNIST AD		CIFAR-10 AD	
Subset	AAT	AT[15–21]	AAT	AT[15–21]	AAT	AT[15–21]
1	187.2	300	194.2	500	100	100
2	165	300	182.8	500	100	100
3	247.6	500	193.4	500	100	100
4	163.2	500	186.4	500	100	100
5	113.2	300	183.8	500	100	100
6	188	300	189.2	500	100	100
7	111.6	300	177	500	100	100
8	184.6	300	192.6	500	100	100
9	188.8	300	189	500	100	100
10	321.8	500	186.8	500	100	100
11	127.6	300	–	–	–	–
12	300	300	–	–	–	–
13	140.4	300	–	–	–	–
14	96.2	300	–	–	–	–
15	300	300	–	–	–	–
Average	189.0	340	187.5	500	100	100

In this section, we conducted extensive experiments to evaluate the proposed method, and each experiment was repeated 5 times to obtain averaged results, ensuring a more robust and reliable evaluation. The experimental results clearly demonstrate that, compared to existing methods, AAT achieves superior AUROC scores and significantly reduces the required training time.

5 Conclusion

In this literature, we investigate the role of the auxiliary subnetwork in training autoencoder networks. In some circumstances, the auxiliary subnetwork will interfere with the later training of the autoencoder. To alleviate the problem, a novel adaptive auxiliary training for autoencoders (AAT) is proposed. The core idea of AAT is to supervise the training process and adaptively remove the auxiliary subnetwork when the auxiliary subnetwork no longer contributes to the training of the autoencoder. AAT ensures that the auxiliary subnetwork constantly plays an active role in the training of the autoencoder. Thus, a better model can be achieved through the proposed supervision mechanism. Experimental results show that AAT achieves superior performance and significantly reduces the required training time.

In addition to image anomaly detection, AAT has potential applications in fields such as speech emotion recognition [17, 26] and top-N recommendation [18]. Taking speech emotion recognition [17] as an example, if oscillations in the autoencoder's accuracy are observed during the training of the autoencoder network, AAT can be considered as a potential solution. To employ AAT, it is essential to first establish a metric that quantifies the similarity between generated speech and the original input speech. Subsequently, the AAT training mechanism can be implemented according to the methodology and procedures outlined in this literature.

Acknowledgement. This work is funded by Zhejiang "Lingyan" Research and Development Program (No. 2022C03121). We are grateful for the support of this program.

References

1. Zhang, K., et al.: History matching of naturally fractured reservoirs using a deep sparse autoencoder. SPE J. **26**(04), 1700–1721 (2021)
2. Lai, Z., Liu, S., Efros, A.A., Wang, X.: Video autoencoder: self-supervised disentanglement of static 3d structure and motion. In: Proceedings of the IEEE/CVF International Conference on Computer Vision 2021, pp. 9730–9740 (2021)
3. Parmar, G., Li, D., Lee, K., Tu, Z.: Dual contradistinctive generative autoencoder. In: Proceedings of the IEEE/CVF Conference on Computer Vision and Pattern Recognition, pp. 823–832 (2021)
4. Nguyen, H., Tran, K.P., Thomassey, S., Hamad, M.: Forecasting and anomaly detection approaches using LSTM and LSTM autoencoder techniques with the applications in supply chain management. Int. J. Inf. Manage. **57**, 102282 (2021)
5. Kim, J., Kong, J., Son, J.: Conditional variational autoencoder with adversarial learning for end-to-end text-to-speech. In: International Conference on Machine Learning PMLR, pp. 5530–5540 (2021)
6. Shao, H., Xia, M., Wan, J., de Silva, C.W.: Modified stacked autoencoder using adaptive morlet wavelet for intelligent fault diagnosis of rotating machinery. IEEE/ASME Trans. Mechatron. **27**(1), 24–33 (2021)
7. Wang, C., Lucey, S.: Paul: procrustean autoencoder for unsupervised lifting. In: Proceedings of the IEEE/CVF Conference on Computer Vision and Pattern Recognition, pp. 434–443 (2021)
8. Ngairangbam, V.S., Spannowsky, M., Takeuchi, M.: Anomaly detection in high-energy physics using a quantum autoencoder. Phys. Rev. D **105**(9), 095004 (2022)
9. Le, V.-T., Kim, Y.-G.: Attention-based residual autoencoder for video anomaly detection. Appl. Intell. **53**(3), 3240–3254 (2023)
10. Chen, X., et al.: Context autoencoder for self-supervised representation learning University of Trento, pp. 1–16. Trento Italy University of Amsterdam, Amsterdam The Netherlands (2023)

11. Yan, S., Shao, H., Xiao, Y., Liu, B., Wan, J.: Hybrid robust convolutional autoencoder for unsupervised anomaly detection of machine tools under noises. Robot. Comput.-Integr. Manufactur. **79**, 102441 (2023)
12. Fanai, H., Abbasimehr, H.: A novel combined approach based on deep autoencoder and deep classifiers for credit card fraud detection. In: College of Hydroelectric and Digitalization Engineering, Huazhong University of Science and Technology, Hubei Province 430074 Wuhan, China; Dispatch and Communication Center, Hunan Electric Power Company, Hunan Province C, vol. 217, p. 119562 (2023)
13. Liang, Y., Liang, W.: ResWCAE: Biometric Pattern Image Denoising Using Residual Wavelet-Conditioned Autoencoder arXiv preprint arXiv:2307.12255 (2023)
14. Kuzmanovic, M., Hatt, T., Feuerriegel, S.: Deconfounding temporal autoencoder: estimating treatment effects over time using noisy proxies. In: Presented at the Proceedings of Machine Learning for Health, Proceedings of Machine Learning Research (2021). https://proceedings.mlr.press/v158/kuzmanovic21a.html
15. Akcay, S., Atapour-Abarghouei, A., Breckon, T.P.: Ganomaly: Semi-supervised anomaly detection via adversarial training. In: Asian Conference on Computer Vision, pp. 622–637. Springer (2018)
16. Akçay, S., Atapour-Abarghouei, A., Breckon, T.P.: Skip-ganomaly: Skip connected and adversarially trained encoder-decoder anomaly detection. In: 2019 International Joint Conference on Neural Networks (IJCNN), pp. 1–8. IEEE (2019)
17. Latif, S., Rana, R., Khalifa, S., Jurdak, R., Epps, J., Schuller, B.W.: Multi-task semi-supervised adversarial autoencoding for speech emotion recognition. In: IEEE Transactions on Affective computing (2020)
18. Yuan, F., Yao, L., Benatallah, B.: Adversarial collaborative auto-encoder for top-n recommendation. In: 2019 International Joint Conference on Neural Networks (IJCNN), pp. 1–8. IEEE (2019
19. Zhong, G., Gao, W., Liu, Y., Yang, Y., Wang, D.-H., Huang, K.: Generative adversarial networks with decoder–encoder output noises. Neural Netw. **127**, 19–28 (2020)
20. Zhang, Z., Chen, Y., Wagner, D.: Seat: Similarity encoder by adversarial training for detecting model extraction attack queries. In: Proceedings of the 14th ACM Workshop on Artificial Intelligence and Security, pp. 37–48 (2021)
21. Blance, A., Spannowsky, M., Waite, P.: Adversarially-trained autoencoders for robust unsupervised new physics searches. J. High Energy Phys. **2019**(10), 1–19 (2019)
22. Bergmann, P., Fauser, M., Sattlegger, D., Steger, C.: MVTec AD--A comprehensive real-world dataset for unsupervised anomaly detection. In Proceedings of the IEEE/CVF Conference on Computer Vision and Pattern Recognition, pp. 9592–9600 (2019)
23. Kimura, D., Chaudhury, S., Narita, M., Munawar, A., Tachibana, R.: Adversarial discriminative attention for robust anomaly detection. In: Proceedings of the IEEE/CVF Winter Conference on Applications of Computer Vision, pp. 2172–2181 (2020)
24. Marafioti, A., Majdak, P., Holighaus, N., Perraudin, N.: GACELA: a generative adversarial context encoder for long audio inpainting of music. IEEE J. Select. Top. Sign. Process. **15**(1), 120–131 (2020)
25. Li, C., et al.: Fusing convolutional generative adversarial encoders for 3D printer fault detection with only normal condition signals. Mech. Syst. Signal Process. **147**, 107108 (2021)
26. Sahu, S., Gupta, R., Sivaraman, G., AbdAlmageed, W., Espy-Wilson, C.: Adversarial auto-encoders for speech based emotion recognition, arXiv preprint arXiv:1806.02146 (2018)
27. Chouchane, O., et al.: Differentially private adversarial auto-encoder to protect gender in voice biometrics. In: Proceedings of the 2023 ACM Workshop on Information Hiding and Multimedia Security, pp. 127–132 (2023)
28. Wang, Z., Bovik, A.C., Sheikh, H.R., Simoncelli, E.P.: Image quality assessment: from error visibility to structural similarity. IEEE Trans. Image Process. **13**(4), 600–612 (2004)

29. Goodfellow, I., Bengio, Y., Courville, A.: Deep Learning. MIT Press (2016)
30. LeCun, Y.: The MNIST database of handwritten digits. http://yann.lecun.com/exdb/mnist/ (1998)
31. Krizhevsky, A., Hinton, G.: Learning multiple layers of features from tiny images (2009)
32. Gong, D., et al.: Memorizing normality to detect anomaly: Memory-augmented deep autoencoder for unsupervised anomaly detection. In: Proceedings of the IEEE/CVF International Conference on Computer Vision, pp. 1705–1714 (2019)
33. Zong, B., et al.: Deep autoencoding gaussian mixture model for unsupervised anomaly detection. In: International Conference on Learning Representations (2018)
34. Zhai, S., Cheng, Y., Lu, W., Zhang, Z.: Deep structured energy based models for anomaly detection. In: presented at the Proceedings of the 33rd International Conference on Machine Learning, Proceedings of Machine Learning Research (2016). https://proceedings.mlr.press/v48/zhai16.html
35. Scholkopf, B., Williamson, R., Smola, A., Shawe-Taylor, J., Platt, J.: Support vector method for novelty detection. Adv. Neural. Inf. Process. Syst. **12**(3), 582–588 (2000)
36. Kingma, D.P., Welling, M.: Auto-encoding variational bayes, arXiv preprint arXiv:1312.6114 (2013)
37. Van den Oord, A., Kalchbrenner, N., Espeholt, L., Vinyals, O., Graves, A.: Conditional image generation with pixelcnn decoders. Adv. Neural Inf. Process. Syst. 4790–4798 (2016)

Matrix Contrastive Learning for Short Text Clustering

Zhengzhong Zhu, Jiankuo Li, Xuejie Zhang, Jin Wang, and Xiaobing Zhou[✉]

School of Information Science and Engineering, Yunnan University, Kunming 650500, Yunnan, China
zhouxb@ynu.edu.cn

Abstract. Recently, many studies have combined contrastive learning with clustering to address this issue and achieved excellent clustering results. However, traditional contrastive learning methods suffer from class conflict. We propose a new framework called Matrix Contrastive Learning (MCL) for text clustering to address this issue. Firstly, data augmentation techniques are utilized to generate pairs of positive and negative instances for all anchor examples. These pairs are mapped into a feature space, where the rows of the matrix represent soft labels for individual instances, and the columns represent cluster representations. We perform contrastive learning at both the instance and cluster levels using these rows and columns. To further improve the cluster allocation in unsupervised clustering tasks and alleviate the class conflict problem caused by instance-level contrastive learning in unsupervised conditions, the K-Nearest Neighbors algorithm is used to filter out negative instances. We conducted extensive experiments on eight challenging text datasets and compared MCL with six existing clustering methods. The results show that MCL significantly outperforms the competing methods. The code is available at https://github.com/2251821381/MCL.

Keywords: Short text clustering · Deep clustering · Contrastive learning

1 Introduction

With the rapid development of contrastive learning, unsupervised learning has experienced significant improvements in performance [3]. Instance-level contrastive learning combined with deep clustering has emerged as a basis for clustering methods and achieved the most advanced results. For instance, SCCL [28] applied instance-level contrastive learning to solve the problem of short text clustering and achieved good results by reducing the problem of frequent overlap between different categories. However, this method still has certain limitations that instance-level contrastive learning can result in the class conflict problem [24], which is from sampling a sufficient number of negative instances inevitably produces negative instance pairs with similar semantics that should be closer in

© The Author(s), under exclusive license to Springer Nature Singapore Pte Ltd. 2024
B. Luo et al. (Eds.): ICONIP 2023, CCIS 1961, pp. 541–554, 2024.
https://doi.org/10.1007/978-981-99-8126-7_42

the embedded space rather than being separated by the contrastive loss. This problem is referred to as the class conflict problem and has been shown to harm representation learning. In essence, instance-level learning preserves only local smoothness around each instance in the embedded space and largely ignores the global semantic structure of the dataset.

To address this issue, we propose a new Matrix Contrastive Learning (MCL) framework to learn instance and cluster representations, as shown in Fig. 1. MCL first constructs a feature matrix for data pairs using data augmentation. The rows of the feature matrix can be treated as soft labels, indicating the probability that each instance belongs to each cluster. Meanwhile, the columns reflect the distribution of clusters over the instances. Then, we perform instance-level and cluster-level contrastive learning by pulling together positive and pushing away negative pairs in the row and column spaces of the feature matrix. In the instance contrastive learning level, we use K-Nearest Neighbors contrastive learning [29] to filter out a large number of negative samples and improve the clustering effect. In the cluster contrastive learning level, we project the features of instances into a subspace with a dimension equal to the number of clusters and treat the labels as a special representation. In this sense, we can regard samples from the same cluster as positive, and samples from different clusters as negative, which implicitly alleviates the class conflict problem. Thus, MCL combines instance-level contrastive learning and cluster-level contrastive learning, which is more beneficial for performing clustering tasks. Our contributions are as follows.

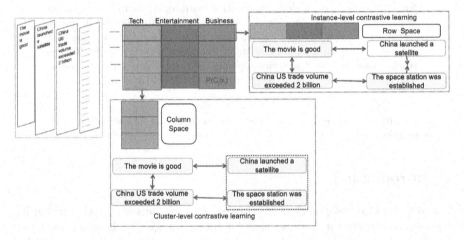

Fig. 1. By treating the rows of the feature matrix as instance-level soft labels ($P(c_j$ —x_i) represents the probability of sample i belonging to cluster j, the columns can be viewed as cluster-level representations distributed across the datasets. Consequently, instance- and cluster-level contrastive learning can be performed in the row and column spaces of the feature matrix, respectively.

- In the field of text clustering, we present a novel approach that maps instance and cluster representations to the rows and columns of a feature matrix, respectively, allowing for easy integration of deep clustering with contrastive learning.
- We effectively address the class conflict problem inherent in traditional instance-level contrastive learning by incorporating K-Nearest Neighbors contrastive learning and treating samples from different clusters as negative samples in our framework, making our approach significant for combining contrastive learning and deep clustering.
- We analyze and verify the effectiveness of MCL through extensive experiments. MCL performs best on most datasets Compared to the state-of-the-art short text clustering models.

2   Related Work

2.1   Contrastive Learning

Various contrastive learning methods have been developed to address a wide range of tasks [3, 5, 23], as contrastive learning has shown great promise in representation learning in latent spaces, including text representation [17], knowledge graph [14], and few-shot classification [25]. Specifically designed for cluster tasks, Prototype Contrastive [12] have made significant contributions to the field by generating different "views" of data, developing new contrastive loss functions and updating strategies. ProPos [8] introduces a new deep clustering method that combines the strengths of both contrastive and non-contrastive approaches to promote compactness within clusters and separability between clusters. However, these methods only focus on performing contrastive learning at the instance level and aim to learn a general-purpose representation that can be applied to various downstream tasks.

In contrast, MCL simultaneously incorporates contrastive learning at both instance and cluster levels, guided by the concept of "label as representation" [1]. Our method differs significantly from these contrastive learning methods in simultaneously performing both row and column-wise contrastive learning. This elegant idea is that the rows and columns of the feature matrix correspond to the instance and cluster representations, respectively.

2.2   Short Text Clustering

Short text clustering poses challenges compared to general text clustering problems because of the lack of contextual information and a limited number of words. Early work utilized neural networks to learn richer representations to address these issues, where word embeddings [16] were adopted to further improve performance. Meanwhile, contextualized word embeddings have achieved great success [22]. With the emergence of the BERT model using the Transformer architecture to capture contextual information, the representation of each word vector is not only related to the word itself, but also includes its contextual information in the entire sentence, which has been extensively explored for short text clustering owing to it better represent the meaning of each word.

Recently, several methods have been proposed to perform representation learning using contrastive learning and combine with clustering learning to improve clustering performance. For example, SCCL uses self-supervised contrastive clustering to perform representation learning and achieved clustering using high-confidence cluster assignments. IIC [9] based on the information bottleneck theory directly maximizes the mutual information between the original and augmented samples. PICA [4] learns the most semantically plausible data separation by maximizing the partition confidence of the clustering solution. SPICE [18] uses pseudo-labels generated by preliminary clustering, i.e., self-labeling, further to improve the clustering performance in a multi-stage manner. DACL [13] adopts a joint training method for clustering and auxiliary tasks, optimizes the contrastive loss function with pseudo-labels, and designs a new adjustment function to dynamically adjust the weights of the auxiliary task loss and clustering loss to the total loss during the training process.

Compared with most works focusing solely on contrastive learning, our MCL alleviates the impact of negative sample pairs, which helps the model learn more beneficial representations for clustering. Our method represents an improvement over existing contrastive learning for clustering tasks.

3   Methodology

Our network framework is mainly divided into three parts, the first is the pair construction backbone, the other part is the instance-level contrastive learning head with K-Nearest Neighbors and the cluster-level contrastive learning head, and the last part is the clustering head.

3.1   Pair Construction Backbone

The pair construction backbone consists of two parts. The first part is the construction of data pairs, where positive and negative pairs are built with the help of data augmentation. Specifically, given a data instance x_i, we apply data transformations T to generate two related samples(we choose Bertbase and Roberta to generate the augmented pairs), $x_i^a = T^a\ (x_i)$ and $x_i^b = T^b\ (x_i)$. The other part is the pre-trained model DistillBERT [21]. The proposed method is theoretically architecture-agnostic. We use DistillBERT as a shared neural network to extract features from the augmented samples $h_a^i = f(T^a), h_b^i = f(T^b)$ (Fig. 2).

3.2   Instance-Level Contrastive Learning Head with K-Nearest Neighbors

The proposed approach is based on K-Nearest Neighbors contrastive learning, a modified version of instance-level contrastive learning to alleviate the class

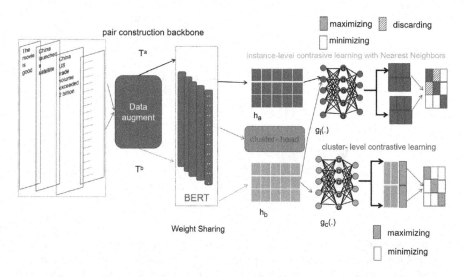

Fig. 2. The framework of matrix contrastive learning. Given data pairs, a shared deep neural network is used to extract features from augmented data pairs. Two separate MLP layers are used to project the features into row and column spaces for instance- and cluster-level contrastive learning, respectively. A single clustering head is used to determine which cluster an instance belongs to. At the instance level, the K-Nearest Neighbor (KNN) algorithm is used to discard some negative samples.

conflict problem. Instead of using negative examples sampled from the entire dataset, K-Nearest Neighbors contrastive learning only considers a small subset of examples closest to the query instance in the feature space. This reduces the chance of selecting negative examples that are semantically similar to the query instance and thus avoids creating class conflict in the training process. Specifically, given a query instance x_i, the KNN algorithm searches for the k-nearest neighbors to form a neighborhood N_i. Theoretically, the instances in N_i should have a similar semantic with x_i. During the training process, we sampled a small batch of instances $B = \{x_i\}_{i=1}^M$. For each instance $x_i \in B$, we uniformly sample a neighbor x_i^a from its neighborhood N_i and periodically update the neighborhood of each sample. Then, we use data augmentation to generate $\hat{x}_i$ and $\hat{x}_i^a$ for x_i and x_i^a, respectively. Here, we consider $\hat{x}_i$ and $\hat{x}_i^a$ as two views of x_i, forming a positive pair. Next, we will input them into the encoder to obtain feature vectors h_a^i and h_b^i and an augmented batch $B' = \{x_i\}_{i=1}^M$, for all generated samples. To compute the contrastive loss, we constructed an adjacency matrix M for B', which is a $2M \times 2M$ binary matrix where 1 represents a positive relation, and 0 represents a negative relation. To alleviate the information loss induced by contrastive loss, we do not directly conduct contrastive learning on the feature matrix. A two-layer nonlinear MLP_{g_I} to map the feature matrix to a subspace via $z_i^a = g_I\ (h_a^i\ )$, $z_i^b = g_I\ (h_b^i\ )$where the instance-level contrastive loss is applied. The pair-wise similarity is measured by cosine distance (Fig. 3). We can write the contrastive loss [10] as

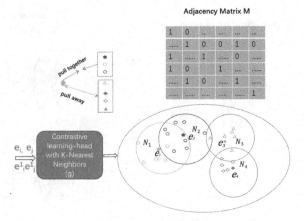

Fig. 3. Contrastive learning-head with K-Nearest Neighbors. For example, a batch of four training instances $=\{x_i\}_{i=1}^4$ (let $\{e_i\}_{i=1}^4$ be the sample feature) and their respective neighborhoods $=\{N_i\}_{i=1}^4$ are plotted (represented by hollow markers within large circles). Since x_2 falls within N_1, x_2 and its neighbors are positive instances for x_1 (but not vice versa since $x1$ is not in N_2). We also show an example of an adjacency matrix M

$$l_i = -\frac{1}{|\mathcal{C}_i|} \sum_{j \in \mathcal{C}_i} \log \frac{\exp\left(\text{sim}\left(\tilde{z}_i, \tilde{z}_j\right)/\tau_I\right)}{\sum_{k \neq i}^{2M} \exp\left(\text{sim}\left(\tilde{z}_i, \tilde{z}_k\right)/\tau_I\right)} \tag{1}$$

$$\mathcal{L}_{\text{ins}} = \frac{1}{2M} \sum_{i=1}^{2M} l_i \tag{2}$$

3.3 Cluster-Level Contrastive Learning Head

According to the concept of label representation, when mapping the features of a sample to a feature space whose dimensionality is equal to the number of clusters, the i-th dimension of the feature vector can be interpreted as the probability of belonging to the i-th cluster. Therefore, the feature vector can be regarded as a soft label for the sample. Especially, if $Y^a \in R^{N \cdot E}$ represents (and Y^b the second augmentation), Y^a represents the output of the augmented data, where $Y^a_{n,d}$ can be interpreted as the probability that sample n belongs to cluster d, with N being the batch size and E being the number of clusters. Since each sample belongs to only one cluster, the feature vectors of Y^a ideally tend towards a one-hot vector. Therefore, cluster-level contrastive learning can pull samples from the same cluster closer together and push samples from different clusters further apart, which alleviates the class conflict issue to some extent. We use another two-layer MLP_{g_C} to map the feature matrix to a D-dimensional feature space. Here, $\hat{y}_i^a$ represents the soft label of sample x_i^a (the i-th row of Y^a), $\hat{y}_i^b$ is the representation obtained from the second type of data augmentation for cluster i. Then, we construct positive cluster example pairs $\hat{y}_i^a, \hat{y}_i^b$ and leave all other $2E$-2 cluster pairs as negatives. Similarly, we use cosine distance to measure the similarity between cluster pairs,

$$s\left(\hat{y}_i^{k_1}, \hat{y}_j^{k_2}\right) = \frac{\left(\hat{y}_i^{k_1}\right)^{\top}\left(\hat{y}_j^{k_2}\right)}{\left\|\hat{y}_i^{k_1}\right\|\left\|\hat{y}_j^{k_2}\right\|} \tag{3}$$

where $k_1, k_2 \in \{a, b\}$ and $i, j \in [1, E]$. To differentiate between cluster $\hat{y}_i^a$ and all other clusters except $\hat{y}_i^b$, we use the following loss function,

$$\hat{\ell}_i^a = -\log \frac{\exp\left(s\left(\hat{y}_i^a, \hat{y}_i^b\right)/\tau_C\right)}{\sum_{j=1}^{E}\left[\exp\left(s\left(\hat{y}_i^a, \hat{y}_j^a\right)/\tau_C\right) + \exp\left(s\left(\hat{y}_i^a, \hat{y}_j^b\right)/\tau_C\right)\right]} \tag{4}$$

where τ_C is the cluster-level temperature parameter controlling the softness of clustering. By iterating through all clusters, we calculate the entropy of the cluster assignment probabilities of $\hat{y}_i^a$ and $\hat{y}_i^b$, which helps to avoid the trivial solution where most instances are assigned to the same cluster [7], in each mini-batch under each data augmentation. The final cluster-level contrastive loss is calculated,

$$\mathcal{L}_{clu} = \frac{1}{2E}\sum_{i=1}^{E}\left(\hat{\ell}_i^a + \hat{\ell}_i^b\right) - H(Y) \tag{5}$$

in which the entropy of the cluster assignment probability needs to be calculated by

$$H(Y) = -\sum_{i=1}^{E}\left[P\left(\hat{y}_i^a\right)\log P\left(\hat{y}_i^a\right) + P\left(\hat{y}_i^b\right)\log P\left(\hat{y}_i^b\right)\right] \tag{6}$$

$$P\left(\hat{y}_i^k\right) = \frac{1}{N}\sum_{t=1}^{N}Y_{ti}^k \tag{7}$$

3.4   Clustering Head

Assuming there are K clusters, we first use the K-means [2] algorithm to calculate the cluster centers, denoted as $u_k, k \in \{1, 2, 3..., k\}$. Let e_j be the sample feature of sample x_i. According to SCCL, we first apply the Student's t-distribution [15] to calculate the assignment distribution q_{jk}, which denotes assigning each sample representation e_j to the k-th cluster.

$$q_{jk} = \frac{\left(1 + \|e_j - \mu_k\|_2^2/\alpha\right)^{-\frac{\alpha+1}{2}}}{\sum_{k'=1}^{K}\left(1 + \|e_j - \mu_{k'}\|_2^2/\alpha\right)^{-\frac{\alpha+1}{2}}} \tag{8}$$

where α represents the degrees of freedom of the t-distribution. Specifically, let p_{jk} denote the auxiliary probability defined as here $f_k = \sum_{j=1}^{m} q_{jk}$ can be interpreted as a soft cluster frequency approximation within a small batch. To approximate the centroids of each cluster, we utilize a linear layer as the clustering head and update the centroids iteratively by leveraging a target distribution p_{jk} [26].

$$p_{jk} = \frac{q_{jk}^2/f_k}{\sum_{k'} q_{jk}^2/f_{k'}} \tag{9}$$

To encourage the clustering assignment probabilities to converge towards the target distribution, we optimize their KL divergence.

$$\ell_j^C = \text{KL}\,[p_j\|q_j] = \sum_{k=1}^{K} p_{jk} \log \frac{p_{jk}}{q_{jk}} \tag{10}$$

Then, the clustering loss is as follows.

$$\mathcal{L}_{\text{Cluster}} = \sum_{j=1}^{M} \ell_j^C/M \tag{11}$$

4 Experiment

4.1 Dataset

We assess the performance of the proposed MCL model on eight benchmark datasets for short text clustering, i.e., SearchSnippets [27], StackOverflow [27], Biomedical [19], AgNews [19], Biomedical [20], Tweet [20], GoogleNews [19]. Among them, GoogleNews contains GoogleNews-TS, GoogleNews-T and GoogleNews-S.

4.2 Baseline Models

We demonstrate that our model can outperform or achieve highly competitive results on clustering short texts. To compare our results, we consider the following baseline methods.

- STCC [27]: This method has three independent stages. It first pretrains a Word2Vec word embedding on a large in-domain corpus. Then, a convolutional neural network is optimized to enrich the representations that are fed into K-means for the final clustering stage.
- Self-Train [6]: This method enhances the pretrained word embeddings. It adopts an autoencoder obtained by layer-wise pretraining, which is further fine-tuned with a clustering objective and updates the target distribution through carefully chosen intervals that vary across datasets. In contrast, our method updates it per iteration and achieves significant improvement.

- SimCTC [11] proposes a simple contrast learning method of text clustering to improve clustering performance.
- HAC-SD [20]: This method applies hierarchical agglomerative clustering on top of a sparse pairwise similarity matrix obtained by zeroing out similarity scores lower than a chosen threshold value.
- BoW TF-IDF: This method applies K-means on top of the associated features with a dimension of 1500.
- SCCL [28]: This method proposes a novel framework called Supporting Clustering with Contrastive Learning (SCCL) to leverage contrastive learning to promote better separation.
- DACL [13]: This method improves SCCL from the perspective of pseudo-label.

4.3   Main Result

We report the comparison results in Table 1. Our MCL model performs significantly better on most datasets.

Visualization. To demonstrate qualitatively how the proposed MCL has good clustering performance, we use t-SNE [15] to visualize the embedding of SCCL and our proposed MCL on the test set. It can be seen that although the distributions of representations from SCCL have achieved good separation, there is still overlap between clusters. On the other hand, there is a clear difference between the representations produced by our proposed MCL, with less overlap (Fig. 4 and 5).

Effect of Two Contrastive Head. To validate the effectiveness of the two contrastive learning heads, we conduct ablation studies. Our approach involves removing the cluster-level contrastive learning(cch) head from the model, leaving only the K-Nearest-Neighbor contrastive learning head(Kch). After verifying the effectiveness of the nearest-neighbor contrastive learning head, we add the cluster-level contrastive learning head back into the model to evaluate its promotion effect.

As shown in Table 2, when two clustering heads are combined, the model performs better, which indicates the joint effect of the two contrastive learning heads to some extent.

4.4   Ablation Experiment

K **Value Analysis of Contrastive Learning-Head with K-Nearest Neighbors.** Different K values affect the performance of the K-Nearest Neighbor contrastive learning head. The horizontal axis represents the K values, while the vertical axis represents the ACC or NMI. It is worth noting that when the number of nearest neighbors is 0, we simply double the same instances as in traditional contrastive learning. MCL still performs very well, demonstrating the

Table 1. The clustering performances on eight object text benchmarks.'-' represents that there are no experimental results on this dataset

	AgNews		SearchSnippets		StackOverflow		Biomedical	
	ACC	NMI	ACC	NMI	ACC	NMI	ACC	NMI
BoW	27.6	2.6	24.3	9.3	18.5	14	14.3	9.2
TF-IDF	34.5	11.9	31.5	19.2	58.4	58.7	28.3	23.2
STCC	-	-	77	63.2	51.1	49	43.6	38.1
Self-Train	-	-	77.1	56.7	59.8	54.8	**54.8**	**47.1**
SimCTC	-	-	85.4	71.1	78.3	75.4	-	-
HAC-SD	81.8	54.6	82.7	63.8	64.8	59.5	40.1	33.5
SCCL	88.2	68.2	85.2	71.1	75.5	74.5	46.2	41.5
DACL	88.6	69.0	86.1	73.6	77.5	76.0	48.6	40.3
MCL	**89.9**	**70.2**	**88.4**	**83.6**	**80.8**	**78.6**	49.2	42.5

	GoogleNews-TS		GoogleNews-T		GoogleNews-S		Tweet	
	ACC	NMI	ACC	NMI	ACC	NMI	ACC	NMI
BoW	57.5	81.9	49.8	73.2	49.0	73.5	49.7	73.6
TF-IDF	68	88.9	58.9	79.3	61.9	83	57	80.7
STCC	-	-	-	-	-	-	-	-
Self-Train	-	-	-	-	-	-	-	-
SimCTC	**90.9**	96.1	-	-	-	-	84.7	**93.4**
HAC-SD	85.8	88	81.8	84.2	80.6	83.5	**89.6**	85.2
SCCL	89.8	94.9	75.8	88.3	83.1	90.4	78.2	89.2
DACL	89.6	94.4	80.5	**89.8**	**84.0**	91.7	82.0	90.6
MCL	90.8	**96.2**	**81.4**	89.5	83.2	**92.2**	84.8	92.4

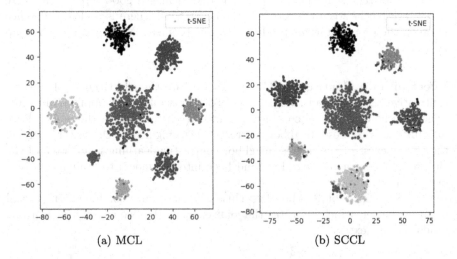

(a) MCL   (b) SCCL

Fig. 4. Embedding space in testing sets of SearchSnippets.

effectiveness of the cluster-level contrastive learning head of MCL. Besides, as the value of K increases, we can observe a phenomenon where both ACC and NMI increase initially and then decrease. This pattern can be easily explained: when the value of K exceeds a certain threshold, it may cause over-sampling in

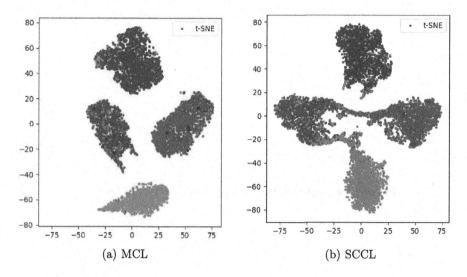

(a) MCL   (b) SCCL

Fig. 5. Embedding space in testing sets of AgNews.

the KNN algorithm. Oversampling can adversely affect feature representation, leading to a decrease in model performance (Fig. 6 and 7).

Table 2. Effectiveness of two contrastive heads.cch(cluster-level contrastive learning head), Kch(K-Nearest-Neighbor contrastive learning head)

Dataset	Contrastive head	ACC	NMI
AgNews	cch+Kch	89.9	70.2
	Kch	88.2	69.4
SearchSnippets	cch+Kch	88.4	83.6
	Kch	87.8	83.2
StackOverflow	cch+Kch	80.8	78.6
	Kch	80.7	77.8
Biomedical	cch+Kch	49.2	42.5
	Kch	48.8	41.4
GoogleNews-TS	cch+Kch	90.8	96.2
	Kch	89.2	92.5
GoogleNews-T	cch+Kch	81.4	89.5
	Kch	79.6	87.8
GoogleNews-S	cch+Kch	83.2	92.2
	Kch	82.4	91.8
Tweet	cch+Kch	84.8	92.4
	Kch	81.4	90.2

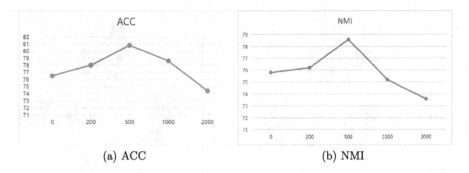

(a) ACC   (b) NMI

Fig. 6. The number of the Nearest Neighbors on Stackoverflow.

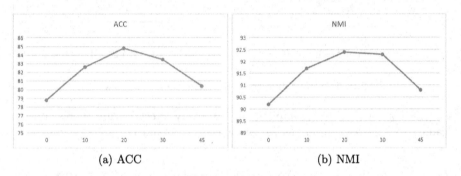

(a) ACC   (b) NMI

Fig. 7. The number of the Nearest Neighbors on Tweety.

5   Conclusion

We propose a feature matrix-based learning method, Matrix Contrastive Learning (MCL), which represents rows and columns as instances and clusters, and performs contrastive learning at both the instance and cluster levels in a unified framework and shows good clustering performance. Our future plans involve expanding the application of MCL to other tasks, such as semi-supervised learning and clustering of images.

Acknowledgement. This work was supported by the Natural Science Foundation of China under Grants No. 61966038 and No. 62266051.

References

1. Abati, D., Tomczak, J., Blankevoort, T., et al.: Conditional channel gated networks for task-aware continual learning. In: Proceedings of the IEEE/CVF Conference on Computer Vision and Pattern Recognition, pp. 3931–3940 (2020)
2. Boyd, D.M., Ellison, N.B.: Social network sites: definition, history, and scholarship. J. Comput.-Mediat. Commun. **13**(1), 210–230 (2007)

3. Chen, T., Kornblith, S., Norouzi, M., et al.: A simple framework for contrastive learning of visual representations. In: International Conference on Machine Learning, pp. 1597–1607. PMLR (2020)
4. Dang, Z., Deng, C., Yang, X., et al.: Doubly contrastive deep clustering. arXiv preprint arXiv:2103.05484 (2021)
5. Gao, T., Yao, X., Chen, D.: SimCSE: simple contrastive learning of sentence embeddings. In: 2021 Conference on Empirical Methods in Natural Language Processing, EMNLP 2021, pp. 6894–6910. Association for Computational Linguistics (ACL) (2021)
6. Hadifar, A., Sterckx, L., Demeester, T., et al.: A self-training approach for short text clustering. In: Proceedings of the 4th Workshop on Representation Learning for NLP (RepL4NLP-2019), pp. 194–199 (2019)
7. Hu, W., Miyato, T., Tokui, S., et al.: Learning discrete representations via information maximizing self-augmented training. In: International Conference on Machine Learning, pp. 1558–1567. PMLR (2017)
8. Huang, Z., Chen, J., Zhang, J., et al.: Learning representation for clustering via prototype scattering and positive sampling. IEEE Trans. Pattern Anal. Mach. Intell. **45**(6), 7509–7524 (2022)
9. Ji, X., Henriques, J.F., Vedaldi, A.: Invariant information clustering for unsupervised image classification and segmentation. In: Proceedings of the IEEE/CVF International Conference on Computer Vision, pp. 9865–9874 (2019)
10. Khosla, P., Teterwak, P., Wang, C., et al.: Supervised contrastive learning. Adv. Neural. Inf. Process. Syst. **33**, 18661–18673 (2020)
11. Li, C., Yu, X., Song, S., et al.: Simctc: A simple contrast learning method of text clustering (student abstract). In: Proceedings of the AAAI Conference on Artificial Intelligence, vol. 36, pp. 12997–12998 (2022)
12. Li, J., Zhou, P., Xiong, C., Hoi, S.: Prototypical contrastive learning of unsupervised representations. In: International Conference on Learning Representations (2020)
13. Li, R., Wang, H.: Clustering of short texts based on dynamic adjustment for contrastive learning. IEEE Access **10**, 76069–76078 (2022)
14. Ma, Y., Zhang, X., Gao, C., et al.: Enhancing recommendations with contrastive learning from collaborative knowledge graph. Neurocomputing **523**, 103–115 (2023)
15. Van der Maaten, L., Hinton, G.: Visualizing data using t-SNE. J. Mach. Learn. Res. **9**, 2579–2605 (2008)
16. Mikolov, T., Chen, K., Corrado, G.S., et al.: Efficient estimation of word representations in vector space (2013)
17. Moukafih, Y., Sbihi, N., Ghogho, M., et al.: SuperConText: supervised contrastive learning framework for textual representations. IEEE Access **11**, 16820–16830 (2023)
18. Niu, C., Shan, H., Wang, G.: Spice: Semantic pseudo-labeling for image clustering. IEEE Trans. Image Process. **31**, 7264–7278 (2022)
19. Phan, X.H., Nguyen, L.M., Horiguchi, S.: Learning to classify short and sparse text & web with hidden topics from large-scale data collections. In: Proceedings of the 17th International Conference on World Wide Web, pp. 91–100 (2008)
20. Rakib, M.R.H., Zeh, N., Jankowska, M., Milios, E.: Enhancement of short text clustering by iterative classification. In: Métais, E., Meziane, F., Horacek, H., Cimiano, P. (eds.) NLDB 2020. LNCS, vol. 12089, pp. 105–117. Springer, Cham (2020). https://doi.org/10.1007/978-3-030-51310-8_10

21. Sanh, V., Debut, L., Chaumond, J., et al.: Distilbert, a distilled version of bert: smaller, faster, cheaper and lighter (2019)
22. Sarzynska-Wawer, J., Wawer, A., Pawlak, A., et al.: Detecting formal thought disorder by deep contextualized word representations. Psychiatry Res. **304**, 114135 (2021)
23. Sun, K., Yao, T., Chen, S., et al.: Dual contrastive learning for general face forgery detection. In: Proceedings of the AAAI Conference on Artificial Intelligence, vol. 36, pp. 2316–2324 (2022)
24. Tejankar, A., Koohpayegani, S.A., Pillai, V., et al.: ISD: self-supervised learning by iterative similarity distillation. In: Proceedings of the IEEE/CVF International Conference on Computer Vision, pp. 9609–9618 (2021)
25. Tian, R., Shi, H.: Momentum memory contrastive learning for transfer-based few-shot classification. Appl. Intell. **53**(1), 864–878 (2023)
26. Xie, J., Girshick, R., Farhadi, A.: Unsupervised deep embedding for clustering analysis. In: International Conference on Machine Learning, pp. 478–487. PMLR (2016)
27. Xu, J., Xu, B., Wang, P., et al.: Self-taught convolutional neural networks for short text clustering. Neural Netw. **88**, 22–31 (2017)
28. Zhang, D., Nan, F., Wei, X., et al.: Supporting clustering with contrastive learning. In: Proceedings of the 2021 Conference of the North American Chapter of the Association for Computational Linguistics: Human Language Technologies, pp. 5419–5430 (2021)
29. Zhang, Y., Zhang, H., Zhan, L.M., Wu, X.M., Lam, A.: New intent discovery with pre-training and contrastive learning. In: Proceedings of the 60th Annual Meeting of the Association for Computational Linguistics (Volume 1: Long Papers), pp. 256–269 (2022)

Sharpness-Aware Minimization for Out-of-Distribution Generalization

Dongqi Li[1], Zhu Teng[1]([⊠]), Qirui Li[2], and Ziyin Wang[2]

[1] Beijing Jiaotong University, Beijing, China
{dongqili,zteng}@bjtu.edu.cn
[2] AFCtech, Kunshan, China
{lqr,wzy}@afctech.com.cn

Abstract. Machine learning models often suffer from a significant decline in performance when they encounter out-of-distribution (OOD) data that differs from the training distribution. The distribution shift can be broadly categorized into diversity shift and correlation shift. While seeking a flat minima in optimization has been shown to improve a neural network's generalization performance with the assumption of independent and identical distribution (IID), it also has been shown to be an effective strategy for improving OOD generalization. However, previous studies potentially focused on addressing diversity shift, leaving the relationship between flat minima and correlation shift unresolved. To address the issue, we propose Sharpness-aware Invariant Risk Minimization (SIRM) as a novel approach to enhance generalization under correlation shift. Our method combines two parts: (1) Invariant risk minimization (IRM), which learns invariant relationships across multiple training environments, and (2) Sharpness-aware minimization (SAM), which finds a flat minima. Our analysis reveals that IRM does not guarantee flat minima and SAM does not improve the generalization in OOD. Moreover, we also analyze the relationship between flat minima and OOD data under correlation shift. Through extensive experiments conducted on image classification datasets, we demonstrate that our proposed method outperforms other methods with a competitive margin.

Keywords: Sharpness-aware minimization · Flat minima · Out-of-distribution generalization · Invariant risk minimization

1 Introduction

In conventional machine learning, training data is assumed to be independently and identically distributed (IID) as the test data. However, this assumption can hardly be satisfied when the test distribution deviates from the training distribution in real cases. Machine learning models may suffer from a sharp drop under distribution shift, which is systematically discussed as the out-of-distribution (OOD) generalization problem. Distribution shift generally can be decomposed into *diversity shift* (a.k.a covariate shift) and *correlation shift* (a.k.a. concept

© The Author(s), under exclusive license to Springer Nature Singapore Pte Ltd. 2024
B. Luo et al. (Eds.): ICONIP 2023, CCIS 1961, pp. 555–567, 2024.
https://doi.org/10.1007/978-981-99-8126-7_43

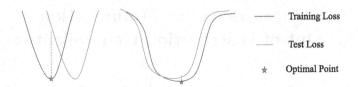

Fig. 1. An illustrative example of training and test loss landscape when achieving a sharp minima (left) and flat minima (right). The sharp optimal point in training loss is not equally optimal in test loss.

shift) [31]. Seeking a flat minima, which finds neural network's parameters in low-loss neighborhoods, has been intuitively assumed to be related to generalization [10,11] in IID cases and shown its performance as training strategies over gradient-based optimizers [14]. In terms of out-of-distribution (OOD) generalization, there have been claims [25,26] that employing optimization strategies to obtain a local flat minima can lead to remarkable performance on OOD datasets. However, these methods primarily focus on addressing diversity shift and are not equally effective for correlation shift. The relationship between flatness and correlation shift remains unresolved.

A notable line of research on the correlation shift problem is learning features with invariant conditioned label distribution across training environments[1] [1], which has been termed as invariant risk minimization (IRM). In contrast to empirical risk minimization (ERM [30]), IRM aims to learn an invariant correlation across multiple training environments, which expects to elicit an invariant predictor. However, our analysis reveals that IRM does not guarantee flatness. And Sharpness-aware minimization (SAM) [8] directly finds flat minima by minimizing the maximum of the loss neighborhood. However, SAM lacks the ability to generalize the OOD data under correlation shift. By examining the flatness in the test distribution and evaluating the generalization bound, our analysis provides insights into the impact of flatness on OOD generalization. These results demonstrate the efficacy of our proposed method, SIRM, in addressing the challenges posed by correlation shift and improving generalization performance. In summary, our contributions are summarized in three-fold:

- We find and analyze the relation between OOD data under correlation shift and seeking flat minima.
- Based on our analysis, we propose a method called Sharpness-aware Invariant Risk Minimization (SIRM) for improving the OOD generalization under correlation shift.
- We conduct comprehensive experiments, including image classification datasets, such as Colored-MNIST, CelebA, and Colored-Object. Experimental results demonstrate that the proposed method outperforms existing methods by a competitive margin.

---

[1] The meanings of environment, group, and domain are all referred to the hierarchy of a dataset, which can be seen as the same definition in this paper.

2   Related Works

2.1   Invariant Risk Learning

IRM [1] proposes a training objective for learning the invariant representations, under the assumption that the Bayes optimal conditional probability remains invariant across environments with overlap. Some IRM variants are proposed like the variance of penalization or loss gradients across training environments [4, 19,24]. However, recent works [16,27] revealed the theoretical pitfall of IRM in the non-linear setting and some other scenarios. Practical works also investigate that the performance of IRM relies on model size [20] and dataset type [9], which means IRM can suffer from overfitting.

2.2   Out-of-Distribution Generalization

Out-of-distribution generalization is a task of generalizing under distribution shift between training and test distribution. Besides IRM, another important line for generalizing data under distribution shift is distributionally robust optimization (DRO) [7,12]. DRO methods propose to optimize the worst-case error over a set of distributions that are required to cover the test distribution. However, DRO methods can encounter the over-pessimism problem, where the training distribution set is often overwhelmingly large, making the learned model make decisions with fairly low confidence. And a notable method called group DRO [28] optimizes the worst-case error by sharing importance weights across training examples and can provide a reasonable region for robust optimization.

2.3   Seeking Flat Minima

Despite the excellent ability to generalize unseen data with neural networks, optimization to low training error does not always mean a better generalization [17]. Intuitively, it is stated that [2,10,11,23] seeking a flat minima is better for generalization compared to a sharp minima, as it can make the deep neural networks (DNNs) models resilient to data noise and model parameter perturbation. Weight averaging (WA) [3,14,25] approach finds flat minima by updating the model parameters for a weight ensemble, which shows its spectacular results under diversity shift. However, it is still not clear if the WA strategy can work under correlation shift. Sharpness-Aware Minimization [8,18] directly seeks flat minima by minimizing the maximum loss around a neighborhood. It is stated [13,26] that SAM can lead to gains in performance for OOD problems. Similarly, they reserve judgment on the relation between correlation shift and flatness.

3   Method

In this section, we first formalize the problem definition and describe the notations used in our paper. Then we briefly introduce the invariant risk minimization [1] and sharpness-aware minimization [8]. At last, we focus on our proposed method Sharpness-aware IRM (SIRM) to handle the OOD data under correlation shift.

3.1 Preliminaries

Throughout the paper, We denote X and Y as random variables and x, y as corresponding samples and labels. We assume that there is a set of multiple environments, $\mathcal{E}_{supp}$, where the data can be extracted from. We can access to a collection of training environments, $\mathcal{E}_{tr} \subset \mathcal{E}_{supp}$ and each training environment denotes $e \in \mathcal{E}_{tr}$. And we indicate the environment index e with the variable X^e and Y^e. Let $\mathcal{X}$ and $\mathcal{Y}$ be the space of X and Y. Our goal is to learn a function $f : \mathcal{X} \rightarrow \mathcal{Y}$, which predicts Y conditioned on X. For DNNs, f can consist of a classifier and a feature extractor parameterized with w and Φ, denoted as $\theta := (w \cdot \Phi)$.

To define the out-of-distribution generalization problem under correlation shift, we consider the feature extracted by feature extractor Φ. We assume that the feature $\Phi(X)$ mainly contains some features that are spurious to labels (like backgrounds in images) called *spurious features*, and other features are truly correlated to labels (like object shapes) called *invariant features*. Our goal is to find a model whose representation $\Phi(X)$ is focused on invariant features and discards spurious features. In this way, we can define the goal of the out-of-distribution problem, learning a robust and invariant model that minimizes the objective over all the considered environments by Eq. 1.

$$\sup_{e \in \mathcal{E}_{supp}} \mathbb{E}_{X^e, Y^e} [l(f(X^e), Y^e)] \tag{1}$$

3.2 The IRM Framework and IRMv1

To handle Eq. 1, IRM [1] Framework aims to solve it by a bi-level optimization. Since the bi-level problem is challenging, a practical variant of IRM called IRMv1 [1] can be formulated by Eq. 2.

$$\min_{\Phi: \mathcal{X} \rightarrow \mathcal{Y}} \sum_{e \in \mathcal{E}_{tr}} R^e(\Phi) + \lambda \cdot \left\| \nabla_{w|w=1.0} R^e(w \cdot \Phi) \right\|^2 \tag{2}$$

where the per-environment risk is formulated as $R^e = \mathbb{E}_{X^e, Y^e} [l(f(X^e), Y^e)]$. In this formulation, additional environment labels are required to regularize the training process and learn invariant predictors. λ is a constant for adjusting the two terms of the formulation. The classifier is "dummy" when $w = 1.0$, which is assumed as a linear model. We continue to make this assumption in the following discussion. Besides IRMv1, there are some other IRM variants recently [5,19,20,34]. For space constraints, we consider IRMv1 as a baseline for the following discussion.

3.3 Sharpness-Aware Minimization

Formally, we discuss the flatness in DNNs mainly about the largest loss change when model weight θ is perturbed with ϵ, then the objective of Sharpness-aware Minimization [8] is given by Eq. 3.

$$\min_{\theta} \max_{\|\epsilon\| \leq \rho} R(\theta + \epsilon) \tag{3}$$

where R is the risk to be minimized and $\rho \geq 0$ is the hyperparameter that controls the maximum norm of ϵ. For obtaining the solution of inner maximization objective is non-trivial, SAM maximizes the first order Taylor expansion approximation to find the worst-case perturbation ϵ^* in Eq. 4.

$$\epsilon^*(\theta) \approx \arg\min_{\|\epsilon\| \leq \rho} R(\theta) + \epsilon^T \nabla_\theta R(\theta) = \rho \nabla_\theta R(\theta) / \|\nabla_\theta R(\theta)\|_2 \tag{4}$$

The current model parameters θ are updated by $\theta + \epsilon^*$, where ϵ^* is obtained by the first backward pass in model training. Then the second backward pass applies Eq. 3 to update the model parameters for seeking a flat minima.

3.4 Sharpness-Aware IRM

Since DNNs have an inductive bias in using spurious features for predicting [6,29] under correlation shift, we can not directly employ the strategy of seeking flatness to ERM models, because we suppose that flatness alone can not avoid it (See more details in Sect. 4.2). To handle the problem of seeking flat minima under correlation shift, we employ IRM for obtaining the invariant features. Then we apply SAM to seek a flat minima, as shown in Eq. 5.

$$\min_{\Phi:\mathcal{X} \to \mathcal{Y}} \sum_{e \in \mathcal{E}_{tr}} R^e(\Phi_{\theta+\epsilon^*}) + \lambda \cdot \left\| \nabla_{w|w=1.0} R^e(w \cdot \Phi_{\theta+\epsilon^*}) \right\|^2 \tag{5}$$

where Φ is the parameters of the feature extractor and ϵ^* is obtained by Eq. 4. We estimate ϵ^* under ERM risk instead of IRMv1 risk in practice. We treat IRMv1 penalty and parameter perturbation as two types of regularizers and balance them with their weight. This helps us find a flat minima under correlation shift and improve generalization under OOD data. The pseudo-code of our method can be shown in Algorithm 1.

4   Relation Between Flatness and OOD Generalization

In this section, we focus on the relationship between flatness and OOD generalization with both theoretical and experimental analysis, especially under correlation shift.

Algorithm 1. Sharpness-aware IRM (exampled with IRMv1)

Input: Training dataset $S = \{(x_i, y_i, e_i)\}^N$, Loss function l, Neighborhood size ρ, Learning rate η, Step size T.

1: Initialize weights Φ_0
2: **for** $t \in 1, ..., T$ **do**
3:    Compute training loss $l_S(\Phi_t)$
4:    Compute $\epsilon_t^*(\Phi_t)$ via Eq. 4
5:    Compute perturbed weights: $\Phi_t^* = \Phi_t + \epsilon_t^*(\Phi_t)$
6:    Compute gradient approximation g for SIRM loss via Eq. 5
7:    Return to Φ_t from Φ_t^*
8:    Update weights: $\Phi_{t+1} = \Phi_t - \eta g$
9: **end for**

4.1   Experimental Analysis

Flatness is a property of model parameter (θ) space and can be analyzed by the Hessian of risk $H = \nabla_\theta^2 R(\theta)$. However, the computation of H is expensive, and thus we can use the maximum eigenvalue of Hessian (λ_{max}) and the trace $Tr(H)$ as a proxy [26] to measure flatness. We assume that low λ_{max} and $Tr(H)$ lead to a flatter loss landscape.

Table 1. The largest Hessian eigenvalue λ_{max} evaluation for different environments on the Colored-MNIST and CelebA dataset.

	Colored-MNIST				CelebA			
	$\mathcal{E}_{tr_1}$	$\mathcal{E}_{tr_2}$	$\mathcal{E}_{te}$	Avg.	$\mathcal{E}_{tr_1}$	$\mathcal{E}_{tr_2}$	$\mathcal{E}_{te}$	Avg.
ERM	70	49	98	72	388	368	467	407
SAM	5	4	8	5	84	98	133	105
IRMv1	11	11	10	10	207	188	193	196
SIRM	5	5	5	**5**	80	81	79	**80**

In Table 1, we have two training environments ($\mathcal{E}_{tr_1}$, $\mathcal{E}_{tr_2}$) and one test environment ($\mathcal{E}_{te}$) on both Colored-MNIST and CelebA datasets (The details of datasets can be seen in Sect. 5.1). We train our DNNs with four methods. It shows that ERM, which can be seen as the vanilla algorithm, has the largest λ_{max} compared to other methods, which means ERM can not obtain flat minima under correlation shift. And IRMv1, which is not employed for finding flat minima, has surprisingly reached lower λ_{max} than ERM. It motivates us to find the relation between IRMv1 and flat minima. The table also shows that IRMv1 can not achieve a similar flat minima level to SAM. Thus, we conjecture that IRMv1 just reaches a sub-optimal flat minima compared to SAM.

We observe that SAM achieves flat minima in training environments but not in the test environment ($\lambda_{max}(\mathcal{E}_{tr_1}) \approx \lambda_{max}(\mathcal{E}_{tr_2}) < \lambda_{max}(\mathcal{E}_{te})$), while IRMv1 has similar but sub-optimal λ_{max} across all environments. We hypothesize that flatness in training environments does not imply flatness in test environments under correlation shift. Motivated by this, we propose our SIRM method to avoid spurious features and seek flat minima for better OOD generalization. Table 1 shows that our method achieves flat mima compared to other methods in the test distribution in both datasets.

4.2 Theoretical Analysis

Let $\hat{R}_{tr}(\theta) := \frac{1}{En} \sum_{e=1}^{E} \sum_{i=1}^{n} l(f(x_i^e), y_i^e)$ be the empirical risk for training environments, where E is the number of training environments and n is the number of samples for each environment. $R_{te}(\theta)$ is denoted as test environemnt risk. Following the definition in [3], we consider the empirical risk based on flatness by the worst empirical risk within neighborhoods in the parameter space as $\hat{R}_{tr}^{\rho}(\theta) := \max_{\|\epsilon\| \leq \rho} \hat{R}_{tr}(\theta + \epsilon)$. Therefore, minimizing $\hat{R}_{tr}^{\rho}(\theta)$ is equal to solving the problem of Eq. 3.

As shown in Fig. 1, which may hold in IID cases, we generally hope to find a flat minima because flat optimal points in the training loss may be equally optimal in the test loss. However, the situation can be broken under distribution shift, since we can not obtain similar loss landscapes between training loss and test loss in OOD cases. Despite the issue, we can still establish the relationship between flatness and OOD generalization theoretically. We follow the proof of [3] and show the relation between $\hat{R}_{tr}^{\rho}(\theta)$ and $R_{te}(\theta)$ with PAC Bayesian generalization bound [22] in OOD generalization.

Theorem 1. *Consider a set of N covers $\{\Theta_k\}_{k=1}^{N}$ such that the parameter space $\Theta \subset \cup_k^N \Theta_k$ where $diag(\Theta) := sup_{\theta,\theta' \in \Theta} \|\theta - \theta'\|$, $N := \lceil (diam(\Theta)/\rho)^d \rceil$ and d is the dimension of Θ. Let v_k be a VC dimension of each Θ_k. Then, for any $\theta \in \Theta$, the following bound holds with probability at least $1 - \delta$.*

$$R_{te}(\theta) < \hat{R}_{tr}^{\rho}(\theta) + \frac{1}{E} \sum_{e=1}^{E} Div(p_{tr_e}, p_{te}) + \max_{k \in [1,N]} \sqrt{\frac{v_k ln(m/v_k) + ln(N/\delta)}{m}} \quad (6)$$

where $Div(p_{tr_e}, p_{te})$ is the divergence of distribution for the test environment and each training environment, and $m = nE$ is the number of training samples.

From Theorem 1, we can observe that the risk of the test environment is bounded by three terms: (1) the worst empirical risk of training environments $\hat{R}_{tr}^{\rho}(\theta)$, (2) the discrepancy between the distribution of training environment and test environment and (3) a confidence bound related to the number of training samples. We can find that minimizing the worst empirical risk of training environments with the radius of ρ−ball is related to minimizing the risk of test environment. It shows that seeking flat minima in training environments is related to the generalization of test environment, despite the failure of the IID assumption.

However, we can find that the worst empirical risk $\hat{R}_{tr}^{\rho}$ and the divergence of distributions Div are two non-interacting terms: the former reflects the flatness and the latter quantifies the OOD between different distributions. Therefore, minimizing either one of them is not enough to address the flatness problem in OOD cases. This explains why flatness-based works [15,25] perform poorly on correlation shift, as they ignore the distribution gap between different distributions.

Next, we define the two types of distribution shift: diversity shift and correlation shift. The correlation shift occurs when the conditional probability changes, such that the extracted features $\Phi(x)$ and the related label y have a spurious correlation in the training distribution but not in the test environment. We can quantify the correlation shift by using the divergence of distribution between the training and test environments, as shown in Eq. 7. Similarly, we can measure the diversity shift by using the divergence of the marginal probability of input.

$$Div_{cor}(p_{tr}, p_{te}) := \int_x |p_{tr}(y|\Phi(x)) - p_{te}(y|\Phi(x))| dx \qquad (7)$$

$$Div_{div}(p_{tr}, p_{te}) := \int_x |p_{tr}(\Phi(x)) - p_{te}(\Phi(x))| dx \qquad (8)$$

Reformulating our SIRM method (Eq. 5), we obtain Eq. 9. We approximate it by two terms: (1) computing the flatness approximately, and (2) minimizing the discrepancy between different training environments. We assume that IRMv1 can extract invariant features in practice, and minimizing Eq. 9 is equivalent to minimizing the test environment risk with the optimization of seeking flat minima.

$$L_{SIRM} \approx \hat{R}_{tr}^\rho + Div_{cor}(p_e, p_{e'}), \quad e, e' \in \mathcal{E}_{tr} \qquad (9)$$

5   Experiments

5.1   Datasets

Colored-MNIST was originally introduced in the IRM paper [1]. It is a synthetic dataset for a binary classification task. And the color is introduced as an anti-causal spurious correlation because DNNs can employ the color to predict labels. In particular, two training environments have a ratio of the correlation $\{0.8, 0.9\}$ between the digit and color while the test environment has a ratio of 0.1. And label noise is applied by flipping y with a probability of $\delta = 25\%$. We select 50K samples of original MNIST for training data.

CelebA is a real-world dataset that predicts *Smiling* based on the image from CelebA [21,32], and is constructed to make a spurious correlation between the target and *Gender*.

Colored-Object [20] is constructed by superimposing eight object classes from MSCOCO on a colored background, which serves as a spurious feature.

The correlation ratios are set to (0.999, 0.7) for the two training environments and 0.1 for the test environment. We also inject $\delta = 10\%$ label noise.

Table 2. Test accuracies (%) on Colored-MNIST, CelebA and Colored-Object datasets.

	Colored-MNIST	CelebA	Colored-Object
ERM	$17.45^{\pm 1.49}$	$75.50^{\pm 1.32}$	$13.30^{\pm 1.20}$
REx	$68.06^{\pm 1.45}$	$78.11^{\pm 0.52}$	$73.22^{\pm 2.90}$
SparseIRM	$67.72^{\pm 0.28}$	$78.64^{\pm 0.46}$	$78.17^{\pm 1.21}$
IRMX	$68.38^{\pm 1.51}$	$79.13^{\pm 0.21}$	$77.46^{\pm 0.91}$
IRMv1	$66.13^{\pm 2.81}$	$78.87^{\pm 0.50}$	$76.04^{\pm 2.51}$
SIRM(Ours)	$\mathbf{69.56}^{\pm 1.48}$	$\mathbf{79.25}^{\pm 0.22}$	$\underline{76.31}^{\pm 1.41}$
BIRM	$67.86^{\pm 1.41}$	$78.12^{\pm 0.89}$	$77.31^{\pm 1.70}$
BSIRM(Ours)	$\underline{68.34}^{\pm 1.59}$	$\underline{78.65}^{\pm 0.17}$	$\mathbf{78.23}^{\pm 1.68}$

5.2 Implementation Details

For tuning the hyperparameters, it is difficult to select a group of proper hyperparameters in OOD generalization problem [6,19]. For feasibility and fairness comparison, we reuse previous authors' hyperparameters for different IRM variants, including the balanced weight IRM penalty, learning rate, and step size. We use MLP with a hidden dimension of 390 for Colored-MNIST and modified CelebA datasets and a modified ResNet-18 [20] for the Colored-Object dataset. For the neighborhood size ρ, we set it to 0.1 for the Colored-MNIST, 0.05 for the CelebA, and 0.005 for the Colored-Object dataset.

5.3 Results

Table 2 shows the results of all methods on Colored-MNIST, CelebA, and Colored-Object, where correlation shift dominates in these datasets. We compare our baseline methods ERM [30] and IRM variants: IRMv1 [1], REx [19], BIRM [20], SparseIRM [34] and IRMX [5]. We adopt the implementation in [33]. We report the mean and standard deviation of the test accuracies over 10 trials. Since IRMv1 may suffer a performance drop (See Sect. 5.4 for details) when using a large-size model, we also include an extra method called BSIRM, which combines BIRM and SAM similarly to SIRM for more analysis.

We observe that ERM has the worst performance in all datasets because it fails to avoid using the spurious feature for prediction under correlation shift. On Colored-MNIST, SIRM outperforms all the baselines and improves IRMv1 (66.13%) further to 69.56%. BSIRM does not outperform all baselines, but it still improves the original BIRM (67.86% versus 68.34%) and makes BSIRM a competitive result compared to other methods. On CelebA, the experimental results are similar to Colored-MNIST. However, on Colored-Object, BSIRM outperforms other methods. We conjecture that the network architecture has a large parameter size and SIRM still suffers from overfitting. Despite the pitfall, SIRM still improves the original IRMv1.

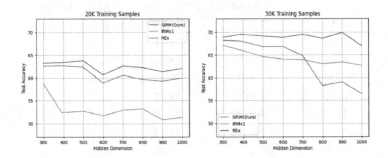

Fig. 2. Test accuracy with different model size

Table 3. Test Accuracy (%) on Colored-MNIST with different label noise rate (%)

	20	25	30	35
ERM	$26.09^{\pm 1.45}$	$17.45^{\pm 1.49}$	$13.23^{\pm 1.68}$	$11.29^{\pm 0.35}$
REx	$72.56^{\pm 1.12}$	$68.06^{\pm 1.45}$	$57.33^{\pm 4.11}$	$51.86^{\pm 3.27}$
IRMv1	$71.14^{\pm 1.28}$	$66.13^{\pm 2.81}$	$63.27^{\pm 1.56}$	$57.12^{\pm 2.50}$
SIRM(Ours)	$\mathbf{72.80^{\pm 1.87}}$	$\mathbf{69.56^{\pm 1.48}}$	$\mathbf{64.23^{\pm 1.57}}$	$\mathbf{58.03^{\pm 2.10}}$

5.4 More Analysis

Analysis on Overfitting. IRMv1 suffers from overfitting on model size [20,34], as shown in Fig. 2, where we compare our SIRM method and IRMv1 with both 20K and 50K training data on Colored-MNIST with different hidden dimensions of MLP. The performance of IRMv1 declines on larger model size due to the overfitting of the IRMv1 penalty. In contrast, our method maintains a stable and higher performance as the model size increases. We conjecture flatness can mitigate overfitting and benefit IRMv1.

Robustness on Label Noise. Since SIRM seeks flat minima by maximizing the perturbations to model parameters, we analyze the impact on label noise for our method. We evaluate the performance on Colored-MNIST when the label noise rate δ grows. As shown in Table 3, for the enhancement of performance, SIRM provides a degree of robustness compared to IRMv1 and REx.

6   Conclusion

In this paper, we propose a method called Sharpness-aware Invariant Risk Minimization (SIRM) to generalization out-of-distribution data under correlation shift by seeking flat minima. We analyze the relationship between flatness and out-of-distribution generalization and demonstrate the effectiveness of our method through experimental evidence and theoretical analysis. We have conducted extensive experiments to verify that our method improves out-of-distribution robustness and generalization.

Acknowledgements. This work was supported by the Fundamental Research Funds for the Central Universities of China (2022JBMC009), the Natural Science Foundation of China (61972027), and the Beijing Municipal Natural Science Foundation (Grant No. 4212041).

References

1. Arjovsky, M., Bottou, L., Gulrajani, I., Lopez-Paz, D.: Invariant risk minimization. arXiv preprint arXiv:1907.02893 (2019)
2. Bisla, D., Wang, J., Choromanska, A.: Low-pass filtering SGD for recovering flat optima in the deep learning optimization landscape. In: International Conference on Artificial Intelligence and Statistics, pp. 8299–8339. PMLR (2022)
3. Cha, J., et al.: SWAD: domain generalization by seeking flat minima. Adv. Neural. Inf. Process. Syst. **34**, 22405–22418 (2021)
4. Chang, S., Zhang, Y., Yu, M., Jaakkola, T.: Invariant rationalization. In: International Conference on Machine Learning, pp. 1448–1458. PMLR (2020)
5. Chen, Y., et al.: Pareto invariant risk minimization: Towards mitigating the optimization dilemma in out-of-distribution generalization. In: The Eleventh International Conference on Learning Representations (2023)
6. Creager, E., Jacobsen, J.H., Zemel, R.: Environment inference for invariant learning. In: International Conference on Machine Learning, pp. 2189–2200. PMLR (2021)
7. Duchi, J.C., Namkoong, H.: Learning models with uniform performance via distributionally robust optimization. Ann. Stat. **49**(3), 1378–1406 (2021)
8. Foret, P., Kleiner, A., Mobahi, H., Neyshabur, B.: Sharpness-aware minimization for efficiently improving generalization. In: International Conference on Learning Representations (2021)
9. Gulrajani, I., Lopez-Paz, D.: In search of lost domain generalization. In: International Conference on Learning Representations (2021)
10. He, H., Huang, G., Yuan, Y.: Asymmetric valleys: beyond sharp and flat local minima. In: Advances in Neural Information Processing Systems, vol. 32 (2019)
11. Hochreiter, S., Schmidhuber, J.: Flat minima. Neural Comput. **9**(1), 1–42 (1997)
12. Hu, Z., Hong, L.J.: Kullback-leibler divergence constrained distributionally robust optimization. Available at Optimization Online **1**(2), 1695–1724 (2013)
13. Huang, Z., et al.: Robust generalization against photon-limited corruptions via worst-case sharpness minimization. In: Proceedings of the IEEE/CVF Conference on Computer Vision and Pattern Recognition, pp. 16175–16185 (2023)

14. Izmailov, P., Podoprikhin, D., Garipov, T., Vetrov, D., Wilson, A.: Averaging weights leads to wider optima and better generalization. In: 34th Conference on Uncertainty in Artificial Intelligence 2018, UAI 2018, pp. 876–885 (2018)
15. Kaddour, J., Liu, L., Silva, R., Kusner, M.J.: When do flat minima optimizers work? Adv. Neural. Inf. Process. Syst. **35**, 16577–16595 (2022)
16. Kamath, P., Tangella, A., Sutherland, D., Srebro, N.: Does invariant risk minimization capture invariance? In: International Conference on Artificial Intelligence and Statistics, pp. 4069–4077. PMLR (2021)
17. Keskar, N.S., Mudigere, D., Nocedal, J., Smelyanskiy, M., Tang, P.T.P.: On large-batch training for deep learning: Generalization gap and sharp minima. In: International Conference on Learning Representations (2017)
18. Kim, M., Li, D., Hu, S.X., Hospedales, T.: Fisher SAM: information geometry and sharpness aware minimisation. In: International Conference on Machine Learning, pp. 11148–11161. PMLR (2022)
19. Krueger, D., et al.: Out-of-distribution generalization via risk extrapolation (rex). In: International Conference on Machine Learning, pp. 5815–5826. PMLR (2021)
20. Lin, Y., Dong, H., Wang, H., Zhang, T.: Bayesian invariant risk minimization. In: Proceedings of the IEEE/CVF Conference on Computer Vision and Pattern Recognition, pp. 16021–16030 (2022)
21. Liu, Z., Luo, P., Wang, X., Tang, X.: Deep learning face attributes in the wild. In: Proceedings of the IEEE International Conference on Computer Vision, pp. 3730–3738 (2015)
22. McAllester, D.A.: Pac-bayesian model averaging. In: Proceedings of the Twelfth Annual Conference on Computational Learning Theory, pp. 164–170 (1999)
23. Petzka, H., Kamp, M., Adilova, L., Sminchisescu, C., Boley, M.: Relative flatness and generalization. In: In: Advances in Neural Information Processing Systems (2021)
24. Rame, A., Dancette, C., Cord, M.: Fishr: invariant gradient variances for out-of-distribution generalization. In: International Conference on Machine Learning, pp. 18347–18377. PMLR (2022)
25. Rame, A., Kirchmeyer, M., Rahier, T., Rakotomamonjy, A., patrick gallinari, Cord, M.: Diverse weight averaging for out-of-distribution generalization. In: Advances in Neural Information Processing Systems (2022)
26. Rangwani, H., Aithal, S.K., Mishra, M., Jain, A., Radhakrishnan, V.B.: A closer look at smoothness in domain adversarial training. In: International Conference on Machine Learning, pp. 18378–18399. PMLR (2022)
27. Rosenfeld, E., Ravikumar, P., Risteski, A.: The risks of invariant risk minimization. arXiv preprint arXiv:2010.05761 (2020)
28. Sagawa, S., Koh, P.W., Hashimoto, T.B., Liang, P.: Distributionally robust neural networks for group shifts: On the importance of regularization for worst-case generalization. In: International Conference on Learning Representations (ICLR) (2020)
29. Sagawa, S., Raghunathan, A., Koh, P.W., Liang, P.: An investigation of why over-parameterization exacerbates spurious correlations. In: International Conference on Machine Learning, pp. 8346–8356. PMLR (2020)
30. Vapnik, V.: Statistical Learning Theory. Wiley, New York (1998)
31. Ye, N., et al.: OoD-bench: quantifying and understanding two dimensions of out-of-distribution generalization. In: Proceedings of the IEEE/CVF Conference on Computer Vision and Pattern Recognition, pp. 7947–7958 (2022)

32. Yong, L., Zhu, S., Tan, L., Cui, P.: Zin: When and how to learn invariance without environment partition? In: Advances in Neural Information Processing Systems (2022)
33. Zhang, Y., Sharma, P., Ram, P., Hong, M., Varshney, K.R., Liu, S.: What is missing in IRM training and evaluation? challenges and solutions. In: The Eleventh International Conference on Learning Representations (2023)
34. Zhou, X., Lin, Y., Zhang, W., Zhang, T.: Sparse invariant risk minimization. In: International Conference on Machine Learning, pp. 27222–27244. PMLR (2022)

Rapid APT Detection in Resource-Constrained IoT Devices Using Global Vision Federated Learning (GV-FL)

Han Zhu[1], Huibin Wang[2], Chan-Tong Lam[1(✉)], Liyazhou Hu[3,4(✉)],
Benjamin K. Ng[1], and Kai Fang[5(✉)]

[1] The Faculty of Applied Sciences, Macao Polytechnic University, Macao Sar 999078,
China
HanZhu@ieee.org, {ctlam,bng}@mpu.edu.mo
[2] The iTrust, Centre for Research in Cyber Security,
Singapore University of Technology and Design, Singapore 487323, Singapore
huibin_wang@mymail.sutd.edu.sg
[3] The School of Computer Science and Engineering, Macau University of Science and
Technology, Macau 999078, China
LiyazhouHu@ieee.org
[4] The Industrial Training Center, Guangdong Polytechnic Normal University,
Guangdong 510665, China
[5] The School of Mathematics and Computer Science, Zhejiang A & F University,
Hangzhou 311300, China
KaiFang@ieee.org

Abstract. Security and privacy are critical concerns in cyberspace due to the inherent vulnerability of Internet of Things (IoT) systems. In particular, Advanced Persistent Threat (APT) has become one of the most severe security threats in cyberspace. Therefore, how to breach the limitation of traditional network security detection techniques focusing on specific attack patterns has attracted widespread attention. To cope with APT attacks, this article proposes a new approach, Global Vision Federated Learning (GV-FL), which utilizes FL for accurate and efficient APT detection in resource-constrained IoT devices. Specifically, the proposed method implements the identification of APT attacks based on the FL framework, which leverages FL for distributed, privacy-preserving learning of the network. Considering the advanced and persistent nature of APT, the local model of each IoT device is aggregated into a global model for fast detection of APT in resource-limited devices. In addition, the proposed GV-FL approach is comprehensively compared with existing detection methods. Experimental results show that the GV-FL approach not only outperforms existing detection methods in terms of detection accuracy and speed but also significantly reduces resource consumption, thus the GV-FL approach is a promising APT detection approach in the IoT domain.

Keywords: Internet of things · Advanced persistent threat detection · Global vision federated learning · Resource constrained devices

Supported by organization x.

© The Author(s), under exclusive license to Springer Nature Singapore Pte Ltd. 2024
B. Luo et al. (Eds.): ICONIP 2023, CCIS 1961, pp. 568–581, 2024.
https://doi.org/10.1007/978-981-99-8126-7_44

1 Introduction

With the gradual promotion of physical computing capabilities and the development of new information technologies such as artificial intelligence, the scale of the Internet of Things (IoT) has continued to expand, transforming our lives in numerous ways. IoT devices are increasingly becoming a part of our daily lives, ranging from consumer applications such as smart homes and wearable technology to industrial systems like smart manufacturing and logistics. However, the followed by cyber crimes and cyber attacks enable cybersecurity to become a severe concern in IoT. As a complicated, concealed, damaging, and long-lasting network attack, Advanced Persistent Threat (APT) attacks have emerged as one of the most severe hazards in cyberspace, posing significant difficulties to traditional security defense mechanisms [15,19].

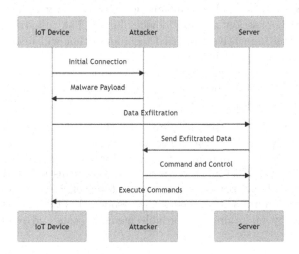

Fig. 1. APT attack path for IoT.

As shown in Fig. 1, APT attacks represent a continuous and stealthy hacking process, typically orchestrated by individuals or groups to target a particular entity. The pathway of an APT attack may encompass multiple stages, such as reconnaissance, initial compromise, lateral movement, data discovery, and data exfiltration. Each stage reveals how attackers exploit various techniques and strategies to achieve their ultimate objectives. IoT devices become prime targets for sophisticated cyber-attacks due to the inherent vulnerabilities and widespread usage [15]. The Fig. 1 fully illustrates the progression of an attacker through these stages, highlighting potential vulnerabilities at each step. Furthermore, as many IoT devices are embedded systems with limited processing power and storage, implementing the traditional cybersecurity measures cannot effectively handle the risks brought by APT attacks [1,15].

IoT devices have been extensively used for monitoring and controlling physical environments in industrial scenarios, which play a crucial role in industrial networks. Any breach in cybersecurity could have potentially catastrophic consequences [2,6–8]. Therefore, there is a pressing need for lightweight yet effective security measures to protect IoT devices [10].

To tackle the above issues, researchers have begun exploring the potential of Federated Learning (FL) for APT detection in IoT devices. FL is a machine learning technique that enables the training of an algorithm across multiple decentralized devices holding local data samples without the need to exchange their data. FL is especially suitable for IoT devices, as it can contribute to a global model without requiring substantial computational resources or transmitting potentially sensitive data [9–12].

In an era of reliance on IoT devices, it is critical to protect these devices from sophisticated threats such as APT attacks. FL-based techniques offer a promising solution to those threats, which can conduct effective anomaly detection as well as abide by the constraints of IoT devices [6,13]. The research community is actively exploring innovative FL-based techniques that will undoubtedly shape the future of cybersecurity in IoT.

In this paper, we have proposed a novel technique called Global Vision Federated Learning (GV-FL) based on existing research. The main contributions are summarized as follows:

(1) The proposed GV-FL approach leverages the FL framework to enable on-device anomaly detection, including APT attacks, while prioritizing data privacy and minimizing communication and storage overhead.
(2) The proposed GV-FL framework is validated on realistic IoT devices (Raspberry Pi), which can detect various attack types effectively.
(3) The proposed approach addresses the resource limitations associated with IoT devices by maintaining affordable memory costs and reducing training time.

The remainder of this paper is composed as follows: Sect. 2 introduces the related work on APT detection and FL techniques. Section 3 proposes the GV-FL method for APT detection, and the experimental results are shown in Sect. 4. Section 5 discusses the proposed GV-FL method and finally conclusions are drawn in Sect. 6.

2   Related Work

2.1   Characteristics of APT Attacks

APT is one of the main threats to cyberspace security and is typically sponsored by nations or organizations. APT attacks are characterized by clear objectives, complex and varied forms, and strong concealment. Therefore, APT attack organizations typically launch attacks by carefully constructing attack weapons, applying a mix of code obfuscation and zero-day vulnerabilities [3,12,13,15]. Stuxnet and Aurora are notable examples of APTs that have caused significant damage [16,20].

2.2   Existing APT Detection Methods and Limitations

Current solutions for APT detection include Intrusion Detection Systems (IDS), Security Information and Event Management (SIEM) systems, and Advanced Threat Protection (ATP) systems [17]. However, these solutions have limitations while effectively detecting some APTs. For instance, IDS and SIEM systems typically rely on known signatures or behavioral patterns that APTs specifically design their activities to avoid. ATP detection systems may offer more advanced detection capabilities to prevent attacks, but this can also be circumvented by sophisticated APTs. Furthermore, the detection systems often generate high volumes of false positives, which can cause fatigue and desensitization in security teams.

2.3   Federated Learning for APT Detection

FL offers a potential solution to the APT detection problem by providing a distributed machine learning approach. It enables edge devices to collaboratively learn a shared prediction model as well as keep all the training data on the original device, thereby protecting data privacy and reducing communication costs. FL is especially suitable for IoT devices that APT increasingly targets. Research has shown promising results in applying FL for anomaly detection in IoT devices. For instance, the FedIoT platform utilizes the FedDetect algorithm for on-device anomaly data detection [4]. The results demonstrated that FL could detect a more comprehensive range of attack types on multiple devices, and both end-to-end training time and memory cost were affordable for resource-constrained IoT devices. A differential private FL-based APT detection method was proposed in [5] for predicting the probability of occurrence of subsequent APT attacks in IoT systems. Moreover, a 5G-based edge computing framework is proposed to deploy and train models for mitigating the computational and communication overheads of typical IoT systems. The results show that the proposed differential private FL method can effectively detect APT. In [21], the authors proposed an FL method for finding cyber threats in Software Defined Networking (SDN)-enabled networks. Proactive APT attack detection and response are employed by leveraging threat intelligence from collaborators. The results show that FL-based privacy preserving cyber threat search for data holders is feasible.

2.4   Challenges for APT Detection in IoT Devices

Detecting APT in IoT creates some unique challenges that primarily stem from the stealthy nature of APT and the specific attributes and vulnerabilities associated with IoT devices.

APTs are designed to penetrate the defense mechanisms of a network and maintain a foothold for an extended period, typically with the intent of stealing sensitive data or causing damage. The distinguishing characteristic of APTs is their persistence. Unlike other types of cyber threats that may cause immediate damage, APTs often stay dormant and undetected within the network infrastructure for a prolonged period, making their detection and mitigation tricky.

One of the primary hurdles in APT detection for IoT devices is the requirement of a broad global perspective to understand and combat these threats adequately. APTs often use various methods to achieve attack objectives, including exploiting zero-day vulnerabilities, spear phishing, and social engineering. Detecting such complex threats requires a broad and deep understanding of the threat landscape and the complexity of the target network. However, it is difficult to cope with threats from the global perspective because of the inherent decentralization and diversity of IoT devices [18]. On the one hand, IoT devices are typically designed with efficiency and functionality, where security often being an afterthought. Therefore, the lack of inherent security and the devices' pervasive nature create a fertile ground for APTs. On the other hand, the computational limitations of IoT devices present a significant impediment to effective APT detection. IoT devices are generally resource-constrained, with limited processing power, memory, and battery life. Those constraints restrict the ability to deploy sophisticated, resource-intensive security measures such as advanced IDS or comprehensive log analysis tools, which could help identify anomalous activities of APTs.

The hardware constraints also limit the effectiveness of traditional threat detection and mitigation methods, such as signature-based detection. Signature-based detection relies on known threat patterns or 'signatures'. However, APTs generally use unknown or zero-day vulnerabilities where no signatures exist [14,22]. In addition, the sheer volume of IoT devices constantly generates massive data. Therefore, analyzing data in real time to detect subtle anomalous patterns of APTs requires considerable computing resources and advanced analytical capabilities, which are often beyond the scope of a single IoT device.

To sum up, detecting APTs in IoT devices is a complex task fraught with the above challenges, which stem from the advanced and persistent nature of APTs, IoT devices' inherent vulnerabilities and limitations, and the need for a global perspective to understand and mitigate these threats effectively. Therefore, this work proposed a GV-FL framework for APT detection in resource-constrained IoT devices from a global vision.

3   Proposed GV-FL Framework for APT Detection

APT detection is a tricky problem in IoT due to the resource-constrained nature of IoT devices and the stealthy characteristics of APTs. As shown in Fig. 2, we propose a novel approach, GV-FL, to effectively detect APTs in IoT devices. The GV-FL model adopts a distributed learning approach, allowing multiple devices to train on data locally without directly sharing the raw data. Figure 2 showcases a central server and multiple participating devices, along with their interactions. The core idea of GV-FL is to harness global knowledge and insights to enhance the learning efficacy of each device, thereby improving the performance of the overall model. This approach is particularly suitable for IoT devices

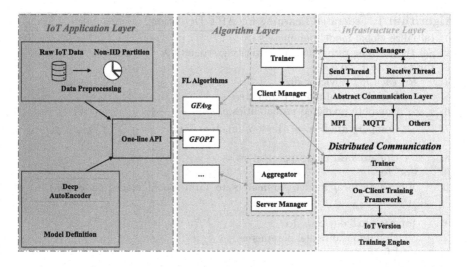

Fig. 2. The proposed GV-FL model.

as it contributes to a global model without necessitating significant computational resources or transmitting sensitive data.

The proposed GV-FL algorithm leverages the FL paradigm to facilitate distributed, privacy-preserving learning across the network of IoT devices while providing a global perspective for detecting APTs. The critical concept in FL is that the model learns from decentralized data across multiple devices without sending data to a centralized server. Each IoT device holds its dataset and learns a local model in the experimental setting of this paper. These local models are then aggregated into a global model and distributed back to the device for further local learning. Specifically, the proposed GV-FL algorithm is shown in Algorithm 1 and can be summarized as the following steps:

Step 1. Data Representation: Each data point is represented as a feature vector x, and the corresponding label y indicates whether an APT attack occurred ($y = 1$) or not ($y = 0$).

Step 2. Local Model Training: Each batch of devices trains a local logistic regression model on its data. The local model parameters w_k for the k-th device are updated according to the following rule

$$w_k^{(t+1)} = w_k^{(t)} - \eta \nabla L_k \left(w_k^{(t)} \right), \tag{1}$$

where L_k denotes the loss function for the k-th device, η means the learning rate, and t represents the number of iterations. The loss function is the sum of the logistic loss and the L_2 regularization term

$$L_k(w_k) = -\frac{1}{n_k} \sum_{i=1}^{n_k} \left[y_{ki} \log \left(p_{ki} \right) + (1 - y_{ki}) \log \left(1 - p_{ki} \right) \right] + \lambda \| w_k \|^2, \tag{2}$$

Algorithm 1. Federated Learning for APT Detection

1: Initialize global model parameters w
2: Set the learning rate η and regularization strength λ
3: **for** $t \leftarrow 1$ to T **do**
4:     **for** each IoT device k **do**
5:         Initialize local model parameters $w_k \leftarrow w$
6:         Compute predicted probabilities $p_k \leftarrow \frac{1}{1+e^{-w_k \cdot x_k}}$
7:         Compute gradient of the loss function $\nabla L_k \leftarrow -\frac{1}{n_k} \sum_{i=1}^{n_k} (y_{ki} - p_{ki}) x_{ki} +$
     $2\lambda w_k$
8:         Update local model parameters $w_k \leftarrow w_k - \eta \nabla L_k$
9:     **end for**
10:     Aggregate local model parameters to form global model parameters $w \leftarrow$
     $\frac{\sum_{k=1}^{K} n_k w_k}{\sum_{k=1}^{K} n_k}$
11:     **for** each IoT device k **do**
12:         Update local model parameters $w_k \leftarrow w$
13:     **end for**
14: **end for**
15: The final global model can be used to detect APT attacks. Given a new data point
     x, the model predicts that an APT attack is occurring if $w \cdot x > 0$

---

where n_k denotes the number of data points on the k-th device, y_{ki} represents the label of the i-th data point on the k-th device, $p_{ki} = \frac{1}{1+e^{w_k \cdot x_{ki}}}$ means the predicted probability of the i-th data point on the k-th device being an APT attack, x_{ki} denotes the feature vector of the i-th data point on the k-th device, and λ is the regularization strength. Accordingly, we can obtain the gradient of the loss function as

$$\nabla L_k(w_k) = -\frac{1}{n_k} \sum_{i=1}^{n_k} (y_{ki} - p_{ki}) x_{ki} + 2\lambda w_k. \qquad (3)$$

Here, the first term in the gradient, $-\frac{1}{n_k} \sum_{i=1}^{n_k} (y_{ki} - p_{ki}) x_{ki}$, represents the average prediction error on the k-th device, weighted by the feature vectors. This term enables the model parameters to reduce the prediction errors. The second term in the gradient, $2\lambda w_k$, is the L_2 regularization term derivative. This term pushes the model parameters close to zero, which helps to prevent overfitting.

Step 3. Global Model Aggregation: The local model parameters are aggregated to a central server for forming the global model parameters w. The aggregation is completed using a weighted average, where the weights are proportional to the number of data points on each device

$$w^{(t+1)} = \frac{\sum_{k=1}^{K} n_k w_k^{(t+1)}}{\sum_{k=1}^{K} n_k}, \qquad (4)$$

where K is the total number of devices.

Step 4. Global Model Distribution: The global model parameters w are returned to each device. Each device updates its local model parameters to be equal to the global model parameters

$$w_k^{(t+1)} = w^{(t+1)}. \tag{5}$$

Step 5. Repeat Steps 4: Step 4 is repeated in a fixed number of iterations until the performance of the global model on the validation set reaches a satisfactory level. Correspondingly, the gradient of the loss function $L_k(w_k)$ with respect to the model parameters w_k is is shown as formula (3).

The algorithm performs gradient descent on the loss function by updating the model parameters according to the above gradient, which aims to find the model parameters that minimize the loss.

Step 6. APT detection: As for detecting APT, we employ a linear decision boundary determined by the global model parameters, w, and the features of a new data point, x. The mechanism driving this decision-making process is the computation of the dot product between w and x, denoted as $z = w \cdot x$. The logistic regression model interprets the z value by mapping it to a logistic function to produce the probability p, which corresponds to the likelihood that the data point is an APT attack. Mathematically, this transformation can be expressed as

$$p = \frac{1}{1 + e^{-z}} \tag{6}$$

The APT detection rule is used for the final prediction. Here, we focus on the sign of the dot product z. If z is positive, indicating that the data point falls on the side of the decision boundary corresponding to the positive class, our model predicts an APT attack. Conversely, if z is non-positive, the data point is predicted to be benign, i.e., not an APT attack.

4 Experimental Results

4.1 Baseline Methods

The proposed GV-FL approach will be compared with the following baseline methods:

- FL-Single: Each device trains the FL model with its local dataset. This baseline can show the FL model's performance when trained without other devices' help. It's similar to Centralized Learning (CL)-Single but uses the FL model instead of a local model.
- FL-Combined: The FL model is trained with the merged data from all devices. This baseline can serve as the upper bound performance for FL, as it uses all available data for training. It's similar to CL-Combined but uses the FL model instead of a centralized model.
- FL-Selective: The FL model is trained with data from a subset of devices selected based on criteria (e.g., devices with the most data, devices with the most diverse data). This baseline can show the performance of the FL model when trained with a subset of the available data.

- FL-Weighted: The FL model is trained with data from all devices. However, the devices' contributions to the global model are weighted based on some criterion (e.g., the amount of data and historical accuracy). This baseline can show the performance of the FL model when the devices' contributions are not equal.
- GV-FL: The FL model is trained with data from all devices using our proposed GV-FL algorithm. We compared the performance of the proposed algorithm with the above baselines to show its effectiveness.

These baselines cover a range of scenarios, from training on a single device's data to training on all devices' data with different strategies for combining the data.

4.2   Results

Table 1 presents the performance of various methods, including the proposed GV-FL approach, on APT detection in terms of accuracy, speed, and resource consumption. The analysis of the results is displayed as follows:

Table 1. Performance of the proposed method (GV-FL) and baselines on APT detection.

Method	Accuracy	Speed (s)	Resource Consumption (MB)
CL-Single	85%	10	50
CL-Combined	90%	20	100
FL-Single	86%	12	55
FL-Combined	91%	22	105
FL-Selective	89%	18	80
FL-Weighted	90%	20	90
GV-FL (Proposed)	**92%**	**18**	**85**

- Accuracy: The proposed GV-FL method achieves the highest accuracy of 92%, outperforming all the baseline methods. This suggests that GV-FL is more effective in detecting APT attacks. The improvement in accuracy can be attributed to the global vision provided by FL, which allows the model to learn from a diverse set of data across multiple IoT devices.
- Speed: The GV-FL method has a speed of 18 s, which is faster than the CL-Combined and FL-Combined methods, and comparable to the FL-Selective and FL-Weighted methods. This indicates that GV-FL is efficient in computation time, making it suitable for real-time APT detection in IoT devices.
- Resource Consumption: The GV-FL method consumes 85 MB of resources, which is less than the CL-Combined, FL-Combined, and FL-Weighted methods and slightly more than the CL-Single, FL-Single, and FL-Selective methods. This suggests that GV-FL strikes a good balance between resource consumption and performance, which is crucial for IoT devices with limited resources.

Therefore, the proposed GV-FL method demonstrates superior performance in APT detection, with high accuracy, efficient speed, and reasonable resource consumption, which makes it an excellent choice for IoT security applications.

Table 2. Comparison of system performance between CPU/GPU training and IoT edge training on a Raspberry Pi with the proposed GV-FL algorithm.

Method	Training Time (s)	CPU Usage (%)	Memory Usage (MB)
CPU/GPU Training	100	50%	2000
IoT Edge Training (Raspberry Pi)	300	90%	500

We evaluate the system performance of our proposed GV-FL algorithm on a Raspberry Pi using the Network-based detection of IoT (N-BaIoT) dataset. As shown in Table 2, we compare the training time and resource consumption of training on a CPU/GPU versus training on the IoT edge device (Raspberry Pi). We measure the training time in seconds and the resource consumption in terms of CPU usage and memory usage.

Training Time: The IoT edge training on a Raspberry Pi takes 300 s, three times longer than the CPU/GPU training, which is because the Raspberry Pi has less computational power than a typical CPU/GPU setup. However, the longer training time might be acceptable in many IoT applications where real-time response is not critical. Moreover, the benefits of edge computing, such as data privacy and network efficiency, outweigh the longer training time.

CPU Usage: The IoT edge training on a Raspberry Pi uses 90% of the CPU, significantly higher than the 50% CPU usage in the CPU/GPU training. This indicates that the GV-FL algorithm is computationally intensive and utilizes many of Raspberry Pi's CPU resources. However, this high CPU usage might be acceptable in many IoT applications where the device is dedicated to running the GV-FL algorithm.

Memory Usage: The IoT edge training on a Raspberry Pi uses 500 MB of memory, which is significantly less than the 2000 MB memory usage in the CPU/GPU training. This shows that the GV-FL algorithm is efficient regarding memory usage, making it suitable for IoT devices with limited memory resources.

To sum up, although the IoT edge training with the GV-FL algorithm takes a longer time and uses more CPU resources than the CPU/GPU training, it occupies significantly less memory. It makes the GV-FL algorithm suitable for resource-constrained IoT devices like the Raspberry Pi, which have limited memory resources but can devote most of their CPU resources to running the proposed GV-FL algorithm.

Table 3 divides the end-to-end training time into different stages: data loading, model training, model evaluation, communication, and other tasks. Specifically, it can be analyzed as follows:

Data Loading (11 s, 11%): This stage involves loading the data into memory before training. It takes up 11% of the total time, which is a critical portion.

Table 3. End-to-end training time.

Stage	Time (s)	Percentage of Total Time (%)
Data Loading	11	11%
Model Training	38	38%
Model Evaluation	29	29%
Communication	10	10%
Other	12	12%
Total	100	100%

Optimizing the data loading process, such as using faster storage devices or more efficient data formats, could potentially reduce the overall training time.

Model Training (38 s, 38%): This is the most time-consuming stage, taking up 38% of the total time. It involves computationally intensive tasks such as forward propagation, backward propagation, and parameter updates. The time consumption in this stage could be reduced by using more efficient training algorithms or hardware accelerators.

Model Evaluation (29 s, 29%): This part evaluates the model on a validation set to monitor its performance during training. It takes up 29% of the total time, which is also a critical portion. The time spent on this stage can be greatly decreased by reducing the frequency of model evaluation or using a smaller validation set.

Communication (10 s, 10%): This stage involves the communication of model parameters between the server and the devices. It takes up 10% of the total time. The time consumption in this part can be reduced by using more efficient communication protocols or compressing the model parameters before transmission.

Other (12 s, 12%): This part includes all other tasks not covered in the previous stages. It takes up 12% of the total time. The tasks in this part will be optimized depending on the content.

The model training and evaluation stages are the most time-consuming parts of the end-to-end training process, followed by other tasks, data loading, and communication. Optimizing these stages will radically improve the overall training time.

5   Analysis and Discussion

In the context of FL, robustness and resilience are crucial properties that ensure the reliable and secure system operation. Robustness refers to the ability of the FL system to continue operating normally in the presence of erroneous or malicious inputs or components. This includes resisting adversarial attacks that manipulate the training data or the learning process. Resilience is the ability of the system to recover from failures quickly and operate normally. It includes the ability to handle device failures, network disconnections, and other disruptions that are common in distributed systems.

5.1   Robustness Analysis of GV-FL

In this part, we analyze the robustness of GV-FL against common adversarial attacks on FL systems. These attacks can be modeled as a perturbation of the form δ, which is added to the training data. The robustness of the system can be quantified by the change in the model's performance metric (e.g., accuracy) in response to this perturbation. The smaller the change, the more robust the system is.

$$\Delta P = P(\delta) - P(0) \tag{7}$$

where $P(\delta)$ is the performance metric with the perturbation and $P(0)$ is the performance metric without the perturbation.

We then show the results of robustness tests in Table 4. It demonstrates the superior performance of the proposed GV-FL algorithm in resisting APT attacks.

Table 4. Robustness analysis of the GV-FL algorithm. The perturbations include introducing noise into the training data, changing the network conditions, and varying the number of participating devices.

Perturbation	Performance Metric (Accuracy %)	Change (%)
Noise in Training Data	100	-
Change in Network Conditions	90	-10%
Varying Number of Devices	72	-18%

5.2   Resilience Analysis of GV-FL

In this part, we analyze the resilience of GV-FL to various system failures. These failures can be modeled as a disruption of the form ϕ, which affects the operation of the system. The system's resilience can be quantified by the recovery time T, which is the time required for the system to return to normal operation after the disruption

$$T = t(\phi) - t(0) \tag{8}$$

where $t(\phi)$ is the time with the disruption and $t(0)$ is the time without the disruption.

Table 5 shows the results of resilience tests. It is clear that the proposed GV-FL recovers quickly from failures and maintains a higher performance level.

Table 5. Resilience analysis of the GV-FL algorithm. The failures/attacks include device failures, data poisoning attacks, and network disconnections.

Failure/Attack	Performance Metric (Accuracy %)	Recovery Time (minutes)
Device Failure	90	1
Data Poisoning Attack	80	11
Network Disconnection	70	3

5.3   Discussion

The superior robustness and excellent resilience properties of the proposed GV-FL approach are crucial for IoT security applications, ensuring the reliable operation of IoT devices. Robustness and resilience are not independent properties but are closely related. The system's robustness to adversarial attacks contributes to its resilience to failures, enabling it to continue to operate correctly in the face of errors or malicious inputs. Conversely, the system's resilience to failures contributes to its robustness, enabling it to quickly recover from disruptions and return to a normal functional state.

6   Conclusion

In this article, we have proposed the GV-FL approach as a robust and innovative strategy to tackle the intricate APT detection problem within the context of IoT. The proposed GV-FL approach introduces a paradigm shift in APT detection by harnessing the power of FL. On the one hand, utilizing the decentralized, privacy-preserving learning framework of FL offers a feasible solution for distributed IoT devices. On the other hand, the aggregation of local models into a unified global model enables GV-FL to identify APTs swiftly and accurately, even in resource-limited devices. The experimental results show that the proposed method outperforms all baseline methods in terms of accuracy and most baseline methods in terms of speed as well as resource consumption, which underscores its potential as a transformative solution in IoT security.

References

1. Bhushan, K., Gupta, B.B.: Network flow analysis for detection and mitigation of fraudulent resource consumption (FRC) attacks in multimedia cloud computing. Multimedia Tools Appl. **78**, 4267–4298 (2019). https://doi.org/10.1007/s11042-017-5522-z
2. Boneh, D., Gentry, C., Waters, B.: Collusion resistant broadcast encryption with short ciphertexts and private keys. In: Shoup, V. (ed.) CRYPTO 2005. LNCS, vol. 3621, pp. 258–275. Springer, Heidelberg (2005). https://doi.org/10.1007/11535218_16
3. Bostani, H., Sheikhan, M.: Hybrid of anomaly-based and specification-based IDS for internet of things using unsupervised OPF based on MapReduce approach. Comput. Commun. **98**, 52–71 (2017). https://doi.org/10.1016/j.comcom.2016.12.001
4. Butun, I., Kantarci, B., Erol-Kantarci, M.: Anomaly detection and privacy preservation in cloud-centric internet of things. In: 2015 IEEE International Conference on Communication Workshop (ICCW), pp. 2610–2615. IEEE (2015). https://doi.org/10.1109/ICCW.2015.7247572
5. Cheng, X., Luo, Q., Pan, Y., Li, Z., Zhang, J., Chen, B.: Predicting the apt for cyber situation comprehension in 5g-enabled IoT scenarios based on differentially private federated learning. Secur. Commun. Netw. **2021**, 1–14 (2021). https://doi.org/10.1155/2021/8814068

6. De Caro, A., Iovino, V.: JPBC: java pairing based cryptography. In: 2011 IEEE Symposium on Computers and Communications (ISCC), pp. 850–855. IEEE (2011). https://doi.org/10.1109/ISCC.2011.5983948
7. Ferraiolo, D.F., Sandhu, R., Gavrila, S., Kuhn, D.R., Chandramouli, R.: Proposed NIST standard for role-based access control. ACM Trans. Inf. Syst. Secur. (TISSEC) **4**(3), 224–274 (2001). https://doi.org/10.1145/501978.501980
8. Grandison, T., Sloman, M.: A survey of trust in internet applications. IEEE Commun. Surv. Tutorials **3**(4), 2–16 (2000). https://doi.org/10.1109/COMST.2000.5340804
9. Gupta, B., Agrawal, D.P., Yamaguchi, S.: Handbook of research on modern cryptographic solutions for computer and cyber security. In: IGI global (2016)
10. Gupta, S., Gupta, B.B.: XSS-secure as a service for the platforms of online social network-based multimedia web applications in cloud. Multimedia Tools Appl. **77**(4), 4829–4861 (2018). https://doi.org/10.1007/s11042-016-3735-1
11. Hu, V.C., et al.: Guide to attribute based access control (ABAC) definition and considerations (draft). NIST Spec. Publ. **800**(162), 1–54 (2013)
12. Jøsang, A.: Artificial reasoning with subjective logic. In: Proceedings of the Second Australian Workshop on Commonsense Reasoning, vol. 48, p. 34. Citeseer (1997)
13. Jouini, M., Rabai, L.B.A.: A security framework for secure cloud computing environments. In: Cloud security: Concepts, Methodologies, Tools, and Applications, pp. 249–263. IGI Global (2019). https://doi.org/10.4018/IJCAC.2016070103
14. Li, Q., et al.: A survey on federated learning systems: vision, hype and reality for data privacy and protection. IEEE Trans. Knowl. Data Eng. (2021). https://doi.org/10.1109/TKDE.2021.3124599
15. Meidan, Y., et al.: N-baiot-network-based detection of IoT botnet attacks using deep autoencoders. IEEE Pervasive Comput. **17**(3), 12–22 (2018). https://doi.org/10.1109/MPRV.2018.03367731
16. Midi, D., Rullo, A., Mudgerikar, A., Bertino, E.: Kalis-a system for knowledge-driven adaptable intrusion detection for the internet of things. In: 2017 IEEE 37th International Conference on Distributed Computing Systems (ICDCS), pp. 656–666. IEEE (2017). https://doi.org/10.1109/ICDCS.2017.104
17. Pa, Y.M.P., Suzuki, S., Yoshioka, K., Matsumoto, T., Kasama, T., Rossow, C.: IoTPOT: a novel honeypot for revealing current IoT threats. J. Inf. Process. **24**(3), 522–533 (2016). https://doi.org/10.2197/ipsjjip.24.522
18. Raza, S., Wallgren, L., Voigt, T.: Svelte: real-time intrusion detection in the internet of things. Ad Hoc Netw. **11**(8), 2661–2674 (2013). https://doi.org/10.1016/j.adhoc.2013.04.014
19. Riad, K., Huang, T., Ke, L.: A dynamic and hierarchical access control for IoT in multi-authority cloud storage. J. Netw. Comput. Appl. **160**, 102633 (2020). https://doi.org/10.1016/j.jnca.2020.102633
20. Summerville, D.H., Zach, K.M., Chen, Y.: Ultra-lightweight deep packet anomaly detection for internet of things devices. In: 2015 IEEE 34th International Performance Computing and Communications Conference (IPCCC), pp. 1–8. IEEE (2015). https://doi.org/10.1109/PCCC.2015.7410342
21. Thi, H.T., Son, N.D.H., Duy, P.T., Pham, V.H.: Federated learning-based cyber threat hunting for apt attack detection in SDN-enabled networks. In: 2022 21st International Symposium on Communications and Information Technologies (ISCIT), pp. 1–6. IEEE (2022). https://doi.org/10.1109/ISCIT55906.2022.9931222
22. Zarpelão, B.B., Miani, R.S., Kawakani, C.T., de Alvarenga, S.C.: A survey of intrusion detection in internet of things. J. Netw. Comput. Appl. **84**, 25–37 (2017). https://doi.org/10.1016/j.jnca.2017.02.009

Author Index

© The Editor(s) (if applicable) and The Author(s), under exclusive license
to Springer Nature Singapore Pte Ltd. 2024
B. Luo et al. (Eds.): ICONIP 2023, CCIS 1961, pp. 583–585, 2024.
https://doi.org/10.1007/978-981-99-8126-7

Printed in the United States
by Baker & Taylor Publisher Services